about this series at http://www.springer.com/series/10283

Springer INdAM Se

More informatio

Volume 16

Laurent Gosse • Roberto Natalini
Editors

Innovative Algorithms and Analysis

Springer

Editors
Laurent Gosse
Istituto per le Applicazioni del Calcolo
CNR
Rome, Italy

Roberto Natalini
Istituto per le Applicazioni del Calcolo
CNR
Rome, Italy

ISSN 2281-518X ISSN 2281-5198 (electronic)
Springer INdAM Series
ISBN 978-3-319-49261-2 ISBN 978-3-319-49262-9 (eBook)
DOI 10.1007/978-3-319-49262-9

Library of Congress Control Number: 2017934879

Printed on acid-free paper

This Springer imprint is published by Springer Nature
The registered company is Springer International Publishing AG
The registered company address is: Gewerbestrasse 11, 6330 Cham, Switzerland

Foreword

This book attempts to provide a fresh perspective on the well-posedness of hyperbolic models by highlighting some open questions and how some of them appear or are treated in similar instances. These issues were discussed at a workshop entitled *Innovative Algorithms and Analysis* held at INdAM (Rome) between 17th and 20th May 2016 and organized by Laurent Gosse and Roberto Natalini. The hope of the organizers was that the diversity of the invited speakers would help to identify common themes in the study of nonclassical solutions to degenerate hyperbolic problems, balance laws with nonlocal source terms, and weak admissible solutions to conservation laws in multiple space dimensions.

The theory of hyperbolic conservation laws has matured significantly since the groundbreaking work of Bressan and Bianchini on the vanishing viscosity method; the stability analysis of Bressan, Liu, and Yang; and the study of viscous profiles by Zumbrun. Moreover, the work of Qiang and Ambrosio has improved our understanding of the functional framework necessary for discussion of singularities in transport problems. Numerics for conservation laws has also become quite sophisticated, although 1D formulations still serve as the basis for most multi-D discretizations. Many of the better known open questions relate to multiscale effects in first-order hyperbolic problems, macroscopic limits, and Riemann problems in 2D. Nevertheless, a few problems in these subjects still persist in these subjects, particularly in 2D, such as the role of diffusion and entropy and the nature of singularities.

In one space dimension, the theory of conservation laws is well understood, yet there remain a few unanswered issues. For example, in chapter "Analysis and Simulation of Nonlinear and Nonlocal Transport Equations", Lagoutière and Vauchelet discuss measure-valued solutions appearing in nonlocal balance laws modeling aggregation, from both the analytical and the numerical perspective. Matthes and Söllner study drift-diffusion with nonlocal aggregation effects using the global Wasserstein gradient flow, which again leads to solutions outside the classical framework. Karlsen, Risebro, and Storrøsten present a few examples of degenerate conservation laws for which the observed numerical convergence rate for high-order schemes is greater than the best estimates given by Kružkov's stability theory. These results indicate that despite the success of the established stability theories, there remain phenomena just beyond the classical BV framework.

Christoforou provides in chapter "On Hyperbolic Balance Laws and Applications" a concise and modern summary of the theory of balance laws in 1D, emphasizing necessary well-posedness conditions for the source terms. The discussion identifies the strengths and weaknesses of different techniques and could be useful to graduate students entering the field. These results are then applied to the Gauss-Codazzi system for the metric of isometric immersions of 2D surfaces in $\mathbb{R}^3$, particularly for metrics with minimal regularity. Relativity theory provides a number of interesting problems for balance laws, as is further highlighted by the contribution of Bini, Damour, and Geralico on perturbations of black hole spacetimes.

Established stability theories for conservation laws, although able to identify asymptotic profiles, lack precise information concerning previous system states and proceed with rough approximations of wave interactions. Gosse and Zuazua present in chapter "Filtered Gradient Algorithms for Inverse Design Problems of One-Dimensional Burgers Equation" an open problem from aerospace engineering that requires the identification of all initial profiles leading to a given asymptotic profile, which is something akin to a multivalued adjoint problem. Deeper understanding of such a problem could help in the characterization of loss of information through shocks.

In two space dimensions and beyond, the numerics of conservation laws have largely been done under the assumption that high-order schemes in 1D, combined with dimensional flux-splitting, are sufficient to generate numerical approximations which converge to a class of unique solutions, something akin to the a posteriori conclusion of the Lax-Richtmyer theorem. Unfortunately, the absence of a general well-posedness theory in multi-D does not warrant such a casual attitude toward numerical methods in 2D. In chapter "A Numerical Glimpse at Some Non-standard Solutions to Compressible Euler Equations", Chiodaroli and Gosse attempt to numerically reproduce nonuniqueness in 2D solutions for isentropic flows by exploiting earlier explicit constructions of nonuniqueness by De Lellis and Szekelyhidi. As was already pointed out by Volker Elling, the nonuniqueness is caused by unbounded vorticity, but even when entropy admissibility is satisfied.

Numerical MHD developers, who work almost exclusively in multi-D, have always been strong proponents for the use of multi-D solvers, that is, solvers without dimensional splitting. In chapter "A Well-Balanced Scheme for the Euler Equations with Gravitation", Käppeli presents a well-balanced scheme that preserves the isothermal hydrostatic equilibrium. His work requires the standard local reconstruction procedure to be re-examined in order to ensure that the equilibrium is preserved, in conjunction with compatible discretizations of the potential. Gosse also presents, in chapter "Viscous Equations Treated with $\mathscr{L}$-Splines and Steklov-Poincaré Operator in two Dimensions", a re-interpretation of many classical solvers using local approximations that are solutions to local analogs of the global problem. In both cases, the authors find fault in reconstructions based on unphysical criteria.

As the authors of this book have made clear, some fundamental insights into hyperbolic problems are still lacking. It was the hope of the workshop organizers that by collecting these works into a monograph, this would encourage the community to re-examine some of these problems. It is this authors' belief that this monograph will go a long way toward reaching those goals.

Montréal, Canada
September 2016

Marc Laforest

Preface

Back in 1978, nothing worked. Now in 2013, almost everything works, but nothing works particularly well.

Professor of Aerospace Engineering
PHILIP L. ROE

Historically, algorithm development may be seen as an iterative optimization process subject to three constraints, severe and perhaps contradictory with each other:

- Accuracy, that is, the maximum deviation between the numerical approximation and the exact solution of the continuous model (assuming it is well posed, we shall come back to that issue later!) the user is willing to tolerate
- Stability, which usually refers to the ability of a constructive process not to diverge when the mesh size (or another "small parameter") is arbitrarily decreased
- Efficiency, or simply the cost of execution, in terms of CPU flops

A cornerstone result proved by Peter Lax, the so-called *"Equivalence Theorem"*, states that satisfying the first two criteria is a sufficient condition for convergence of the sequence of approximations. The third criterion has a meaning which varies a lot according to the epoch: nowadays, "costly" refers to tasks, the complexity of which would have been unthinkable[1] during Courant's, Richtmyer's, or Samarskii's time. A concrete manifestation of "cost constraints" was the introduction of so-called *variational finite-differences*, as a branch of Ritz-type approximation methods endowed with a sufficiently small stencil, so as to keep on allowing a rather cheap inversion of the resulting (sparse) mass matrix, even at high orders of accuracy.

Despite certain refinements, like uniform-type estimates originally developed for convection-dominated transport (nowadays rephrased as "asymptotic-preserving"

[1]New areas of research, like "Big Data," are partly spawned by contemporary computing power.

properties), the notion of "stability" is rather easy to apprehend; it can, though, be an arduous task to pick the right norm in which uniform bounds are to be sought. However, what really lies behind the term "accuracy" can be more puzzling and context dependent. The more elementary notion of accuracy relies on both truncation errors and Taylor's expansions, that is, on the density of polynomials in reasonable spaces of smooth, at least continuous, functions (Stone-Weierstrass theorem). However, when it comes to handling differential operators endowed with either weak solutions or advanced geometric features, such a notion has many shortcomings; researchers henceforth developed alternative, more elaborate notions of accuracy, including:

1. Symplectic integrators, typically for Hamiltonian systems in which longtime preservation of certain qualitative features of phase-space trajectories is crucial
2. Mimetic finite differences for multidimensional diffusion equations, where, given a strongly nonuniform gridding of a complex geometry, one is led to define compatible (possibly high-order) discretizations of ubiquitous differential operators, like div, ∇, and curl, in such a way that duality pairings (like Stokes formula), holding at the continuous level, are preserved at a discrete one
3. Consistency with both the original equation and a supplementary differential operator, of frequent use in continuum mechanics, where physically relevant (weak) solutions are believed to be identified thanks to an "entropy inequality"

These simple examples (to which might have been appended, e.g., hypocoercivity estimates) show that, through the passing of time, researchers switched from a notion of accuracy, as being measured mostly through remaining terms induced by finite-difference approximations of partial derivatives, to the notion of **preservation of either qualitative or quantitative features displayed by the exact solution of the underlying problem**, which is really more challenging, in the sense that it requires a thorough knowledge of the exact solution's peculiarities in order to wisely decide which ones are so salient as to be necessary in its discrete counterpart.

Accuracy $\rightarrow$ Exact Solution's Features Preservation

Clearly, in a context of, say, diffusion equations with rough, position-dependent coefficients, possibly admitting weak solutions endowed with sharp transition layers, usual truncation estimates give little, if any, useful information; the reason is that one really needs error bounds depending as less as possible on the coefficients' smoothness. Hence, the need of substituting usual polynomial bases (in both finite elements and finite differences) with more suitable, problem-fit functions was already well identified by Trefftz in his seminal 1926 paper: *"Instead of using trial functions that satisfy the boundary conditions but violate the differential equation, we approximate the solution by selecting functions that violate the boundary conditions but satisfy the differential equation. We can view this procedure as a generalization of series expansions using particular solutions and the new approach in fact contains the conventional series expansions as a special case."* Such an idea lies at the roots of many modern algorithms for elliptic or parabolic equations, like Scharfetter-Gummel's exponential fit (sometimes called "harmonic averaging"), $\mathscr{L}$-splines, patch functions, tailored schemes, or virtual elements methods. Moreover, it can be set up for treating both hyperbolic and kinetic models, too, as it corresponds to so-called "well-balanced" or "Caseology" approaches, respectively.

Accordingly, the book aims at presenting contributions, written by several leading researchers, which address specific and challenging applications along with some corresponding approximation strategies:

- *Lagrangian schemes and wavefront-tracking*. Lagrangian discretizations are usually a good alternative to standard, "Eulerian" ones when equations preserving both positivity and the total mass are posed in an unbounded domain. Then it is possible to "compactify" the computational domain by tracking the location of particles, instead of solving for the repartition of their mass density. Two contributions deal with such topics, both aiming at different objectives:
  - The one proposed by Amadori et al. concerns a wavefront-tracking algorithm designed for solving the Kuramoto-Sakaguchi model of synchronization in a simple setting where a limited number of frequencies is present.
  - The text of Matthes and Söllner is concerned with a Lagrangian scheme approximating nonlinear drift-diffusion with nonlocal aggregation terms. A thorough and quantitative study of such an algorithm, involving a minimization process at each time step, is produced, along with practical validations.
- *Astrophysics and post-Newtonian computations*. Practical numerical computation of atmospheres surrounding white dwarf, i.e., rather dense, stars is a challenging task. Especially, computer codes should be able to efficiently preserve isothermal hydrostatic steady states in the presence of powerful gravity. Moreover, Käppeli asks them to be reliable on dynamics like the well-known Rayleigh-Taylor instability, inducing possibly strong velocity fields. At the other end of the spectrum lie the amazing results by Bini et al., aiming at very specific applications, like gravitational wave detection and identification, in the realm of which the weakness of observable signals implies that the algorithms must be endowed with an extremely high accuracy (in terms of significant figures). Since a general relativistic computation of the trajectory of test particles in a black hole (Schwarzschild or Kerr) background is still mostly unfeasible, one relies on post-Newtonian approximations, but extreme care is needed so as to ensure that cumulative truncations errors remain below the scale of the observables to be retrieved.
- *Corrugated isometric embeddings in 2D*. To isometrically embed an abstract Riemannian manifold into an Euclidean space is a problem with a long, vivid, and prestigious story: from Hilbert stating that the 2D hyperbolic half-plane cannot be embedded into $\mathbb{R}^3$ to the amazing Nash-Kuiper algorithm, producing a

"corrugated" surface isometric to the flat torus in ambient space.[2] It turns out that, thanks to results previously accumulated in the realm of 1D hyperbolic systems of balance laws, an astute reformulation of Gauss-Codazzi equations involving "fluid dynamics variables" allows to prove, existence of such new embeddings of low regularity. Concrete examples are provided in Christoforou's contribution.

- *"Caseology" and $\mathscr{L}$-spline interpolation.* Accurate numerical discretizations can sometimes be derived by working out local exact solutions of the differential operator showing up in the PDEs under consideration; this idea is at the basis of Trefftz's method, the so-called homogeneous schemes of Tikhonov and Samarskii, the famous Scharfetter-Gummel's scheme for drift-diffusion equations, or, more recently, the well-balanced method for hyperbolic systems of (position-dependent) balance laws. It turns out that such ideas yield powerful approximations of solutions of (mostly $(1 + 1)$-dimensional) kinetic models as well. A spectacular example is provided by Calvez et al., who set up the method of "Case's elementary solutions" in the context of both an existence proof of traveling chemotactic waves of bacteria in micro-channels and their practical computation in a particularly hazardous context of bistability for certain ranges of parameters. A related method of "elementary solutions" is set up by Lorenzani, who deals with a challenging linearized BGK (so-called MacCormack's) kinetic model of binary diluted gaseous mixtures. It is remarkable that such a model, apparently holding in $1_x + 1_v$ dimensions, takes into account collisions involving three-dimensional velocities, albeit at the price of an intricate algebra. However, this intricacy doesn't prevent the production of numerical results of practical interest, especially relevant for MEMS (microelectromechanical systems) applications. Finally, when it comes to macroscopic, strictly parabolic equations, and by analogy with Scharfetter-Gummel's original ideas, Gosse presents a seemingly original manner of systematically building accurate numerical fluxes by means of the Steklov-Poincaré (also called "Dirichlet-to-Neumann," "voltage-to-current") pseudo-differential operator; in 2D, this formulation allows to derive efficient approximations of drift-diffusion equations, even in a context where their solutions may develop boundary, or internal, layers.
- *Specific solutions of conservation laws.* Establishing well-posedness in the sense of Hadamard for the equations of multidimensional fluid mechanics witnessed a dramatic twist when, in 2010, De Lellis and Székelyhidi proved that 2D incompressible Euler equations satisfy Gromov's h-principle. Such a breakthrough allows Nash-Kuiper algorithms to be invoked, in order to prove the existence of an infinity of weak, oscillating, and bounded solutions, hence ill-posedness. Later, it was the task of Chiodaroli to extend their proofs to 2D compressible, isentropic Euler equations admitting shock waves. In such a context, it was proved that for specific initial data, beyond shock onset, an infinity of weak, highly oscillating, entropy-dissipating bounded solutions do

[2]See V. Borrelli, S. Jabrane, F. Lazarus, B. Thibert, Flat tori in three-dimensional space and convex integration. Proc. Natl. Acad. Sci. USA **109** (2012) 7218–7223.

exist. What triggers curiosity is that the roadmap leading to these existence results was furnished by differential geometry techniques, instead of a more standard method of vanishing viscosity, which proved successful for establishing well-posedness in the one-dimensional case. So, the question of the ability of numerical finite-difference schemes, inevitably replete of artificial viscosity, to produce any approximation of these so-called nonstandard solutions came quickly at the forefront. Accordingly, a first step was, for Chiodaroli and Gosse, the derivation of convenient, possibly Lipschitz, initial data, for which former theorems ensure a multiplicity of oscillating solutions after shock onset. Then, a convenient numerical setup was built (on Cartesian grids), exploiting Euler equation's rotational invariance through a specific commutator, for which it resulted that oscillations, apparently compatible with nonstandard existence theory, were detected at the expected place.

– *Entropy dissipation, irreversibility, and inverse design*. Issues related to sonic-boom optimization for aircrafts gave rise to harsh control problems for inviscid systems of conservation laws. As a preliminary stepping stone, a simplification of compressible Euler equations yield a Burgers-type conservation law involving possibly nonlocal terms, for which specific pressure profiles, so-called F-functions, are particularly relevant. Then one faces the so-called "inverse design problem", consisting in, given a certain target profile at time $T > 0$, deducing the (set of) initial data able to generate it under the action of the solution operator. Gosse and Zuazua prove that such a set of initial designs is necessarily a convex of L^∞. Being L^2-dissipative because of entropy inequalities, hence mostly irreversible even in the presence of admissible targets, it can be fairly delicate to set up a numerical iterative process because of nonuniqueness of initial profiles. One way to visualize in practice several types of initial design is filtering, especially into scale-limited subspaces of L^2, by means of "father wavelet" functions.
– *Convergence rate estimation*. Stability theory of the underlying continuous equation usually permits proofs of rigorous asymptotics for the global error of corresponding numerical schemes with the grid-size parameter set to zero. A famous example is the Kuznetsov $\frac{1}{2}$-rate governing most entropy-consistent discretizations, proof of which relies heavily on Kružkov's L^1 stability results. Yet, the sharpness of such rates can be quite delicate to assess. Risebro et al. study such issues in the context of degenerate parabolic equations, meaning that they cannot be reduced as "a dominant diffusion process perturbed by lower-order convection terms," in both one and two space dimensions. One difficulty lies in the weak regularity of solutions, which may display discontinuities. Even worse, in the contribution by Lagoutière and Vauchelet, convergence rates are rigorously obtained for aggregation equations admitting measure solutions possibly involving Dirac masses! Obviously, the choice of the metric in which such a convergence process can be observed is delicate, and the authors found out that the so-called Wasserstein metric is the correct one in order to recover a Kuznetsov-type $\frac{1}{2}$-rate.

Unfortunately, certain speakers at the workshop were not able to contribute. Among them were T. Foglizzo, who may have offered a valuable perspective on specific experimental shallow water models for neutron star accretion, and B. Després, whose results on polynomial moment methods for uncertainty quantification in scalar conservation laws would have been complementary to several other contributions.

Better luck next time!

Rome, Italy Laurent Gosse
Rome, Italy Roberto Natalini
November 2016

Contents

Contributors

Debora Amadori DISIM, University of L'Aquila, L'Aquila, Italy

Donato Bini Istituto per le Applicazioni del Calcolo (CNR), Rome, Italy

Vincent Calvez Institut Camille Jordan, Université de Lyon 1, Lyon, France
Inria, Project-Team NUMED, Lyon, France

Elisabetta Chiodaroli École Polytechnique Fédérale de Lausanne, Lausanne, Switzerland

Cleopatra Christoforou Department of Mathematics and Statistics, University of Cyprus, Nicosia, Cyprus

Thibault Damour Institut des Hautes Etudes Scientifiques, Bures-sur-Yvette, France

Andrea Geralico Astrophysical Observatory of Torino, INAF, Pino Torinese, Italy

Laurent Gosse Istituto per le Applicazioni del Calcolo (CNR), Rome, Italy

Seung-Yeal Ha Department of Mathematical Sciences, Seoul National University, Seoul, Korea

Roger Käppeli ETH Zürich, Seminar for Applied Mathematics, Zürich, Switzerland

Kenneth H. Karlsen Department of Mathematics, University of Oslo, Oslo, Norway

Marc Laforest École Polytechnique de Montréal, Montréal, QC, Canada

Frédéric Lagoutiere Institut Camille Jordan, Université Claude Bernard Lyon 1, Villeurbanne, France

Silvia Lorenzani Politecnico di Milano, Milan, Italy

Daniel Matthes Technische Universität München, Garching, Germany

Jinyeong Park Departamento de Matemática Aplicada, Universidad de Granada, Granada, Spain

Nils H. Risebro Department of Mathematics, University of Oslo, Oslo, Norway

Benjamin Söllner Technische Universität München, Garching, Germany

E.B. Storrøsten Department of Mathematics, University of Oslo, Oslo, Norway

Monika Twarogowska Unité de Mathématiques Pures et Appliquées, Ecole Normale Supérieure de Lyon, Lyon, France
Inria, Project-Team NUMED, Lyon, France

Nicolas Vauchelet LAGA - UMR 7539, Institut Galilée, Université Paris 13, Villetaneuse, France

Enrique Zuazua DeustoTech - Fundación Deusto, Bilbao, Basque Country, Spain
Departamento de Matemáticas, Universidad Autónoma de Madrid, Madrid, Spain
Facultad Ingeniería, Universidad de Deusto, Bizkaia, Spain

A Nonlocal Version of Wavefront Tracking Motivated by Kuramoto-Sakaguchi Equation

Debora Amadori, Seung-Yeal Ha, and Jinyeong Park

Abstract In this paper, we present a modified wave-front tracking algorithm which is suitable for the analysis of scalar conservation laws with nonlocal terms. This method has been first employed in Shen and Zhang (Arch Ration Mech Anal 204:837–879, 2012) to analyze a nonlocal Hamilton-Jacobi equation related to a granular flow and later used in other works. Such an approach leads to a possibly simpler analysis in obtaining rigorous quantitative estimates on approximate solutions, compared to a classical iteration procedure based on the recomputation of the nonlocal term at each time step. Here, we delineate this method for a nonlocal equation namely "*the Kuramoto-Sakaguchi equation*" arising from the kinetic modeling of collective motion of large ensemble of Kuramoto oscillators, for which BV-weak solutions and their large time behavior are investigated in Amadori et al. (J Differ Equ **262**, 978–1022, 2017).

1 Introduction

The jargon "*synchronization*" denotes the phenomenon in which rhythms of coupled oscillators are adjusted due to their weak interactions in the sense that the magnitude is invariant, whereas the phases are coupled. In this paper, we present a modified wave-tracking algorithm adapted for the scalar conservation law with a nonlocal flux. To fix the idea, we consider a kinetic model arising from the study of collective dynamics of many weakly coupled Kuramoto oscillators. We refer to [1]

D. Amadori (✉)
DISIM, University of L'Aquila, L'Aquila, Italy
e-mail: amadori@univaq.it

S.-Y. Ha
Department of Mathematical Sciences and Research Institute of Mathematics, Seoul National University, Seoul 08826, Korea

Korea Institute for Advanced Study, Hoegiro 85, Seoul 02455, Korea
e-mail: syha@snu.ac.kr

J. Park
Departamento de Matemática Aplicada, Universidad de Granada, Granada 18071, Spain
e-mail: jpark@ugr.es

L. Gosse, R. Natalini (eds.), *Innovative Algorithms and Analysis*,
Springer INdAM Series 16, DOI 10.1007/978-3-319-49262-9_1

for a general review of Kuramoto phase model. We consider a large ensemble of Kuramoto oscillators whose state is effectively approximated by moments of some one-oscillator probability density function, $f(\theta, \Omega, t)$, at phase θ, natural frequency Ω and time t. Dynamics follow the Kuramoto-Sakaguchi (KS) equation [12],

$$\begin{aligned} &\partial_t f + \partial_\theta(\omega[f]f) = 0 \quad \theta \in \mathbf{T} := \mathbf{R}/2\pi\mathbf{Z}, \ \ \Omega \in \mathbf{R}, \ \ t \in \mathbf{R}_+, \\ &\omega[f](\theta, \Omega, t) = \Omega - KL[f](\theta), \\ &L[f(\cdot, t)](\theta) := \int_{\mathbf{T}\times\mathbf{R}} \sin(\theta - \theta_*) f(\theta_*, \Omega, t) d\theta_* d\Omega \end{aligned} \tag{1}$$

subject to suitable initial datum:

$$\begin{aligned} &f(\theta, \Omega, 0) = f_0(\theta, \Omega), \qquad \int_{\mathbf{T}} f_0(\theta, \Omega)\, d\theta = g(\Omega), \\ &\int_{\mathbf{R}} \Omega g(\Omega)\, d\Omega = 0, \qquad \int_{\mathbf{R}} g(\Omega)\, d\Omega = 1, \end{aligned} \tag{2}$$

where K and $g(\Omega)$ are positive coupling strength and density function for natural frequencies, respectively. Thanks to (1) and to the periodicity of f and $L[f]$, the density function $g(\Omega) = \int_{\mathbf{T}} f(\theta, \Omega, t)\, d\theta$ is independent on time. Notice also that the operator L can be expressed in terms of the density

$$\rho(\theta, t) := \int_{\mathbf{R}} f(\theta, \Omega, t) d\Omega$$

by

$$L[\rho(\cdot, t)](\theta) := \int_{\mathbf{T}} \sin(\theta - \theta_*) \rho(\theta_*, t) d\theta_*\,,$$

so that the equation in (1) rewrites as a kinetic model

$$\boxed{\partial_t f + \Omega \partial_\theta f - K\partial_\theta(L[\rho]f) = 0\,.} \tag{3}$$

We refer to [3] for detailed mathematical description for (1)–(2). The purpose of this paper is to present analytical tools to approximate bounded weak solutions to (1)–(2), mainly for simple frequency distributions $\Omega \mapsto f(\theta, \Omega, t)$. They possibly lead to a quantitative analysis on the large-time behavior of solutions, and also serve as a numerical method to simulate weak solutions to Eq. (1).

In Sect. 2, a particular situation, already considered in e.g. [5, §3], [6, §4], [4, §9.1], and involving an ensemble of Kuramoto oscillators oscillating at identical frequency, is considered. In such a case, a simple phase-shift $\theta \to \theta + \Omega t$ allows to eliminate the transport term in (3) (because $g(\Omega)$ is independent of time), yielding somehow paradoxically a synchronization process for oscillators at zero frequency. Accordingly, **the natural frequency measure $g(\Omega)$ is monokinetic**,

$$g(\Omega) = \int_{\mathbf{T}} f(\theta, \Omega, t)\, d\theta = \delta(\Omega).$$

Integrating (3) in Ω gives a macroscopic equation for $\rho(\theta, t) = f(\theta, 0, t)$,

$$\partial_t \rho - K\partial_\theta(L[\rho]\rho) = 0, \quad (\theta, t) \in \mathbf{T} \times \mathbf{R}. \tag{4}$$

Equation (4) is a scalar conservation law with a nonlocal flux on a periodic domain **T**, which is structurally analogous to the equations dealt in [2, 7, 8, 11, 14] where the spatial domain is the whole space **R** or $\mathbf{R}^n$. We shall discuss the front-tracking algorithm describing the continuous-time coupled dynamics of phase fronts and heights, and we shall also study the well-ordering principle of fronts and dynamics of order parameters associated with front-height dynamics. In Sect. 3, a special case of (1) for non-identical oscillators is rendered by means of a bi-modal (or "bi-kinetic", by opposition to the previous mono-kinetic) density function g,

$$g(\Omega) = m^- \delta_{-\bar{\Omega}} + m^+ \delta_{\bar{\Omega}}, \qquad \bar{\Omega} \in (0, \infty),$$

which yields a "two-stream approximation" of (3), with obvious notation,

$$\begin{aligned} &\partial_t \rho^- - \bar{\Omega}\partial_\theta \rho^- - K\partial_\theta(L[m^+\rho^+ + m^-\rho^-]\rho^-) = 0, \\ &\partial_t \rho^+ + \bar{\Omega}\partial_\theta \rho^+ - K\partial_\theta(L[m^+\rho^+ + m^-\rho^-]\rho^+) = 0. \end{aligned}$$

For the time being, a complete analysis for the non-identical case in the frame work front-tracking algorithm is still lacking, hence our analysis in Sect. 3 will provide a first stepping stone toward the general theory of the KS equation for nonidentical oscillators. In Sect. 4, we present several numerical examples to illustrate our front-tracking scheme and compare them with analytical results in Sects. 2 and 3 (we refer to [3] for more details). Finally, Sect. 5 is devoted to the brief summary of this paper and discussion of possible future directions.

2 A Special Case of Identical Oscillators, $\Omega = 0$

In this section, we consider (1) for identical oscillators, so that $g(\Omega) = \delta(\Omega)$:

$$\rho(\theta, t) = \int_{\mathbf{T}} f(\theta, \Omega, t) g(\Omega) d\Omega = f(\theta, 0, t).$$

Then, the local phase density ρ satisfies the Cauchy problem for the continuity equation with nonlocal velocity field:

$$\begin{aligned} &\partial_t \rho - K\partial_\theta (L[\rho]\rho) = 0, \quad \theta \in \mathbf{T},\ t > 0, \\ &L[\rho(\cdot,t)](\theta) := \int_{\mathbf{T}} \sin(\theta - \theta_*)\rho(\theta_*,t)d\theta_*, \\ &\rho(\theta,0) = \rho_0(\theta) \in BV(\mathbf{T}), \quad \rho_0(\theta) \geq 0, \quad \int_{\mathbf{T}} \rho_0 \, d\theta = 1. \end{aligned} \tag{5}$$

Weak solutions of (5) is understood in distributional sense.

Kuramoto Order Parameters We next introduce order parameters $R(t)$ and $\varphi(t)$ as the amplitude and phase of the centroid of $e^{i\theta}$:

$$R(t)\, \mathrm{e}^{i\varphi(t)} := \int_{\mathbf{T}} \mathrm{e}^{i\theta} \rho(\theta,t)\, d\theta. \tag{6}$$

In (6), we compare the real and imaginary parts to find

$$R(t)\cos\varphi = \int_{\mathbf{T}} \cos(\theta)\rho(\theta,t)d\theta, \qquad R(t)\sin\varphi = \int_{\mathbf{T}} \sin(\theta)\rho(\theta,t)\, d\theta. \tag{7}$$

and recalling the definition (5) of L:

$$L[\rho](\theta) = R(t)\,(\sin\theta\cos\varphi - \cos\theta\sin\varphi) = R(t)\sin(\theta - \varphi(t)). \tag{8}$$

This yields an alternative form for the Kuramoto-Sakaguchi equation:

$$\partial_t \rho - K\partial_\theta (R(t)\sin(\theta - \varphi(t))\,\rho) = 0.$$

By direct calculation from (7), we have the dynamical systems for (R,φ):

$$\begin{aligned} \dot{R} &= KR \int_{\mathbf{T}} \rho \sin^2(\theta - \varphi)\, d\theta \geq 0, \\ R\dot{\varphi} &= -KR \int_{\mathbf{T}} \rho \sin(\theta - \varphi)\cos(\theta - \varphi) d\theta. \end{aligned}$$

The Front-Tracking Algorithm Here we present a procedure to define approximate solutions to problem (5) in two steps. Note that the continuity equation (5) for ρ can be rewritten in a quasilinear form, i.e. a transport equation with a source term:

$$\boxed{\partial_t \rho - KL[\rho]\partial_\theta \rho = K\rho(\partial_\theta L[\rho]), \quad (\theta,t) \in \mathbf{T}\times\mathbf{R}.} \tag{9}$$

Let $(\theta(t), \rho(t)) = (\theta(t; \theta_0), \rho(t; \rho_0)$ be a forward characteristics associated with (9) which is a solution to the characteristic system:

$$\begin{aligned} &\dot{\theta} = -KL[\rho], \quad \dot{\rho} = K\rho(\partial_\theta L[\rho]), \quad t > 0, \\ &(\theta, \rho)(0) = (\theta_0, \rho_0). \end{aligned} \tag{10}$$

- *Step A: Sampling of the initial data* Let $N \in \mathbf{N}$, and set

$$\Delta\theta = \frac{2\pi}{N}, \quad \theta_{0j} = j\Delta\theta, \quad j = 1, \cdots, N-1 \tag{11}$$

  We choose ρ_{0j}, the initial value of ρ in the interval $(\theta_{0j}, \theta_{0(j+1)})$, as follows:

$$\rho_{0j} := \rho_0(\theta_{0i}+), \qquad j = 0, \ldots, N-1.$$

  where we set $\theta_{N+1} = \theta_1 + 2\pi, \theta_0 = \theta_N - 2\pi$.
- *Step B: Approximate solutions* We define an approximate solution $\rho_N = \rho(\theta, t)$ on a periodicity interval $(\theta_0(t), \theta_N(t))$ as a function of the form

$$\rho(\theta, t) = \sum_{i=0}^{N-1} \rho_i(t)\chi_{(\theta_i(t), \theta_{i+1}(t))}(\theta), \qquad \theta \in (\theta_0(t), \theta_N(t)). \tag{12}$$

  Motivated by the characteristic system (10), we consider the following discrete system of characteristic equations:

$$\begin{aligned} &\dot{\theta}_j = -KL[\rho](\theta_j), \quad t > 0, \\ &\dot{\rho}_j = K\rho_j \frac{L[\rho](\theta_{j+1}) - L[\rho](\theta_j)}{\theta_{j+1} - \theta_j} \end{aligned} \tag{13}$$

  subject to initial data:

$$\theta_j(0) = \theta_{0j}, \qquad \rho_j(0) = \rho_{0j}, \qquad j = 0, \ldots, N-1.$$

Let $I(t)$ be the L^1 norm of $\rho(\cdot, t)$ on the periodic domain $\mathbf{T}$, that is

$$I(t) := \|\rho(\cdot, t)\|_{L^1(\mathbf{T})} = \sum_{j=0}^{N-1} \rho_j \left(\theta_{j+1} - \theta_j\right)$$

Dynamics of the Front-Height System (13) In this part, we provide two a priori estimates "*well-ordering of fronts*" and "*monotonicity of R*" associated with the dynamics of the front-height system.

Proposition 1 (See [3]) *Let $(\rho_i(t), \theta_i(t))$ be a solution to the characteristic system (13). Then, the following assertions hold:*

1. *The set of discontinuities satisfies*

$$\theta_0(t) < \theta_1(t) < \theta_2(t) < \cdots < \theta_N(t), \quad \forall\, t > 0.$$

2. *The local masses are conserved:*

$$\rho_j(t)\left(\theta_{j+1}(t) - \theta_j(t)\right) = \rho_j(0)\left(\theta_{j+1}(0) - \theta_j(0)\right), \quad j = 0, \ldots, N-1$$

and hence the total mass is conserved: $I(t) = I(0)$.

Both the properties stated in Proposition 1 hold for the discrete ODE system (13), which itself approximates the continuous macroscopic equation (4). Such a conservation law involving a nonlocal flux is well handled by so-called "Lagrangian schemes", see [9, 13], in which one tracks the time-evolution of $\theta(t, \rho_j)$. Especially, the "well-ordering of fronts" corresponds to the monotonicity-preserving property shown in [9, §2], and hereafter in [13].

The Kuramoto Order Parameters For the approximate solution $\rho(\theta, t)$ given in (12) (we omit the dependence on N for simplicity), we define the Kuramoto order parameters $R(t)$ and $\varphi(t)$ as in (6). Using (8), we rewrite the system (13) as

$$\boxed{\dot{\theta}_j = -KR\sin\left(\theta_j - \varphi\right), \qquad \dot{\rho}_j = KR\rho_j \frac{\sin\left(\theta_{j+1} - \varphi\right) - \sin\left(\theta_j - \varphi\right)}{\theta_{j+1} - \theta_j}.} \tag{14}$$

Note that θ_j is piecewise monotonic:

$$\dot{\theta}_j < 0 \quad \text{if } \varphi < \theta_j < \varphi + \pi, \qquad \dot{\theta}_j > 0 \quad \text{if } \varphi - \pi < \theta_j < \varphi.$$

The sign of $\left(\theta_j - \varphi\right)$ may change in time. Moreover one has, for $j < \ell$,

$$\begin{aligned} \dot{\theta}_\ell - \dot{\theta}_j &= -KR\left(\sin\left(\theta_\ell - \varphi\right) - \sin\left(\theta_j - \varphi\right)\right) \\ &= -2KR\sin\left(\frac{\theta_\ell - \theta_j}{2}\right)\cos\left(\frac{\theta_\ell + \theta_j}{2} - \varphi\right). \end{aligned}$$

Note that $0 < \theta_\ell - \theta_j < 2\pi$, so that the term $\sin(\ldots)$ in the line above is positive. In the following lemma we deduce the evolution equations for R and ϕ.

Lemma 1 *Let* (θ_j, ρ_j) *be a solution to the characteristic system (14). Then, the following identities hold:*

$$\begin{aligned}
\dot{R} &= KR\sum_{j=0}^{N-1} \rho_j \mathscr{L}(\theta_j - \varphi, \theta_{j+1} - \varphi), \\
R\dot{\varphi} &= KR\sum_{j=0}^{N-1} \rho_j \mathscr{F}(\theta_j - \varphi, \theta_{j+1} - \varphi),
\end{aligned} \tag{15}$$

where the functions $\mathscr{L}$ *and* $\mathscr{F}$ *are defined as follows.*

$$\mathscr{L}(x, y) := \frac{(\sin y - \sin x)^2}{y - x} - [\cos y \sin y - \cos x \sin x] \tag{16}$$

$$\mathscr{F}(x, y) := -(\sin y - \sin x)\Big(\frac{(\cos y - \cos x)}{y - x} + (\sin y + \sin x)\Big). \tag{17}$$

Proof

(i) It follows from (6) that one obtains

$$R(t) = \int_{\mathbf{T}} e^{i(\theta - \varphi(t))} \rho(\theta, t)\, d\theta.$$

This yields

$$\begin{aligned}
R(t) &= \int_{\mathbf{T}} \cos(\theta - \varphi(t)) \rho(\theta, t)\, d\theta, \\
0 &= \int_{\mathbf{T}} \sin(\theta - \varphi(t))\, \rho(\theta, t)\, d\theta.
\end{aligned} \tag{18}$$

or equivalently

$$\begin{aligned}
R &= \sum_{j=0}^{N-1} \rho_j \left(\sin\left(\theta_{j+1} - \varphi\right) - \sin\left(\theta_j - \varphi\right)\right), \\
0 &= \sum_{j=0}^{N-1} \rho_j \left(\cos\left(\theta_{j+1} - \varphi\right) - \cos\left(\theta_j - \varphi\right)\right).
\end{aligned}$$

By using (18) and the above identities, we find that

$$\begin{aligned}\dot{R} &= \frac{d}{dt}\sum_{j=0}^{N-1}\left[\rho_j \int_{\theta_j}^{\theta_{j+1}} \cos(\theta - \varphi(t))d\theta\right] \\ &= \underbrace{\dot{\varphi}\int_{\mathbf{T}} \sin(\theta - \varphi(t))\,\rho(\theta,t)\,d\theta}_{=0} + \sum_{j=0}^{N-1} I_j \\ &= \sum_{j=0}^{N-1} I_j,\end{aligned}$$

where the term I_j is defined as follows.

$$I_j = \dot{\rho}_j\left(\sin(\theta_{j+1}-\varphi) - \sin(\theta_j-\varphi)\right) + \rho_j\left(\cos(\theta_{j+1}-\varphi)\theta'_{j+1} - \cos(\theta_j-\varphi)\dot{\theta}_j\right).$$

By means of (14), the term I_j rewrites as

$$\begin{aligned}I_j &= KR\rho_j\left\{\frac{\left(\sin(\theta_{j+1}-\varphi) - \sin(\theta_j-\varphi)\right)^2}{\theta_{j+1}-\theta_j}\right. \\ &\qquad \left. - \left[\cos(\theta_{j+1}-\varphi)\sin(\theta_{j+1}-\varphi) - \cos(\theta_j-\varphi)\sin(\theta_j-\varphi)\right]\right\} \\ &= KR\rho_j \mathscr{L}(\theta_j-\varphi, \theta_{j+1}-\varphi)\end{aligned}$$

with $\mathscr{L}$ as in (16).

(ii) An equation for $\dot{\varphi}$ is deduced from the second identity in (18). Indeed, by taking its time derivative, one gets

$$\dot{\varphi}(t)R(t) = \sum_{j=0}^{N-1} \dot{\rho}_j \int_{\theta_j}^{\theta_{j+1}} \sin(\theta-\varphi)d\theta - \sum_{j=0}^{N-1} \rho_j\left(\theta'_{j+1}\sin(\theta_{j+1}-\varphi) - \theta'_j\sin(\theta_j-\varphi)\right)$$

so that

$$\begin{aligned}R\dot{\varphi} &= KR\sum_{j=0}^{N-1}\rho_j\left[\frac{\sin(\theta_{j+1}-\varphi) - \sin(\theta_j-\varphi)}{\theta_{j+1}-\theta_j}\int_{\theta_j}^{\theta_{j+1}} \sin(\theta-\varphi)\,d\theta\right. \\ &\qquad \left. - \left(\sin^2(\theta_{j+1}-\varphi) - \sin^2(\theta_j-\varphi)\right)\right] \\ &= KR\sum_{j=0}^{N-1}\rho_j \mathscr{F}(\theta_j-\varphi, \theta_{j+1}-\varphi)\end{aligned}$$

with $\mathscr{F}$ as in (17). Hence (15) is proved. □

In the following lemma, we study the sign of the quantity $\mathscr{L}$ in (16).

Lemma 2 *If* $x < y \leq x + \frac{\pi}{2}$, *then,*

$$\mathscr{L}(x,y) = \frac{(\sin y - \sin x)^2}{y-x} - [\cos y \sin y - \cos x \sin x] > 0\,.$$

Proof The argument of this proof refines slightly the one of Lemma 4.3 in [3]. Indeed, let $x < y$ with $x \in [-\pi, \pi]$. Then, we have

$$\begin{aligned}(y-x)\mathscr{L} &= (\sin y - \sin x)^2 - (y-x)\,[\cos y \sin y - \cos x \sin x] \qquad (19)\\ &= \left(\int_x^y \cos s\, ds\right)^2 + (y-x)\int_x^y (1 - 2\cos^2 s)\, ds\\ &= \left(\int_x^y \cos s\, ds\right)^2 + (y-x)^2 - 2(y-x)\int_x^y \cos^2 s\, ds\,.\end{aligned}$$

Two possible cases occur.

(1) If $\cos s$ does not change its sign on the interval (x, y), as for $-\pi/2 \leq x < y \leq \pi/2$ or $\pi/2 \leq x < y \leq 3\pi/2$, then

$$\begin{aligned}\left(\int_x^y \cos s\, ds\right)^2 + (y-x)^2 &> 2(y-x)\left|\int_x^y \cos s\, ds\right| = 2(y-x)\int_x^y |\cos s|\, ds\\ &> 2(y-x)\int_x^y \cos^2 s\, ds\,,\end{aligned}$$

so that $\mathscr{L} > 0$.

(2) On the other hand, if $x \in [0, \pi/2]$ and $y \in [\pi/2, \pi]$, then $\cos x \sin x \geq 0$ and $\cos y \sin y \leq 0$. The same conclusion holds if $x \in [-\pi, -\pi/2]$ and $y \in [-\pi/2, 0]$. Hence from (19) one finds that $\mathscr{L} \geq 0$ (being $\mathscr{L} = 0$ only if $x = 0$, $y = \pi$).

In particular, let $x \in [-\pi, \pi]$ and $x < y \leq x + \frac{\pi}{2}$. If the condition in **(1)** is not satisfied, it means that $\cos s$ changes its sign on the interval $[x, y] \subset [x, x + \pi/2]$. Then the argument in **(2)** can be applied. □

In the next proposition, we study the strict monotonicity of the order parameter R.

Proposition 2 *For a positive constant* $T \in (0, \infty]$, *let* (θ_i, ρ_i) *be a solution to the front-height system (13) in the time-interval* $[0, T)$ *satisfying*

$$\sup_{0 \leq t < T}\ \max_{0 \leq j \leq N-1} |\theta_{j+1}(t) - \theta_j(t)| \leq \frac{\pi}{2}. \qquad (20)$$

Then, R *is monotonically increasing in* $t \in [0, T)$.

Proof It follows from (20) and Lemma 2 that

$$\mathscr{L}(\theta_j - \varphi, \theta_{j+1} - \varphi) > 0.$$

Again, it follows from the equation for R in Lemma 1 that

$$\dot{R} = KR \sum_{j=0}^{N-1} \rho_j \mathscr{L}(\theta_j - \varphi, \theta_{j+1} - \varphi) \geq 0.$$

Hence R is monotonically increasing at time $t \in [0, T)$. □

As a corollary, we have the following estimate.

Corollary 1 *If* $R(0) > 0$ *and* $N \geq 4$, *then* $\dot{R}(0) > 0$.

Proof Recalling (11), one has $\theta_{0(j+1)} - \theta_{0j} = 2\pi/N \leq \pi/2$. Then, thanks to Lemma 2, each term in the expression for $\dot{R}$ [see (15)] is positive. □

As an extension of the analysis covered in this section, we can consider a generalized KS equation by letting the coupling strength depend on the spatial position:

$$\begin{aligned} &\partial_t \rho - \partial_\theta (a(L[\rho])\rho) = 0, \quad (x, \theta, t) \in \mathbf{R} \times \mathbf{T} \times \mathbf{R}_+ \\ &L[\rho](\theta, t) := \int_{\mathbf{T}} \sin(\theta - \theta_*)\rho(\theta_*, t) d\theta_*, \end{aligned} \tag{21}$$

where

$$a = a(x) \in C^1(\mathbf{R}), \quad 0 \leq a'(x) \leq \bar{a} \quad \forall x \in \mathbf{R}, \tag{22}$$

as in [11]. We will not further investigate Eqs. (21)–(22) in this note.

3 Two Different Oscillators with Opposite Frequencies

In this section, we consider the ensemble of non-identical oscillators where the density function $g(\Omega)d\Omega$ is not a Dirac measure. In this case, fronts θ_j in (11) may not preserve their order along the dynamics (13) at least in finite-time interval. However, the crossing between fronts are at most countable. Hence we can define approximate solutions accordingly. This issue will be treated in our future work. As an immediate generalization of identical oscillator case, we consider bi-modal distribution function for $g(\Omega)d\Omega$:

$$g(\Omega)d\Omega = m^- \delta_{-\bar{\Omega}} + m^+ \delta_{\bar{\Omega}}, \qquad \bar{\Omega} \in (0, \infty), \quad m^- + m^+ = 1.$$

In this case, recalling that the distribution $gd\Omega$ is preserved in time, it is reasonable to introduce the partial densities $\rho^{\pm}$:

$$\rho^-(\theta,t) := f(\theta,-\bar{\Omega},t), \qquad \rho^+(\theta,t) := f(\theta,\bar{\Omega},t).$$

Then, it is easy to see that

$$\begin{aligned}\rho(\theta,t) &= \int_{\mathbf{R}} f(\theta,\Omega,t)g(\Omega)d\Omega = \int_{\mathbf{R}} f(\theta,\Omega,t)\Big(m^-\delta_{-\bar{\Omega}} + m^+\delta_{\bar{\Omega}}\Big)d\Omega \\ &= m^-\rho^-(\theta,t) + m^+\rho^+(\theta,t),\end{aligned}$$

along with,

$$\begin{aligned}L[\rho(\cdot,t)](\theta) &= \int_{\mathbf{T}} \sin(\theta-\theta_*)(m^+\rho^+ + m^-\rho^-)(\theta_*,t)d\theta_* \\ &= m^+L[\rho^+] + m^-L[\rho^-].\end{aligned}$$

Recalling Eq. (3), after integration in $d\Omega$ and localization around $\pm\bar{\Omega}$, one obtains that the local density functions $\rho^{\pm}$ satisfy the coupled 2×2 system:

$$\begin{aligned}&\partial_t\rho^- - \bar{\Omega}\partial_\theta\rho^- - K\partial_\theta(L[m^+\rho^+ + m^-\rho^-]\rho^-) = 0, \quad (\theta,t)\in\mathbf{T}\times\mathbf{R}_+, \\ &\partial_t\rho^+ + \bar{\Omega}\partial_\theta\rho^+ - K\partial_\theta(L[m^+\rho^+ + m^-\rho^-]\rho^+) = 0.\end{aligned} \tag{23}$$

Notice that the two equations are coupled through the term $L[\rho] = m^+L[\rho^+] + m^-L[\rho^-]$. Similar to (13) for identical oscillators, we introduce a dynamical system for the fronts $\theta_j^{\pm}$ and heights $\rho_j^{\pm}$: for $j = 0,\cdots,N-1$ and $t>0$,

$$\begin{aligned}\dot{\theta}_j^+ &= \bar{\Omega} - KL[\rho](\theta_j^+), \\ \dot{\theta}_j^- &= -\bar{\Omega} - KL[\rho](\theta_j^-), \\ \dot{\rho}_j^+ &= K\rho_j^+\frac{L[\rho](\theta_{j+1}^+) - L[\rho](\theta_j^+)}{\theta_{j+1}^+ - \theta_j^+}, \\ \dot{\rho}_j^- &= K\rho_j^-\frac{L[\rho](\theta_{j+1}^-) - L[\rho](\theta_j^-)}{\theta_{j+1}^- - \theta_j^-},\end{aligned} \tag{24}$$

where $\theta_N^{\pm}(t) = \theta_0^{\pm}(t) + 2\pi$. The functions $\rho^{\pm}(\theta,t)$ are defined by

$$\rho^{\pm}(\theta,t) = \sum_{j=0}^{N-1}\rho_j^{\pm}(t)\chi_{(\theta_j^{\pm}(t),\theta_{j+1}^{\pm}(t))}(\theta), \qquad \theta\in(\theta_0^{\pm}(t),\theta_N^{\pm}(t)).$$

Hence, the function $\rho(\theta,t) = m^-\rho^-(\theta,t) + m^+\rho^+(\theta,t)$ is recovered over the period $\mathbf{T}$ and the integral term $L[\rho]$, needed in (24), is naturally defined as $\int_{\mathbf{T}} \sin(\theta - \theta_*)\rho(\theta_*,t)d\theta_*$. In this sense, L can be written as a function of the all set of unknowns, $L[\theta_j^\pm, \rho_j^\pm]$.

We have the following properties.

Lemma 3 *Let $(\theta_j^\pm, \rho_j^\pm)$ be a solution to system (24). Then each local mass is conserved:*

$$\rho_j^\pm(t)(\theta_{j+1}^\pm(t) - \theta_j^\pm(t)) = \rho_j^\pm(0)(\theta_{j+1}^\pm(0) - \theta_j^\pm(0)), \quad j = 0, \cdots, N-1.$$

In particular, the masses of ρ^+ and ρ^- are conserved:

$$\frac{d}{dt}\int_{\mathbf{T}} \rho^+(\theta,t)d\theta = 0, \qquad \frac{d}{dt}\int_{\mathbf{T}} \rho^-(\theta,t)d\theta = 0, \quad t > 0$$

and, consequently, the quantity $\int_{\mathbf{T}} \rho(\theta,t) = m^- \int_{\mathbf{T}} \rho^-(\theta,t)d\theta + m^+ \int_{\mathbf{T}} \rho^+(\theta,t)d\theta$ is constant in time. Moreover, having set $I_0 = \int_{\mathbf{T}} \rho(\theta,t)d\theta$, then

$$\rho_j^\pm(0)e^{-KI_0t} \le \rho_j^\pm(t) \le \rho_j^\pm(0)e^{KI_0t}, \qquad j = 0, \ldots, N-1. \tag{25}$$

Proof For the local mass conservation, we take the time derivative:

$$\begin{aligned}
&\frac{d}{dt}\rho_j^+(\theta_{j+1}^+ - \theta_j^+) \\
&= \dot\rho_j^+(\theta_{j+1}^+ - \theta_j^+) + \rho_j^+(\dot\theta_{j+1}^+ - \dot\theta_j^+) \\
&= K\rho_j^+\Big[L[\rho](\theta_{j+1}^+) - L[\rho](\theta_j^+)\Big] - K\rho_j^+\Big[L[\rho](\theta_{j+1}^+) - L[\rho](\theta_j^+)\Big] \\
&= 0.
\end{aligned}$$

Similarly, we can obtain $\frac{d}{dt}\Big(\rho_j^-(\theta_{j+1}^- - \theta_j^-)\Big) = 0$. About the estimate in $\rho_j^\pm(t)$, it follows from the equation for $\rho_j^\pm$ in (24) and the bound on $\partial_\theta L[\rho(\cdot,t)]$:

$$|\partial_\theta L[\rho(\cdot,t)](\theta)| \le \int_{\mathbf{T}} |\cos(\theta - \theta_*)|\rho(\theta_*,t)d\theta_* \le I_0.$$

Then, by Gronwall lemma, (25) is proved. □

Lemma 4 *Let $(\theta_j^\pm, \rho_j^\pm)$ be a solution to system (24). For $\theta_j^\pm \in \mathbf{T}$, consider a covering $\{\tilde{\theta}_n\}_{n\in\mathbf{Z}}$, $\tilde{\theta}_n^\pm \in \mathbf{R}$, such that*

$$\tilde{\theta}_n^\pm = \theta_j^\pm + 2k\pi \quad \text{for} \quad n = kN + j, \quad 0 \le j < N.$$

Then, the coverings of fronts $\{\tilde{\theta}_j^+\}$ and $\{\tilde{\theta}_j^-\}$ are well ordered in the sense that

$$\cdots < \tilde{\theta}_0^-(t) < \tilde{\theta}_1^-(t) < \cdots < \tilde{\theta}_N^-(t) < \cdots,$$
$$\cdots < \tilde{\theta}_0^+(t) < \tilde{\theta}_1^+(t) < \cdots < \tilde{\theta}_N^+(t) < \cdots, \quad t \ge 0.$$

Proof We only consider the well-ordering of $\tilde{\theta}_i^+$; the other case can be treated similarly. By definition, the initial fronts $\tilde{\theta}_i^+(0)$ are well-ordered in the sense that

$$\cdots < \tilde{\theta}_0^+(0) < \tilde{\theta}_1^+(0) < \cdots < \tilde{\theta}_N^+(0) < \cdots.$$

We claim that the trajectories θ_i do not meet in any finite future time, i.e.,

$$\tilde{\theta}_i^+(t) < \tilde{\theta}_{i+1}^+(t), \quad i \in \mathbf{Z}, \ t > 0.$$

Suppose not, i.e., there exists $i_0 \in \{0, \cdots, N-1\}$ and $t_* < \infty$ such that

$$\tilde{\theta}_{i_0}^+(t) < \tilde{\theta}_{i_0+1}^+(t), \quad t < t_* \quad \text{and} \quad \tilde{\theta}_{i_0}^+(t_*) = \tilde{\theta}_{i_0+1}^+(t_*). \tag{26}$$

Without loss of generality, we assume that only these two fronts meet at time t_*. By the continuity of the function $L[\rho](\theta) = \int_{\mathbf{T}} \sin(\theta - \theta_*)\rho(\theta_*, t)\, d\theta_*$, we have $L[\rho](\tilde{\theta}_{i_0}^+(t_*)) = L[\rho](\tilde{\theta}_{i_0+1}^+(t_*))$. By the uniqueness of ODE solution (see [3, proof of Lemma 3.1] for details) near $t = t_*$, this contradicts (26).

Indeed, it is possible to obtain an ODE system for $\{\rho_j^\pm\}_{j=0}^{N-1}$, $\{\theta_j^-\}_{j=0}^{N-1}$, $\{\theta_k^+\}$ with $k \ne i_0, i_0 + 1$ and θ_* such that $\dot{\theta}_* = \bar{\Omega} - KL[\rho](\theta_*)$, for which the initial conditions at $t = t_*$ coincide with the one for (24). Hence two distinct solutions for $t < t_*$, $|t - t_*|$ suitably small, are deduced: one with $\theta_{i_0}^+(t) = \theta_{i_0+1}^+(t) = \theta_*(t)$ and the one with $\theta_{i_0}^+(t) < \theta_{i_0+1}^+(t)$, leading to a contradiction to uniqueness. □

For a solution $(\theta_j^\pm, \rho_j^\pm)$ to system (24) which is well-separated:

$$\theta_0^- \le \cdots \le \theta_{N-1}^- \le \theta_0^+ \le \cdots \le \theta_{N-1}^+,$$

we set two functionals measuring the diameter of whole configurations $\Theta := (\Theta^-, \Theta^+)$ and gap between two sub-ensembles Θ^- and Θ^+, respectively:

$$D(\Theta)(t) := \theta^+_{N-1}(t) - \theta^-_0(t), \qquad \Delta(\Theta) := \theta^+_0(t) - \theta^-_{N-1}(t).$$

In the following lemma, we provide a sufficient condition leading to

$$D(\Theta) < \pi, \quad \Delta(\Theta) > 0, \quad t > 0.$$

We set Δ^∞ and choose a small positive constant δ to satisfy the relations:

$$\Delta^\infty := \frac{2\bar{\Omega}}{K\big((m^+)^2 + (m^-)^2\big)}, \qquad \frac{\pi}{4}\Big(1-\sqrt{1-2\Delta^\infty}\Big) < \delta < \frac{\pi}{4}\Big(1+\sqrt{1-2\Delta^\infty}\Big).$$

Lemma 5 *Suppose that the following conditions hold:*

1. *The coupling strength K is sufficiently large to satisfy $\Delta^\infty < \frac{1}{2}$.*
2. *The initial fronts and heights satisfy a separation property in the sense that*

$$-\frac{\pi}{2} + \delta < \theta^-_0(0) < \cdots < \theta^-_{N-1}(0) \le 0 \le \theta^+_0(0) < \cdots < \theta^+_{N-1}(0) < \frac{\pi}{2} - \delta,$$
$$\Delta(\Theta(0)) := \theta^+_0(0) - \theta^-_{N-1}(0) < \Delta^\infty,$$
$$\rho^\pm_i \ge 0 \quad \text{for} \quad i = 1, \cdots, N-2, \quad \text{and} \quad \rho^\pm_{N-1} = 0.$$

Then, for any front-tracking approximate solutions to system (24), we have

$$D(\Theta(t)) < \pi - 2\delta \quad \text{and} \quad \Delta(\Theta(t)) > \Delta(\Theta(0)), \quad t \ge 0.$$

Proof

(i) (Estimate on $D(\Theta)$): We first consider the dynamics of $D(\Theta(t)) := \theta^+_{N-1}(t) - \theta^-_0(t)$:

$$\begin{aligned}
\frac{d}{dt}D(\Theta) &= 2\bar{\Omega} - K\Big[m^+L[\rho^+](\theta^+_{N-1}) + m^-L[\rho^-](\theta^+_{N-1}) \\
&\qquad - m^+L[\rho^+](\theta^-_0) - m^-L[\rho^-](\theta^-_0)\Big] \\
&= 2\bar{\Omega} - K\Big[m^+\Big\{L[\rho^+](\theta^+_{N-1}) - L[\rho^+](\theta^-_0)\Big\} \\
&\qquad + m^-\Big\{L[\rho^-](\theta^+_{N-1}) - L[\rho^-](\theta^-_0)\Big\}\Big] \\
&=: 2\bar{\Omega} - K\Big[m^+\mathcal{I}_+ + m^-\mathcal{I}_-\Big].
\end{aligned}$$

Since $\rho^{\pm}_{N-1} = 0$,

$$\begin{aligned}
\mathscr{I}_{\pm} &= \sum_{j=0}^{N-1} \rho_j^{\pm} \Big\{ \cos(\theta^{+}_{N-1} - \theta^{\pm}_{j+1}) - \cos(\theta^{+}_{N-1} - \theta^{\pm}_{j}) \\
&\qquad\qquad - \cos(\theta^{-}_{0} - \theta^{\pm}_{j+1}) + \cos(\theta^{-}_{0} - \theta^{\pm}_{j}) \Big\} \\
&= \sum_{j=0}^{N-2} \rho_j^{\pm} \Big\{ \Big(\cos(\theta^{+}_{N-1} - \theta^{\pm}_{j+1}) - \cos(\theta^{-}_{0} - \theta^{\pm}_{j+1}) \Big) \\
&\qquad\qquad - \Big(\cos(\theta^{+}_{N-1} - \theta^{\pm}_{j}) - \cos(\theta^{-}_{0} - \theta^{\pm}_{j}) \Big) \Big\} \\
&= \sum_{j=0}^{N-2} -2\rho_j^{\pm} \Big\{ \sin\Big(\frac{\theta^{+}_{N-1} - \theta^{-}_{0}}{2} \Big) \sin\Big(\frac{\theta^{+}_{N-1} + \theta^{-}_{0}}{2} - \theta^{\pm}_{j+1} \Big) \\
&\qquad\qquad - \sin\Big(\frac{\theta^{+}_{N-1} - \theta^{-}_{0}}{2} \Big) \sin\Big(\frac{\theta^{+}_{N-1} + \theta^{-}_{0}}{2} - \theta^{\pm}_{j} \Big) \Big\} \\
&= \sum_{j=0}^{N-2} 2\rho_j^{\pm} \sin\frac{D(\Theta)}{2} \Big\{ \sin\Big(\frac{\theta^{+}_{N-1} + \theta^{-}_{0}}{2} - \theta^{\pm}_{j} \Big) - \sin\Big(\frac{\theta^{+}_{N-1} + \theta^{-}_{0}}{2} - \theta^{\pm}_{j+1} \Big) \Big\}
\end{aligned}$$

If $0 < D(\Theta) < \pi - 2\delta$, we claim that

$$\sin\frac{D(\Theta)}{2} > \frac{D(\Theta)}{\pi} > 0. \tag{27}$$

Indeed, notice that $x = \frac{D(\Theta)}{2} \in (0, \frac{\pi}{2})$. Since the function $x \mapsto \frac{\sin x}{x}$ is strictly decreasing on $(0, \frac{\pi}{2}]$, then one has

$$\frac{\sin x}{x} > \frac{\sin(\pi/2)}{\pi/2} = \frac{2}{\pi}, \qquad x \in (0, \frac{\pi}{2}).$$

By substitution of $x = \frac{D(\Theta)}{2}$ in the inequality above, we obtain (27). The claim is proved. Now, since

$$-\frac{\pi}{2} + \delta < \frac{\theta^{+}_{N-1} + \theta^{-}_{0}}{2} - \theta^{\pm}_{j} < \frac{\pi}{2} - \delta \quad \text{for} \quad j = 0, \cdots, N-1,$$

we have

$$\sin\left(\frac{\theta_{N-1}^{+}+\theta_0^{-}}{2}-\theta_j^{\pm}\right)-\sin\left(\frac{\theta_{N-1}^{+}+\theta_0^{-}}{2}-\theta_{j+1}^{\pm}\right)\geq(\theta_{j+1}^{\pm}-\theta_j^{\pm})\sin\delta$$

for $j=0,\cdots,N-2$. Thus, we have

$$\mathscr{I}_{\pm}\geq 2\sin\frac{D(\Theta)}{2}\sin\delta\sum_{j=0}^{N-2}\rho_j^{\pm}(\theta_{j+1}^{\pm}-\theta_j^{\pm})\geq\frac{2m^{\pm}\sin\delta}{\pi}D(\Theta(t))$$

This yields a Gronwall's inequality for $D(\Theta)$:

$$\frac{d}{dt}D(\Theta)\leq 2\bar{\Omega}-\frac{2K\big((m^{+})^2+(m^{-})^2\big)\sin\delta}{\pi}D(\Theta),$$

and we have

$$\begin{aligned}D(\Theta(t))\leq&\frac{\pi\bar{\Omega}}{K\big((m^{+})^2+(m^{-})^2\big)\sin\delta}\\&+\left[D(\Theta(0))-\frac{\pi\bar{\Omega}}{K\big((m^{+})^2+(m^{-})^2\big)\sin\delta}\right]e^{-\frac{2K\left((m^{+})^2+(m^{-})^2\right)\sin\delta}{\pi}t}\end{aligned}$$

Since $\frac{\pi}{4}\left(1-\sqrt{1-2\Delta^{\infty}}\right)<\delta<\frac{\pi}{4}\left(1+\sqrt{1-2\Delta^{\infty}}\right)$, we have

$$(\pi-2\delta)\sin\delta>(\pi-2\delta)\frac{2}{\pi}\delta>\frac{\pi\Delta^{\infty}}{2}.$$

Thus, we have

$$D(\Theta(t))\leq\pi-2\delta.$$

(ii) (Estimate on $\Delta(\Theta)$): We next consider the dynamics of $\Delta(\Theta):=\theta_0^{+}-\theta_{N-1}^{-}$:

$$\begin{aligned}\frac{d}{dt}\Delta(\Theta)&=2\bar{\Omega}-K\Big[m^{+}L[\rho^{+}](\theta_0^{+})+m^{-}L[\rho^{-}](\theta_0^{+})\\&\qquad-m^{+}L[\rho^{+}](\theta_{N-1}^{-})-m^{-}L[\rho^{-}](\theta_{N-1}^{-})\Big]\\&=2\bar{\Omega}-K\Big[m^{+}\big\{L[\rho^{+}](\theta_0^{+})-L[\rho^{+}](\theta_{N-1}^{-})\big\}\\&\qquad+m^{-}\big\{L[\rho^{-}](\theta_0^{+})-L[\rho^{-}](\theta_{N-1}^{-})\big\}\Big]\\&=:2\bar{\Omega}-K\Big[m^{+}\mathscr{L}_{+}+m^{-}\mathscr{L}_{-}\Big].\end{aligned}$$

Since $\rho^{\pm}_{N-1} = 0$, from the definition of $L[\rho](\theta)$, we have

$$
\begin{aligned}
\mathscr{L}_{\pm} &= \sum_{j=0}^{N-1} \rho_j^{\pm} \Big\{ \cos(\theta_0^{+} - \theta_{j+1}^{\pm}) - \cos(\theta_0^{+} - \theta_j^{\pm}) \\
&\qquad\qquad - \cos(\theta_{N-1}^{-} - \theta_{j+1}^{\pm}) + \cos(\theta_{N-1}^{-} - \theta_j^{\pm}) \Big\} \\
&= \sum_{j=0}^{N-2} \rho_j^{\pm} \Big\{ \Big(\cos(\theta_0^{+} - \theta_{j+1}^{\pm}) - \cos(\theta_{N-1}^{-} - \theta_{j+1}^{\pm}) \Big) \\
&\qquad\qquad - \Big(\cos(\theta_0^{+} - \theta_j^{\pm}) - \cos(\theta_{N-1}^{-} - \theta_j^{\pm}) \Big) \Big\} \\
&= \sum_{j=0}^{N-2} 2\rho_j^{\pm} \sin\Big(\frac{\theta_0^{+} - \theta_{N-1}^{-}}{2} \Big) \\
&\qquad\qquad \times \Big\{ \sin\Big(\frac{\theta_0^{+} + \theta_{N-1}^{-}}{2} - \theta_j^{\pm} \Big) - \sin\Big(\frac{\theta_0^{+} + \theta_{N-1}^{-}}{2} - \theta_{j+1}^{\pm} \Big) \Big\} \\
&\le \sum_{j=0}^{N-2} \rho_j^{\pm} \Delta(\Theta)(\theta_{j+1}^{\pm} - \theta_j^{\pm}) = \Delta(\Theta) m^{\pm}.
\end{aligned}
$$

Thus, we have

$$
\frac{d}{dt}\Delta(\Theta) \ge 2\bar{\Omega} - K\Delta(\Theta)\big((m^{+})^2 + (m^{-})^2\big).
$$

This yields

$$
\begin{aligned}
\Delta(\Theta(t)) \ge & \frac{2\bar{\Omega}}{K\big((m^{+})^2 + (m^{-})^2\big)} \\
& + \Big(\Delta(\Theta(0)) - \frac{2\bar{\Omega}}{K\big((m^{+})^2 + (m^{-})^2\big)} \Big) e^{-K\big((m^{+})^2 + (m^{-})^2\big)t}.
\end{aligned}
$$

Since $\Delta(\Theta(0)) < \frac{2\bar{\Omega}}{K\big((m^{+})^2 + (m^{-})^2\big)}$, we have

$$
\Delta(\Theta(t)) > \Delta(\Theta(0)), \quad t > 0.
$$

Remark 1 Lemma 5 explains that once the fronts are aligned and segregated, then it is preserved, i.e., there is no collision henceforward.

4 A Few Numerical Examples

Yet, we provide several numerical simulations corresponding to analytical results in Sects. 2 and 3. Throughout this section, we implement numerical simulations using the fourth order Runge-Kutta method with time step $h = 0.001$.

4.1 *Identical Oscillators: $\Omega = 0$*

Here, we deal with identical Kuramoto oscillators. Without loss of generality, we may assume that $g(\Omega) = \delta_0$. We set the coupling strength $K = 10$ and set the number of fronts to be 100. In Fig. 1, we choose the initial data as follows.

$$\theta_0 = -\frac{\pi}{2}, \quad \theta_{j+1} = \theta_j + \frac{\pi}{N} \quad \text{where } j = 0, \cdots, N-2, \quad \theta_{N-1} = \frac{\pi}{2},$$

$$\rho_i = \frac{1}{\pi} \quad \text{for} \quad i = 0, \cdots, N-2, \quad \rho_{N-1} = 0.$$

We can see the emergence of concentration in Fig. 1a at $\theta = 0$. The exponential growth of $\|\rho(\cdot, t)\|_\infty$ is shown in Fig. 1c, d. As we observed in Corollary 1, the order parameter R monotonically increase in Fig. 1b. Further numerical examples for identical oscillators are presented in [3].

4.2 *Nonidentical Oscillators*

Next, we display numerical simulations for non-identical oscillators case with two symmetric natural frequencies. To fix the idea, we consider a symmetric case:

$$m^+ = m^- = \frac{1}{2}, \qquad g(\Omega) = \frac{1}{2}\delta_{-\Omega} + \frac{1}{2}\delta_\Omega,$$

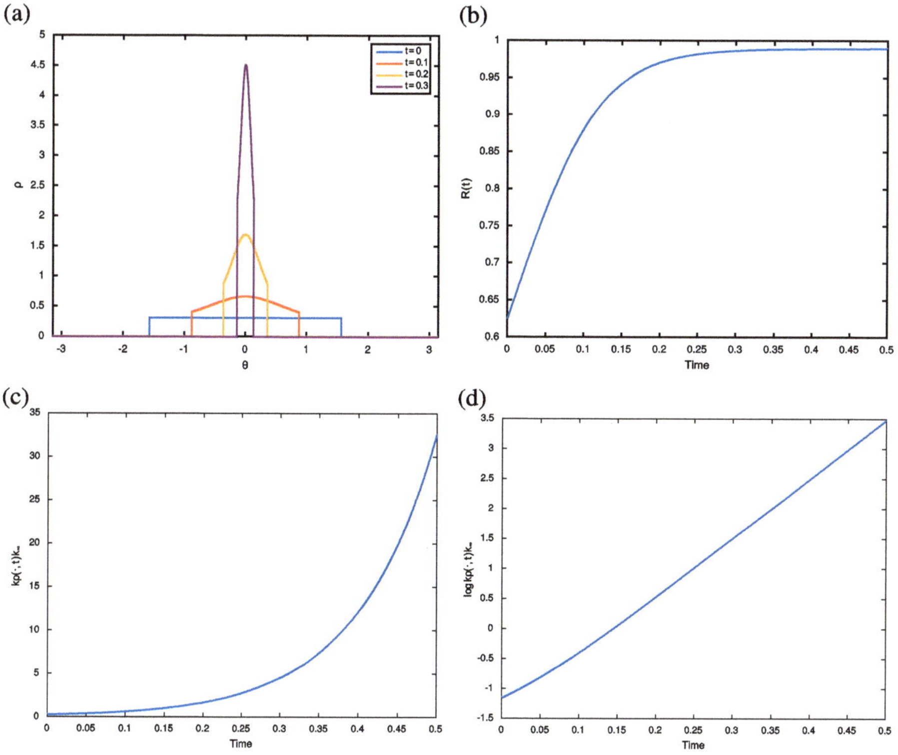

Fig. 1 Identical oscillators. (**a**) Time evolution of synchronization. (**b**) Order parameter $R(t)$. (**c**) $\|\rho(\cdot,t)\|_\infty$. (**d**) $\log\|\rho(\cdot,t)\|_\infty$

where $\Omega = 1$. We set the initial configuration $\rho^+(\theta,0)$ and $\rho^-(\theta,0)$ to satisfy

$$
\begin{aligned}
&0 < \theta_0^+(0) < \cdots, < \theta_{N-1}^+(0) < \frac{\pi}{2},\\
&\theta_j^-(0) = -\theta_{N-1-j}^+(0) \quad \text{for} \quad j = 0,\cdots,N-1,\\
&\theta_{j+1}^+(0) - \theta_j^+(0) = \Delta\theta \quad \text{where} \quad \Delta\theta = \frac{\theta_{N-1}^+(0) - \theta_0^+(0)}{N-1},\\
&\rho_j^+(0) = \rho_j^-(0) = \frac{1}{2\Delta\theta(N-2)} \quad \text{for} \quad j = 0,\cdots,N-2,\\
&\rho_{N-1}^+(0) = \rho_{N-1}^-(0) = 0.
\end{aligned}
\tag{28}
$$

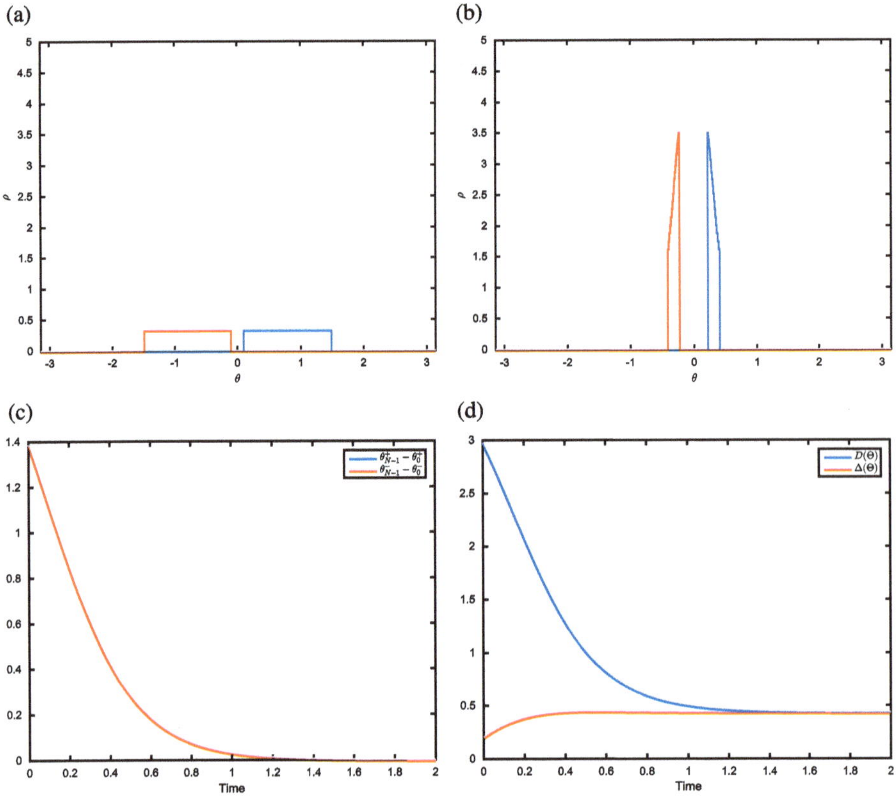

Fig. 2 Bi-modal case I. (**a**) Initial configuration with $N = 20$. (**b**) Configuration at time $t = 0.6$. (**c**) $\theta^{\pm}_{N-1} - \theta^{\pm}_{0}$. (**d**) $D(\Theta)$ and $\Delta(\Theta)$

For $N = 20$, we set $K = 10$ and the initial configuration (28) is described in Fig. 2a. As time $t \to \infty$, the fronts of the approximate solution converge to the same point so that $\theta^{\pm}_{j+1} - \theta^{\pm}_{j}$ goes to zeros for $j = 0, \cdots, N-2$. This concentration shows the synchronization of oscillators. Once the fronts are segregated like the initial data, the fronts do not have collision, i.e., $\Delta(\Theta(t)) = \theta^{+}_{0}(t) - \theta^{-}_{N-1}(t) > 0$ for $t > 0$ as in Fig. 2d. A non-symmetric case has different weights: $m^{+} = 0.2$ and $m^{-} = 0.8$.

$$g(\Omega) = 0.8\delta_{-\Omega} + 0.2\delta_{\Omega}$$

We set the initial configuration as given in Fig. 3a. In this case, although the diameter of support $\theta^{\pm}_{N-1} - \theta^{\pm}_{0}$ may not decrease monotonically, it converges to zero eventually. As we can see in Fig. 3d, the segregation of fronts is preserved. The system with multiple natural frequencies has a distribution $g(\Omega)$ like,

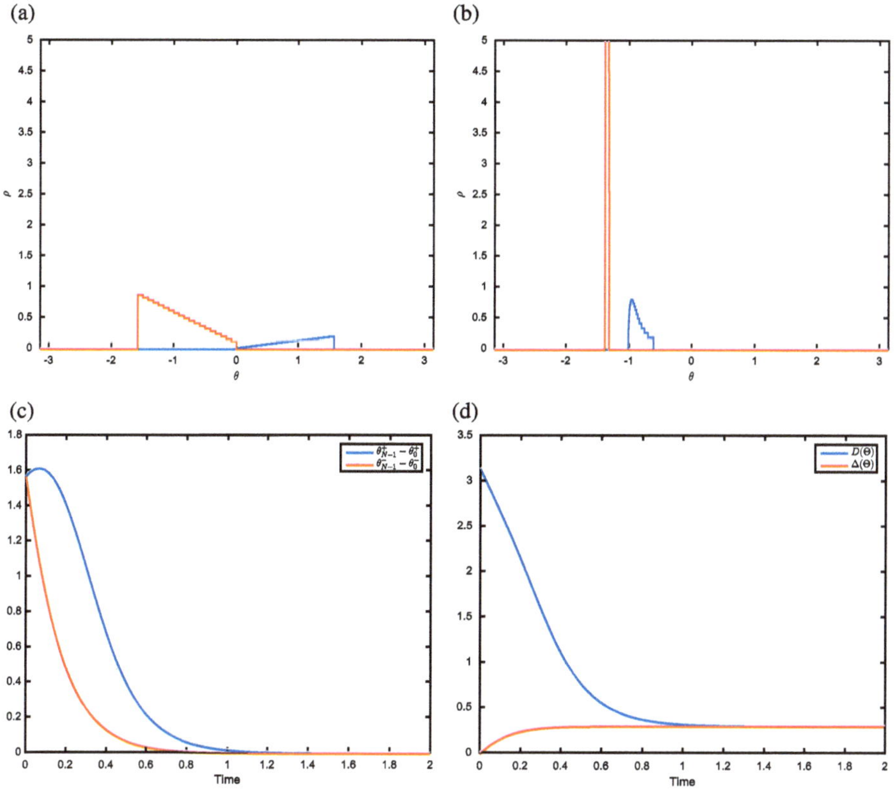

Fig. 3 Bi-modal case II. (**a**) Initial configuration with $N = 20$. (**b**) Configuration at time $t = 0.5$. (**c**) $\theta^{\pm}_{N-1} - \theta^{\pm}_0$. (**d**) $D(\Theta)$ and $\Delta(\Theta)$

$$g(\Omega) = \sum_{i=1}^{5} m_i \delta_{\Omega_i},$$

$$(m_1, m_2, m_3, m_4, m_5) = (0.13, 0.2, 0.3, 0.12, 0.25),$$

$$(\Omega_1, \Omega_2, \Omega_3, \Omega_4, \Omega_5) = (-1, -0.3, 0.1, 0.5, 1)$$

We set the coupling strength $K = 10$ and take initial data to satisfy

$$\text{supp } \rho^1 \subset \left[-\frac{\pi}{2}, -\frac{3\pi}{10}\right], \text{ supp } \rho^2 \subset \left[-\frac{3\pi}{10}, -\frac{\pi}{10}\right], \text{ supp } \rho^3 \subset \left[-\frac{\pi}{10}, \frac{\pi}{10}\right],$$

$$\text{supp } \rho^4 \subset \left[\frac{\pi}{10}, \frac{3\pi}{10}\right], \text{ supp } \rho^5 \subset \left[\frac{3\pi}{10}, \frac{\pi}{2}\right],$$

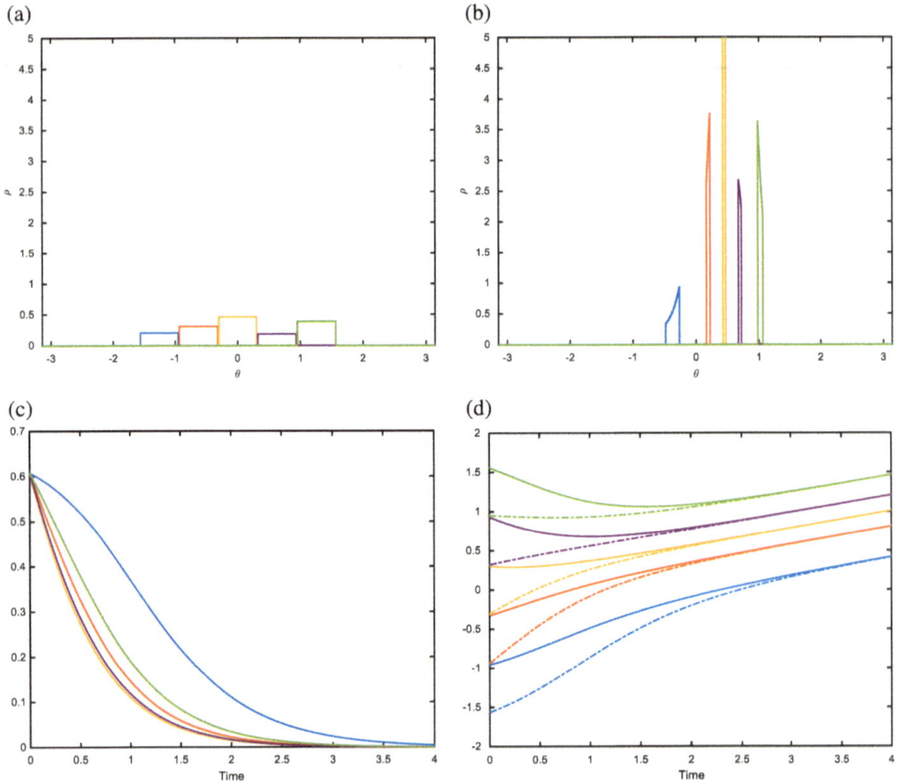

Fig. 4 Multiple natural frequencies: $(m_1, m_2, m_3, m_4, m_5) = (0.13, 0.2, 0.3, 0.12, 0.25)$, $(\Omega_1, \Omega_2, \Omega_3, \Omega_4, \Omega_5) = (-1, -0.3, 0.1, 0.5, 1)$. (**a**) Initial configuration with $N = 20$. (**b**) Configuration at time $t = 1.5$. (**c**) $\theta^i_{N-1} - \theta^i_0$. (**d**) $\theta^i_0, \theta^i_{N-1}$

and ρ^i's are uniform in each intervals. In Fig. 4b, c, the numerical results show the emergence of concentration for each distribution. In Fig. 4d, the solid lines represent the location of largest front θ^i_{N-1} and dash-dot line is the location of smallest front θ^i_0 for each distribution $i = 1, \cdots, 5$. This shows the segregated fronts do not collide in finite time.

5 Conclusion

In this paper, we have discussed a modified front-tracking algorithm used in the synchronization problem of the kinetic Kuramoto-Sakaguchi equation. The front-tracking algorithm has been very successfully used in the realm of hyperbolic conservation laws to construct entropic BV weak solutions for the system of 1D hyperbolic conservation laws in BV framework and recently, it has been further generalized in the study of a scalar conservation law with nonlocal flux arising from the modeling of granular and pedestrian flows, etc. In the context of synchronization modeling of many coupled oscillators, a scalar conservation law with nonlocal flux appears naturally in the kinetic modeling of large ensemble of weakly coupled oscillators. For the ensemble of identical Kuramoto oscillators, the modified front-tracking algorithm was applied for the construction of BV weak solutions in authors' recent work [3] (see also [10]). In contrast, for distributed natural frequencies, a general and complete analysis based on the front-tracking algorithm is not available yet. In this paper, we address the dynamics of order parameters and consider a simplest case where the set of natural frequencies has only two elements, i.e., bi-modal case. In this case, we explained how the front-tracking algorithm can be adopted for the large-time dynamics of the Kuramoto-Sakaguchi equation. Of course, the analysis provided in this paper only provides preliminary results for the large-time behavior of the entropic BV weak solution. A general treatment of the Kuramoto-Sakaguchi equation with distributed natural frequencies will be considered in future investigations in the front-tracking algorithm.

Acknowledgements The first author would like to thank the organizers of the *INdAM Workshop on Innovative Algorithms and Analysis* (Rome, May 17–20 2016) for their kind invitation and hospitality.

The work of D. Amadori is supported by the Miur-PRIN 2012 Project *Nonlinear Hyperbolic Partial Differential Equations, Dispersive and Transport Equations: theoretical and applicative aspects*. The work of S.-Y. Ha is partially supported by a National Research Foundation of Korea Grant (2014R1A2A2A05002096) funded by the Korean government, and the work of J. Park was supported by NRF (National Research Foundation of Korea) Grant funded by Korean Government (NRF-2014-Fostering Core Leaders of the Future Basic Science Program).

References

1. Acebrón, J.A., Bonilla, L.L., Pérez Vicente, C.J.P., Ritort, F., Sigler, R.: The Kuramoto model: a simple paradigm for synchronization phenomena. Rev. Mod. Phys. **77**, 137–185 (2005)
2. Amadori, D., Shen, W.: Front tracking approximations for slow erosion. Discrete Contin. Dyn. Syst. **32**, 1481–1502 (2012)
3. Amadori, D., Ha, S.-Y., Park, J.: On the global well-posedness of BV weak solutions for the Kuramoto-Sakaguchi equation. J. Differ. Equ. **262**, 978–1022 (2017)
4. Balmforth, N.J., Sassi, R.: A shocking display of synchrony. Physica D **143**, 21–55 (2000)
5. Benedetto, D., Caglioti, E., Montemagno, U.: On the complete phase synchronization for the Kuramoto model in the mean-field limit. Commun. Math. Sci. **13**, 1775–1786 (2015)
6. Carrillo, J.A., Choi, Y.-P., Ha, S.-Y., Kang, M.-J., Kim, Y.: Contractivity of transport distances for the kinetic Kuramoto equation. J. Stat. Phys. **156**, 395–415 (2014)
7. Colombo, R.M., Herty, M., Mercier, M.: Control of the continuity equation with a non local flow. ESAIM: Control Optim. Calc. Var. **17**, 353–379 (2011)
8. Crippa, G., Lécureux-Mercier, M.: Existence and uniqueness of measure solutions for a system of continuity equations with non-local flow. NoDEA Nonlin. Differ. Equ. Appl. **20**, 523–537 (2013)

9. Gosse, L., Toscani, G.: Lagrangian numerical approximations to one-dimensional convolution-diffusion equations. SIAM J. Sci. Comput. **28**, 1203–1227 (2006)
10. Ha, S.-Y., Ko, D., Park, J., Zhang, X.: Collective synchronization of classical and quantum oscillators. EMS Surv. Math. Sci. **3**(2), 209–267 (2016)
11. James, F., Vauchelet, N.: Numerical methods for one-dimensional aggregation equations. SIAM J. Numer. Anal. **53**, 895–916 (2015)
12. Lancellotti, C.: On the Vlasov limit for systems of nonlinearly coupled oscillators without noise. Transp. Theory Stat. Phys. **34**, 523–535 (2005)
13. Matthes, D., Söllner, B.: Convergent Lagrangian discretization for drift-diffusion with nonlocal aggregation. In: Gosse, L., Natalini, R. (eds.) Innovative Algorithms and Analysis. Springer INdAM Series, vol. 16. Springer International Publishing, Cham (2017)
14. Shen, W., Zhang, T.: Erosion profile by a global model for granular flow. Arch. Ration. Mech. Anal. **204**, 837–879 (2012)

High-Order Post-Newtonian Contributions to Gravitational Self-force Effects in Black Hole Spacetimes

Donato Bini, Thibault Damour, and Andrea Geralico

Abstract The explicit analytical computation of first-order metric perturbations in black hole spacetimes is described in the case of a perturbing mass moving on an equatorial circular orbit. The perturbation equations can be separated into an angular part and a radial part. The latter satisfies a single inhomogeneous radial Schrödinger-like equation with a Dirac-delta singular source term, whose solutions are built up through Green's function techniques. Various types of approximate analytical homogeneous solutions (and corresponding Green's functions) can be constructed: Post-Newtonian solutions (expanded in powers of $1/c$), Mano-Suzuki-Takasugi solutions (expanded in series of hypergeometric functions), Wentzel-Kramers-Brillouin (WKB) solutions (large l expansion). The perturbed black-hole metric constructed by suitably combining these different kind of solutions can then be used to compute, in analytical form, gauge-invariant quantities. These include several "potentials" entering the effective-one-body formalism (shortly reviewed here). The latter formalism is a new way of describing the gravitational interaction of two masses which has played a crucial role in the recent detection of gravitational waves.

1 Introduction

The two-body problem is one of the arenas where general relativity (GR) can be accurately confronted to experiment, either via the accurate radio timing of binary pulsars, or via the detection of the gravitational wave signals emitted by coalescing compact binaries. The first detection of gravitational-wave signals has concerned

---

D. Bini (✉)
Istituto per le Applicazioni del Calcolo, 00185 Rome, Italy
e-mail: donato.bini@gmail.com

T. Damour
Institut des Hautes Etudes Scientifiques, 91440 Bures-sur-Yvette, France
e-mail: damour@ihes.fr

A. Geralico
Astrophysical Observatory of Torino, INAF, 10025 Pino Torinese, Italy
e-mail: andrea.geralico@gmail.com

L. Gosse, R. Natalini (eds.), *Innovative Algorithms and Analysis*,
Springer INdAM Series 16, DOI 10.1007/978-3-319-49262-9_2

binary systems made of (spinning) black holes [1, 2]. This spectacular discovery gives a strong incentive for further improving our analytical knowledge of the relativistic gravitational interaction in binary systems.

Solving the two-body problem in the context of the full GR theory is, however, a tall order. Henceforth, over the years several analytical-relativity methods have been actively pursued: post-Newtonian (PN) theory (see, e.g., [3, 4] and references therein), gravitational self-force (GSF) theory (see, e.g., [5] and references therein), and the Effective-One-Body (EOB) formalism [6–9]. In addition, numerical relativity techniques have allowed, since 2005 [10], to stably evolve coalescing binary black holes. Recent years have witnessed a useful synergy between both the various analytic methods, and the results of full numerical relativity simulations (see, e.g., [11–13] and references therein).

In this paper we shall focus on recently developed analytical approaches to GSF theory, i.e. to the dynamics of a small mass orbiting around a large mass, and on the use of GSF theory for informing the EOB formalism. Here, we shall consider, for simplicity, the following situation: A small mass m_1 is orbiting a (Schwarzschild or Kerr) black hole of large mass m_2 on a spatially circular equatorial path (helical in the spacetime). The background metric is modified (perturbed) by the particle, and one can analytically construct the perturbed metric by solving the associated (first-order) perturbation equations. This task is simplified by the presence of several symmetries: (1) the temporal and axial Killing-vector symmetries of the background metric, together with the Killing-tensor symmetry linked to separability; and (2) the helical Killing-vector symmetry of the perturbed spacetime (associated with the helical nature of the worldline of m_1).

Gravitational perturbations of Schwarzschild and Kerr black holes are governed by the Regge-Wheeler-Zerilli [14, 15] equation and the Teukolsky [16–18] equation, respectively. Due to the background spacetime symmetries, and the monochromatic nature of the circular orbit of m_1, the angular part and the radial part of the perturbation can be separated, so that one is left with a single inhomogeneous ordinary differential equation for the radial function with Dirac-delta source terms having support at the particle radius $r = r_0$.

In the case of a Schwarzschild black hole, for example, the main steps are listed below.

1. One has first to explicitly determine solutions to the homogeneous radial equation, for any given multipole order l. This can be done either numerically, or by means of analytical approximations. Here we focus on the construction of analytical solutions to the homogeneous radial equation. These can be obtained in (at least) two ways: simple PN solutions (which, however, neither properly take into account the matching between the near zone and the wave zone, nor the presence of the black hole horizon) and "radiative" solutions (which cure the shortcomings of the PN solutions), following a method developed by Mano, Suzuki and Takasugi [19, 20]. A third kind of solutions, Wentzel-Kramers-Brillouin (WKB) solutions, corresponding to the large l expansion, is also useful in the regularization process, see below.

2. Next one uses the Green's function method to solve the associated inhomogeneous equation.
3. Each multipolar component (for a given l) of the perturbed metric can then be computed at the location r_0 of the perturbing particle, since it is finite there. Let $h_l(r_0)$ denote such a component. It is analytically obtained (after PN-expanding the "radiative" solutions) as an expansion in powers of r_0, or, better, $u \equiv Gm_2/(c^2 r_0)$. [In the following, we often use units where $G = c = 1$.] However, the fact that the self-field of the particle (computed at the location of its point-like source) is singular implies that the sum over l of $h_l(r_0)$ is a divergent series.
4. Regularization techniques have then been developed to subtract the "singular part" of the field, so as to obtain its "regular part" only. When working (as we do) with multipolar expansions, the regularization consists in subtracting from $h_l(r_0)$ a certain regularization piece (or subtraction term), say $B_l(r_0)$ [21, 22]. In our analytical approach, we can both analytically predict the value of the subtraction term (by using the WKB solutions to the RW equation) and directly compute it from our analytic solutions (so that a first check consists in showing that these two quantities agree).
5. By using the regular part of the perturbed metric (eventually "cured" also for the presence of discontinuities or jumps) one then construct, from the convergent series $\sum_l (h_l(r_0) - B_l(r_0))$, gauge-invariant quantities which can then be transformed into quantities of direct dynamical significance.

None of the above points is trivial to implement. Of great help is the implementation of all the calculations into symbolic algebra manipulator systems, like Mathematica™ or Maple™.

Finally, some modifications to the above procedure are necessary when dealing with the Teukolsky equation in the case of a perturbed Kerr black hole.

2 Effective-One-Body (EOB) Formalism

The EOB formalism [6–9] plays an important role for transcribing our GSF results into quantities of direct dynamical significance. This formalism maps the general relativistic dynamics of a two-body system onto the dynamics of a single body moving within an effective metric. Let us briefly summarize here the EOB formalism in the simplest case of non-spinning bodies.

In GR a test particle ($m_1/m_2 \to 0$) moving in a given gravitational background of a (spinning/non-spinning) black hole (of mass m_2) follows a geodesic worldline in the metric $g_{\mu\nu}(m_2)$. However, when m_1 is comparable to m_2, the GR situation is more complicated: each body follows a geodesic in a metric which is modified by the presence of both masses. When the ratio $q = m_1/m_2$ is a small parameter, GSF theory can compute the dynamics of the small mass m_1 to first order in q, i.e. taking into account the first-order back-reaction of m_1 on itself. Before the conception of the EOB formalism, the only analytical way of representing the dynamics of comparable-mass ($q \sim 1$) binary systems was the PN approximation method. The EOB formalism provided (starting in 1999) a new way of representing the dynamics of comparable-mass systems. This formalism starts from dynamical results extracted either from PN calculations or from GSF calculations and transcribes them into a new representation of the relative two-body dynamics. This new representation consists (modulo some additional corrections) in deriving the *relative* two-body dynamics as the dynamics of an *effective* body of mass $\mu \equiv m_1 m_2/(m_1+m_2)$ following a geodesic in some *effective* metric $g_{\mu\nu}^{\rm eff}(M, \nu)$, where $M \equiv m_1 + m_2$, and where

$$\nu \equiv \frac{\mu}{M} = \frac{m_1 m_2}{(m_1 + m_2)^2}. \tag{1}$$

In the EOB formalism, the symmetric mass ratio ν plays the role of a deformation parameter, continuously connecting the test-particle case $\nu = 0$ to the equal-mass case $\nu = \frac{1}{4}$. Note that ν is not used as an expansion parameter. The GSF dynamics corresponds to the case where one works to first order in ν in the $\nu-$deformed effective metric $g_{\mu\nu}^{\rm eff}(M, \nu)$.

The basic aim of the EOB formalism was to resum (using specific resummation techniques [8, 23]) the results of PN calculations of the motion and radiation of binary systems so as to be able to analytically describe the last few orbits, the merger and the post-merger gravitational-wave signal of coalescing binary black holes. As early as 2000 [7], i.e. 5 years before the existence of numerical simulations of binary black-hole coalescences, it made predictions about the late dynamics, and the complete gravitational-wave signal of coalescing binary black holes. These predictions have been broadly confirmed by subsequent numerical simulations [10, 24]. The recent discovery, interpretation and data analysis of the first gravitational-wave signals by LIGO [1, 2] has been crucially helped by the development of the EOB formalism, improved by extracting additional non-perturbative information from a relatively small sample of numerical relativity simulations [25, 26].

To give a flavor of the EOB formalism, let us review the definition of the EOB Hamiltonian for non-spinning bodies, at the second post-Newtonian (2PN) level, from the corresponding PN-expanded Hamiltonian, as obtained in [6] (see, respectively, [8] and [27] for the corresponding 3PN-level and 4PN-level results). The μ-rescaled 2PN Hamiltonian (written in ADM coordinates and in the center-of-mass system) reads

$$\hat{H}(\mathbf{p}, \mathbf{q}) = \hat{H}_{\rm N}(\mathbf{p}, \mathbf{q}) + \frac{1}{c^2}\hat{H}_{\rm 1PN}(\mathbf{p}, \mathbf{q}) + \frac{1}{c^4}\hat{H}_{\rm 2PN}(\mathbf{p}, \mathbf{q}) + \dots , \tag{2}$$

with

$$
\begin{aligned}
\hat{H}_{\rm N}(\mathbf{p},\mathbf{q}) &= \frac{1}{2}p^2 - \frac{1}{q}\,,\\
\hat{H}_{\rm 1PN}(\mathbf{p},\mathbf{q}) &= -\frac{1}{8}(1-3\nu)p^4 - \frac{1}{2q}[(3+\nu)p^2 + \nu(\mathbf{n}\cdot\mathbf{p})^2] + \frac{1}{2q^2}\,,\\
\hat{H}_{\rm 2PN}(\mathbf{p},\mathbf{q}) &= \frac{(1-5\nu+5\nu^2)}{16}p^6\\
&\quad + \frac{1}{8q}[(5-20\nu-3\nu^2)p^4 - 2\nu^2p^2(\mathbf{n}\cdot\mathbf{p})^2 - 3\nu^2(\mathbf{n}\cdot\mathbf{p})^4]\\
&\quad + \frac{1}{2q^2}[(5+8\nu)p^2 + 3\nu(\mathbf{n}\cdot\mathbf{p})^2] - \frac{1+3\nu}{4q^2}\,,
\end{aligned}
\tag{3}
$$

where we recall the notation

$$
M = m_1 + m_2\,, \qquad \mu = \frac{m_1 m_2}{M}\,, \qquad \nu = \frac{\mu}{M}\,. \tag{4}
$$

In the center-of-mass frame, the momenta of the two bodies are such that $\mathbf{P}_1 = \mathbf{P} = -\mathbf{P}_2$ and their relative (ADM) position is $\mathbf{Q} = \mathbf{q}_1 - \mathbf{q}_2$. The rescaled relative positions and momenta entering the μ-rescaled Hamiltonian above are

$$
\mathbf{q} = \frac{\mathbf{Q}}{GM}\,, \qquad \mathbf{p} = \frac{\mathbf{P}}{\mu}\,, \tag{5}
$$

with $q = |\mathbf{q}|$ and $\mathbf{n} = \mathbf{q}/q$. In addition, time and energy are rescaled into

$$
\hat{t} = \frac{t}{GM}\,, \quad \hat{H} = \frac{H^{\rm NR}}{\mu} = \frac{H^{\rm R} - Mc^2}{\mu}\,. \tag{6}
$$

The relative-motion Hamiltonian is invariant under time translations and space rotations, implying the following conservation laws for energy and angular momentum

$$
\mathscr{E}^{\rm NR} = \mu\hat{H}\,, \qquad \mathscr{J}_{\rm cm} = \mu GM\mathbf{j} = \mu GM\mathbf{q}\times\mathbf{p}\,. \tag{7}
$$

It is convenient to solve the motion by using the Hamilton-Jacobi approach, i.e., writing the (rescaled) action in the separated form

$$
\hat{S} = \frac{S}{\mu GM} = -\hat{\mathscr{E}}^{\rm NR}\hat{t} + j\phi + \hat{S}_r(r, \hat{\mathscr{E}}^{\rm NR}, j)\,, \tag{8}
$$

and adopting a polar coordinate system in the plane of the relative motion

$$
q^x = r\cos\phi\,, \qquad q^y = r\sin\phi\,, \qquad q^z = 0\,. \tag{9}
$$

The radial action $\hat{S}_r(r, \hat{\mathscr{E}}^{\rm NR}, j)$ is then obtained by a quadrature:

$$\hat{S}_r(r, \hat{\mathscr{E}}^{\rm NR}, j) = \int dr \sqrt{\mathscr{R}(r, \hat{\mathscr{E}}^{\rm NR}, j)}\,. \tag{10}$$

This allows one to explicitly compute the corresponding integrated radial action variable:

$$i_r = \frac{I_R}{\mu G M} \equiv \frac{2}{2\pi} \int_{r_{\rm min}}^{r_{\rm max}} dr \sqrt{\mathscr{R}(r, \hat{\mathscr{E}}^{\rm NR}, j)}\,. \tag{11}$$

At the 2PN level, following [28], one finds

$$\begin{aligned} i_r = & \frac{1}{\sqrt{-2\hat{\mathscr{E}}^{\rm NR}}}\left[1 + \frac{15-\nu}{4}\frac{\hat{\mathscr{E}}^{\rm NR}}{c^2} + \frac{35+30\nu+3\nu^2}{32}\frac{\hat{\mathscr{E}}^{\rm NR2}}{c^4}\right] \\ & -j + \frac{1}{c^2 j}\left[3 + \frac{15-6\nu}{2}\frac{\hat{\mathscr{E}}^{\rm NR}}{c^2}\right] + \frac{35-10\nu}{4}\frac{1}{c^4 j^2}\,. \end{aligned} \tag{12}$$

Introducing the "principal action variable"

$$\mathscr{N} = I_R + \mathscr{J} = G\mu M(i_r + j) = G\mu M n \tag{13}$$

and restoring the relativistic energy $\mathscr{E}^{\rm R} = \mathscr{E}^{\rm NR} + Mc^2$, one finds the 2PN "Delaunay Hamiltonian" (i.e. the energy expressed in terms of the action variables $n = i_r + j$ and j)

$$\begin{aligned} \hat{\mathscr{E}}^{\rm R}(n,j) = & \frac{1}{\nu} - \frac{1}{2n^2}\left[1 + \left(\frac{6}{nj} - \frac{1}{4}\frac{15-\nu}{n^2}\right)\frac{1}{c^2}\right. \\ & \left. + \left(\frac{5}{2}\frac{7-2\nu}{nj^3} + \frac{27}{n^2 j^2} - \frac{3}{2}\frac{35-4\nu}{n^3 j} + \frac{145-15\nu+\nu^2}{8n^4}\right)\frac{1}{c^4}\right], \end{aligned} \tag{14}$$

where $\hat{\mathscr{E}} = \mathscr{E}/(\mu c^2)$, cn and cj are dimensionless.

This (Delaunay-type) version of the Hamiltonian, Eq. (14), has the great advantage of being *invariant* under coordinate transformations, and, more generally under canonical transformations. Remember that, in quantum mechanics, the Bohr-Sommerfeld quantization rules states that the discrete spectrum of the two-body bound states is obtained by requiring that the action variables are quantized in units of $\hbar$. More precisely, $\mathscr{N}/\hbar$ is the "principal quantum number" and $\mathscr{J}/\hbar$ is the "total angular momentum quantum number." This result suggests that the map between the usual PN dynamics and the corresponding EOB dynamics should be defined by identifying the real action variables to their effective EOB counterparts. Let us see how this can be implemented in practice.

We wish to associate with the real, conservative, relative two-body dynamics the dynamics of a single body, of mass m_0, moving in an effective external pseudo-Riemannian spacetime

$$ds^2_{\rm eff} = g^{\rm eff}_{\mu\nu}(x^\lambda_{\rm eff}) dx^\mu_{\rm eff} dx^\nu_{\rm eff} . \tag{15}$$

The geodesic action describing this effective one-body dynamics is

$$S_{\rm eff}(z_0^\mu) = -\int m_0 c \sqrt{-g^{\rm eff}_{\mu\nu}(z_0^\lambda) dz_0^\mu dz_0^\nu} . \tag{16}$$

In the simplest situation of two non-spinning bodies, the rotational symmetry of the relative dynamics implies that the effective metric must be spherically symmetric. It can then be parametrized as:

$$ds^2_{\rm eff} = -A(R_{\rm eff})c^2 dt^2_{\rm eff} + B(R_{\rm eff}) dR^2_{\rm eff} + R^2_{\rm eff} C(R_{\rm eff})[d\theta^2_{\rm eff} + \sin^2\theta_{\rm eff} d\varphi^2_{\rm eff}] , \tag{17}$$

where we can choose the effective coordinates so that $C(R_{\rm eff}) = 1$, as we shall henceforth do.

Writing the effective Hamilton-Jacobi equation

$$g^{\rm eff}_{\mu\nu} \frac{\partial S}{\partial x^\mu_{\rm eff}} \frac{\partial S}{\partial x^\nu_{\rm eff}} + m_0^2 c^2 = 0 \tag{18}$$

and separating the variables (for equatorial plane motion) as

$$S_{\rm eff} = -\mathcal{E}_0 t_{\rm eff} + \mathcal{J}_0 \varphi_{\rm eff} + S^0_R(R_{\rm eff}, \mathcal{E}_0, \mathcal{J}_0) \tag{19}$$

leads to an *effective* radial integral given by

$$I^0_R(\mathcal{E}_0, \mathcal{J}_0) = \frac{2}{2\pi} \int_{R_{\rm min}}^{R_{\rm max}} dR_{\rm eff} \sqrt{\mathcal{R}_0(R_{\rm eff}, \mathcal{E}_0, \mathcal{J}_0)} , \tag{20}$$

with

$$\mathcal{R}_0(R_{\rm eff}, \mathcal{E}_0, \mathcal{J}_0) = \frac{B(R_{\rm eff})}{A(R_{\rm eff})} \frac{\mathcal{E}_0^2}{c^2} - B(R_{\rm eff}) \left(m_0 c^2 + \frac{\mathcal{J}_0^2}{R^2_{\rm eff}} \right) . \tag{21}$$

Introducing the effective principal action variable $\mathcal{N}_0 = I^0_R + \mathcal{J}_0$ and performing the definite radial action integral (at 2PN) allows one to compute the (coordinate invariant) effective energy levels in a form which depends on the coefficients a_n, b_n

entering the PN expansion of the two effective metric potentials $A(R)$ and $B(R)$:

$$A(R) = 1 + \frac{a_1}{c^2 R} + \frac{a_2}{c^4 R^2} + \dots,$$
$$B(R) = 1 + \frac{b_1}{c^2 R} + \frac{b_2}{c^4 R^2} + \dots. \tag{22}$$

It remains to define the dictionary between the real and the effective Delaunay Hamiltonians. As already said, it is natural to identify

$$\mathscr{N} = \mathscr{N}_0, \qquad \mathscr{J} = \mathscr{J}_0, \tag{23}$$

since both of them are quantized in units of $\hbar$. On the other hand, the precise relation between the real energy ($\mathscr{E}^{\rm R}_{\rm real} = (m_1 + m_2)c^2 + \mathscr{E}^{\rm NR}_{\rm real}$) and the effective one ($\mathscr{E}_0 = m_0 c^2 + \mathscr{E}^{\rm NR}_0$) is a priori unclear. The EOB formalism only requires that there should exist a one-to-one map f between the real and effective "levels", i.e.

$$\mathscr{E}_0(\mathscr{N}_0, \mathscr{J}_0) = f[\mathscr{E}^{\rm R}_{\rm real}(\mathscr{N}, \mathscr{J})], \tag{24}$$

or, in PN-expanded form

$$\frac{\mathscr{E}^{\rm NR}_0}{m_0 c^2} = \frac{\mathscr{E}^{\rm NR}_{\rm real}}{\mu c^2}\left[1 + \alpha_1 \frac{\mathscr{E}^{\rm NR}_{\rm real}}{\mu c^2} + \alpha_2 \left(\frac{\mathscr{E}^{\rm NR}_{\rm real}}{\mu c^2}\right)^2 + \dots\right], \tag{25}$$

with the coefficients α_i parametrizing the unknown map f.

At the 2PN level [6], it was found that the natural requirements

$$m_0 = \mu, \qquad a_1 = -2GM, \qquad b_1 = 2GM \tag{26}$$

lead to a unique solution for all the other 2PN-level unknowns, i.e. $\alpha_1, \alpha_2, a_2, a_3$ and b_2. This solution yields remarkably simple results both for the energy map f and for the effective metric potentials $A(R)$ and $B(R)$. Namely $\alpha_1 = \frac{1}{2}$ and $\alpha_2 = 0$, i.e.

$$\mathscr{E}_0 = \frac{(\mathscr{E}^{R}_{\rm real})^2 - m_1^2 c^4 - m_2^2 c^4}{2(m_1 + m_2)c^2}, \tag{27}$$

and

$$a_2 = 0, \qquad a_3 = 2\nu(GM)^3 \qquad b_2 = (4 - 6\nu)(GM)^2. \tag{28}$$

This then gives for the main EOB potentials at the 2PN accuracy

$$A(R) = 1 - 2\frac{GM}{c^2R} + 2\nu\frac{(GM)^3}{c^6R^3} + \dots,$$
$$B(R) = 1 + 2\frac{GM}{c^2R} + (4 - 6\nu)\frac{(GM)^2}{c^4R^2} + \dots. \tag{29}$$

Passing to the dimensionless inverse radial variable

$$u = \frac{GM}{c^2R}, \tag{30}$$

we have then

$$\begin{aligned} A(u) &= 1 - 2u + 2\nu u^3 + O(u^4) \\ &\equiv 1 - 2u + \nu a_{1\mathrm{SF}}(u) + \nu^2 a_{2\mathrm{SF}}(u) + O(\nu^3), \\ B(u) &= 1 + 2u + (4 - 6\nu)u^2 + O(u^3) \\ &\equiv (1-2u)^{-1} + \nu b_{1\mathrm{SF}}(u) + \nu^2 b_{2\mathrm{SF}}(u) + O(\nu^3), \end{aligned} \tag{31}$$

where the first lines give the PN expansions of $A(u, \nu)$ and $B(u, \nu)$ (i.e. their expansions in powers of u), while the second lines give their GSF expansions (i.e. their expansions in powers of ν).

We see on the above expressions how the EOB formalism can naturally incorporate, in a complementary manner, both information coming from PN theory (expansion in powers of u) and information coming from GSF theory (expansion in powers of ν). From PN results, one currently knows A and B to the 4PN level, i.e. modulo $O(u^6)$ in $A(u, \nu)$ and modulo $O(u^5)$ in $B(u, \nu)$ [27]. On the other hand, from first-order GSF theory, it is only possible to access the linear-in-ν pieces $a_{1\mathrm{SF}}(u)$ of $A(u, \nu)$ and $b_{1\mathrm{SF}}(u)$ of $B(u, \nu)$ (together with the linear-in-ν piece of the higher-than-quadratic correction $Q(u, p_r, \nu)$ to the effective Hamilton-Jacobi equation that starts entering at the 3PN level [8]). For instance, concerning $a_{1\mathrm{SF}}(u)$, one currently knows from GSF theory both its global functional value, with high numerical accuracy, in the strong-field interval $0 < u < \frac{1}{3}$ [29], and its 22.5PN-accurate weak-field expansion [30].

3 Detweiler's Gauge-Invariant Function and the EOB Main Radial Potential

A useful gauge-invariant quantity for perturbations induced by a particle orbiting a Schwarzschild black hole was suggested by Detweiler in 2008 [31]. Let $g^{(0)}_{\alpha\beta}(x^\alpha)$ denote the background metric written in standard coordinates $(x^\alpha) = (t, r, \theta, \phi)$

with associated line element

$$ds^2 = -f(r)dt^2 + f(r)^{-1}dr^2 + r^2(d\theta^2 + \sin^2\theta d\phi^2), \qquad f(r) = 1 - \frac{2m_2}{r}, \tag{32}$$

and let $q\,h_{\mu\nu}(x^\alpha)$ (with $q = m_1/m_2$) denote the perturbation, so that the perturbed metric is $g_{\alpha\beta} = g^{(0)}_{\alpha\beta} + q\,h_{\alpha\beta} + O(q^2)$. A particle moving along an equatorial circular orbit has 4-velocity

$$U = \Gamma k = U^\alpha \partial_\alpha, \qquad k = \partial_t + \Omega\partial_\phi, \tag{33}$$

with $U \cdot U = -1$ (computed within the *perturbed* metric along the circular orbit) and Γ being a normalization factor

$$\Gamma = \frac{1}{\sqrt{1 - \frac{2m_2}{r} - \Omega^2 r^2}}\left[1 + \frac{h^R_{kk}}{2\left(1 - \frac{2m_2}{r} - \Omega^2 r^2\right)}\right], \tag{34}$$

where $h^R_{kk} \equiv \left[h_{\mu\nu}k^\mu k^\nu\right]^{\rm reg}$ is the *regularized* value of $h_{kk} = h_{\mu\nu}k^\mu k^\nu$. Here the gauge-invariant (and constant) orbital frequency Ω is used to parametrize the one-parameter family of circular orbits at different radii. Denoting $z_1 \equiv \Gamma^{-1}$, Detweiler's redshift function is defined as the correspondence $\Omega \to z_1$.

The assumption that the vector field k is a Killing field for the perturbed metric implies that

$$h_{\alpha\beta} = h_{\alpha\beta}(\bar\phi, r, \theta), \qquad \bar\phi = \phi - \Omega t. \tag{35}$$

The conditions for geodesic motion are

$$\Omega = \sqrt{\frac{m_2}{r^3}}\left(1 - \frac{r}{4m_2} r\partial_r h_{kk}\right), \qquad \partial_{\bar\phi} h_{kk} = 0. \tag{36}$$

The former equation can be inverted to express r as a function of Ω, or equivalently of the gauge-invariant dimensionless frequency parameter $y \equiv (m_2\Omega)^{2/3}$ as

$$\frac{m_2}{r} = y\left(1 + \frac{m_2}{6y^2}\partial_r h_{kk}\right). \tag{37}$$

Substituting this into Eq. (34) and evaluating at the particle position finally gives

$$z_1(y) = \sqrt{1-3y} - \frac{h^R_{kk}(y)}{2\sqrt{1-3y}}. \tag{38}$$

The first law of binary black hole mechanics [32, 33] allows one to extract the dynamical information contained in Detweiler's redshift function $z_1(y)$. In particular, there is a direct, simple link between $h^R_{kk}(y)$ and the first-order self-force correction $a_{1SF}(u)$ to the main radial potential $A(u, \nu) = 1-2u+\nu a_{1SF}(u)+O(\nu^2)$ entering the EOB formalism, namely [34]

$$a_{1SF}(x) = -\frac{1}{2}h^R_{kk} - \frac{x(1-4x)}{\sqrt{1-3x}}\,, \tag{39}$$

x being here a generic variable (reducing to u or y when $\nu \to 0$). The PN expansion of $a_{1SF}(x)$ starts as [35, 36]

$$\begin{aligned}
a_{1SF}(x) &= 2x^3 + \left(\frac{94}{3} - \frac{41}{32}\pi^2\right)x^4 \\
&+ \left(-\frac{4237}{60} + \frac{2275}{512}\pi^2 + \frac{256}{5}\ln(2) + \frac{128}{5}\gamma + \frac{64}{5}\ln(x)\right)x^5 \\
&+ \left(-\frac{1066621}{1575} + \frac{246367}{3072}\pi^2 - \frac{31736}{105}\ln(2) - \frac{7004}{105}\ln(xe^{2\gamma})\right. \\
&+ \left.\frac{243}{7}\ln(3)\right)x^6 + \frac{13696}{525}\pi x^{13/2} + O(x^7)\,.
\end{aligned} \tag{40}$$

The present knowledge of this function reaches the 22.5PN order level [30] and exhibits an interesting increasing transcendental structure involving logarithmic terms, Euler gamma constant, Riemann zeta values, etc.

4 GSF Calculations in a Schwarzschild Spacetime: A Review of Important Details

We review in this section some of the technically important details of our recent works [35–41] which are either unpublished or are hidden in the appendices of several different published papers.

One has first to decompose the perturbed metric in tensor harmonics in some gauge; in our case, we use the Regge-Wheeler (RW) gauge [14], following the Regge-Wheeler-Zerilli approach [15]. As a result, the metric functions of different parities satisfy different equations: odd-parity waves satisfy the simple RW equation with odd-parity source terms; even-parity waves satisfy the more complicated Zerilli equation with even-parity source terms. Next, using results by Chandrasekhar [42], the Zerilli equation can be mapped onto a RW equation with modified source terms so that one is left with solving in all cases a RW equation, with appropriate (odd-parity or even-parity) source terms. [The latter source terms are singular at the location $r = r_0$ of the particle, namely $S^{(\mathrm{even/odd})}_{lm\omega}(r) = \sum_k c_k(r_0)\delta^{(k)}(r-r_0)$, with

$\delta^{(k)}(r - r_0)$ denoting the kth-order derivative of the Dirac delta function.] We then have:

$$\mathscr{L}^{(r)}_{(\mathrm{RW})}[R^{(\mathrm{even/odd})}_{lm\omega}] = S^{(\mathrm{even/odd})}_{lm\omega}(r)\,. \tag{41}$$

Here $\mathscr{L}^{(r)}_{(\mathrm{RW})}$ denotes the RW operator

$$\begin{aligned}\mathscr{L}^{(r)}_{(\mathrm{RW})} &= \left(1 - \frac{2M}{r}\right)^2 \frac{d^2}{dr^2} + \frac{2M}{r^2}\left(1 - \frac{2M}{r}\right)\frac{d}{dr} + [\omega^2 - V_{(\mathrm{RW})}(r)] \\ &= \frac{d^2}{dr_*^2} + [\omega^2 - V_{(\mathrm{RW})}(r)]\,,\end{aligned} \tag{42}$$

with $d/dr_* = f(r)d/dr$, and the RW potential

$$V_{(\mathrm{RW})}(r) = \left(1 - \frac{2M}{r}\right)\left(\frac{l(l+1)}{r^2} - \frac{6M}{r^3}\right)\,. \tag{43}$$

Note that here, as well as in all GSF sections below, the large mass is denoted as M (instead of m_2) and the small mass by μ (instead of m_1), to follow a standard notation in perturbation theory. One should keep in mind that, in all EOB-related results, we instead use the standard PN-EOB notation m_1, m_2.

The first step is to explicitly determine analytical solutions to the homogeneous RW equation, using the techniques outlined above.

4.1 PN Solutions to the Homogeneous RW Equation

Restoring the factors of c, the homogeneous RW equation $\mathscr{L}^{(r)}_{(\mathrm{RW})}X_{l\omega}(r) = 0$ reads

$$\begin{aligned}X''_{l\omega}(r) &+ \frac{2M\eta^2}{r^2\left(1 - \frac{2M\eta^2}{r}\right)}X'_{l\omega}(r) - \frac{1}{\left(1 - \frac{2M\eta^2}{r}\right)^2 r^2}\Bigg[- \omega^2\eta^2 r^2 \\ &+ \left(1 - \frac{2M\eta^2}{r}\right)\left(L - \frac{6M\eta^2}{r}\right)\Bigg]X_{l\omega}(r) = 0\,,\end{aligned} \tag{44}$$

where $L = l(l+1)$ and $\eta = 1/c$. By definition the PN expansion is an expansion in powers of η. Expanding the above equation in series of η and solving order by order one can obtain two independent PN-type solutions: an "in" PN solution (defined as having r^{l+1} as leading order term) and an "up" PN solution (defined as having r^{-l} as leading order term), respectively. Actually, one a priori only needs to compute the PN "in" solution, because the PN "up" solution directly follows from the "in" solution simply by the replacement $l \to -l - 1$. We find

$$X_{l\omega}^{\text{in(PN)}}(r) = r^{l+1}\left(1 + A_2^{\text{in}\,(PN,l)}\eta^2 + A_4^{\text{in}\,(PN,l)}\eta^4 + \dots\right), \tag{45}$$

where the first few coefficients $A_k^{\text{in}\,(PN,l)}$ are listed below using the notation $X_1 = M/r$, $X_2 = (\omega r)^2$:

$$\begin{aligned}
A_2^{\text{in}\,(PN,l)} &= -\frac{(l-2)(l+2)}{l}X_1 - \frac{1}{2(2l+3)}X_2\,,\\
A_4^{\text{in}\,(PN,l)} &= \frac{(l-2)(l-3)(l+2)(l+1)}{(-1+2l)l}X_1^2 + \frac{(l^3-5l^2-14l-12)}{2(l+1)(2l+3)l}X_2X_1\\
&\quad+\frac{1}{8(2l+5)(2l+3)}X_2^2\,,\\
A_6^{\text{in}\,(PN,l)} &= -\frac{(l-2)(l-3)(l-4)(l+2)(l+1)}{3(-1+2l)(l-1)}X_1^3\\
&\quad-\frac{2(15l^4+30l^3+28l^2+13l+24)}{(-1+2l)(2l+1)(l+1)l(2l+3)}X_1^2X_2\ln(r/R)\\
&\quad-\frac{(3l^4-27l^3-154l^2-220l-120)}{24(l+1)l(2l+5)(l+2)(2l+3)}X_1X_2^2\\
&\quad-\frac{1}{48(2l+5)(2l+7)(2l+3)}X_2^3\,,\\
A_8^{\text{in}\,(PN,l)} &= \frac{(l-2)(l-3)(l-4)(-5+l)(l+2)(l+1)}{6(-1+2l)(2l-3)}X_1^4\\
&\quad-\frac{\bar{B}_8^{\ln}\ln(r/R)+\bar{B}_8}{6(-1+2l)(2l+1)l^3(2l+3)(l+1)(l-1)}X_2X_1^3\\
&\quad-\frac{\bar{C}_8^{\ln}\ln(r/R)+\bar{C}_8}{24l(-1+2l)(2l+1)(2l+3)^3(l+1)(2l+5)(l+2)}X_2^2X_1^2\\
&\quad+\frac{(5l^5-60l^4-645l^3-1788l^2-1928l-840)}{240l(l+3)(2l+3)(l+1)(2l+5)(2l+7)(l+2)}X_1X_2^3\\
&\quad+\frac{1}{384(2l+9)(2l+3)(2l+5)(2l+7)}X_2^4\,,
\end{aligned} \tag{46}$$

with

$$\begin{aligned}
\bar{B}_8^{\ln} &= -12l(l-1)(l-2)(l+2)(15l^4+30l^3+28l^2+13l+24)\,,\\
\bar{B}^8 &= 4l^{10}-36l^9-141l^8-135l^7+207l^6-765l^5-1042l^4\\
&\quad-132l^3+1728l^2-1344l+576\,,
\end{aligned}$$

$$\bar{C}_8^{\ln} = -24(2l+5)(l+2)(2l+3)(15l^4+30l^3+28l^2+13l+24)\,,$$
$$\bar{C}_8 = 24l^9+156l^8-1994l^7-16283l^6-44315l^5-52967l^4$$
$$-14363l^3+35794l^2+45708l+20520\,. \tag{47}$$

The length (and complexity) of these coefficients increases with higher powers of η. This is why we have only displayed the first few ones here. They can only be stored in some mathematical manipulation software, like for example Mathematica^{TM} or Maple^{TM}.

As exhibited in the coefficient $A_6^{\mathrm{in}\,(PN,\,l)}$, $A_8^{\mathrm{in}\,(PN,\,l)}$, etc., there are logarithms ("PN-logs") entering these coefficients which involve a priori arbitrary length scales (here denoted as R, R_1, etc.). When such PN-logs first appear in the in-PN solution they are associated with r-independent combinations of the type $\eta^6 X_1^2 X_2 = (\eta^2 M/r)^2(\eta\omega r)^2$, and its higher powers (for R_1, etc.). [The scale R first appears at order η^6, the scale R_1 first appears at order η^{12}, ..., with new scales associated with higher powers of $\eta^6 X_1^2 X_2 = (\eta^2 M/r)^2(\eta\omega r)^2$.] However, the dependence of $X_{l\omega}^{\mathrm{in(PN)}}$ on the choice of these scales is spurious, because changing these scales only modify $X_{l\omega}^{\mathrm{in(PN)}}(r)$ by a constant multiplicative factor. In other words, the PN in-solution $X_{l\omega}^{\mathrm{in(PN)}}(r)$ satisfies "Renormalization Group Equations", with respect to the group of rescalings of $R, R_1, \ldots$, similar to those appearing in Quantum Field Theory. For simplicity, let us identify all log scales, i.e. $R = R_1 = \ldots$. We then find that the logarithmic derivative of $X_{l\omega}^{\mathrm{in(PN)}}$ with respect to the logarithm of the scale R, is a mere constant:

$$\frac{\partial_{\ln R} X_{l\omega}^{\mathrm{in(PN)}}}{X_{l\omega}^{\mathrm{in(PN)}}} = C_l M^2\omega^2\eta^6 + D_l M^4\omega^4\eta^{12} + O(\eta^{18})\,, \tag{48}$$

where

$$C_l = \frac{2(28l^2+30l^3+13l+15l^4+24)}{l(2l+3)(l+1)(-1+2l)(1+2l)} \tag{49}$$

$$D_l = \frac{2d_l}{(-3+2l)(5+2l)(-1+l)(2+l)l^3(1+2l)^3(-1+2l)^3(l+1)^3(2l+3)^3}\,,$$

with

$$d_l = 18480l^{16}+147840l^{15}+456120l^{14}+605640l^{13}+8295l^{12}-1096830l^{11}$$
$$-1678310l^{10}-1520455l^9-1355518l^8-1397512l^7-1217380l^6-733273l^5$$
$$+675625l^4+1855326l^3+850608l^2-102816l-51840\,.$$

One can then work, without loss of generality, with a "minimal PN solution" where all the log scales are fixed to, say, the value $R = GM/c^2$. See below the discussion for the additional, physically different logarithms entering the up-PN solution.

Let us further note that: (1) the PN in solution cannot be used when dealing with $l = 0, 1$ multipoles, since many of the above coefficients are singular for these values of l. This is physically related to the fact that the $l = 0$ and $l = 1$ solutions correspond to gauge terms which must be separately added and which correspond to changes in, respectively, the background mass (monopole, $l = 0$) and angular momentum (dipole, $l = 1$); (2) When $l = 2$ the PN *up* solution cannot be used as is at (and beyond) order η^8. Indeed, the coefficient $A_8^{\text{in}\,(PN,l)}$ contains a factor $\sim 1/(l+3)$ which, when converted into the corresponding up coefficient (by replacing $l \to -l-1$), gives a term $\sim 1/(l-2)$ which is singular for $l = 2$. This situation is quite general in the sense that $A_9^{\text{in}\,(PN,l)}$ contains a factor $\sim 1/(l+4)$ which, when converted into the corresponding up coefficient (by replacing $l \to -l-1$), gives a potentially singular term $\sim 1/(l-3)$ etc. These formally singular terms $\sim 1/(l-l_0)$ entering the up-PN solution signal the appearance of new types of logarithmic terms entering the up-PN solution. For instance, the term in $A_8^{\text{in}\,(PN,l)}$ that contains a factor $\sim 1/(l+3)$ is proportional to $X_1 X_2^3$ and formally generates in $X^{\text{up(PN)}}_{(l=2)\omega}(r)$ (when expanding it in Laurent series in $l-2$) a term

$$X^{\text{up(PN) singular}}_{(l=2)\omega}(r) \propto \frac{r^{-l}}{l-2} X_1 X_2^3 \propto \eta^8 \left(c_1 \frac{r^3}{l-2} + c_2 r^3 \ln(r/R') + O(l-2) \right). \tag{50}$$

We see on this example how new logarithms arise in the up-PN solution. The first such logarithm, exhibited in the above equation, corresponds to a 4PN correction having a quadrupolar ($l = 2$) structure, and being proportional to the corresponding in-PN solution $\sim r^{l+1} = r^3$. The ambiguity in this logarithmic contribution linked to the arbitrariness in the new log scale R' corresponds to the tail-transported hereditary quadrupolar nonlocal breakdown, arising at the 4PN level, in the PN-expanded nearzone metric pointed out in [43]. The latter nonlocal ambiguity (which corresponds to mixing[1] some in-PN solution within the up-PN one) cannot be fixed from the sole PN-based solutions of the RW equation. [In spite of the spurious dependence of the in-PN solutions on the "in" scales R, R_1 etc., there is also an ambiguity in the correct determination of the in-solution linked to the boundary condition at the horizon. However, the latter ambiguity arises at a higher PN order, namely, when $l = 2$, $O(\eta^{13})$ for dissipative effects [44, 45], and $O(\eta^{16})$ for conservative effects; see below.] One needs to resort to the MST hypergeometric-expansion solution (which incorporates nonlocal information necessary for a correct

[1] Before the admixtion of this 4PN-level nonlocal contribution in the up $l = 2$ solution, there arises a 2.5PN-level admixtion of an in-type $l = 2$ solution $\propto r^3$, which corresponds to radiation damping. However, the latter is a dissipative effect which does not contribute to the conservative GSF effects we are interested in. See Sect. 4.3 below for a more detailed discussion of these 2.5PN dissipative, and 4PN conservative, admixtions.

matching between the near zone and the wave zone) in order to resolve the types of ambiguities arising in the up-PN solutions. [The nonlocally constructed MST solutions also provide a correct matching at the horizon for the in- (and up-) solutions.] However, our analytical understanding of these ambiguities, as being associated with tail-transported nonlocal effects, shows that their PN order is linked (and actually increases) with the value of their multipolarity l. More precisely, they arise at order $\eta^{(2l+4)}$ (($l+2$)-PN order) in the even-parity up-PN solution, and at order $\eta^{(2l+6)}$ (($l+3$)-PN order) in the odd-parity up-PN solution [36]. Because of these links, at any given PN order, one will only need MST technology to compute a finite number of low-multipolarity solutions. For the higher-multipolarity ones, it is enough to use the PN solutions. For instance, if we are looking for a 4PN-accurate result (as in [35]), it is enough to compute the $l=2$ solutions with MST technology, and the $l>2$ ones with PN technology.

4.2 MST Solutions to the Homogeneous RW Equation

The physical Green's function we are interested in is the *retarded* one. It is obtained, by choosing for $X^{\rm in}_{l\omega}$ a solution that is incoming from $r=+\infty$ (and purely ingoing on the horizon), and for $X^{\rm up}_{l\omega}$ a solution that is upgoing from the horizon (and purely outgoing at infinity). Mano, Suzuki and Takasugi [19, 20] gave analytic expressions for the so-defined $X^{\rm in}_{l\omega}$ and $X^{\rm up}_{l\omega}$ in the form of series of hypergeometric (HG) functions (of the usual Gauss-type for the "in" solution and of the confluent type for the "up" solution), namely,

$$X^{\rm in(HG)}_{l\omega}(r) = C^{\nu}_{\rm (in)}(x)\sum_{n=-\infty}^{\infty} a^{\nu}_n \times \\ \times \overline{F}(n+\nu-1-i\epsilon, -n-\nu-2-i\epsilon, 1-2i\epsilon; x]\,, \tag{51}$$

$$X^{\rm up(HG)}_{l\omega}(r) = C^{\nu}_{\rm (up)}(z)\sum_{n=-\infty}^{\infty} a^{\nu}_n (-2iz)^n \times \\ \times \overline{\Psi}(n+\nu+1-i\epsilon, 2n+2\nu+2; -2iz)\,. \tag{52}$$

Here, $x = 1-c^2r/2GM$, $z=\omega r/c$, $\epsilon = 2GM\omega/c^3 = 2mGM\Omega/c^3$,

$$\nu = l + \frac{1}{2l+1}\Bigg[-2-\frac{s^2}{l(l+1)} + \frac{[(l+1)^2-s^2]^2}{(2l+1)(2l+2)(2l+3)} \\ - \frac{(l^2-s^2)^2}{(2l-1)2l(2l+1)}\Bigg]\epsilon^2 + O(\epsilon^4) \tag{53}$$

is an ϵ-modified avatar of l [19, 20] (with $s^2 = 4$ in the present spin 2 case), and

$$
\begin{aligned}
C^{\nu}_{(\text{in})}(x) &= c_{(\text{in})} e^{i\epsilon[(x-1)-\ln(-x)]}(1-x)^{-1}\,, \\
C^{\nu}_{(\text{up})}(z) &= c_{(\text{up})} e^{iz} z^{\nu+1}\left(1-\frac{\epsilon}{z}\right)^{-i\epsilon} 2^{\nu} e^{-\pi\epsilon} e^{-i\pi(\nu+1)}\,, \\
\overline{F}(a,b,c;x) &= \frac{\Gamma(a)\Gamma(b)}{\Gamma(c)} \quad F(a,b,c;x)\,, \\
\overline{\Psi}(a,b;\zeta) &= \frac{\Gamma(a-2)\Gamma(a)}{\Gamma(a^*)\Gamma(a^*+2)}\Psi(a,b;\zeta)\,,
\end{aligned}
\tag{54}
$$

with a^* denoting the complex conjugate of a and Ψ the second Kummer function. Here the quantities $c_{(\text{in})} = \eta^{\alpha^{(l)}_{(\text{in})}}$ and $c_{(\text{up})} = \eta^{\alpha^{(l)}_{(\text{up})}}$ are some l-dependent powers of $\eta = 1/c$ that are defined so that $C^{\nu}_{(\text{in})}(x)$ and $C^{\nu}_{(\text{up})}(z)$ both start with zeroth order in η. For example, for $l = 2$ we take $c_{(\text{in})} = \eta^{-2}$ and $c_{(\text{up})} = \eta^{-3}$ so that

$$
\begin{aligned}
C_{(\text{in})}(r) &= \frac{2M}{r} - 2iM\omega\eta - Mr\omega^2\eta^2 + O(\eta^3)\,, \\
C_{(\text{up})}(r) &= -4\omega^3 r^3 - 4i\omega^4 r^4 \eta + 2\omega^5 r^5 \eta^2 + O(\eta^3)\,.
\end{aligned}
$$

Finally, the two-sided sequence of coefficients a^{ν}_n entering both series (51) and (52) are obtained by solving a three-term recursion relation, $\alpha^{\nu}_n a^{\nu}_{n+1} + \beta^{\nu}_n a^{\nu}_n + \gamma^{\nu}_n a^{\nu}_{n-1} = 0$, obtained by Mano et al. (see Eqs. (2.5)–(2.8) in [19]). To solve this three-term relation between $n = -N$ and $n = +N$ (included) one initiates the recursion with $a_{N+1} = 0$ and $a_{-N-1} = 0$ and chooses $a_0 = 1$ to have a fully determined set of algebraic equations.

The expansions of $X^{\text{in(HG)}}_{l\omega}$ and $X^{\text{up(HG)}}_{l\omega}$ in powers of η thus have the general form

$$
X^{\text{in(HG)}}_{l\omega}(r) = \tilde{c}_{\text{in}} r^{l+1} \sum_{k=0}^{k_{\max}} A^{\text{in}(l)}_k \eta^k\,, \tag{55}
$$

$$
X^{\text{up(HG)}}_{l\omega}(r) = \tilde{c}_{\text{up}} r^{-l} \sum_{k=0}^{k_{\max}} A^{\text{up}(l)}_k \eta^k\,, \tag{56}
$$

where $\tilde{c}_{\text{in}}$ and $\tilde{c}_{\text{up}}$ are some normalization coefficients, where $A^{\text{in}(l)}_0 = 1 = A^{\text{up}(l)}_0$, and where the logarithmic dependence in η is absorbed in the expansion coefficients A_k. The A_k's are functions of the two quantities $X_1 = GM/r$ (linked to the weak-field expansion), and $\sqrt{X_2} = \omega r$ (linked to the near-zone expansion). Modulo some logarithmic dependence in r, the A_k's are polynomials in X_1 and $\sqrt{X_2}$.

Let us indicate the structure of the η expansion of the hypergeometric *up* solutions (52) for the particular case $l = 2$. We derived similar explicit results for $l = 3, 4\ldots$. The expansion is extended up to some preassigned accuracy $\eta^{k_{\max}}$, with $k_{\max} \geq 12$ ($k_{\max} = 12$ corresponding to the (fractional) 6PN level). When using $c_{\rm up} = c^3 = \eta^{-3}$ in the definition of $C^{\nu}_{(\rm up)}(z)$ in Eq. (52), the η expansion of the hypergeometric "up" solution for $l = 2$, i.e., $X^{HG,l=2}_{(\rm up)}$, reads

$$X^{HG,l=2}_{(\rm up)} = -\frac{i}{16\omega^2 r^2}\sum_k A_k^{{\rm up}(HG,l=2)}\eta^k\,, \tag{57}$$

where HG stands for "hypergeometric" and the explicit values of the first coefficients $A_k^{\rm up}$ read

$$
\begin{aligned}
A_0^{{\rm up}\,(HG,l=2)} &= 1\,,\\
A_1^{{\rm up}\,(HG,l=2)} &= 0\,,\\
A_2^{{\rm up}\,(HG,l=2)} &= \frac{1}{6}X_2 + \frac{5}{3}X_1\,,\\
A_3^{{\rm up}\,(HG,l=2)} &= \left(6i\gamma - 2\pi - \frac{43}{6}i\right)X_1\sqrt{X_2}\,,\\
A_4^{{\rm up}\,(HG,l=2)} &= \frac{7}{6}X_1X_2 + \frac{1}{24}X_2^2 + \frac{20}{7}X_1^2\,,\\
A_5^{{\rm up}\,(HG,l=2)} &= \left[\left(10i\gamma - \frac{10}{3}\pi - \frac{215}{18}i\right)X_1^2\right.\\
&\quad\left.+\left(-\frac{43}{36}i + i\gamma - \frac{1}{3}\pi\right)X_1X_2 + \frac{1}{45}iX_2^2\right]\sqrt{X_2}\,,\\
A_6^{{\rm up}\,(HG,l=2)} &= 5X_1^3 + \left(-18\gamma^2 + \frac{7}{3}\pi^2 - 12i\pi\gamma - \frac{272551}{8820} + \frac{214}{105}\ln(2\omega r\eta)\right.\\
&\quad\left.+\frac{4729}{105}\gamma + \frac{43}{3}i\pi\right)X_1^2X_2 + \frac{7}{24}X_1X_2^2 - \frac{1}{144}X_2^3\,,\\
A_7^{{\rm up}\,(HG,l=2)} &= \left[\left(-\frac{430}{21}i + \frac{120}{7}i\gamma - \frac{40}{7}\pi\right)X_1^3 + \left(-\frac{301}{36}i - \frac{7}{3}\pi + 7i\gamma\right)X_2X_1^2\right.\\
&\quad\left.+\left(-\frac{1}{12}\pi - \frac{43}{144}i + \frac{1}{4}i\gamma\right)X_2^2X_1 - \frac{1}{630}iX_2^3\right]\sqrt{X_2}\,,
\end{aligned}
$$

$$
\begin{aligned}
A_8^{{\rm up}\,(HG,l=2)} &= \frac{80}{9}X_1^4 + \left(-20i\pi\gamma + \frac{215}{9}i\pi + \frac{4729}{63}\gamma - 30\gamma^2 + \frac{35}{9}\pi^2 - \frac{1477681}{26460}\right.\\
&\quad\left.+\frac{214}{63}\ln(2\omega r\eta)\right)X_2X_1^3 + \left(-\frac{76169}{17640} + \frac{7}{18}\pi^2 + \frac{43}{18}i\pi - 2i\pi\gamma\right.
\end{aligned}
$$

$$
+\frac{4729}{630}\gamma - 3\gamma^2 + \frac{107}{315}\ln(2\omega r\eta)\Big)X_2^2X_1^2
$$
$$
+\left(\frac{3547}{10800} - \frac{4}{45}\ln(2\omega r\eta) - \frac{2}{9}\gamma\right)X_2^3X_1 + \frac{1}{3456}X_2^4,
$$

$$
\begin{aligned}
A_9^{\mathrm{up}\,(HG,\,l=2)} = {} & \Bigg[\left(-10\pi - \frac{215}{6}i + 30i\gamma\right)X_1^4 + \Bigg(-2\pi^3 - 36i\gamma^3 + \frac{4943}{35}i\gamma^2 \\
& + \frac{428}{35}i\gamma\ln(2\omega r\eta) + \frac{1691881}{17640}i - 8i\zeta(3) + 14i\pi^2\gamma + \frac{272551}{4410}\pi \\
& - \frac{4601}{315}i\ln(2\omega r\eta) - \frac{428}{105}\pi\ln(2\omega r\eta) + 36\gamma^2\pi - \frac{9458}{105}\pi\gamma \\
& - \frac{882067}{4410}i\gamma - \frac{9251}{630}i\pi^2\Bigg)X_2X_1^3 + \left(\frac{7}{4}i\gamma - \frac{7}{12}\pi - \frac{301}{144}i\right)X_2^2X_1^2 \\
& + \left(\frac{121}{3360}i - \frac{1}{24}i\gamma + \frac{1}{72}\pi\right)X_2^3X_1 + \frac{1}{22680}iX_2^4\Bigg]\sqrt{X_2},
\end{aligned}
$$

$$
\begin{aligned}
A_{10}^{\mathrm{up}\,(HG,\,l=2)} = {} & 16X_1^5 + \Bigg(\frac{20}{3}\pi^2 - \frac{240}{7}i\pi\gamma + \frac{860}{21}i\pi - \frac{220798}{2205} + \frac{856}{147}\ln(2\omega r\eta) \\
& + \frac{18916}{147}\gamma - \frac{360}{7}\gamma^2\Bigg)X_2X_1^4 + \Bigg(\frac{301}{18}i\pi + \frac{4729}{90}\gamma - 14i\pi\gamma - 21\gamma^2 \\
& + \frac{107}{45}\ln(2\omega r\eta) - \frac{33862}{945} + \frac{49}{18}\pi^2\Bigg)X_2^2X_1^3 + \Bigg(-\frac{45901}{70560} + \frac{43}{72}i\pi \\
& + \frac{107}{1260}\ln(2\omega r\eta) + \frac{4729}{2520}\gamma - \frac{1}{2}i\pi\gamma - \frac{3}{4}\gamma^2 + \frac{7}{72}\pi^2\Bigg)X_2^3X_1^2 \\
& + \left(\frac{1}{63}\gamma - \frac{108721}{4233600} + \frac{2}{315}\ln(2\omega r\eta)\right)X_2^4X_1 - \frac{1}{172800}X_2^5. \qquad (58)
\end{aligned}
$$

Note that this "up" solution contains logarithmic terms, $\ln(2\omega r\eta)$, starting at the 3PN level (i.e., A_6^{up}), and even squared logarithms, $(\ln(2\omega r\eta))^2$, at the 6PN level A_{12}^{up}. These logarithms have various physical meanings: some are gauge effects, some are related to far-zone effects and enter the relation between algorithmic multipole moments and source variables, some (with imaginary coefficients) are linked to tail modifications of radiation-reaction effects, and some are linked to tail modifications of conservative effects.

In the case $l = 2$ (as an example), an accurate expression for the generalized angular momentum variable ν reads

$$\nu = 2 - \frac{107}{210}\epsilon^2 - \frac{1695233}{9261000}\epsilon^4 - \frac{76720109901233}{480698687700000}\epsilon^6 - \frac{71638806585865707261481}{389235629236738284000000}\epsilon^8 - \frac{27036066493983382155489949365364 3}{11256262345978018393784760000000 00}\epsilon^{10} + O(\epsilon^{12}) . \quad (59)$$

Furthermore, the *leading-order* solution for the system of coefficients a_n^ν of Eqs. (51) and (52) in the case $l = 2$ when taking $N = 20$ is the following

$$\begin{aligned}
a^\nu_{-20} &= 0 ,\\
a^\nu_{-19} &= \frac{2097152}{742411353415124 25}M^{18}\omega^{18}\eta^{54} ,\\
a^\nu_{-18} &= \frac{3670016}{102097316853353 25}iM^{17}\omega^{17}\eta^{51} ,\\
a^\nu_{-17} &= -\frac{3670016}{869617582393725}M^{16}\omega^{16}\eta^{48} ,\\
a^\nu_{-16} &= -\frac{28672}{631530560925}iM^{15}\omega^{15}\eta^{45} ,\\
a^\nu_{-15} &= \frac{229376}{516706822575}M^{14}\omega^{14}\eta^{42} ,\\
a^\nu_{-14} &= \frac{8192}{2096781309}iM^{13}\omega^{13}\eta^{39} ,\\
a^\nu_{-13} &= -\frac{57344}{1874899689}M^{12}\omega^{12}\eta^{36} ,\\
a^\nu_{-12} &= -\frac{14336}{68316183}iM^{11}\omega^{11}\eta^{33} ,\\
a^\nu_{-11} &= \frac{200704}{162083493}M^{10}\omega^{10}\eta^{30} ,\\
a^\nu_{-10} &= \frac{50176}{8186035}iM^{9}\omega^{9}\eta^{27} ,\\
a^\nu_{-9} &= -\frac{250880}{10201059}M^{8}\omega^{8}\eta^{24} ,\\
a^\nu_{-8} &= -\frac{7840}{103041}iM^{7}\omega^{7}\eta^{21} ,
\end{aligned}$$

$$
\begin{aligned}
a_{-7}^{\nu} &= \frac{1920}{11449} M^6 \omega^6 \eta^{18}\,,\\
a_{-6}^{\nu} &= \frac{7840}{34347} i M^5 \omega^5 \eta^{15}\,,\\
a_{-5}^{\nu} &= -\frac{1568}{11449} M^4 \omega^4 \eta^{12}\,,\\
a_{-4}^{\nu} &= -\frac{28}{107} i M^5 \omega^5 \eta^{15}\,,\\
a_{-3}^{\nu} &= -\frac{56}{963} M^4 \omega^4 \eta^{12}\,,\\
a_{-2}^{\nu} &= \frac{22}{1605} M^4 \omega^4 \eta^{12}\,,\\
a_{-1}^{\nu} &= -\frac{2}{5} i M^3 \omega^3 \eta^{9}\,,\\
a_{0}^{\nu} &= 1\,,\\
a_{1}^{\nu} &= -\frac{5}{3} i M \omega \eta^{3}\,,\\
a_{2}^{\nu} &= -\frac{60}{49} M^2 \omega^2 \eta^{6}\,,\\
a_{3}^{\nu} &= \frac{5}{9} i M^3 \omega^3 \eta^{9}\,,\\
a_{4}^{\nu} &= \frac{160}{891} M^4 \omega^4 \eta^{12}\,,\\
a_{5}^{\nu} &= -\frac{32}{715} i M^5 \omega^5 \eta^{15}\,,\\
a_{6}^{\nu} &= -\frac{128}{14157} M^6 \omega^6 \eta^{18}\,,\\
a_{7}^{\nu} &= \frac{64}{41769} i M^7 \omega^7 \eta^{21}\,,\\
a_{8}^{\nu} &= \frac{256}{1146327} M^8 \omega^8 \eta^{24}\,,\\
a_{9}^{\nu} &= -\frac{256}{8973909} i M^9 \omega^9 \eta^{27}\,,\\
a_{10}^{\nu} &= -\frac{1024}{315918225} M^{10} \omega^{10} \eta^{30}\,,\\
a_{11}^{\nu} &= \frac{128}{386122275} i M^{11} \omega^{11} \eta^{33}\,,
\end{aligned}
$$

$$
\begin{aligned}
a^{\nu}_{12} &= \frac{16384}{531690372675} M^{12}\omega^{12}\eta^{36}\,,\\
a^{\nu}_{13} &= -\frac{16384}{6242302541475} iM^{13}\omega^{13}\eta^{39}\,,\\
a^{\nu}_{14} &= -\frac{65536}{317740905907425} M^{14}\omega^{14}\eta^{42}\,,\\
a^{\nu}_{15} &= \frac{32768}{2178417568201875} iM^{15}\omega^{15}\eta^{45}\,,\\
a^{\nu}_{16} &= \frac{65536}{64045476505135125} M^{16}\omega^{16}\eta^{48}\,,\\
a^{\nu}_{17} &= -\frac{65536}{1004831410292584875} iM^{17}\omega^{17}\eta^{51}\,,\\
a^{\nu}_{18} &= -\frac{262144}{67041355167702790875} M^{18}\omega^{18}\eta^{54}\,,\\
a^{\nu}_{19} &= \frac{65536}{2961732457975830288 75} iM^{19}\omega^{19}\eta^{57}\,,\\
a^{\nu}_{20} &= 0\,.
\end{aligned}
\tag{60}
$$

It is evident that the full solution of the a^{ν}_{n} coefficients cannot be displayed here and it can be only stored in the associated Maple™ or Mathematica™ generating code.

4.3 Comparison Between PN and MST Solutions for $l = 2$

To clarify the admixtion (discussed above in terms of a matching between the near zone and the wave zone) of in-type PN multipolar solutions in the exact up-type HG solution (and reciprocally), we can compare (in the case $l = 2$) the PN solutions $X_{\rm in}^{(l=2,{\rm PN})}$ and $X_{\rm up}^{(l=2,{\rm PN})}$ with the corresponding MST solutions $X_{\rm in}^{(l=2,{\rm HG})}$, $X_{\rm up}^{(l=2,{\rm HG})}$, limiting the comparison to the order $O(\eta^{16})$ (included). Since both PN and MST solutions form complete sets of solutions, one can express one pair in term of the other, through linear relations of the type

$$
\begin{aligned}
X_{\rm up}^{(l=2,{\rm HG})} &= A_{\rm up}X_{\rm up}^{(l=2,{\rm PN})} + B_{\rm in}X_{\rm in}^{(l=2,{\rm PN})}\\
&= A_{\rm up}\left(X_{\rm up}^{(l=2,{\rm PN})} + \beta_2 X_{\rm in}^{(l=2,{\rm PN})}\right)\,,\\
X_{\rm in}^{(l=2,{\rm HG})} &= \tilde{A}_{\rm up}X_{\rm up}^{(l=2,{\rm PN})} + \tilde{B}_{\rm in}X_{\rm in}^{(l=2,{\rm PN})}\\
&= \tilde{B}_{\rm in}\left(\chi_2 X_{\rm up}^{(l=2,{\rm PN})} + X_{\rm in}^{(l=2,{\rm PN})}\right)\,,
\end{aligned}
\tag{61}
$$

where $A_{\rm up}$, $B_{\rm in}$, $\tilde{A}_{\rm up}$ and $\tilde{B}_{\rm in}$ are constant coefficients (computed as series of η, e.g., $A_{\rm up} = \sum_k A_{\rm up}^{(k)} \eta^k$). For convenience, we have also introduced the ratios

$$\beta_2 = \frac{B_{\rm in}}{A_{\rm up}}, \qquad \chi_2 = \frac{\tilde{A}_{\rm up}}{\tilde{B}_{\rm in}}. \tag{62}$$

By using the notation

$$X = \ln(2e^{\gamma}\eta\omega) \equiv Z + \gamma, \qquad Y = \ln\left(2M\eta^2\right), \tag{63}$$

we find for the real and imaginary parts of the latter ratios

$$\begin{aligned}
\text{Re}\,(\beta_2) &= \left(-\frac{4}{45}X + \frac{109}{720}\right)M\omega^6\eta^8 - \frac{214}{4725}\pi\omega^7 M^2\eta^{11} \\
&+ \left(-\frac{677812}{496125}X + \frac{856}{4725}X^2 - \frac{16}{45}\zeta(3) - \frac{214}{2835}\pi^2\right. \\
&\left.+\frac{4862286281}{2222640000}\right)\omega^8 M^3\eta^{14}, \\
\text{Im}\,(\beta_2) &= \frac{1}{45}\omega^5\eta^5 + \frac{2}{45}\omega^6 M\pi\eta^8 + \left(\frac{169453}{496125} - \frac{428}{4725}X + \frac{4}{135}\pi^2 -\right)\omega^7 M^2\eta^{11} \\
&+ \left(-\frac{856}{4725}\pi X + \frac{338906}{496125}\pi\right)\omega^8 M^3\eta^{14},
\end{aligned} \tag{64}$$

and

$$\begin{aligned}
\text{Re}\,(\chi_2) &= -\frac{32}{15}M^7\omega^2\,(12Y - 25)\,\eta^{16}, \\
\text{Im}\,(\chi_2) &= \frac{64}{5}M^6\omega\eta^{13}.
\end{aligned} \tag{65}$$

Note the structure of β_2: $\beta_2 = i\omega^5\eta^5/45 + O(\eta^8)$. Here, the purely imaginary coefficient $i\omega^5\eta^5/45$ corresponds to the admixtion of the (Burke-Thorne) quadrupolar radiation damping within the up solution. This has no effect on the conservative dynamics, so that, when computing (as we do) conservative effects, the admixtion of in-PN solution within the up solution starts at the 4PN order, $O(\eta^8)$.

Concerning the structure of the other mixing coefficient χ_2, namely $\chi_2 = i\frac{64}{5}M^6\omega\eta^{13} + O(\eta^{16})$, it shows that the horizon boundary condition introduces a dissipative effect at the η^{13} level [44, 45], and starts having a conservative effect only at the η^{16} level.

For completeness, let us also give the expressions of the coefficients $A_{\rm up}$, $B_{\rm in}$, $\tilde{A}_{\rm up}$ and $\tilde{B}_{\rm in}$. The nonvanishing coefficients of the η-expansion of $A_{\rm up}$ and $B_{\rm in}$ are given by

$$
\begin{aligned}
A^{(0)}_{\rm up} &= -\frac{i}{16\omega^2}\,,\\
A^{(3)}_{\rm up} &= \frac{M}{8\omega}\left(-\frac{43}{12}+3\gamma+i\pi\right)\,,\\
A^{(6)}_{\rm up} &= \frac{M^2}{4}\left(\frac{43}{12}\pi-\frac{7}{12}i\pi^2-\frac{107}{210}iZ+\frac{9}{2}i\gamma^2-\frac{4729}{420}i\gamma-3\pi\gamma+\frac{272551}{35280}i\right)\,,\\
A^{(9)}_{\rm up} &= \frac{M^3\omega}{2}\left[\left(-\frac{4601}{2520}+\frac{107}{70}\gamma+\frac{107}{210}i\pi\right)Z\right.\\
&\quad -\frac{9}{2}\gamma^3+\left(-\frac{9}{2}i\pi+\frac{4943}{280}\right)\gamma^2+\left(-\frac{882067}{35280}+\frac{7}{4}\pi^2+\frac{4729}{420}i\pi\right)\gamma\\
&\quad \left.-\frac{9251}{5040}\pi^2-\zeta(3)+\frac{1}{4}i\pi^3-\frac{272551}{35280}i\pi+\frac{1691881}{141120}\right]\,,
\end{aligned}
$$

$$
\begin{aligned}
A^{(12)}_{\rm up} &= \omega^2M^4\left\{-\frac{11449}{88200}iZ^2+\left[\frac{321}{140}i\gamma^2+\left(-\frac{506003}{88200}i-\frac{107}{70}\pi\right)\gamma+\frac{4601}{2520}\pi\right.\right.\\
&\quad \left.+\frac{19861979}{5292000}i-\frac{107}{360}i\pi^2\right]Z+\frac{359}{560}\pi^3+\frac{1291}{420}i\zeta(3)+\pi\zeta(3)\\
&\quad -\frac{27}{8}i\gamma^4+\left(\frac{9}{2}\pi+\frac{5157}{280}i\right)\gamma^3+\left(-\frac{4943}{280}\pi-\frac{14243011}{352800}i+\frac{21}{8}i\pi^2\right)\gamma^2\\
&\quad +\left(-\frac{3}{4}\pi^3-3i\zeta(3)-\frac{29251}{5040}i\pi^2+\frac{420397183}{10584000}i+\frac{882067}{35280}\pi\right)\gamma\\
&\quad \left.-\frac{4815277630548731}{356258075040000}i-\frac{1691881}{141120}\pi-\frac{121}{1440}i\pi^4+\frac{7469477}{2116800}i\pi^2\right\}\,,
\end{aligned}
$$

$$
\begin{aligned}
A^{(15)}_{\rm up} &= \omega^3M^5\left\{\left(\frac{11449}{44100}i\pi-\frac{492307}{529200}+\frac{11449}{14700}\gamma\right)Z^2+\left[-\frac{321}{70}\gamma^3\right.\right.\\
&\quad +\left(\frac{528901}{29400}-\frac{321}{70}i\pi\right)\gamma^2+\left(\frac{506003}{44100}i\pi+\frac{107}{60}\pi^2-\frac{64509007}{2646000}\right)\gamma\\
&\quad \left.+\frac{107}{420}i\pi^3-\frac{989857}{529200}\pi^2-\frac{107}{105}\zeta(3)+\frac{49467121}{4536000}-\frac{19861979}{2646000}i\pi\right]Z\\
&\quad +\frac{81}{20}\gamma^5+\left(-\frac{16113}{560}+\frac{27}{4}i\pi\right)\gamma^4+\left(-\frac{21}{4}\pi^2+\frac{5100271}{58800}-\frac{5157}{140}i\pi\right)\gamma^3
\end{aligned}
$$

$$
\begin{aligned}
&+\left(-\frac{1390209563}{10584000}+\frac{30749}{1680}\pi^2+\frac{14243011}{176400}i\pi-\frac{9}{4}i\pi^3+9\zeta(3)\right)\gamma^2\\
&+\left(-\frac{420397183}{5292000}i\pi+\frac{121}{240}\pi^4+6i\pi\zeta(3)-\frac{4877629}{211680}\pi^2-\frac{4087}{210}\zeta(3)\right.\\
&\left.+\frac{16388410196014663}{178129037520000}+\frac{689}{168}i\pi^3\right)\gamma-\frac{102497}{302400}\pi^4+\frac{11}{240}i\pi^5\\
&+\frac{68928373}{7056000}\pi^2+\frac{130897}{9800}\zeta(3)+\frac{4815277630548731}{178129037520000}i\pi-\frac{4577}{2400}i\pi^3\\
&\left.-\frac{7}{6}\pi^2\zeta(3)-\frac{46037484216446383}{2137548450240000}-\frac{1291}{210}i\zeta(3)\pi+\frac{6}{5}\zeta(5)\right\}, \qquad (66)
\end{aligned}
$$

and

$$
\begin{aligned}
B_{\rm in}^{(5)} &= \frac{1}{720}\omega^3,\\
B_{\rm in}^{(8)} &= \frac{M\omega^4}{72}i\left(-\frac{671}{480}+\gamma+\frac{2}{5}Z\right),\\
B_{\rm in}^{(11)} &= \frac{M^2\omega^5}{30}\left[\left(\frac{233}{210}-\frac{1}{3}i\pi-\gamma\right)Z-\frac{7}{4}\gamma^2+\left(-\frac{1}{3}i\pi+\frac{30941}{6720}\right)\gamma+\frac{13157}{20160}i\pi\right.\\
&\left.-\frac{1}{72}\pi^2-\frac{3243467}{1209600}\right],\\
B_{\rm in}^{(14)} &= M^3\omega^6\left\{\left[-\frac{1}{10}i\gamma^2+\left(\frac{1}{15}\pi+\frac{233}{1050}i\right)\gamma+\frac{7}{540}i\pi^2-\frac{257899}{3024000}i\right.\right.\\
&\left.-\frac{43}{540}\pi\right]Z+\left(-\frac{751853}{1209600}i-\frac{63551}{302400}\pi+\frac{11}{1080}i\pi^2\right)\gamma\\
&+\left(\frac{1}{15}\pi+\frac{12799}{22400}i\right)\gamma^2-\frac{3}{20}i\gamma^3+\frac{1}{90}i\zeta(3)+\frac{565751}{3628800}\pi-\frac{6803}{403200}i\pi^2\\
&\left.+\frac{14496690547}{106686720000}i\right\}, \qquad (67)
\end{aligned}
$$

respectively.

There are only two nonvanishing "up" coefficients

$$
\begin{aligned}
\tilde{A}_{\rm up}^{(13)} &= -\frac{M^2}{30},\\
\tilde{A}_{\rm up}^{(16)} &= -\frac{iM^3\omega}{15}Y+\frac{1439}{6320}iM^3\omega, \qquad (68)
\end{aligned}
$$

while the nonvanishing "in" coefficients are

$$
\begin{aligned}
\tilde{B}_{\rm in}^{(0)} &= \frac{i}{384M^4\omega}\,,\\
\tilde{B}_{\rm in}^{(3)} &= \frac{47}{6720M^3}\,,\\
\tilde{B}_{\rm in}^{(6)} &= \frac{107i}{20160M^2}\omega Y - \frac{41869i}{1881600M^2}\omega\,,\\
\tilde{B}_{\rm in}^{(9)} &= \frac{\omega^2}{M}\left[\frac{107}{30240}\pi^2 + \frac{5029}{352800}Y - \frac{2105693}{59270400}\right],\\
\tilde{B}_{\rm in}^{(12)} &= i\omega^3\left[\frac{11449}{2116800}Y^2 - \frac{6707783}{177811200}Y + \frac{107}{2520}\zeta(3) - \frac{107}{18144}\pi^2\right.\\
&\quad\left.+\frac{129666772886101}{8550193800960000}\right],\\
\tilde{B}_{\rm in}^{(15)} &= M\omega^4\left[-\frac{420907}{27216000}\pi^2 - \frac{107}{113400}\pi^4 + \frac{5029}{44100}\zeta(3) - \frac{115405993}{2222640000}Y\right.\\
&\quad\left.-\frac{38361833226400711}{987547384010880000} + \frac{538103}{37044000}Y^2 + \frac{11449}{1587600}\pi^2 Y\right]. \qquad (69)
\end{aligned}
$$

4.4 Confluent Heun Formal Solutions

Let us also note (though this is not very useful in practice), that, formally, the general solution of the RW equation can be expressed in terms of the so-called Heun confluent functions as

$$X_{l\omega}(r) = r^3 e^{i\omega r}\left[C_1(r-2M)^{2i\omega M}H_1 + C_2(r-2M)^{-2i\omega M}H_2\right], \qquad (70)$$

where

$$
\begin{aligned}
H_1 &= \mathrm{HeunC}\left[-4iM\omega, 4i\omega M, 4, -8\omega^2M^2, 4-L+8\omega^2M^2, 1-\frac{r}{2M}\right]\\
H_2 &= \mathrm{HeunC}\left[-4iM\omega, -4iM\omega, 4, -8\omega^2M^2, 4-L+8\omega^2M^2, 1-\frac{r}{2M}\right].
\end{aligned} \qquad (71)
$$

However, to use these solutions one would need to expand these Heun functions, which are expressed in terms of the "large variable" (at infinity)

$$x = 1 - \frac{r}{2M}\,, \qquad (72)$$

into manageable functions of a "small" variable, such as

$$y = \frac{2M}{r} = \frac{1}{1-x}\,. \tag{73}$$

In the case of hypergeometric functions, there exist transformation laws which can be used for re-expressing hypergeometric functions of x into hypergeometric functions of $y = 1/(1-x)$. No such convenient transformation laws are available yet for the Heun functions. The MST technology is equivalent to expanding the Heun confluent functions in infinite series of hypergeometric functions for which there exist simple, convergent series expansion in terms of the small variable y.

4.5 Green's Function

The next step consists of using the Green's function method to solve the associated inhomogeneous equation (41). The retarded Green's function is expressed in terms of the two independent *homogeneous* solutions $X^{\rm in}_{l\omega}$ and $X^{\rm up}_{l\omega}$ of the RW operator (discussed above) as

$$\begin{aligned} G(r,r') &= \frac{1}{W}\Big[X^{\rm in}_{l\omega}(r)X^{\rm up}_{l\omega}(r')H(r'-r) + X^{\rm in}_{l\omega}(r')X^{\rm up}_{l\omega}(r)H(r-r')\Big] \\ &\equiv G_{\rm (in)}(r,r')H(r'-r) + G_{\rm (up)}(r,r')H(r-r')\,, \end{aligned} \tag{74}$$

where $W_{l\omega}$ denotes the (constant) Wronskian

$$W_{l\omega} = \left(1-\frac{2M}{r}\right)\left[X^{\rm in}_{l\omega}(r)\frac{d}{dr}X^{\rm up}_{l\omega}(r) - \frac{d}{dr}X^{\rm in}_{l\omega}(r)X^{\rm up}_{l\omega}(r)\right] = \text{const.} \tag{75}$$

and $H(x)$ is the Heaviside step function. Both even-parity and odd-parity solutions are then given by integrals over the corresponding (distributional) sources as

$$R^{\rm (even/odd)}_{lm\omega}(r) = \int dr'\,\frac{G(r,r')}{f(r')}S^{\rm (even/odd)}_{lm\omega}(r')\,. \tag{76}$$

4.6 Source Terms

In the case of odd-parity metric perturbations the source terms are given by

$$S^{\rm (odd)}_{lm\omega} = s_0^{\rm odd}\delta(r-r_0) + s_1^{\rm odd}\delta'(r-r_0)\,, \tag{77}$$

with

$$
\begin{aligned}
s_0^{\rm odd} &= 16 i\pi\mu \left(1-\frac{3M}{r_0}\right)^{-1/2}\left(1-\frac{2M}{r_0}\right)\frac{m\Omega^2}{l(l+1)(l-1)(l+2)}\,Y^{*\prime}_{lm}\left(\frac{\pi}{2},0\right)\,,\\
s_1^{\rm odd} &= -r_0\left(1-\frac{2M}{r_0}\right)s_0^{\rm odd}\,.
\end{aligned}
\tag{78}
$$

In the case of even-parity metric perturbations instead we have

$$
S^{\rm (even)}_{lm\omega} = s_0^{\rm even}\delta(r-r_0)+s_1^{\rm even}\delta'(r-r_0)+s_2^{\rm even}\delta''(r-r_0)\,,
\tag{79}
$$

with

$$
\begin{aligned}
s_0^{\rm even} &= -8\mu\pi\frac{(-r_0+2M)N_0}{\lambda(\lambda+1)(r_0\lambda+3M)r_0^4}\left(1-\frac{3M}{r_0}\right)^{-1/2} Y^*_{lm}\left(\frac{\pi}{2},0\right)\,,\\
s_1^{\rm even} &= -8\mu\pi\frac{(r_0-2M)^2N_1}{\lambda(\lambda+1)(r_0\lambda+3M)r_0^4}\left(1-\frac{3M}{r_0}\right)^{-1/2} Y^*_{lm}\left(\frac{\pi}{2},0\right)\,,\\
s_2^{\rm even} &= -24\mu\pi M\frac{(r_0-2M)^4}{r_0^3(r_0\lambda+3M)(\lambda+1)}\left(1-\frac{3M}{r_0}\right)^{-1/2} Y^*_{lm}\left(\frac{\pi}{2},0\right)\,,
\end{aligned}
\tag{80}
$$

where $\lambda=(l-1)(l+2)/2$, $\omega=m\Omega=m\sqrt{M/r_0^3}$ and

$$
\begin{aligned}
N_0 &= 9r_0^3\Omega^2m^2M^2-2r_0^5\Omega^2\lambda^3+3r_0^2M\lambda+3r_0^2M\lambda^2-24r_0M^2\lambda-14r_0M^2\lambda^2\\
&\quad -r_0^5\Omega^2\lambda^2-3r_0^4\Omega^2M\lambda-6r_0^4\Omega^2M\lambda^2-15r_0^3\Omega^2M^2\lambda+r_0^5\Omega^2m^2\lambda^2\\
&\quad -2r_0^3\lambda^3-r_0^3\lambda^2+3r_0^4\Omega^2m^2M\lambda+r_0^5\Omega^2m^2\lambda^3-6M^2\Omega^2\lambda^2r_0^3-3Mr_0^4\Omega^2\lambda^3\\
&\quad +6M^2r_0^3\Omega^2m^2\lambda+3Mr_0^4\Omega^2m^2\lambda^2-\lambda^4r_0^3+24M^3\lambda^2-r_0^5\Omega^2\lambda^4+4M^2\lambda^3r_0\\
&\quad +2M\lambda^4r_0^2-9r_0^3\Omega^2M^2+2r_0^2M\lambda^3+36M^3\lambda\,,\\
N_1 &= 2r_0^2M\lambda^3-r_0^3\lambda^3-r_0^2M\lambda^2-3r_0^4\Omega^2M\lambda^2+6r_0M^2\lambda^2-r_0^3\lambda^2+36M^3\lambda\\
&\quad +3r_0^4\Omega^2m^2M\lambda-3r_0^2M\lambda-12r_0M^2\lambda-3r_0^4\Omega^2M\lambda-9r_0^3\Omega^2M^2\lambda\\
&\quad -9r_0^3\Omega^2M^2+9r_0^3\Omega^2m^2M^2\,.
\end{aligned}
\tag{81}
$$

4.7 Summation over m

Another (nontrivial) task is that of performing the summation over the magnetic number $-l \leq m \leq l$. To this end, i.e. to compute the quantities

$$S_{k,l} = \sum_{m=-l}^{l} m^k \left| Y_{lm}\left(\frac{\pi}{2}, 0\right) \right|^2 , \tag{82}$$

and

$$S'_{k,l} = \sum_{m=-l}^{l} m^k \left| Y'_{lm}\left(\frac{\pi}{2}, 0\right) \right|^2 , \tag{83}$$

one uses the following "generating functions" identities (see Appendix F in [46])

$$\begin{aligned} \Gamma_l(z) &= \sum_{m=-l}^{l} e^{mz} \left| Y_{lm}\left(\frac{\pi}{2}, 0\right) \right|^2 \\ &= \frac{2l+1}{4\pi} e^{lz} F\left(\frac{1}{2}, -l, 1; 1 - e^{-2z}\right) \end{aligned} \tag{84}$$

and

$$\begin{aligned} \Delta_l(z) &= \sum_{m=-l}^{l} e^{mz} \left| Y'_{lm}\left(\frac{\pi}{2}, 0\right) \right|^2 \\ &= \frac{2l+1}{4\pi^2} e^{(l-1)z} \frac{\Gamma(l+\frac{1}{2})\Gamma(\frac{3}{2})}{\Gamma(l)} F\left(\frac{3}{2}, -l+1, -l+\frac{1}{2}; e^{-2z}\right) . \end{aligned} \tag{85}$$

4.8 WKB Approximation for $X^{\text{in}}_{l\omega}$ and $X^{\text{up}}_{l\omega}$

The analytical computation of the subtraction constant B (needed to compute the regularized value of h_{kk}) depends on the large l behavior of the homogeneous in and up solutions. As noted in [37], this behavior can be computed by applying the Wentzel-Kramers-Brillouin (WKB) approximation to the RW equation, which is just a 1-dimensional Schrödinger equation of the form

$$\frac{d^2}{dx^2}\Psi = \frac{Q(x)}{\hbar^2}\Psi(x) , \tag{86}$$

with $\hbar^2 \sim 1/(l(l+1)) \ll 1$.

The standard WKB expansion of the solutions of Eq. (86) is of the form

$$\Psi(x) = e^{\frac{S_0}{\hbar}+S_1+O(\hbar)}, \tag{87}$$

where we will neglect the correction $O(\hbar)$ (which is of fractional order $O(\hbar^2)$ with respect to the leading-order term). At this approximation we have two independent solutions

$$\Psi_\pm(x) = C_\pm \frac{e^{\pm\int p dx}}{\sqrt{p}}, \qquad p \equiv \sqrt{\frac{Q}{\hbar^2}}, \tag{88}$$

corresponding to

$$\frac{S_0}{\hbar} = \pm\int p dx, \qquad S_1 = -\frac{1}{2}\ln p. \tag{89}$$

Notice that the choice $C_\pm = 1/\sqrt{2}$ implies that the Wronskian of these solutions is 1

$$W = \Psi_-\Psi'_+ - \Psi_+\Psi'_- = 1. \tag{90}$$

Let us apply this approximation to the homogeneous RW equation

$$\frac{d}{dr^*}X = \left[f(r)\left(\frac{l(l+1)}{r^2} - \frac{6M\eta^2}{r^3}\right) - \eta^2\omega^2\right]X, \tag{91}$$

with $\omega = m\Omega$ and

$$dr^* = \frac{1}{f(r)}dr, \qquad f(r) = 1 - 2M\eta^2/r. \tag{92}$$

Let

$$\mathscr{L} \equiv l + \frac{1}{2}, \tag{93}$$

so that, in the large l limit we have

$$l(l+1) = \mathscr{L}^2 - \frac{1}{4} \approx \mathscr{L}^2. \tag{94}$$

We can then use as small expansion parameter

$$\hbar \equiv \frac{1}{\mathscr{L}} \approx \frac{1}{\sqrt{l(l+1)}}. \tag{95}$$

In the RW equation, considered in the large-l limit with fixed ratio

$$W \equiv m/\mathscr{L}\,, \tag{96}$$

the potential Q_* follows after the factorization of $l(l+1) \approx \mathscr{L}^2$. Indeed, we have (neglecting terms of fractional order $O(\hbar^2) = O(\mathscr{L}^{-2})$, such as the term $6M\eta^2/r^3$)

$$\begin{aligned} \frac{Q_*}{\hbar^2} &= l(l+1)\left[f(r)\frac{1}{r^2} - \eta^2\frac{m^2}{l(l+1)}\Omega^2\right] + \dots \\ &= \mathscr{L}^2\left[f(r)\frac{1}{r^2} - \eta^2 W^2\Omega^2 + O(\mathscr{L}^{-2})\right]. \end{aligned} \tag{97}$$

Denoting

$$\Delta(r) = f(r) - \eta^2 W^2 \Omega^2 r^2 = 1 - \frac{2M\eta^2}{r} - \eta^2 W^2 \Omega^2 r^2\,, \tag{98}$$

we have

$$p_*^2 = Q_* = \mathscr{L}^2\frac{\Delta(r)}{r^2}\,, \qquad p_* = \mathscr{L}\frac{\sqrt{\Delta(r)}}{r}\,. \tag{99}$$

Therefore

$$\frac{S_0}{\hbar} = \pm\int p_* dr_* = \pm\int p_*\frac{dr}{f} = \pm\mathscr{L}\int\frac{\sqrt{\Delta(r)}}{f(r)}\frac{dr}{r} \tag{100}$$

and

$$S_1 = -\frac{1}{2}\ln p_*\,. \tag{101}$$

In the present case

$$\sqrt{p_*} = \sqrt{\mathscr{L}}\frac{\Delta^{1/4}}{\sqrt{r}} \tag{102}$$

and

$$\frac{C_\pm}{\sqrt{p_*}} = \tilde{C}_\pm\frac{\sqrt{r}}{\Delta^{1/4}}\,, \tag{103}$$

where we have re-absorbed the factor $\sqrt{\mathscr{L}}$ in the constant $C_\pm$ [$\tilde{C}_\pm = C_\pm/\sqrt{\mathscr{L}}$]. Indicating the $O(\mathscr{L}^{-2})$ approximation (keeping W fixed) by the symbol $\approx$, we end up with two explicit approximate solutions of the homogeneous RW equations given by

$$X_\pm \approx \tilde{C}_\pm \frac{\sqrt{r}}{\Delta^{1/4}} e^{\pm \mathscr{L} \int \frac{\sqrt{\Delta}}{f}\frac{dr}{r}}\,. \tag{104}$$

To leading PN approximation, we have $\Delta \approx 1 \approx f$ so that the leading-order PN approximations of the WKB solutions read $X_\pm \approx \tilde{C}_\pm r^{\frac{1}{2}} e^{\pm \mathscr{L} \ln r} \approx \tilde{C}_\pm r^{\frac{1}{2} \pm \mathscr{L}}$, in which we recognize the leading-order approximation of the two PN solutions: r^{l+1} and r^{-l}.

This agreement between the PN solutions and the WKB ones formally persists to all PN orders. Indeed, one finds (up to fractional $1/\mathscr{L}^2$ error) that there exists a constant $\widehat{C}$ (which depends only on $\mathscr{L}$ and the arbitrary log scales $R, R_1, \ldots$ entering the "in" PN solution) such that

$$X^{\rm in/up}_{l\omega}(r) \approx e^{\pm \mathscr{L} \widehat{C}} \frac{\sqrt{r}}{\Delta^{1/4}} e^{\pm \mathscr{L} \int \frac{\sqrt{\Delta}}{f}\frac{dr}{r}}\,. \tag{105}$$

In other words, writing our PN "in" solution in the form

$$X^{\rm in}_{l\omega}(r) = r^{l+1}\left(1 + A^{l\omega}(r)\right)\,, \tag{106}$$

where $A^{l\omega}(r) = O(\eta^2)$, there exists an $\mathscr{L}$−dependent constant $\widehat{C}$ such that

$$\frac{1}{\mathscr{L}} \ln\left[\Delta^{1/4}\left(1 + A^{l\omega}(r)\right)\right] = \int \frac{dr}{r}\left(\frac{\sqrt{\Delta}}{f} - 1\right) + \widehat{C} + O(\frac{1}{\mathscr{L}^2})\,. \tag{107}$$

Up to η^{16} included, we find

$$\widehat{C}(\mathscr{L}, R, R_1) = c_6^* \eta^6 + c_{12}^* \eta^{12}\,, \tag{108}$$

with coefficients c_6^* and c_{12}^* given by

$$\begin{aligned}
\frac{c_6^*}{M^2\Omega^2W^2} &= \frac{1}{8}\mathscr{L}^2 - \frac{31}{16}\mathscr{L} + \frac{27}{32} + \frac{15}{4}\ln(R) + \frac{235}{64\mathscr{L}} + O\left(\frac{1}{\mathscr{L}^2}\right)\,,\\
\frac{c_{12}^*}{M^4\Omega^4W^4} &= \frac{5}{768}\mathscr{L}^5 - \frac{31}{192}\mathscr{L}^4 + \frac{7735}{9216}\mathscr{L}^3 + \frac{5}{32}\mathscr{L}^2 + \frac{34427}{12288}\mathscr{L} + \frac{15481}{3072}\\
&\quad + \frac{1155}{64}\ln(R_1) + \frac{1511951}{49152\mathscr{L}} + O\left(\frac{1}{\mathscr{L}^2}\right)\,.
\end{aligned} \tag{109}$$

Note that in this derivation R and R_1 have been assumed to be independent on l, and no scale has been introduced in the various logarithms entering the WKB phase $\int dr/r = \ln r$.

Let us now sketch how the WKB solutions lead to an analytic derivation of the subtraction term B. To this end we compute $h_{kk} = h_{kk}^{(\rm even)} + h_{kk}^{(\rm odd)}$ with a Green's function approximated by the WKB solutions, taking the large $\mathscr{L}$ limit (with $m/\mathscr{L} = W$ =fixed). In this limit the discrete summation over m can be replaced by an integration over the continuous variable W. More precisely, writing $m = \mathscr{L}\sin\theta$ (i.e., $W = \sin\theta$), we can use the following relations

$$
\begin{aligned}
\sum_m \frac{4\pi}{2l+1}|Y_{lm}(\pi/2,0)|^2 &= \frac{2}{\pi}\int_0^{\pi/2} d\theta = \frac{2}{\pi}\int_0^1 \frac{dW}{\sqrt{1-W^2}}\,,\\
\sum_m \frac{4\pi}{2l+1}|Y'_{lm}(\pi/2,0)|^2 &= \frac{2}{\pi}\int_0^{\pi/2} \cos^2\theta d\theta = \frac{2}{\pi}\int_0^1 (1-W^2)\frac{dW}{\sqrt{1-W^2}}\,,
\end{aligned}
\tag{110}
$$

which yield ($u = M/r$)

$$
\begin{aligned}
h_{kk}^{(\rm even)\,\rm WKB} &= 2u\Gamma\int_0^1 \frac{dW}{\sqrt{1-W^2}}\frac{(1-u-2uW^2)^2}{\sqrt{1-2u-W^2u}}\,,\\
h_{kk}^{(\rm odd)\rm WKB} &= -8u^2\Gamma\int_0^1 \frac{dW}{\sqrt{1-W^2}}(1-W^2)\sqrt{1-2u-W^2u}\,.
\end{aligned}
\tag{111}
$$

The sum of these terms finally gives $h_{kk}^{\rm WKB} = B$, with

$$
\begin{aligned}
B &= 2u(1-3u)^{3/2}\int_0^1 \frac{dW}{\sqrt{(1-W^2)(1-2u-W^2u)}}\\
&= 2u\frac{(1-3u)^{3/2}}{\sqrt{1-2u}}\frac{2}{\pi}\text{EllipticK}\left(k^2 = \frac{u}{1-2u}\right)\,.
\end{aligned}
\tag{112}
$$

This result agrees with the one obtained by means of local worldline expansions (as in [22]). Note also that the PN expansion of B (i.e. its expansion in powers of u, $B^{\rm PN}(u) = 2u - \frac{13}{2}u^2 + O(u^3)$) can be directly obtained by taking the large l limit of the PN-expanded $h_{kk\,l}^{\rm PN}$. This is a useful feature of our approach, because it shows that one does not need to separately compute the value of the subtraction constant B.

4.9 Computation of the Regularized Value of h_{kk}

Using: (1) the $l = 0, 1$ metric perturbations due to Zerilli [15] (re-expressed in an asymptotically flat gauge [38]); (2) a finite number of MST-computed metric perturbations, corresponding to multipole orders $2 \le l \le l_{\max}$; and (3) the PN-computed high-l metric perturbations for $l > l_{\max}$; one can analytically evaluate the perturbed metric in a given gauge (the RW gauge in our case) and at a generic spacetime point. One can then proceed to analytically evaluate (physically relevant) gauge invariant quantities such as Detweiler's metric contraction

$$h_{kk} = h_{\alpha\beta}k^\alpha k^\beta\,, \qquad k^\alpha = \delta^\alpha_t + \Omega\delta^\alpha_\phi\,, \tag{113}$$

evaluated at the particle's location $r = r_0$, $\theta = \pi/2$, $\phi = \Omega t$. One has at a generic spacetime point

$$\begin{aligned} h_{kk}(t,r,\theta,\phi) &= \sum_{lm}[h_{kk}^{lm\,\text{(even)}}(t,r,\theta,\phi) + h_{kk}^{lm\,\text{(odd)}}(t,r,\theta,\phi)]\,,\\ h_{kk}^{lm\,\text{(even)}}(t,r,\theta,\phi) &= h_{kk}^{lm\,\text{(even)}}(t,r)e^{im\phi}Y_{lm}(\theta)\,,\\ h_{kk}^{lm\,\text{(odd)}}(t,r,\theta,\phi) &= h_{kk}^{lm\,\text{(odd)}}(t,r)e^{im\phi}Y'_{lm}(\theta)\,, \end{aligned} \tag{114}$$

where $Y_{lm}(\theta)$ denotes the value of the usual spherical harmonics at $\phi = 0$, while $Y'_{lm}(\theta)$ denotes its θ-derivative. Once evaluated along the particle's world line the multipolar pieces of the previous quantities turn out to be

$$\begin{aligned} h_{kk}^{lm\,\text{(even)}}(r_0) &= \left|Y_{lm}\left(\frac{\pi}{2},0\right)\right|^2 A_{lm}^{\text{even}}(r_0)J_{\text{in}}(r_0)J_{\text{up}}(r_0)\,,\\ h_{kk}^{lm\,\text{(odd)}}(r_0) &= \left|\partial_\theta Y_{lm}\left(\frac{\pi}{2},0\right)\right|^2 A_{lm}^{\text{odd}}(r_0)\tilde{J}_{\text{in}}(r_0)\tilde{J}_{\text{up}}(r_0)\,, \end{aligned} \tag{115}$$

where $A_{lm}^{\text{even/odd}}(r_0)$ and

$$\begin{aligned} J_{\text{in/up}}(r_0) &= \alpha_{\text{in/up}}(r_0)X_{l\omega}^{\text{in/up}}(r_0) + \beta_{\text{in/up}}(r_0)\left.\frac{dX_{l\omega}^{\text{in/up}}}{dr}\right|_{r=r_0}\,,\\ \tilde{J}_{\text{in/up}}(r_0) &= \tilde{\alpha}_{\text{in/up}}(r_0)X_{l\omega}^{\text{in/up}}(r_0) + \tilde{\beta}_{\text{in/up}}(r_0)\left.\frac{dX_{l\omega}^{\text{in/up}}}{dr}\right|_{r=r_0}\,, \end{aligned} \tag{116}$$

are known functions of r_0. See [35, 36] and references therein for additional details.

Next, the above quantities $h_{kk}^{lm\,\text{(even/odd)}}$ should summed over $m \in [-l, l]$ using the summation formulas mentioned above. This leads to a finite contribution $h_{kk}^l = \sum_m \left(h_{kk}^{lm\,\text{(even)}} + h_{kk}^{lm\,\text{(odd)}}\right)$ at each multipole order l.

The value of $h_{kk}(r_0)$, evaluated at the location of the particle, is then formally given by the sum of the series $\sum_l h_{kk}^l$. However, this series diverges, which corresponds to the fact that $h_{kk}(r)$ is singular as $r \to r_0$. The regularized value, h_{kk}^{R}, of $\lim_{r\to r_0} h_{kk}(r)$ is conveniently computed by summing the series obtained by subtracting from each multipolar piece h_{kk}^l the corresponding multipolar piece of the singular part of $h_{kk}(r)$ [21, 22]. In the case of $h_{kk}(r)$ the l-th multipolar piece of its singular part is simply given by the $l \to \infty$ limit of h_{kk}^l, say $B \equiv \lim_{l\to\infty} h_{kk}^l$.

We saw above how this limit can be computed by using WKB techniques, with the result (first obtained by worldline-expansion techniques [22, 31])

$$B = \frac{4M}{\pi r_0}\left(1 - \frac{3M}{r_0}\right)\sqrt{\frac{r_0 - 3M}{r_0 - 2M}}\mathrm{EllipticK}\left(\sqrt{\frac{M}{r_0 - 2M}}\right), \tag{117}$$

or, denoting $u = M/r_0$,

$$\begin{aligned} B &= \frac{2u(1-3u)^{3/2}}{(1-2u)^{1/2}}\frac{2}{\pi}\mathrm{EllipticK}\left(\sqrt{\frac{u}{1-2u}}\right) \\ &= 2u - \frac{13}{2}u^2 + \frac{9}{32}u^3 + \frac{83}{128}u^4 + \frac{12361}{8192}u^5 + \frac{116163}{32768}u^6 + \frac{4409649}{524288}u^7 \\ &\quad + O(u^8). \end{aligned} \tag{118}$$

As $h^l_{kk} = B + O(l^{-2})$, the series $h^{\rm R}_{kk} = \sum_l (h^l_{kk} - B)$ is convergent. When analytically computing the PN expansion of the latter series, the contribution coming from large values of l is explicitly given by some rational function of l, which can be explicitly summed after decomposing it in partial fractions. This stage involves, in particular, Riemann zeta series, such as $\sum_l 1/(l+1)^2 = \zeta(2) = \pi^2/6$.

Starting in 2013, the above techniques were used to obtain the analytical values of the PN expansion coefficients of $h^{\rm R}_{kk}$ at 4PN [35], and beyond [30, 36, 39, 40], namely

$$\begin{aligned} h^{\rm R}_{kk} &= \sum_{l=0}^{\infty}(h^l_{kk} - B) \\ &= -2u + 5u^2 + \frac{5}{4}u^3 + \left(-\frac{1261}{24} + \frac{41}{16}\pi^2\right)u^4 \\ &\quad + \left(\frac{157859}{960} - \frac{256}{5}\gamma - \frac{128}{5}\ln(u) - \frac{512}{5}\ln(2) - \frac{2275}{256}\pi^2\right)u^5 \\ &\quad + \cdots. \end{aligned} \tag{119}$$

Before these works, $h^{\rm R}_{kk}$ was analytically known only at the 3PN approximation [47] (together with the knowledge of the 4PN [48] and 5PN [49] logarithmic contributions).

Using the first law of black hole mechanics [32–34], the knowledge of the PN expansion of $h^{\rm R}_{kk}$ can be transcribed into the dynamically directly significant knowledge of the $O(\nu)$ correction $a_{1\rm SF}(u)$ to the EOB main radial potential $A(u;\nu)$ [see Eq. (39)]. The so-derived GSF-based 4PN knowledge of $h^{\rm R}_{kk}$ (and $a_{1\rm SF}(u)$) [35] has been crucial for deriving the complete 4PN two-body dynamics [4].

5 GSF Calculations in a Kerr Spacetime

Let us complete our review by giving some of the details of GSF analytical calculations along circular orbits around a *spinning* (Kerr) black hole (of mass M and spin $S = Ma$). In this section, we switch to a mostly negative signature, in order to apply the standard tools of the Newman-Penrose (NP) formalism. The Kerr line element written in standard Boyer-Lindquist coordinates then reads

$$ds^2 = \left(1 - \frac{2Mr}{\Sigma}\right) dt^2 + \frac{4aMr\sin^2\theta}{\Sigma} dt d\phi - \frac{\Sigma}{\Delta} dr^2 \\ - \Sigma d\theta^2 - \left(r^2 + a^2 + \frac{2Mra^2\sin^2\theta}{\Sigma}\right)\sin^2\theta d\phi^2 , \qquad (120)$$

where $\Delta = r^2 - 2Mr + a^2$ and $\Sigma = r^2 + a^2\cos^2\theta$. A principal NP frame is

$$l = \frac{1}{\Delta}[(r^2 + a^2)\partial_t + \Delta\partial_r + a\partial_\phi] ,$$
$$n = \frac{1}{2\Sigma}[(r^2 + a^2)\partial_t - \Delta\partial_r + a\partial_\phi] ,$$
$$m = -\frac{\bar{\rho}}{\sqrt{2}}\left[ia\sin\theta\partial_t + \partial_\theta + \frac{i}{\sin\theta}\partial_\phi\right] , \qquad (121)$$

with nonvanishing spin coefficients

$$\rho = -\frac{1}{r - ia\cos\theta} , \quad \beta = -\frac{\bar{\rho}\cos\theta}{2\sqrt{2}\sin\theta} , \quad \varpi = \frac{ia\sin\theta\rho^2}{\sqrt{2}} , \quad \tau = -\frac{ia\sin\theta}{\sqrt{2}\Sigma} ,$$
$$\mu = \frac{\Delta\rho}{2\Sigma} , \quad \gamma = \mu + \frac{r - M}{2\Sigma} , \quad \alpha = \pi - \bar{\beta} .$$

The associated frame derivatives are denoted by

$$\mathbf{D} = l^\mu\partial_\mu , \qquad \boldsymbol{\Delta} = n^\mu\partial_\mu , \qquad \boldsymbol{\delta} = m^\mu\partial_\mu . \qquad (122)$$

A particle of mass μ moving along an equatorial circular orbit has 4-velocity

$$u = u^t k , \qquad k = \partial_t + \Omega\partial_\phi . \qquad (123)$$

Circular geodesics are characterized by

$$u^t = \frac{r_0^{3/2} \pm M^{1/2}a}{\sqrt{r_0^3 - 3Mr_0^2 \pm 2aM^{1/2}r_0^{3/2}}} , \qquad \Omega = \frac{\pm M^{1/2}}{r_0^{3/2} \pm aM^{1/2}} , \qquad (124)$$

where the upper/lower sign corresponds to co/counter-rotating orbits. The corresponding stress-energy tensor is given by

$$T^{\mu\nu} = \frac{\mu}{u^t r^2} u^\mu u^\nu \delta_3 \,, \qquad \delta_3 = \delta(r - r_0)\delta(\theta - \pi/2)\delta(\phi - \Omega t) \,. \tag{125}$$

General perturbations on a Kerr background due to a field of spin-weight s can be studied using the approach of Teukolsky [16–18]. This formalism has been further developed and applied to many interesting situations (see, e.g., [45] and references therein). We will follow below [50], where all necessary tools to compute the metric perturbations induced by a particle orbiting a rotating black hole have been given.

The Teukolsky master equation reads as

$$\mathscr{T}_s(\psi_s) = 8\pi \Sigma T_s \,, \tag{126}$$

with

$$\begin{aligned} \mathscr{T}_s = & \left[\frac{(r^2+a^2)^2}{\Delta} - a^2 \sin^2\theta\right]\partial_{tt} - 2s\left[\frac{M(r^2-a^2)}{\Delta} - r - ia\cos\theta\right]\partial_t \\ & + \frac{4aMr}{\Delta}\partial_{t\phi} - \Delta^{-s}\partial_r(\Delta^{s+1}\partial_r) \\ & - \frac{1}{\sin\theta}\partial_\theta(\sin\theta\,\partial_\theta) - 2s\left[\frac{a(r-M)}{\Delta} + i\frac{\cos\theta}{\sin^2\theta}\right]\partial_\phi \\ & + \left(\frac{a^2}{\Delta} - \frac{1}{\sin^2\theta}\right)\partial_{\phi\phi} + (s^2\cot^2\theta - s) \,, \end{aligned} \tag{127}$$

and T_s denotes the source term (to be specified below). Separation of variables

$$\psi_s = \sum_{lm\omega} {}_sR_{lm\omega}(r)\ {}_sS_{lm\omega}(\theta)\ e^{i(m\phi-\omega t)} \,, \tag{128}$$

leads to the following angular (homogeneous) and radial (inhomogeneous) equations

$$\mathscr{L}^s_\theta\, {}_sS_{lm\omega}(\theta) = 0 \,, \qquad \mathscr{L}^s_r\, {}_sR_{lm\omega}(r) = -8\pi T_{slm\omega} \,, \tag{129}$$

where

$$\begin{aligned} \mathscr{L}^s_r &\equiv \Delta^{-s}\frac{d}{dr}\left(\Delta^{s+1}\frac{d}{dr}\right) + \left[\frac{K^2 - 2is(r-M)K}{\Delta} + 4is\omega r - \lambda\right] , \\ \mathscr{L}^s_\theta &\equiv \frac{1}{\sin\theta}\frac{d}{d\theta}\left(\sin\theta\frac{d}{d\theta}\right) + \left[\xi^2\cos^2\theta - 2s\xi\cos\theta - \frac{2ms\cos\theta + s^2 + m^2}{\sin^2\theta} + E\right] , \end{aligned} \tag{130}$$

with $\xi = a\omega$, $K = (r^2 + a^2)\omega - ma$, and $\lambda \equiv \lambda_{(l,m,s;\xi)} = E - s(s+1) - 2m\xi + \xi^2$; $E \equiv E_{(l,m,s;\xi)}$ denoting the eigenvalue entering the angular equation.

5.1 Solutions to the Angular Teukolsky Equation

For fixed s and ξ, the angular functions ${}_sS_{lm\omega}$ can be taken as real functions and form a complete and orthogonal system on the sphere. In fact, introducing the "spin-weighted spheroidal harmonics"

$$ {}_s\mathscr{S}_{lm\omega}(\theta,\phi) = {}_sS_{lm\omega}(\theta)e^{im\phi} \,, \tag{131} $$

they satisfy the relations

$$ \delta(\cos\theta - \cos\theta')\delta(\phi-\phi') = \sum_{l,m,s} {}_s\mathscr{S}_{lm\omega}(\theta,\phi)\,{}_s\overline{\mathscr{S}}_{lm\omega}(\theta',\phi') \,, \tag{132} $$

and

$$ \int_0^\pi \int_0^{2\pi} {}_s\mathscr{S}_{lm\omega}(\theta,\phi)\,{}_s\overline{\mathscr{S}}_{l'm'\omega}(\theta,\phi)_s \sin\theta d\theta d\phi = \delta_{ll'}\delta_{mm'} \,, \tag{133} $$

together with the property

$$ {}_s\overline{\mathscr{S}}_{lm\omega} = (-1)^{m+s}{}_{-s}\mathscr{S}_{l,-m,-\omega} \,. \tag{134} $$

In the limit $\xi = 0$ the spin-weighted spheroidal harmonics coincide with the spin-weighted spherical harmonics [51], i.e., ${}_sS_{lm0}(\theta)e^{im\phi} = {}_sY_{lm}(\theta,\phi)$.

The spin-weighted spheroidal harmonics can be decomposed in terms of spin-weighted spherical harmonics as

$$ {}_s\mathscr{S}_{lm\omega}(\theta,\phi) = \sum_{j=l_{\min}}^{\infty} b_j^{lm\omega}\ {}_sY_{jm}(\theta,\phi) \,, \qquad l_{\min} = \max\{|m|,|s|\} \,, \tag{135} $$

where $b_j^{lm\omega}(\xi = 0) = \delta_{j,l}$. It turns out that

$$ \begin{aligned} b_j^{lm\omega} &= \delta_{j,l} + \delta_{j,l+1}P^{(1),+}_{l,m,\omega}(\xi) + \delta_{j,l-1}P^{(1),-}_{l,m,\omega}(\xi) \\ &\quad + \delta_{j,l+2}P^{(2),+}_{l,m,\omega}(\xi) + \delta_{j,l-2}P^{(2),-}_{l,m,\omega}(\xi) + O(\xi^3) \,, \end{aligned} \tag{136} $$

where we have used the notation

$$ P^{(1),+}_{l,m,\omega}(\xi) = a_1\xi + a_2\xi^2 + O(\xi^3) \,, \qquad P^{(2),+}_{l,m,\omega}(\xi) = b_2\xi^2 + O(\xi^3) \,. \tag{137} $$

Explicitly, in the case $s = 2$, we have

$$
\begin{aligned}
{}_2S_{lm\omega}(\theta)e^{im\phi} &= b_l^{lm\omega}\ {}_2Y_{lm}(\theta,\phi) \\
&+ b_{l+1}^{lm\omega}P^{(1),+}_{l,m,\omega}(\xi)\ {}_2Y_{l+1,m}(\theta,\phi) + b_{l-1}^{lm\omega}P^{(1),-}_{l,m,\omega}(\xi)\ {}_2Y_{l-1,m}(\theta,\phi) \\
&+ b_{l+2}^{lm\omega}P^{(2),+}_{l,m,\omega}(\xi)\ {}_2Y_{l+2,m}(\theta,\phi) + b_{l-2}^{lm\omega}P^{(2),-}_{l,m,\omega}(\xi)\ {}_2Y_{l-2,m}(\theta,\phi) \\
&+ O(\xi^3)\,.
\end{aligned} \tag{138}
$$

Similarly, the angular eigenvalue can be written as a series expansion in ξ as

$$
E = l(l+1) + \xi E_1(l,m,s) + \xi^2 E_2(l,m,s) + O(\xi^3)\,. \tag{139}
$$

Substituting Eq. (135) in the Teukolsky angular equation leads to

$$
\sum_j b_j[\xi^2\cos^2\theta - 2\xi s\cos\theta - j(j+1)]_s Y_{jm}(\theta,\phi) = -E\sum_j b_j\ {}_sY_{jm}(\theta,\phi)\,, \tag{140}
$$

which gives a recurrence relation for the $b_j \equiv b_j^{lm\omega}$ coefficients and an algebraic equation for the angular eigenvalue E order by order in ξ. This is accomplished by using the relations

$$
\cos\theta\ {}_sY_{lm} = \sum_{j=l-1}^{l+1} c_{j,l,1}\ {}_2Y_{jm}(\theta,\phi)\,, \qquad \cos^2\theta\ {}_sY_{lm} = \sum_{j=l-2}^{l+2} c_{j,l,2}\ {}_2Y_{jm}(\theta,\phi)\,, \tag{141}
$$

with coefficients (symmetric in the first pair of indices)

$$
c_{l,l-1,1} = \frac{\sqrt{(l^2-m^2)(l^2-s^2)}}{l\sqrt{(2l-1)(2l+1)}} \equiv Q(l)\,, \quad c_{l,l,1} = -\frac{ms}{l(l+1)}\,, \quad c_{l+1,l,1} = Q(l+1) \tag{142}
$$

and

$$
\begin{aligned}
c_{l,l-2,2} &= Q(l)Q(l-1)\,, \\
c_{l,l-1,2} &= -\frac{2ms}{(l-1)(l+1)}Q(l)\,, \\
c_{l,l,2} &= 1 - 2(l+1)[Q(l+1)]^2 + 2l[Q(l)]^2\,, \\
c_{l+1,l,2} &= -\frac{2ms}{l(l+2)}Q(l+1)\,, \\
c_{l+2,l,2} &= Q(l+2)Q(l+1)\,.
\end{aligned} \tag{143}
$$

Multiplying then Eq. (140) by ${}_sY_{km}$ and integrating over the 2-sphere, leads to

$$A_{k-2}b_{k-2} + A_{k-1}b_{k-1} + A_k b_k + A_{k+1}b_{k+1} + A_{k+2}b_{k+2} = -Eb_k\,, \tag{144}$$

where

$$\begin{aligned}
A_{k-2} &= \xi^2 c_{k-2,k,2}\,,\\
A_{k-1} &= \xi^2 c_{k-1,k,2} - 2s\xi c_{k-1,k,1}\,,\\
A_k &= \xi^2 c_{k,k,2} - 2s\xi c_{k,k,1} - k(k+1)\,,\\
A_{k+1} &= \xi^2 c_{k+1,k,2} - 2s\xi c_{k+1,k,1}\,,\\
A_{k+2} &= \xi^2 c_{k+2,k,2}\,,
\end{aligned} \tag{145}$$

The solution for the angular eigenvalue up to the order ξ^2 is given by

$$E_{(l,m,s;\xi)} = l(l+1) - \frac{2s^2 m}{l(l+1)}\xi + [H(l+1) - H(l) - 1]\xi^2 + O(\xi^3)\,, \tag{146}$$

with

$$H(l) = \frac{2(l^2-m^2)(l^2-s^2)^2}{(2l-1)l^3(2l+1)}\,, \qquad l \geq 2\,, \tag{147}$$

related to $Q(l)$ defined above by $[Q(l)]^2/l = H(l)/2(l^2-s^2)$. The nonvanishing b_j coefficients to the same order are the following

$$\begin{aligned}
&b_l = 1\,, \qquad b_{l-1} = \xi A^{(1)} + \frac{1}{2}\xi^2 A^{(2)}\,, \qquad b_{l+1} = \xi B^{(1)} + \frac{1}{2}\xi^2 B^{(2)}\,,\\
&b_{l-2} = \frac{1}{2}\xi^2 C^{(2)}\,, \qquad b_{l+2} = \frac{1}{2}\xi^2 D^{(2)}\,,
\end{aligned} \tag{148}$$

with

$$\begin{aligned}
A^{(1)} &= \frac{s}{l}Q(l)\,,\\
B^{(1)} &= -\frac{s}{l+1}Q(l+1)\,,\\
A^{(2)} &= A^{(1)}\frac{2m}{(l-1)(l+1)}\left(1 - \frac{2s^2}{l^2}\right),\\
B^{(2)} &= B^{(1)}\frac{2m}{l(l+2)}\left(1 - \frac{2s^2}{(l+1)^2}\right),
\end{aligned}$$

$$C^{(2)} = -\frac{1}{2l-1}Q(l)Q(l-1)\left(1-\frac{2s^2}{l}\right),$$

$$D^{(2)} = \frac{1}{2l+3}Q(l+2)Q(l+1)\left(1+\frac{2s^2}{(l+1)}\right). \qquad (149)$$

As a final task, one should normalize these b_j so that $\sum_j b_j^2 = 1$. Therefore, we have to replace $b_j \to \tilde{b}_j = b_j/\sqrt{\sum_j b_j^2}$ in the expansion (135), with $\sum_j b_j^2 = 1 + \xi^2(|A^{(1)}|^2 + |B^{(1)}|^2) + O(\xi^3)$, so that

$$\tilde{b}_l = 1 - \frac{1}{2}\xi^2(|A^{(1)}|^2 + |B^{(1)}|^2), \qquad \tilde{b}_{l\pm 1} = b_{l\pm 1}, \qquad \tilde{b}_{l\pm 2} = b_{l\pm 2}. \qquad (150)$$

Finally, the spin-weighted spherical harmonics ${}_sY_{lm}(\theta,\phi)$ can be straightforwardly expressed in terms of the standard spherical harmonics $Y_{lm}(\theta,\phi) = {}_0Y_{lm}(\theta,\phi)$ and their derivatives [51].

5.2 *Solutions to the Radial Teukolsky Equation*

The Teukolsky radial equation has a source term which depends on the value of the spin-weight parameter s. In the case $s = +2$ (i.e., for $\psi_{s=2} = \psi_0$), we have

$$T_{s=2} = \mathscr{L}_1\left(\mathscr{L}_2(T_{13}) - \mathscr{L}_3(T_{11})\right) + \mathscr{L}_4\left(\mathscr{L}_5(T_{13}) - \mathscr{L}_6(T_{33})\right), \qquad (151)$$

where

$$\mathscr{L}_1 = \boldsymbol{\delta} + \bar{\varpi} - \bar{\alpha} - 3\beta - 4\tau, \quad \mathscr{L}_2 = \mathbf{D} - 2\bar{\rho}, \quad \mathscr{L}_3 = \boldsymbol{\delta} + \bar{\varpi} - 2\bar{\alpha} - 2\beta,$$
$$\mathscr{L}_4 = \mathbf{D} - 4\rho - \bar{\rho}, \quad \mathscr{L}_5 = \boldsymbol{\delta} + 2\bar{\varpi} - 2\beta, \quad \mathscr{L}_6 = \mathbf{D} - \bar{\rho},$$

and where $T_{11} = T_{ll}$, $T_{13} = T_{lm}$, $T_{33} = T_{mm}$ are the frame components of the stress-energy tensor (125) of the particle given by

$$T_{11} = \frac{\mu}{r^2}u^t(1-a\Omega)^2\delta_3,$$

$$T_{13} = \frac{i\mu}{\sqrt{2}r^3}u^t(1-a\Omega)[a-(r^2+a^2)\Omega]\delta_3,$$

$$T_{33} = -\frac{\mu}{2r^4}u^t[a-(r^2+a^2)\Omega]^2\delta_3. \qquad (152)$$

Following the notation of [50], we can write

$$T_{s=2} \equiv T^{(0)} + T^{(1)} + T^{(2)}, \qquad (153)$$

where

$$T^{(0)} = -\mathscr{L}_1\mathscr{L}_3 T_{11}\,, \qquad T^{(1)} = (\mathscr{L}_1\mathscr{L}_2 + \mathscr{L}_4\mathscr{L}_5)T_{13}\,, \qquad T^{(2)} = -\mathscr{L}_4\mathscr{L}_6 T_{33}\,. \tag{154}$$

5.2.1 Green's Function

The (retarded) Green's function of the Teukolsky radial equation is a solution of the equation

$$\mathscr{L}_r^{s=2}(G_{lm}(r,r')) = \frac{1}{\Delta}\delta(r-r')\,. \tag{155}$$

It has the form

$$\begin{aligned} G_{lm}(r,r') &= \frac{(\Delta')^2}{W_{lm}}\left[R_{\rm in}(r)R_{\rm up}(r')H(r'-r) + R_{\rm in}(r')R_{\rm up}(r)H(r-r')\right] \\ &\equiv \frac{(\Delta')^2}{W_{lm}} R_{\rm in}(r_<)R_{\rm up}(r_>)\,, \end{aligned} \tag{156}$$

where $H(x)$ here denotes the Heaviside step function, $\Delta' \equiv \Delta(r')$, $R_{\rm in}(r)$ and $R_{\rm up}(r)$ are two independent solutions to the homogeneous radial Teukolsky equation having (as explained in the Schwarschild case) the appropriate behaviors at the horizon and at infinity, respectively, and W_{lm} is the associated (constant) Wronskian. The full retarded Green's function then turns out to be

$$G(x,x') = \sum_{l,m} \frac{(\Delta')^2}{W_{lm}} R_{\rm in}(r_<)R_{\rm up}(r_>)\,{}_2S_{lm}(\theta){}_2S_{lm}(\theta')e^{im(\phi-\phi')}\,. \tag{157}$$

5.2.2 Source Terms

One can then solve the Teukolsky equation for ψ_0 ($s = 2$), which yields

$$\begin{aligned} \psi_0 &= -8\pi \int \Sigma' T(x', x_0) G(x,x') dr' d(\cos\theta') d\phi' \\ &= -8\pi \int \Sigma' [T^{(0)} + T^{(1)} + T^{(2)}] G(x,x') dr' d(\cos\theta') d\phi' \\ &\equiv \psi_0^{(0)} + \psi_0^{(1)} + \psi_0^{(2)}\,. \end{aligned} \tag{158}$$

The coefficients $\psi_0^{(0,1,2)}$ can be computed straightforwardly and for each of them one has a left part $\psi_0^{(0,1,2)-}$ and a right one $\psi_0^{(0,1,2)+}$, i.e.,

$$\psi_0^{(0,1,2)} = \sum_{lm}[\psi_{0,lm}^{(0,1,2)-}H(r_0 - r) + \psi_{0,lm}^{(0,1,2)+}H(r - r_0)]\,{}_2S_{lm}(\theta)e^{i(m\phi-\omega t)} . \tag{159}$$

They can be conveniently written as

$$\begin{aligned}
\psi_{0,lm}^{(0)-} &= p_0\frac{[\beta_{lm}^{(0)}R_{\rm up}(r_0)]}{W_{lm}}R_{\rm in}(r) ,\\
\psi_{0,lm}^{(0)+} &= p_0\frac{[\beta_{lm}^{(0)}R_{\rm in}(r_0)]}{W_{lm}}R_{\rm up}(r) ,\\
\psi_{0,lm}^{(1)-} &= p_1\frac{[\alpha_{lm}^{(1)}R'_{\rm up}(r_0) + \beta_{lm}^{(1)}R_{\rm up}(r_0)]}{W_{lm}}R_{\rm in}(r) ,\\
\psi_{0,lm}^{(1)+} &= p_1\frac{[\alpha_{lm}^{(1)}R'_{\rm in}(r_0) + \beta_{lm}^{(1)}R_{\rm in}(r_0)]}{W_{lm}}R_{\rm up}(r) ,\\
\psi_{0,lm}^{(2)-} &= p_2\frac{[\alpha_{lm}^{(2)}R'_{\rm up}(r_0) + \beta_{lm}^{(2)}R_{\rm up}(r_0)]}{W_{lm}}R_{\rm in}(r) ,\\
\psi_{0,lm}^{(2)+} &= p_2\frac{[\alpha_{lm}^{(2)}R'_{\rm in}(r_0) + \beta_{lm}^{(2)}R_{\rm in}(r_0)]}{W_{lm}}R_{\rm up}(r) ,
\end{aligned} \tag{160}$$

where

$$p_0 = 4\pi\mu\frac{w_1^2}{r_0^2}\Delta_0^2u^t , \qquad p_1 = 4\pi i\mu\frac{w_1w_2}{r_0^2}u^t , \qquad p_2 = -4\pi\mu\frac{w_2^2}{r_0^2}u^t , \tag{161}$$

with $\Delta_0 = \Delta(r_0)$ and the dimensionless quantities

$$w_1 = 1 - a\Omega , \qquad w_2 = \frac{a - \Omega(r_0^2 + a^2)}{r_0} = \frac{a}{r_0}w_1 - \Omega r_0 . \tag{162}$$

Furthermore,

$$\begin{aligned}
\beta_{lm}^{(0)} &= 2\left(mw_1 - \frac{ia}{r_0}\right)(S_0' + mw_1S_0) + \left[m^2(1 - w_1^2) + 2 - E\right]S_0 ,\\
\alpha_{lm}^{(1)} &= -2\Delta_0^2r_0\left[S_0' + mw_1S_0\right] ,\\
\beta_{lm}^{(1)} &= 2\Delta_0\left[imr_0^2w_2 + 2(a^2 - r_0^2)\right]\left[S_0' + mw_1S_0\right] ,\\
\alpha_{lm}^{(2)} &= 2r_0\Delta_0(Mr_0 - a^2 - imr_0^2w_2)S_0 ,
\end{aligned}$$

$$\beta_{lm}^{(2)} = \left[-2r_0^4 m^2 w_2^2 + \frac{2imw_2 r_0^2(a^2 - r_0^2)}{r_0^2 + a^2}(r_0^2 + 3Mr_0 + a^2) + \lambda_{\rm CH} r_0^2 \Delta_0 \right.$$
$$\left. - \frac{6iamr_0^3 \Delta_0}{r_0^2 + a^2} + 2[(a^2 - r_0^2)^2 - \Delta_0^2]\right] S_0 \,, \tag{163}$$

where $S_0 = {}_2S_{lm}(\pi/2)$, S_0' denotes the derivative of ${}_2S_{lm}$ with respect to θ evaluated at $\theta = \pi/2$, and $\lambda_{\rm CH}$ is the Chandrasekhar constant related to the energy eigenvalue $E = E_{(l,m,2;\xi)}$ by $\lambda_{\rm CH} = E_{(l,m,2;\xi)} + \xi^2 - 2m\xi - 2$.

The harmonic decomposition of $\psi_0^{\pm}$ is then

$$\psi_0^{\pm} = \sum_{lm} {}_2R_{lm\omega}^{\pm}(r)\ {}_2S_{lm\omega}(\theta)\ e^{i(m\phi - \omega t)} \,, \tag{164}$$

with

$${}_2R_{lm\omega}^{\pm}(r) = \psi_{0,lm}^{(0)\pm} + \psi_{0,lm}^{(1)\pm} + \psi_{0,lm}^{(2)\pm} \,, \tag{165}$$

leading to

$${}_2R_{lm\omega}^{-}(r) = \mathscr{A}_{lm\omega,({\rm up})}^{-}(r_0) R_{\rm in}(r) \,, \qquad {}_2R_{lm\omega}^{+}(r) = \mathscr{A}_{lm\omega,({\rm in})}^{+}(r_0) R_{\rm up}(r) \,. \tag{166}$$

The coefficients $\mathscr{A}_{lm\omega,({\rm up})}^{-}$ and $\mathscr{A}_{lm\omega,({\rm in})}^{+}$ can be (formally) expressed as

$$\mathscr{A}_{lm\omega,({\rm up})}^{-} = \frac{1}{W_{lm}}\left[\alpha_{lm}^{-} R_{\rm up}'(r_0) + \beta_{lm}^{-} R_{\rm up}(r_0)\right] \,,$$
$$\mathscr{A}_{lm\omega,({\rm in})}^{+} = \frac{1}{W_{lm}}\left[\alpha_{lm}^{+} R_{\rm in}'(r_0) + \beta_{lm}^{+} R_{\rm in}(r_0)\right] \,, \tag{167}$$

and satisfy the following property $\mathscr{A}_{lm\omega,({\rm up})}^{-}\big|_{l=-l-1} = -\mathscr{A}_{lm\omega,({\rm in})}^{+}$. Consider for instance the left coefficient. The value of each spheroidal harmonics (for given lm) on the equatorial plane vanishes only if the corresponding equatorial value of the ordinary spherical harmonics Y_{lm} also vanishes (and similarly for their θ derivatives). One can therefore numerically re-express, modulo a non-zero numerical factor, the equatorial values of ${}_2S_{lm}(\pi/2)$ (respectively $d\,{}_2S_{lm}(\pi/2)/d\theta$) in terms of $Y_{lm}(\pi/2)$ (respectively $dY_{lm}(\pi/2)/d\theta$). This yields expressions of the form

$$\mathscr{A}_{lm\omega,({\rm up})}^{-} = Y_{lm}^{*} J_{lm\omega}^{(0){\rm up}} + Y_{lm}'{}^{*} J_{lm\omega}^{(1){\rm up}} \,,$$
$$\bar{\mathscr{A}}_{l,-m,-\omega,({\rm up})}^{-} = (-1)^m \left[Y_{lm}^{*} J_{lm\omega}^{(0){\rm up}} - Y_{lm}'{}^{*} J_{lm\omega}^{(1){\rm up}}\right] \,. \tag{168}$$

5.3 Metric Reconstruction

The radiative ($l \geq 2$) perturbed metric (up to parts for which ψ_0 vanishes) is given by

$$h_{\alpha\beta} = \rho^{-4}\{n_\alpha n_\beta D_{nn} + \bar{m}_\alpha \bar{m}_\beta D_{\bar{m}\bar{m}} - n_{(\alpha}\bar{m}_{\beta)} D_{n\bar{m}}\}\Psi + \text{c.c.}, \tag{169}$$

where

$$\begin{aligned} D_{nn} &= (\bar{\boldsymbol{\delta}} - 3\alpha - \bar{\beta} + 5\varpi)(\bar{\boldsymbol{\delta}} - 4\alpha + \varpi), \\ D_{\bar{m}\bar{m}} &= (\boldsymbol{\Delta} + 5\mu - 3\gamma + \bar{\gamma})(\boldsymbol{\Delta} + \mu - 4\gamma), \\ D_{n\bar{m}} &= (\bar{\boldsymbol{\delta}} - 3\alpha + \bar{\beta} + 5\varpi + \bar{\tau})(\boldsymbol{\Delta} + \mu - 4\gamma) \\ &\quad + (\boldsymbol{\Delta} + 5\mu - \bar{\mu} - 3\gamma - \bar{\gamma})(\bar{\boldsymbol{\delta}} - 4\alpha + \varpi), \end{aligned} \tag{170}$$

and where Ψ denotes the Hertz-Debye potential [52–55]. It is related to ψ_0 by

$$\psi_0 = \frac{1}{8}\left[\mathscr{L}^4\bar{\Psi} + 12M\partial_t\Psi\right], \tag{171}$$

with

$$\mathscr{L}^4 = \mathscr{L}_1\mathscr{L}_0\mathscr{L}_{-1}\mathscr{L}_{-2}, \qquad \mathscr{L}_s = -[\partial_\theta - s\cot\theta + i\csc\theta\,\partial_\phi] - ia\sin\theta\,\partial_t. \tag{172}$$

The harmonic decompositions of Ψ and its complex conjugate $\bar{\Psi}$ are given by

$$\begin{aligned} \Psi &= \sum_{lm\omega} {}_2\mathscr{R}_{lm\omega}(r)\ {}_2S_{lm\omega}(\theta)\ e^{i(m\phi-\omega t)}, \\ \bar{\Psi} &= \sum_{lm\omega}(-1)^m\ {}_2\bar{\mathscr{R}}_{l,-m,-\omega}(r)\ {}_{-2}S_{lm\omega}(\theta)\ e^{i(m\phi-\omega t)}, \end{aligned} \tag{173}$$

respectively. The Teukolsky-Starobinski identity

$$\mathscr{L}^4\left({}_{-2}S_{lm\omega}e^{i(m\phi-\omega t)}\right) = D\ \left({}_2S_{lm\omega}e^{i(m\phi-\omega t)}\right), \tag{174}$$

with a real, positive constant D whose square is

$$D^2 = \lambda_{\text{CH}}^2(\lambda_{\text{CH}}+2)^2 + 8a\omega\lambda_{\text{CH}}(m-a\omega)(5\lambda_{\text{CH}}+6) + 48a^2\omega^2[2\lambda_{\text{CH}}+3(m-a\omega)^2], \tag{175}$$

implies

$$\mathscr{L}^4(\bar{\Psi}) = \sum_{l,m}(-1)^m\ {}_2\bar{\mathscr{R}}_{l,-m,-\omega}(r)\ D\ {}_2S_{lm\omega}e^{i(m\phi-\omega t)}. \tag{176}$$

Up to the second order in a we have

$$D = l(l-1)(l+2)(l+1) - 4(l-1)(l+2)m\omega a + \frac{4(l-1)(l+2)}{(2l+3)(2l-1)l^2(l+1)^2}\tilde{D}_2\omega^2 a^2 + O(a^3)\,, \tag{177}$$

with

$$\tilde{D}_2 = l^6+3l^5+5l^4m^2-9l^4-23l^3+10m^2l^3-12l^2+19m^2l^2+14m^2l+12m^2\,. \tag{178}$$

Taking into account that $\partial_t\Psi = -i\omega\Psi$, Eq. (171) thus becomes

$$\psi_0 = \sum_{l,m}\frac{1}{8}\left[(-1)^m D\,{}_2\bar{\mathscr{R}}_{l,-m,-\omega}(r) - 12iM\omega\,{}_2\mathscr{R}_{lm\omega}(r)\right]{}_2S_{lm\omega}e^{i(m\phi-\omega t)}\,. \tag{179}$$

Recalling then the harmonic decomposition (164) of ψ_0 implies

$${}_2R_{lm\omega} = \frac{1}{8}\left[(-1)^m D\,{}_2\bar{\mathscr{R}}_{l,-m,-\omega}(r) - 12iM\omega\,{}_2\mathscr{R}_{lm\omega}(r)\right]\,, \tag{180}$$

which once inverted yields

$${}_2\mathscr{R}_{lm\omega} = 8\frac{(-1)^m D}{D^2+144M^2\omega^2}\,{}_2\bar{R}_{l,-m,-\omega} + 8\frac{12iM\omega}{D^2+144M^2\omega^2}\,{}_2R_{lm\omega}\,. \tag{181}$$

Let us consider the "left" solution ${}_2\mathscr{R}^-_{lm\omega}$, which is given by

$${}_2\mathscr{R}^-_{lm\omega} = 8\frac{(-1)^m D\,\bar{\mathscr{A}}^-_{l,-m,-\omega(\mathrm{up})}}{D^2+144M^2\omega^2}\,\bar{R}^{l,-m}_{\mathrm{in}}(r) + 8\frac{12iM\omega\mathscr{A}^-_{lm\omega,(\mathrm{up})}}{D^2+144M^2\omega^2}\,R^{lm}_{\mathrm{in}}(r)\,, \tag{182}$$

as per Eq. (166). Using the property $\bar{R}^{l,-m}_{\mathrm{in}}(r) = R^{lm}_{\mathrm{in}}(r)$ and recalling Eq. (168) leads to

$${}_2\mathscr{R}^-_{lm\omega} = [\tilde{J}^{(0)\mathrm{up}}_{lm\omega}Y^*_{lm} + \tilde{J}^{(1)\mathrm{up}}_{lm\omega}Y'_{lm}{}^*]R^{lm}_{\mathrm{in}}(r) \equiv \mathscr{C}^-_{(lm\omega)}R^{lm}_{\mathrm{in}}(r)\,, \tag{183}$$

where we have defined

$$\tilde{J}^{(0)\mathrm{up}}_{lm\omega} = \frac{8}{D-12iM\omega}J^{(0)\mathrm{up}}_{lm\omega}\,, \qquad \tilde{J}^{(1)\mathrm{up}}_{lm\omega} = -\frac{8}{D+12iM\omega}J^{(1)\mathrm{up}}_{lm\omega}\,. \tag{184}$$

5.3.1 Non-radiative Multipoles

The contribution of the non-radiative modes $l = 0, 1$ (when seen from outside, i.e. for $r > r_0$) comes from the changes in the mass and angular momentum of the black hole due to the presence of the orbiting particle of mass μ. The Kerr metric perturbed in mass and angular momentum (in Boyer-Lindquist coordinates) acquires the following nonzero components (for $r > r_0$) [50]

$$
\begin{aligned}
h_{tt} &= -\frac{2\delta M}{r}\,, \qquad h_{t\phi} = \frac{2\delta J}{r}\,,\\
h_{rr} &= -\frac{2r^2}{M\Delta^2}[(Mr + a^2)\delta M - a\delta J]\,,\\
h_{\phi\phi} &= \frac{2a}{Mr}[(r + M)a\delta M - (r + 2M)\delta J]\,,
\end{aligned}
\tag{185}
$$

with $\delta M = \mu\tilde{E} = \mu u_t$ and $\delta J = \mu\tilde{L} = -\mu u_\phi$, where $\tilde{E}$ and $\tilde{L}$ denote the conserved energy and angular momentum per unit mass of the particle given by

$$
\begin{aligned}
\tilde{E} &= \frac{1}{r_0^{3/4}} \frac{r_0^{3/2} - 2M\sqrt{r_0} \pm a\sqrt{M}}{\left[r_0^{3/2} - 3M\sqrt{r_0} \pm 2a\sqrt{M}\right]^{1/2}}\,,\\
\tilde{L} &= \pm\frac{\sqrt{M}}{r_0^{3/4}} \frac{r_0^2 \mp 2a\sqrt{Mr_0} + a^2}{\left[r_0^{3/2} - 3M\sqrt{r_0} \pm 2a\sqrt{M}\right]^{1/2}}\,.
\end{aligned}
\tag{186}
$$

5.3.2 Computing h^R_{kk}

Let us now compute the gauge-invariant quantity $h_{kk} = h_{\alpha\beta}k^\alpha k^\beta$. Contracting the radiative part of the perturbed metric (169) twice with $k = \partial_t + \Omega\partial_\phi$ gives

$$
\begin{aligned}
h_{kk}^{(l\geq 2)} &= \rho^{-4}\{(n\cdot k)^2 D_{nn} + (\bar{m}\cdot k)^2 D_{\bar{m}\bar{m}} - (n\cdot k)(\bar{m}\cdot k)D_{n\bar{m}}\}\Psi + \text{c.c.}\\
&\equiv h_{kk}^{(1)} + \text{c.c.}\,,
\end{aligned}
\tag{187}
$$

and substituting the harmonic decompositions (173) of Ψ implies that $h_{kk}^{(1)\,lm\omega}$ has the general form

$$
\begin{aligned}
h_{kk}^{(1)\,lm\omega} &= \left(a_1\,{}_2\mathcal{R}''_{lm\omega} + a_2\,{}_2\mathcal{R}'_{lm\omega} + a_3\,{}_2\mathcal{R}_{lm\omega}\right) S_0\\
&\quad + \left(b_2\,{}_2\mathcal{R}'_{lm\omega} + b_3\,{}_2\mathcal{R}_{lm\omega}\right) S'_0 + c_3\,{}_2\mathcal{R}_{lm\omega} S''_0\,.
\end{aligned}
\tag{188}
$$

Consider the left part $h_{kk}^{(1)\,lm\omega,-}$ (i.e. the limit $r \to r_0^-$). After eliminating the second derivatives through the homogeneous equations and numerically re-expressing (as above) the spheroidal harmonics on the equatorial plane in terms of ordinary spherical harmonics, we formally have

$$
\begin{aligned}
h_{kk}^{(1)\,lm\omega,-} &= \mathscr{C}^-_{(lm\omega)} \left[\left(c_1 R'_{\rm in} + c_2 R_{\rm in}\right) Y_{lm} + \left(d_1 R'_{\rm in} + d_2 R_{\rm in}\right) Y'_{lm}\right] \\
&\equiv \mathscr{C}^-_{(lm\omega)} \left[K^{(0)\rm in}_{lm\omega} Y_{lm} + K^{(1)\rm in}_{lm\omega} Y'_{lm}\right] . \qquad (189)
\end{aligned}
$$

Recalling then the similar decomposition (183) for $\mathscr{C}^-_{(lm)}$, we find (up to mixed terms in Y_{lm} and Y'_{lm} which vanish when summing over m)

$$
h_{kk}^{(1)\,lm\omega,-} = \tilde{J}^{(0)\rm up}_{lm\omega} K^{(0)\rm in}_{lm\omega} |Y_{lm}|^2 + \tilde{J}^{(1)\rm up}_{lm\omega} K^{(1)\rm in}_{lm\omega} |Y'_{lm}|^2 . \qquad (190)
$$

Because of the mild nature of the singularity (at the location of the particle) of the gauge-invariant quantity $h_{kk}(r)$, its lm-th multipolar piece is continuous at r_0: $h_{kk}^{(1)\,lm\omega,+} = h_{kk}^{(1)\,lm\omega,-}$. The above formula therefore gives the common, left and right, contribution coming from the radiative multipoles $l \geq 2$.

One must finally add the contribution of the non-radiative multipoles $l = 0, 1$, i.e.,

$$
h_{kk}^{(l=0,1)} = h_{tt} + 2\Omega h_{t\phi} + \Omega^2 h_{\phi\phi} . \qquad (191)
$$

Its right part ($r > r_0$) is simply evaluated through Eq. (185).

The regularized contraction h^R_{kk} has been analytically computed in [56] through the 8.5PN order and the fourth-order in spin (see also [57], where no assumptions on the magnitude of the black hole rotation parameter are made). In the present spinning black-hole case, the subtraction term $B = \lim_{l\to\infty} h^l_{kk}$, needed for regularization, depends both on the dimensionless frequency parameter $y = (M\Omega)^{2/3}$ (such that $u \equiv M/r = y/(1 - \hat{a}y^{3/2})^{2/3}$) and on the dimensionless spin parameter $\hat{a} = a/M$. Up to $O(y^7)$ and to second order in spin, $B(y, \hat{a})$ is given by

$$
\begin{aligned}
-B &= -2y + \frac{13}{2}y^2 - \frac{9}{32}y^3 - \frac{83}{128}y^4 - \frac{12361}{8192}y^5 - \frac{116163}{32768}y^6 \\
&+ \left[-\frac{4}{3}y^{5/2} - \frac{19}{3}y^{7/2} + \frac{45}{16}y^{9/2} + \frac{1261}{192}y^{11/2} + \frac{191959}{12288}y^{13/2}\right]\hat{a} \\
&+ \left[\frac{1}{2}y^3 + \frac{8}{9}y^4 - \frac{4903}{1152}y^5 - \frac{9653}{1024}y^6\right]\hat{a}^2 + O(y^7, \hat{a}^3) . \qquad (192)
\end{aligned}
$$

Finally, the computation of the regularized Detweiler function $h^R_{kk}(y, \hat{a})$ yields (modulo undisplayed higher-order terms)

$$
\begin{aligned}
-h^R_{kk} &= 2y - 5y^2 - \frac{5}{4}y^3 + \left(\frac{1261}{24} - \frac{41}{16}\pi^2\right)y^4 \\
&+ \left(-\frac{157859}{960} + \frac{2275}{256}\pi^2 + \frac{128}{5}\ln(y) + \frac{256}{5}\gamma + \frac{512}{5}\ln(2)\right)y^5 \\
&+ \left(-\frac{284664301}{201600} - \frac{28016}{105}\gamma - \frac{14008}{105}\ln(y) - \frac{63472}{105}\ln(2) + \frac{246367}{1536}\pi^2\right. \\
&\left. + \frac{486}{7}\ln(3)\right)y^6 + \frac{27392}{525}\pi y^{13/2} \\
&+ \left[-\frac{14}{3}y^{5/2} + \frac{7}{3}y^{7/2} - \frac{103}{4}y^{9/2} + \left(-\frac{1373}{24} - \frac{29}{16}\pi^2\right)y^{11/2}\right. \\
&\left. + \left(-\frac{9370733}{4800} + \frac{3059}{16}\pi^2 - \frac{704}{5}\gamma - \frac{352}{5}\ln(y) - \frac{832}{3}\ln(2)\right)y^{13/2}\right]\hat{a} \\
&+ \left[2y^3 + \frac{64}{9}y^4 + \frac{629}{36}y^5 + \left(\frac{1463}{12} - \frac{593}{256}\pi^2\right)y^6\right]\hat{a}^2 + O(y^7, \hat{a}^3)\,.
\end{aligned}
\tag{193}
$$

6 Concluding Remarks

In the context of gauge-invariant, first-order gravitational self-force computations around black holes, many useful high-order analytical results have been obtained in the last few years (starting from 2013). The simplest case of gravitational self-force redshift effects induced by a particle moving along a circular equatorial orbit around a non-rotating Schwarzschild black hole has been generalized to more complex cases involving: (1) a rotating Kerr background [50, 56–58]; (2) non-circular (i.e., eccentric) orbits [59–66]; (3) the computation of higher-derivative self-force invariants, linked to the dynamics of spinning and/or tidally deformed test bodies [37, 38, 41, 67–69].

The field of first-order gravitational self-force computations is quite active, and mature. Research in second-order gravitational self-force perturbations is also actively starting, see [70] and references therein. The entire field is driven by potential applications to the gravitational wave signals emitted by coalescing binary systems (with comparable, or extreme, mass ratios).

Acknowledgements D.B. and T.D. dedicate this work to the memory of Ulysse Schildge who helped them in many ways during many years. D.B. thanks the organizers of the *INdAM Workshop on Innovative Algorithms and Analysis*, Rome, May 17–20, 2016 for their kind invitation. D.B. and T.D. are grateful to ICRANet for partial support over many years.

References

1. Abbott, B.P., et al., [LIGO Scientific and Virgo Collaborations]: Observation of gravitational waves from a binary black hole merger. Phys. Rev. Lett. **116**, 061102 (2016) [arXiv:1602.03837 [gr-qc]]
2. Abbott, B.P., et al., [LIGO Scientific and Virgo Collaborations]: GW151226: observation of gravitational waves from a 22-solar-mass binary black hole coalescence. Phys. Rev. Lett. **116**, 241103 (2016) [arXiv:1606.04855 [gr-qc]]
3. Blanchet, L.: Gravitational radiation from post-Newtonian sources and inspiralling compact binaries. Living Rev. Rel. **17**, 2 (2014) [arXiv:1310.1528 [gr-qc]]
4. Damour, T., Jaranowski, P., Schäfer, G.: Nonlocal-in-time action for the fourth post-Newtonian conservative dynamics of two-body systems. Phys. Rev. D **89**, 064058 (2014) [arXiv:1401.4548 [gr-qc]]
5. Sago, N., Barack, L., Detweiler, S.L.: Two approaches for the gravitational self force in black hole spacetime: comparison of numerical results. Phys. Rev. D **78**, 124024 (2008) [arXiv:0810.2530 [gr-qc]]
6. Buonanno, A., Damour, T.: Effective one-body approach to general relativistic two-body dynamics. Phys. Rev. D **59**, 084006 (1999) [gr-qc/9811091]
7. Buonanno, A., Damour, T.: Transition from inspiral to plunge in binary black hole coalescences Phys. Rev. D **62**, 064015 (2000) [gr-qc/0001013]
8. Damour, T., Jaranowski, P., Schäfer, G.: On the determination of the last stable orbit for circular general relativistic binaries at the third post-Newtonian approximation. Phys. Rev. D **62**, 084011 (2000) [gr-qc/0005034]
9. Damour, T.: Coalescence of two spinning black holes: an effective one-body approach. Phys. Rev. D **64**, 124013 (2001) [gr-qc/0103018]
10. Pretorius, F.: Evolution of binary black hole spacetimes. Phys. Rev. Lett. **95**, 121101 (2005) [gr-qc/0507014]
11. Hinder, I., et al.: Error-analysis and comparison to analytical models of numerical waveforms produced by the NRAR collaboration. Class. Quant. Grav. **31**, 025012 (2014) [arXiv:1307.5307 [gr-qc]]
12. Le Tiec, A.: The overlap of numerical relativity, perturbation theory and post-Newtonian theory in the binary black hole problem. Int. J. Mod. Phys. D **23**, 1430022 (2014) [arXiv:1408.5505 [gr-qc]]
13. Bernuzzi, S., Nagar, A., Dietrich, T., Damour, T.: Modeling the dynamics of tidally interacting binary neutron stars up to the merger. Phys. Rev. Lett. **114**, 161103 (2015) [arXiv:1412.4553 [gr-qc]]
14. Regge, T., Wheeler, J.A.: Stability of a Schwarzschild singularity. Phys. Rev. **108**, 1063 (1957)
15. Zerilli, F.J.: Gravitational field of a particle falling in a Schwarzschild geometry analyzed in tensor harmonics. Phys. Rev. D **2**, 2141 (1970)
16. Teukolsky, S.A.: Perturbations of a rotating black hole. 1. Fundamental equations for gravitational electromagnetic and neutrino field perturbations. Astrophys. J. **185**, 635 (1973)
17. Press, W.H., Teukolsky, S.A.: Perturbations of a rotating black hole. II. Dynamical stability of the Kerr metric. Astrophys. J. **185**, 649 (1973)
18. Teukolsky, S.A., Press, W.H.: Perturbations of a rotating black hole. III - Interaction of the hole with gravitational and electromagnetic radiation. Astrophys. J. **193**, 443 (1974)

19. Mano, S., Suzuki, H., Takasugi, E.: Analytic solutions of the Regge-Wheeler equation and the post-Minkowskian expansion. Prog. Theor. Phys. **96**, 549 (1996) [gr-qc/9605057]
20. Mano, S., Suzuki, H., Takasugi, E.: Analytic solutions of the Teukolsky equation and their low frequency expansions. Prog. Theor. Phys. **95**, 1079 (1996) [gr-qc/9603020].
21. Barack, L., Mino, Y., Nakano, H., Ori, A., Sasaki, M.: Calculating the gravitational selfforce in Schwarzschild space-time. Phys. Rev. Lett. **88**, 091101 (2002) [gr-qc/0111001]
22. Detweiler, S.L., Messaritaki, E., Whiting, B.F.: Selfforce of a scalar field for circular orbits about a Schwarzschild black hole. Phys. Rev. D **67**, 104016 (2003) [gr-qc/0205079]
23. Damour, T., Iyer, B.R., Nagar, A.: Improved resummation of post-Newtonian multipolar waveforms from circularized compact binaries. Phys. Rev. D **79**, 064004 (2009) [arXiv:0811.2069 [gr-qc]]
24. Buonanno, A., Cook, G.B., Pretorius, F.: Inspiral, merger and ring-down of equal-mass black-hole binaries. Phys. Rev. D **75**, 124018 (2007) [gr-qc/0610122]
25. Taracchini, A., et al.: Effective-one-body model for black-hole binaries with generic mass ratios and spins. Phys. Rev. D **89**, 061502 (2014) [arXiv:1311.2544 [gr-qc]]
26. Damour, T., Nagar, A.: New effective-one-body description of coalescing nonprecessing spinning black-hole binaries. Phys. Rev. D **90**, 044018 (2014) [arXiv:1406.6913 [gr-qc]]
27. Damour, T., Jaranowski, P., Schaefer, G.: Fourth post-Newtonian effective one-body dynamics. Phys. Rev. D **91**, 084024 (2015) [arXiv:1502.07245 [gr-qc]]
28. Damour, T., Schaefer, G.: Higher order relativistic periastron advances and binary pulsars. Nuovo Cim. B **101**, 127 (1988)
29. Akcay, S., Barack, L., Damour, T., Sago, N.: Gravitational self-force and the effective-one-body formalism between the innermost stable circular orbit and the light ring. Phys. Rev. D **86**, 104041 (2012) [arXiv:1209.0964 [gr-qc]]
30. Kavanagh, C., Ottewill, A.C., Wardell, B.: Analytical high-order post-Newtonian expansions for extreme mass ratio binaries. Phys. Rev. D **92**, 084025 (2015) [arXiv:1503.02334 [gr-qc]]
31. Detweiler, S.L.: A consequence of the gravitational self-force for circular orbits of the Schwarzschild geometry. Phys. Rev. D **77**, 124026 (2008) [arXiv:0804.3529 [gr-qc]]
32. Le Tiec, A., Blanchet, L., Whiting, B.F.: The first law of binary black hole mechanics in general relativity and post-Newtonian theory. Phys. Rev. D **85**, 064039 (2012) [arXiv:1111.5378 [gr-qc]]
33. Le Tiec, A., Barausse, E., Buonanno, A.: Gravitational self-force correction to the binding energy of compact binary systems. Phys. Rev. Lett. **108**, 131103 (2012) [arXiv:1111.5609 [gr-qc]]
34. Barausse, E., Buonanno, A., Le Tiec, A.: The complete non-spinning effective-one-body metric at linear order in the mass ratio. Phys. Rev. D **85**, 064010 (2012) [arXiv:1111.5610 [gr-qc]]
35. Bini, D., Damour, T.: Analytical determination of the two-body gravitational interaction potential at the fourth post-Newtonian approximation. Phys. Rev. D **87**, 121501 (2013) [arXiv:1305.4884 [gr-qc]]
36. Bini, D., Damour, T.: High-order post-Newtonian contributions to the two-body gravitational interaction potential from analytical gravitational self-force calculations. Phys. Rev. D **89**, 064063 (2014) [arXiv:1312.2503 [gr-qc]]
37. Bini, D., Damour, T.: Analytic determination of high-order post-Newtonian self-force contributions to gravitational spin precession. Phys. Rev. D **91**, 064064 (2015) [arXiv:1503.01272 [gr-qc]]
38. Bini, D., Damour, T.: Two-body gravitational spin-orbit interaction at linear order in the mass ratio. Phys. Rev. D **90**, 024039 (2014) [arXiv:1404.2747 [gr-qc]]
39. Bini, D., Damour, T.: Analytic determination of the eight-and-a-half post-Newtonian self-force contributions to the two-body gravitational interaction potential. Phys. Rev. D **89**, 104047 (2014) [arXiv:1403.2366 [gr-qc]]
40. Bini, D., Damour, T.: Detweiler's gauge-invariant redshift variable: analytic determination of the nine and nine-and-a-half post-Newtonian self-force contributions. Phys. Rev. D **91**, 064050 (2015) [arXiv:1502.02450 [gr-qc]]

41. Bini, D., Damour, T.: Gravitational self-force corrections to two-body tidal interactions and the effective one-body formalism. Phys. Rev. D **90**, 124037 (2014) [arXiv:1409.6933 [gr-qc]]
42. Chandrasekhar, S.: On the equations governing the perturbations of the Schwarzschild black hole. Proc. R. Soc. Lond. A **343**, 289 (1975)
43. Blanchet, L., Damour, T.: Tail transported temporal correlations in the dynamics of a gravitating system. Phys. Rev. D **37**, 1410 (1988)
44. Poisson, E., Sasaki, M.: Gravitational radiation from a particle in circular orbit around a black hole. 5: Black hole absorption and tail corrections. Phys. Rev. D **51**, 5753 (1995) [gr-qc/9412027]
45. Sasaki, M., Tagoshi, H.: Analytic black hole perturbation approach to gravitational radiation. Living Rev. Relativ. **6**, 6 (2003) [gr-qc/0306120]
46. Nakano, H., Sago, N., Sasaki, M.: Gauge problem in the gravitational selfforce. 2. First postNewtonian force under Regge-Wheeler gauge. Phys. Rev. D **68**, 124003 (2003) [gr-qc/0308027]
47. Blanchet, L., Detweiler, S.L., Le Tiec, A., Whiting, B.F.: Post-Newtonian and numerical calculations of the gravitational self-force for circular orbits in the Schwarzschild geometry. Phys. Rev. D **81**, 064004 (2010) [arXiv:0910.0207 [gr-qc]]
48. Damour, T.: Gravitational self force in a Schwarzschild background and the effective one body formalism. Phys. Rev. D **81**, 024017 (2010) [arXiv:0910.5533 [gr-qc]]
49. Blanchet, L., Detweiler, S.L., Le Tiec, A., Whiting, B.F.: High-order post-Newtonian fit of the gravitational self-force for circular orbits in the Schwarzschild geometry. Phys. Rev. D **81**, 084033 (2010) [arXiv:1002.0726 [gr-qc]]
50. Shah, A.G., Friedman, J.L., Keidl, T.S.: EMRI corrections to the angular velocity and redshift factor of a mass in circular orbit about a Kerr black hole. Phys. Rev. D **86**, 084059 (2012) [arXiv:1207.5595 [gr-qc]]
51. Goldberg, J.N., MacFarlane, A.J., Newman, E.T., Rohrlich, F., Sudarshan, E.C.G.: Spin s spherical harmonics and edth. J. Math. Phys. **8**, 2155 (1967)
52. Chrzanowski, P.L.: Vector potential and metric perturbations of a rotating black hole. Phys. Rev. D **11**, 2042 (1975)
53. Cohen, J.M., Kegeles, L.S.: Electromagnetic fields in curved spaces - a constructive procedure. Phys. Rev. D **10**, 1070 (1974)
54. Kegeles, L.S., Cohen, J.M.: Constructive procedure for perturbations of space-times. Phys. Rev. D **19**, 1641 (1979)
55. Ori, A.: Reconstruction of inhomogeneous metric perturbations and electromagnetic four potential in Kerr space-time. Phys. Rev. D **67**, 124010 (2003) [gr-qc/0207045]
56. Bini, D., Damour, T., Geralico, A.: Spin-dependent two-body interactions from gravitational self-force computations. Phys. Rev. D **92**, 124058 (2015). Erratum: [Phys. Rev. D **93**, 109902 (2016)] [arXiv:1510.06230 [gr-qc]]
57. Kavanagh, C., Ottewill, A.C., Wardell, B.: Analytical high-order post-Newtonian expansions for spinning extreme mass ratio binaries. Phys. Rev. D **93**, 124038 (2016) [arXiv:1601.03394 [gr-qc]]
58. Blanchet, L., Buonanno, A., Le Tiec, A.: First law of mechanics for black hole binaries with spins. Phys. Rev. D **87**, 024030 (2013) [arXiv:1211.1060 [gr-qc]]
59. Barack, L., Sago, N.: Beyond the geodesic approximation: conservative effects of the gravitational self-force in eccentric orbits around a Schwarzschild black hole. Phys. Rev. D **83**, 084023 (2011) [arXiv:1101.3331 [gr-qc]]
60. Akcay, S., Le Tiec, A., Barack, L., Sago, N., Warburton, N.: Comparison between self-force and post-Newtonian dynamics: beyond circular orbits. Phys. Rev. D **91**, 124014 (2015) [arXiv:1503.01374 [gr-qc]]
61. Bini, D., Damour, T., Geralico, A.: Confirming and improving post-Newtonian and effective-one-body results from self-force computations along eccentric orbits around a Schwarzschild black hole. Phys. Rev. D **93**, 064023 (2016) [arXiv:1511.04533 [gr-qc]]

62. Hopper, S., Kavanagh, C., Ottewill, A.C.: Analytic self-force calculations in the post-Newtonian regime: eccentric orbits on a Schwarzschild background. Phys. Rev. D **93**, 044010 (2016) [arXiv:1512.01556 [gr-qc]]
63. van de Meent, M., Shah, A.G.: Metric perturbations produced by eccentric equatorial orbits around a Kerr black hole. Phys. Rev. D **92**, 064025 (2015) [arXiv:1506.04755 [gr-qc]]
64. Le Tiec, A.: First law of mechanics for compact binaries on eccentric orbits. Phys. Rev. D **92**, 084021 (2015) [arXiv:1506.05648 [gr-qc]]
65. Bini, D., Damour, T., Geralico, A.: New gravitational self-force analytical results for eccentric orbits around a Schwarzschild black hole. Phys. Rev. D **93**, 104017 (2016) [arXiv:1601.02988 [gr-qc]]
66. Bini, D., Damour, T., Geralico, A.: High post-Newtonian order gravitational self-force analytical results for eccentric equatorial orbits around a Kerr black hole. Phys. Rev. D **93**, 124058 (2016) [arXiv:1602.08282 [gr-qc]]
67. Barack, L., Damour, T., Sago, N.: Precession effect of the gravitational self-force in a Schwarzschild spacetime and the effective one-body formalism. Phys. Rev. D **82**, 084036 (2010) [arXiv:1008.0935 [gr-qc]]
68. Dolan, S.R., Warburton, N., Harte, A.I., Le Tiec, A., Wardell, B., Barack, L.: Gravitational self-torque and spin precession in compact binaries. Phys. Rev. D **89**, 064011 (2014) [arXiv:1312.0775 [gr-qc]]
69. Dolan, S.R., Nolan, P., Ottewill, A.C., Warburton, N., Wardell, B.: Tidal invariants for compact binaries on quasicircular orbits. Phys. Rev. D **91**, 023009 (2015) [arXiv:1406.4890 [gr-qc]]
70. Bini, D., Damour, T.: Conservative second-order gravitational self-force on circular orbits and the effective one-body formalism. Phys. Rev. D **93**, 104040 (2016) [arXiv:1603.09175 [gr-qc]]

Concentration Waves of Chemotactic Bacteria: The Discrete Velocity Case

Vincent Calvez, Laurent Gosse, and Monika Twarogowska

Abstract The existence of travelling waves for a coupled system of hyperbolic/parabolic equations is established in the case of a finite number of velocities in the kinetic equation. This finds application in collective motion of chemotactic bacteria. The analysis builds on the previous work by the first author (arXiv:1607.00429) in the case of a continuum of velocities. Here, the proof is specific to the discrete setting, based on the decomposition of the population density in special Case's modes. Some counter-intuitive results are discussed numerically, including the co-existence of several travelling waves for some sets of parameters, as well as the possible non-existence of travelling waves.

1 Introduction and Framework

This note is devoted to the analysis of kinetic models for travelling bands of chemotactic bacteria *E. coli* in a microchannel. This builds on the series of papers [7–9, 29, 30]. This series of works is motivated by the following seminal experiment: A population of bacteria *E. coli* is initially located on the left side of a microchannel after centrifugation (approximately 5×10^5 individuals). After short time, a significant fraction of the population moves towards the right side of the channel, at constant speed, within a constant profile [30], see Fig. 1 for a schematic picture. We refer to [2] for the original experiment, and [32] for a thorough review

V. Calvez (✉)
CNRS & Institut Camille Jordan, Université de Lyon 1, Lyon, France

Inria, Project-Team NUMED, Lyon, France
e-mail: vincent.calvez@math.cnrs.fr

L. Gosse
IAC, CNR, via dei Taurini, 19, 00185 Roma, Italy
e-mail: l.gosse@ba.iac.cnr.it

M. Twarogowska
Unité de Mathématiques Pures et Appliquées, Ecole Normale Supérieure de Lyon, Lyon, France

Inria, Project-Team NUMED, Lyon, France
e-mail: monika.twarogowska@ens-lyon.fr

L. Gosse, R. Natalini (eds.), *Innovative Algorithms and Analysis*,
Springer INdAM Series 16, DOI 10.1007/978-3-319-49262-9_3

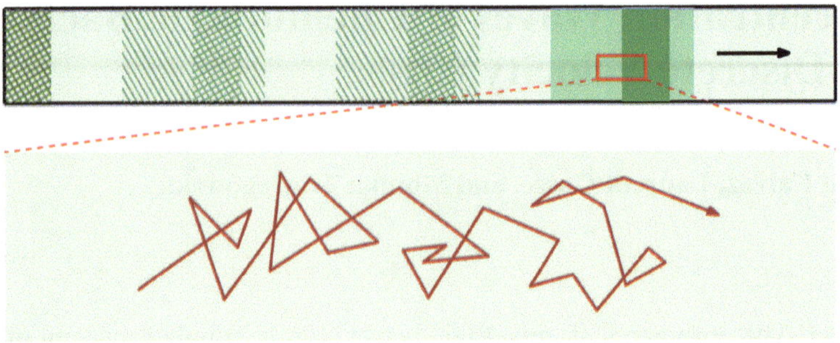

Fig. 1 Cartoon of concentration waves of bacteria as observed in experiments (see [2, 30]). The population of bacteria is initially located on the left hand side of the channel after centrifugation. Shortly, a large fraction of the population detaches and propagate to the right side at constant speed. Individual trajectories follows a run-and-tumble process in first approximation: cells alternate between straight runs and fast reorientation events (tumbles). The duration of run phases is modulated by sensing temporal variations of the chemical gradients in the environment. We refer to [6] for biological aspects of motions of *E. coli*

about the mathematical modelling of collective motion of bacteria in the light of this experiment, initiated by the celebrated work by Keller and Segel [23].

Kinetic models have proven to be well suited to study bacteria locomotion and chemotaxis, which navigate in a liquid medium according to a biased run-and-tumble process [6]. They were first introduced and investigated in the 70s by Stroock [31], then in the early 80s by Alt [4]. We refer to [11, 12, 15, 25, 26, 33] for the description of the run-and-tumble model at multiple scales. In particular, [34] and [18] deals with the modelling of the same experiment, and [3, 14] is about the modelling of the interactions between two strains into the same wave of propagation. We also refer to the recent works [13, 16, 17, 27, 35] concerning numerical simulations of multiscale models of chemotactic bacteria.

Here, we investigate a basic kinetic model, coupled to reaction-diffusion equations for the dynamics of chemical species. The population of bacteria is described by its density $f(t, x, v)$ in the phase space position×velocity. In addition, two chemical species are considered, according to the leading hypothesis in [28–30, 34]. We denote by $N(t, x)$ the concentration of some nutrient, which is distributed homogeneously in the domain at initial time. We also denote by $S(t, x)$ the concentration of some amino-acid which mediates cell-cell communication. The full model is written as follows,

$$\begin{cases} \partial_t f(t,x,v) + v \cdot \nabla_x f(t,x,v) = \displaystyle\int_{v' \in V} \mathbf{T}(t,x,v') f(t,x,v')\, dv(v') - \mathbf{T}(t,x,v) f(t,x,v) \\ \partial_t S(t,x) = D_S \Delta S(t,x) - \alpha S(t,x) + \beta \rho(t,x) \\ \partial_t N(t,x) = D_N \Delta N(t,x) - \gamma \rho(t,x) N(t,x) \frac{1}{2}\,, \end{cases} \tag{1}$$

where ρ denotes the spatial density: $\rho(t,x) = \int f(t,x,v)\,d\nu(v)$. Here, $\alpha, \beta, \gamma, D_S, D_N$ are positive constants. The measure ν is a compactly supported probability measure on the velocity space. The tumbling (scattering) rate $\mathbf{T}(t,x,v)$ expresses temporal sensing of navigating bacteria:

$$\mathbf{T}(t,x,v) = 1 - \chi_S \text{sign}\,(\partial_t S(t,x) + v\nabla_x S(t,x)) - \chi_N \text{sign}\,(\partial_t N(t,x) + v\nabla_x N(t,x)),$$

$$(\chi_S, \chi_N) \in (0, \frac{1}{2}) \times (0, \frac{1}{2}).$$

It is assumed that any single bacteria is influenced by temporal variations of both concentrations S and N along its trajectory with velocity v. Furthermore, it is able to distinguish perfectly between *favourable* directions (with positive variation) and *unfavourable* directions (with negative variation). It modulates the tumbling rate accordingly: runs are relatively longer if direction is favourable (because the tumbling rate is relatively smaller). Finally, we assume that both signal contributions are additive, with possibly two different values for the coefficients χ_S, χ_N.

For the sake of simplicity, we assume that

$$\chi_N \leq \chi_S. \tag{2}$$

This condition appears at some point during the analysis. We believe that our results hold true also in the opposite case $\chi_N > \chi_S$. However, this would require more complicated arguments that we postpone for future work.

As the kinetic equation is conservative, we assume without loss of generality that

$$\iint f(t,x,v)\,d\nu(v)dx = \int \rho(t,x)\,dx = 1. \tag{3}$$

We refer to [7, 30] for a detailed discussion about the relevance of this model.

In this work, we examine the case of a finite number of velocities $N = 2K + 1$, for some integer K. Let (v_k) be the set of discrete velocities, and $\boldsymbol{\omega} = (\omega_k)$ be the corresponding weights. We adopt the following notation: index k ranges from $-K$ to K, with $v_0 = 0$. Let denote $\mathcal{K} = [-K, K]$ the set of indices. The measure ν is defined as follows,

$$\nu = \sum_{k\in\mathcal{K}} \omega_k \delta_{v_k}, \quad \sum_{k\in\mathcal{K}} \omega_k = 1.$$

We assume that nodes and weights are symmetric with respect to the origin:

$$(\forall k \in \mathcal{K}) \quad v_k = -v_{-k}, \quad \omega_k = \omega_{-k}.$$

We seek one-dimensional travelling wave solutions, that we write $f(x-ct,v)$, $S(x-ct), N(x-ct)$ with some slight abuse of notation. Thus, we are reduced to investigate the following problem:

$$\begin{cases} (v_k - c)\partial_z f(z, v_k) = \sum_{k' \in \mathcal{K}} \omega_{k'} T(z, v_{k'} - c) f(z, v_{k'}) - T(z, v_k - c) f(z, v_k) \\ -c\partial_z S(z) - D_S \partial_z^2 S(z) + \alpha S(z) = \beta \rho(z) \\ -c\partial_z N(z) - D_N \partial_z^2 N(z) = -\gamma \rho(z) N(z) , \end{cases} \tag{4}$$

where the speed c is an unknown real number. As the problem is symmetric, we look for a positive value $c > 0$ without loss of generality. The tumbling rate in the moving frame is

$$T(z, v - c) = 1 - \chi_S \mathrm{sign}\left((v - c)\partial_z S(z)\right) - \chi_N \mathrm{sign}\left((v - c)\partial_z N(z)\right) . \tag{5}$$

In [7], the existence of travelling waves is established in the case of a continuum of velocity. Namely, it is assumed that the measure ν is absolutely continuous with respect to the Lebesgue measure, and that the probability density function belongs to L^p for some $p > 1$. Here, we investigate this problem in the case of discrete velocities, which is not contained in [7]. Rather than stating a global result, we present the framework for studying (4). This enables to discuss numerically the possible non existence of travelling waves.

The following methodology is adopted in [7, 29]:

Framework: construction of travelling waves

(i) Assume a priori that N is increasing everywhere, and that S is unimodal, with a single maximum located at $z = 0$, say. This enables to decouple the kinetic equation from the reaction-diffusion equations in (4), since T is then fully characterized by (6), see also Fig. 2.

(ii) Prove that there exists a non trivial density f, which decays exponentially fast as $|z| \to +\infty$. This is an expression of the confinement effect due to the biased modulation of runs, see Sect. 2.

(iii) Prove that the spatial density ρ is also unimodal, with a single maximum located at the transition point $z = 0$, as well. This is the hard task, because f does not share this monotonicity property for all v, but ρ does, as a consequence of compensations in averaging, see Sect. 3.

(iv) Check a posteriori that N is increasing, and that S is unimodal, with a single maximum reached at $z = 0$. The former is unconditionally true, provided that $c > 0$. The latter condition is the crucial one which enables to prescribe the value of c.

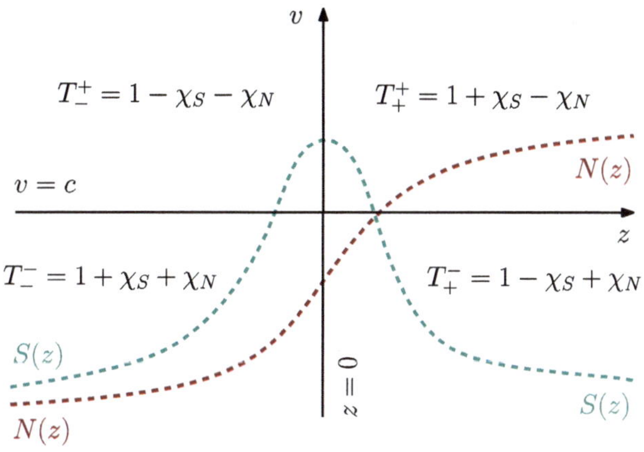

Fig. 2 The tumbling rate $T(z, v - c)$ at a glance, according to the rule of signs expressed in (7)

Suppose that N and S share the appropriate monotonicity conditions, as in (i). Then, T (5) is prescribed as follows:

$$T(z, v - c) = 1 + \chi_S \text{sign}\left((v - c)z\right) - \chi_N \text{sign}\left((v - c)\right) . \tag{6}$$

We adopt the following short-cut notations (see also Fig. 2):

$$\begin{cases} T_-^- = 1 + \chi_S + \chi_N \,, & \text{(direction is doubly unfavourable)}, \\ T_-^+ = 1 - \chi_S - \chi_N & \text{(direction is doubly favourable)}, \\ T_+^- = 1 - \chi_S + \chi_N & \text{(direction is favourable for } S\text{, but unfavourable for} N), \\ T_+^+ = 1 + \chi_S - \chi_N & \text{(direction is unfavourable for } S\text{, but favourable for} N). \end{cases} \tag{7}$$

Suppose that c is given in a suitable interval for confinement purposes (see below). We can associate a probability density f, in a unique way, solution of the first equation in (4). This defines the concentration S, through the spatial density ρ, as the result of the elliptic equation in the second line of (4). The matching condition in point (iv) can be formulated as $\partial_z S(0) = 0$, hence motivates a definition:

Definition 1 Let $c \mapsto \Upsilon(c)$ be the derivative of the concentration $S(z)$ at $z = 0$:

$$\Upsilon(c) = \partial_z S(0) . \tag{8}$$

The ultimate goal of this paper is to prove that this definition makes perfect sense, and that solving the equation $\Upsilon = 0$ in the range of admissible wave speeds is equivalent to solving problem (4). We also discuss **some counter-intuitive**

examples for which the function $\Upsilon(c)$ has multiple roots, or have no admissible root. Section 2 is dedicated to the so-called confinement problem: being given appropriate monotonicity of S and N, prove the existence of a unique normalized density function f solution of the kinetic equation. Here, we follow a numerical analyst's viewpoint, by decomposing the solution into a finite sum of Case's special functions, [1, 10]. Several properties of the solution are also established. Section 3 justifies the framework presented above, as it is proven that the spatial density ρ reaches a unique maximum, so does the chemical concentration S. A careful analysis of the shape of the velocity profiles is required there. Finally, three case studies are presented in Sect. 4, together with numerical simulations of the Cauchy problem.

2 Confinement by Biased Velocity-Jump Processes

We denote by $c_\star < c^\star$ resp. the infimum and the supremum of admissible velocities. Exact definitions are given below, during the course of analysis, see (16)–(19). Roughly speaking, if $c < c_\star$, the cell density is not confined on the right hand side. On the other hand, if $c > c^\star$, the cell density is not confined on the left-hand-side. Let $\mathscr{C} = (c_\star, c^\star) \setminus \{(v_k)_{k\in\mathscr{K}}\}$ be the set of admissible velocities. The **density of tumbling events per unit of time**, which appears together with ρ as a macroscopic quantity in (4), will play a major role in the subsequent analysis:

$$I(z) = \sum_{k\in\mathscr{K}} \omega_k T(z, v_k - c) f(z, v_k)\,, \tag{9}$$

We have the following result, adapted from [7, 8], but in a finite velocity setting.

Theorem 1 *Let $c \in \mathscr{C}$. There exists a unique positive function f with normalization* (3)*, such that for all $z \in \mathbb{R}$, and $k \in \mathscr{K}$,*

$$(v_k - c)\partial_z f(z, v_k) = \sum_{k'\in\mathscr{K}} \omega_{k'} T(z, v_{k'} - c) f(z, v_{k'}) - T(z, v_k - c) f(z, v_k)\,. \tag{10}$$

The functions $z \mapsto f(z, v_k)$ are exponentially decaying on both sides $z < 0$ and $z > 0$. In addition, we have the following asymptotic behaviour: there exist positive numbers $\lambda_+, \mu_+, \lambda_-, \mu_- > 0$, such that

$$\begin{cases} (\forall z > 0) & \dfrac{dI}{dz}(z) = -\mu_+ \exp(-\lambda_+ z)\,(1 + o(z)) \\ (\forall z < 0) & \dfrac{dI}{dz}(z) = \mu_- \exp(\lambda_- z)\,(1 + o(z)) \end{cases} \tag{11}$$

The dependence of the various constants with respect to the parameters is described in the proof. In particular, we keep track of the dependency with respect to the weights $\boldsymbol{\omega}$, as they will vary in Sect. 3. It is crucial to guarantee that the correction terms $o(z)$ in (11) are uniformly small for large $|z|$. In particular, this requires that **the extremal weight ω_K is bounded below by some positive constant**, see (35) below. Note that a similar result in higher dimension was established recently in [24], with a different approach. In the latter work, the description of the stationary distribution f is less explicit.

Proof The proof builds on the numerical analysis developed in [7, Sect. 7]. It is very much inspired from the study of discrete Case's modes for linear kinetic transport equations, see [20] and references therein. Accordingly, we seek the solution as **a combination of Case's modes on each side of the origin** $z = 0$. For this purpose, we define the cutting index $J \geq 0$, such that

$$v_{-K} < v_{-K+1} < (\dots) < v_J < c < v_{J+1} < (\dots) < v_K \,. \tag{12}$$

We make the following ansatz:

$$\begin{cases} (\forall z < 0)\,(\forall v_k) \quad f(z, v_k) = \displaystyle\sum_{j=-K}^{J} a_j \frac{\exp(-\lambda_j(c) z)}{T_-(v_k - c) - \lambda_j(c)(v_k - c)} \\ (\forall z > 0)\,(\forall v_k) \quad f(z, v_k) = \displaystyle\sum_{j=J+1}^{K} b_j \frac{\exp(-\lambda_j(c) z)}{T_+(v_k - c) - \lambda_j(c)(v_k - c)}\,, \end{cases} \tag{13}$$

where (a_j, b_j) are unknown coefficients. It is immediate to check that each mode in (13) is indeed a special solution if, and only if, $\lambda_j(c)$ is a root of the following **dispersion relation,**

$$\boxed{\sum_{k \in \mathcal{K}} \omega_k \frac{T_\pm(v_k - c)}{T_\pm(v_k - c) - \lambda(v_k - c)} = 1\,.} \tag{14}$$

This equation possesses the trivial solution $\lambda = 0$. However, it is excluded since we seek solutions which are integrable over the whole line. Otherwise, (14) is equivalent to

$$\sum_{k \in \mathcal{K}} \omega_k \left(\frac{T_\pm(v_k - c)}{v_k - c} - \lambda \right)^{-1} = 0\,. \tag{15}$$

The latter equation has exactly $K + J + 1$ negative solutions, associated with T_-, which are **interlaced** as follows,

$$- \infty < \frac{T_-^-}{v_J - c} < \lambda_J(c) < \frac{T_-^-}{v_{J-1} - c} < \lambda_{J-1}(c) < \frac{T_-^-}{v_{J-2} - c}$$
$$< (\dots) < \lambda_{-K-1}(c) < \frac{T_-^-}{v_{-K} - c} < \lambda_{-K}(c) < 0 \, .$$

This definition of the exponents $(\lambda_j(c))$ deserves some careful explanation. The fact that there exists a negative root between the last singular value $\frac{T_-^-}{v_{-K}-c}$ and value 0 is a consequence of the increasing monotonicity of (15) with respect to λ, and of the evaluation at $\lambda = 0$, namely

$$\sum_{k \in \mathcal{K}} \omega_k \frac{v_k - c}{T_-(v_k - c)} > 0 \, .$$

The latter expresses the fact that **the mean algebraic run length is positive**, which is obviously required for the confinement phenomenon. This defines an upper value for c: $c < c^\star$, where

$$\sum_{k \in \mathcal{K}} \omega_k \frac{v_k - c^\star}{T_-(v_k - c^\star)} = 0 \, . \tag{16}$$

Intuitively, the speed c cannot be too large, in order to ensure confinement on the left side. Obviously, in the extreme case where $c > v_K$, relative speeds $v_j - c$ are all negative, so there is no possible confinement! Similar conclusions can be drawn on the positive side $z > 0$: (15) has $K - J$ positive solutions, associated with T_+, which are interlaced as follows,

$$0 < \lambda_K(c) < \frac{T_+^+}{v_K - c} < \lambda_{K-1}(c) < \frac{T_+^+}{v_{K-1} - c}$$
$$< (\dots) < \lambda_{J+2}(c) < \frac{T_+^+}{v_{J+2} - c} < \lambda_{J+1}(c) < \frac{T_+^+}{v_{J+1} - c} < +\infty \, . \tag{17}$$

Again, the existence of a positive root below the singular value $\frac{T_+^+}{v_{J+1}-c}$ is guaranteed if, and only if, the mean algebraic run length is negative, namely

$$\sum_{k \in \mathcal{K}} \omega_k \frac{v_k - c}{T_+(v_k - c)} < 0 \, . \tag{18}$$

The latter prescribes a lower value for c: $c > c_\star$, where

$$\sum_{k\in\mathcal{K}} \omega_k \frac{v_k - c_\star}{T_+(v_k - c_\star)} = 0\,. \tag{19}$$

Remark 1 It is a consequence of condition (2) that $c_\star \leq 0$. Indeed, the mean algebraic run length is non positive when $c = 0$. As we seek wave solutions travelling to the right side, it is legitimate to restrict to $c \in (0, c^\star)$ in the present work.

In a second moment, we associate the number of degrees of freedom in (13) with the incoming data on each side of the origin. From (12), we deduce that there are $K + J + 1$ negative relative velocities $(v_j - c)_{j\in[-K,J]}$, and $K - J$ positive relative velocities $(v_j - c)_{j\in[J+1,K]}$. As a consequence, the solution on the left side $f(z, v)|_{z<0}$ is characterized by the incoming data at the origin,[1] i.e. the vector

$$G_- = \begin{pmatrix} f(0, v_{-K}) \\ \vdots \\ f(0, v_J) \end{pmatrix}.$$

The right side $f(z, v)|_{z>0}$ is characterized by incoming data at the origin, i.e. the vector

$$G_+ = \begin{pmatrix} f(0, v_{J+1}) \\ \vdots \\ f(0, v_K) \end{pmatrix}.$$

We may supposedly relate the incoming data, and the degrees of freedom in the decompositions (13), by square matrices, as they are in the same number. Alternatively, we can express compatibility conditions in (13) at $z = 0$ in the form of a transfer operator. For this we must identify both decompositions at $z = 0$:

$$(\forall v_k) \quad \sum_{j=-K}^{J} \frac{a_j}{T_-(v_k - c) - \lambda_j(c)(v_k - c)} = \sum_{j=J+1}^{K} \frac{b_j}{T_+(v_k - c) - \lambda_j(c)(v_k - c)}\,. \tag{20}$$

This is exactly equivalent to developing a fixed point argument on the right inflow data G_+ as in [7, 8]: suppose we are given G_+, we can invert the second part of (13) to find $(b_j)_{j\in[J+1,K]}$ by solving a $(K - J) \times (K - J)$ linear system. This prescribes a left inflow data G_-. Again, we can invert the first part of (13) to find $(a_j)_{j\in[-K,J]}$ by

[1] We refer to [7, 8] for a discussion between this characterization and the Milne problem in radiative transfer theory [5].

solving a $(K+J+1)\times(K+J+1)$ linear system. This yields in turn a right inflow data $\widetilde{G}_+$ that should coincide with G_+. This fixed point procedure is all contained in (20). The existence of a non zero vector such that

$$\begin{pmatrix} \dfrac{1}{T_-(v_k-c)-\lambda_j(c)(v_k-c)} \\ \dfrac{-1}{T_+(v_k-c)-\lambda_j(c)(v_k-c)} \end{pmatrix}_{(k,j)} \begin{pmatrix} a_j \\ \\ b_j \end{pmatrix}_{(j)} = 0\,, \tag{21}$$

is a direct consequence of mass conservation. Indeed, formulation (15) can be rewritten as

$$(\forall j) \quad \sum_{k\in\mathcal{K}} \omega_k \frac{v_k - c}{T_\pm(v_k-c)-\lambda_j(c)(v_k-c)} = 0\,. \tag{22}$$

This means exactly that the row vector $(\omega_k(v_k-c))_{(k)}$ belongs to the kernel of the adjoint problem of (21). We deduce the existence of non trivial coefficients (a_j, b_j). This yields a solution of the stationary problem (10) over the whole line. Positivity and uniqueness of f are both consequences of the ergodicity underlying the linear (decoupled) kinetic equation, see [7, 8]. The rest of the proof of Theorem 1 consists in a series of Lemma that establish appropriate bounds for the solution. This leads ultimately to the quantitative estimate (11). Contrary to the decomposition of the solution in Case's modes (13), we shall now use the **Duhamel formulation along characteristic lines**:

$$(\forall z>0)(\forall v_k<c) \quad f(z,v_k) = \int_0^{+\infty} I(z-s(v_k-c))\exp\left(-sT_+^-\right)ds \tag{23}$$

$$\begin{aligned}(\forall z>0)(\forall v_k>c) \quad f(z,v_k) &= f(0,v_k)\exp\left(-\frac{zT_+^+}{v_k-c}\right) \\ &\quad + \int_0^{\frac{z}{v_k-c}} I(z-s(v_k-c))\exp\left(-sT_+^+\right)ds\,. \end{aligned}\tag{24}$$

Similar formulas hold on the other side $(z<0)$.

Lemma 1 **(L^∞ bound)** *The function f is uniformly bounded, independently of the weight $\boldsymbol{\omega}$.*

Proof Firstly, let recall that the solution f is normalized to have unit mass (3):

$$\int_{\mathbb{R}} \rho(z)\,dz = \int_{\mathbb{R}} \sum_{k\in\mathcal{K}} \omega_k f(z,v_k)\,dz = 1\,.$$

As a by-product, the macroscopic quantity $I(z)$ (9), which is easily comparable with $\rho(z)$, has a uniformly bounded integral over $\mathbb{R}$. We immediately deduce from the Duhamel formula (23), that $f(z, v_k)$ is uniformly bounded for $z > 0$ and $v_k < c$:

$$
\begin{aligned}
f(z, v_k) &= \int_0^{+\infty} I(z+y) \exp\left(-\frac{yT_+^+}{c - v_k}\right) \frac{1}{c - v_k}\, dy \\
&\leq \frac{1}{c - v_k} \int_{\mathbb{R}} I(z)\, dz \\
&\leq \frac{\max T}{c - v_k} \,. \qquad (25)
\end{aligned}
$$

Similar estimate holds true for $z < 0$ and $v_k > c$. Therefore, $f(0, v_k)$ is uniformly bounded for all k, independently of the weight $\boldsymbol{\omega}$.

The same bounds can be propagated to any $z > 0$ (resp. $z < 0$) using (24). □

The decomposition (13) is a nice characterization of the density over $\{z < 0\}$ and $\{z > 0\}$, respectively. The coefficients (a_j, b_j) were not specified in (13). However, they are in relation with the profile at $z = 0$ as in the following lemma.

Lemma 2 (Expression of b_i's) *Each coefficient b_i in* (13) *is given by the following orthogonality formula:*

$$
b_i = \frac{\displaystyle\sum_{k \in \mathcal{K}} \omega_k (v_k - c) f(0, v_k) \frac{T_+(v_k - c)}{T_+(v_k - c) - \lambda_i(c)(v_k - c)}}{\displaystyle\sum_{k \in \mathcal{K}} \omega_k (v_k - c) \frac{T_+(v_k - c)}{\left(T_+(v_k - c) - \lambda_i(c)(v_k - c)\right)^2}} \,. \qquad (26)
$$

Proof We take the scalar product of the second line in (13) and the dual eigenvector indexed by i, $\frac{T_+(v_k - c)}{T_+(v_k - c) - \lambda_i(c)(v_k - c)}$, with the weights $(\omega_k(v_k - c))$.[2] We realize that crossed terms cancel for $j \neq i$:

$$
\begin{aligned}
&\sum_{k \in \mathcal{K}} \omega_k (v_k - c) \frac{T_+(v_k - c)}{\left(T_+(v_k - c) - \lambda_j(v_k - c)\right)\left(T_+(v_k - c) - \lambda_i(v_k - c)\right)} \\
&= \frac{1}{\lambda_i - \lambda_j} \sum_{k \in \mathcal{K}} \omega_k \frac{\left(\lambda_i - \lambda_j\right)(v_k - c) + T_+(v_k - c) - T_+(v_k - c)}{\left(T_+(v_k - c) - \lambda_j(v_k - c)\right)\left(T_+(v_k - c) - \lambda_i(v_k - c)\right)} T_+(v_k - c) \\
&= \frac{1}{\lambda_i - \lambda_j} \left(- \sum_{k \in \mathcal{K}} \omega_k \frac{T_+(v_k - c)}{T_+(v_k - c) - \lambda_j(v_k - c)} + \sum_{k \in \mathcal{K}} \omega_k \frac{T_+(v_k - c)}{T_+(v_k - c) - \lambda_i(v_k - c)} \right) \\
&= 0 \,.
\end{aligned}
$$

[2]Notice that these weights are signed.

The cancellation holds true due to the dispersion relation (14) which is common to λ_i and λ_j. Hence, only the contribution indexed by i remains after multiplication. This yields (26). □

Next, we can better rewrite (26) with positive weights, instead of signed weights, by using the zero flux condition $\sum(v_k - c)f(0, v_k) = 0$, and (22):

$$b_i = \frac{\displaystyle\sum_{k\in\mathcal{K}} \omega_k(v_k - c)f(0, v_k)\frac{\lambda_i(c)(v_k - c)}{T_+(v_k - c) - \lambda_i(c)(v_k - c)}}{\displaystyle\sum_{k\in\mathcal{K}} \omega_k(v_k - c)\frac{\lambda_i(c)(v_k - c)}{(T_+(v_k - c) - \lambda_i(c)(v_k - c))^2}}$$

$$= \frac{\displaystyle\sum_{k\in\mathcal{K}} \omega_k f(0, v_k)\frac{(v_k - c)^2}{T_+(v_k - c) - \lambda_i(c)(v_k - c)}}{\displaystyle\sum_{k\in\mathcal{K}} \omega_k\frac{(v_k - c)^2}{(T_+(v_k - c) - \lambda_i(c)(v_k - c))^2}}. \tag{27}$$

Interestingly, this suggests to use another scalar product, in order to derive appropriate bounds for the coefficients (b_i). This is the purpose of the next lemma.

Lemma 3 (Bound of b_i's) *Each coefficient b_i in* (13) *is bounded in the following way:*

$$(\forall i \in [J+1, K])\ (\forall k \in \mathcal{K}) \qquad \frac{|b_i||v_k - c|}{|T_+(v_k - c) - \lambda_i(c)(v_k - c)|} \le \frac{\max T}{\sqrt{\omega_k}}. \tag{28}$$

Proof For any $z > 0$, let compute the weighted ℓ^2 norm of $f(z, v_k)$ as follows,

$$\|f(z, v_k)\|^2_{\ell^2(\omega_k(v_k-c)^2)} = \sum_{k\in\mathcal{K}} \omega_k(v_k - c)^2 f(z, v_k)^2$$

$$= \sum_{i,j\in[J+1,K]^2} b_i b_j \exp(-(\lambda_i(c) + \lambda_j(c))z)$$

$$\times \sum_{k\in\mathcal{K}} \omega_k(v_k - c)\frac{(v_k - c)}{\left(T_+(v_k - c) - \lambda_j(c)(v_k - c)\right)\left(T_+(v_k - c) - \lambda_i(c)(v_k - c)\right)}$$

$$= \sum_{i\in[J+1,K]} b_i^2 \exp(-2\lambda_i(c)z) \sum_{k\in\mathcal{K}} \omega_k\frac{(v_k - c)^2}{(T_+(v_k - c) - \lambda_i(c)(v_k - c))^2},$$

because the cross terms vanish, exactly as in the proof of Lemma 2.

We deduce from this estimate at $z = 0$, that for all $i \in [J+1, K]$, and all $k \in \mathscr{K}$, we have

$$\sqrt{\omega_k}\frac{|b_i||v_k - c|}{|T_+(v_k - c) - \lambda_i(c)(v_k - c)|} \leq \max_{k' \in \mathscr{K}} \left(f(0, v_{k'})|v_{k'} - c|\right) .$$

This yields (28) by (25). □

A fruitful consequence of the latter estimate concerns the asymptotic behaviour of f, as $z \to +\infty$. Characterization (13) is now of great interest. Indeed, the large space behaviour is characterized by the slowest mode (17). Loosely speaking,

$$(\forall v_k) \quad f(z, v_k) \underset{z \to +\infty}{\sim} \frac{b_K \exp(-\lambda_K(c)z)}{T_+(v_k - c) - \lambda_K(c)(v_k - c)} . \tag{29}$$

Furthermore, the asymptotic monotonicity is also intuitively clear from this decomposition:

$$(\forall v_k) \quad \partial_z f(z, v_k) \underset{z \to +\infty}{\sim} \frac{-\lambda_K(c) b_K \exp(-\lambda_K(c)z)}{T_+(v_k - c) - \lambda_K(c)(v_k - c)} . \tag{30}$$

For upcoming purposes, it is necessary to gain some quantitative control about the prefactors in the r.h.s. of (29) and (30). This is the aim of the next lemma.

Lemma 4 (Bound of b_K from Below) *There exists a constant C, depending on ω, such that $b_K \geq 1/C$.*

Proof We extend easily the orthogonality formula (27) to any $z > 0$:

$$b_i \exp(-\lambda_i(c)z) = \sum_{k \in \mathscr{K}} \left[f(z, v_k)\left(T_+(v_k - c) - \lambda_i(c)(v_k - c)\right)\right] \varphi_{k+}(\lambda_i(c)) ,$$

where the weights $\varphi_{k+}(\lambda_i(c)) > 0$ are such that $\sum \varphi_{k+} = 1$:

$$\varphi_{k+} = \frac{1}{Z_+}\frac{(v_k - c)^2}{(T_+(v_k - c) - \lambda_i(c)(v_k - c))^2} ,$$

$$Z_+ = \sum_{k' \in \mathscr{K}} \omega_{k'} \frac{(v_{k'} - c)^2}{(T_+(v_{k'} - c) - \lambda_i(c)(v_{k'} - c))^2} .$$

For $i = K$, we deduce, after integration over $\mathbb{R}_+$, that

$$\frac{b_K}{\lambda_K(c)} \geq \left(\min_{k \in \mathscr{K}} \frac{T_+(v_k - c) - \lambda_K(c)(v_k - c)}{\omega_k}\right) \int_{\mathbb{R}_+} \rho(z)\, dz . \tag{31}$$

Here, we have used the peculiar property of index K, which is such that $\lambda_K(c)(v_k - c) < T_+(v_k - c)$ for all $k \in \mathscr{K}$ (17). The minimum value is uniformly bounded below by (14). We could conclude from the unit mass normalization, provided that the integral would be taken over $\mathbb{R}$ in (31). To overcome this small issue, it is necessary to connect both sides $\{z > 0\}$ and $\{z < 0\}$ at some point. This is a consequence of the uniform comparison between the two values b_K and a_{-K}. To this end, we notice that the identity (27) can be recast, for $i = K$, as

$$b_K = \| f(0, v_k) \left(T_+(v_k - c) - \lambda_K(c)(v_k - c)\right) \|_{\ell^1(\varphi_+)} \,. \tag{32}$$

On the other side, we have similarly

$$a_{-K} = \| f(0, v_k) \left(T_-(v_k - c) - \lambda_{-K}(c)(v_k - c)\right) \|_{\ell^1(\varphi_-)} \,, \tag{33}$$

with the appropriate choice for the probability weights φ_-:

$$\varphi_{k-} = \frac{1}{Z_-} \frac{\omega_k (v_k - c)^2}{\left(T_-(v_k - c) - \lambda_{-K}(c)(v_k - c)\right)^2} \,,$$

$$Z_- = \sum_{k \in \mathscr{K}} \frac{\omega_k (v_k - c)^2}{\left(T_-(v_k - c) - \lambda_{-K}(c)(v_k - c)\right)^2} \,.$$

To compare the two ℓ^1 norms (32) and (33), we shall establish the following bounds: there exists a constant $C > 0$, depending on $\boldsymbol{\omega}$ such that

$$(\forall k) \quad \frac{1}{C} \leq \frac{\left(T_+(v_k - c) - \lambda_K(c)(v_k - c)\right) \varphi_{k+}}{\left(T_-(v_k - c) - \lambda_{-K}(c)(v_k - c)\right) \varphi_{k-}} \leq C \,.$$

On the one hand, it is easy to bound the quantities $T_\pm(v_k - c) - \lambda_{\pm K}(c)(v_k - c)$ from above and from below, uniformly with respect to k: In fact, it is sufficient to control uniformly the following piece of estimate:

$$\lambda_K(c) < \frac{T_+^+}{v_K - c} \,. \tag{34}$$

But, we deduce from (14) that

$$\omega_K \frac{T_+^+}{T_+^+ - \lambda_K(c)(v_K - c)} \leq 1 \quad \Rightarrow \quad T_+^+ - \lambda_K(c)(v_K - c) \geq \omega_K T_+^+ \,. \tag{35}$$

This gives the required estimate.

Remark 2 Importantly, the bound from below in (35) degenerates only if ω_K vanishes. The fact that it degenerates as ω_K vanishes is quite obvious, as the velocities $v_{\pm K}$ effectively disappear from the problem. The limit system keeps at most $2K-1$ velocities. The crucial point is that it does not depend upon the other weights.

On the other hand, we have

$$\frac{\varphi_{k+}}{\varphi_{k-}} = \frac{Z_-}{Z_+}\frac{(T_+(v_k-c)-\lambda_K(c)(v_k-c))^2}{(T_-(v_k-c)-\lambda_{-K}(c)(v_k-c))^2}\,.$$

It is a consequence of (35) that all terms in this fraction are bounded from above and below. Finally, combining both estimates (31) and the similar one for $\{z<0\}$:

$$\frac{a_{-K}}{\lambda_{-K}(c)} \geq \left(\min_{k\in\mathscr{K}} \frac{T_-(v_k-c)-\lambda_{-K}(c)(v_k-c)}{\omega_k}\right)\int_{\mathbb{R}_-}\rho(z)\,dz\,,$$

together with the normalization $\int_{\mathbb{R}_-}\rho(z)\,dz + \int_{\mathbb{R}_+}\rho(z)\,dz = 1$, we get that there exists a constant C, depending on ω_K, such that

$$\max\,(b_K, a_{-K}) \geq \frac{1}{C}\,.$$

Since b_K, and a_{-K} are uniformly comparable by previous estimates, we conclude that they are both bounded from below by some positive constant. □

Finally, we combine all these estimates to determine the asymptotic behaviour of the derivative of I.

Lemma 5 (Asymptotic Monotonicity of I) *The derivative of the macroscopic quantity I satisfies the following quantitative estimate:*

$$(\forall z>0)\quad \frac{dI}{dz}(z) = -\lambda_K(c)b_K\exp(-\lambda_+(c)z)\,(1+o(z))\,. \tag{36}$$

Proof For $z>0$, we have,

$$\begin{aligned}
\frac{dI}{dz}(z) &= \sum_{k\in\mathscr{K}} \omega_k T_+(v_k-c)\partial_z f(z,v_k)\\
&= -\sum_{k\in\mathscr{K}} \omega_k T_+(v_k-c)\sum_{j=J+1}^{K} b_j\lambda_j(c)\frac{\exp(-\lambda_j(c)z)}{T_+(v_k-c)-\lambda_j(c)(v_k-c)}\\
&= -\lambda_K(c)b_K\exp(-\lambda_K(c)z)\sum_{k\in\mathscr{K}}\omega_k\frac{T_+(v_k-c)}{T_+(v_k-c)-\lambda_K(c)(v_k-c)}\\
&\quad + K(K-J)\exp(-\lambda_{K-1}(c)z)\max_{(k,j)\in\mathscr{K}\times[J+1,K-1]}\left|\frac{\omega_k b_j\lambda_j(c)T_+(v_k-c)}{T_+(v_k-c)-\lambda_j(c)(v_k-c)}\right|\,.
\end{aligned}$$

We deduce from the identity satisfied by $\lambda_K(c)$ (14) (for the first contribution), and from the bound (28) together with the trivial inequality $\omega_k \leq \sqrt{\omega_k}$ (for the second contribution), that

$$\frac{dI}{dz}(z) = -\lambda_K(c) b_K \exp(-\lambda_K(c)z) + \mathscr{O}\left(\exp(-\lambda_{K-1}(c)z)\right) .$$

The estimate (36) follows since $\lambda_K(c)$ and b_K are positive, and also $\lambda_{K-1}(c) > \lambda_K(c)$ (17). □

This lemma concludes the proof of Theorem 1. Before we move to the next Section, let us make the following important observations:

1. The bound $b_K > 0$ was carefully analysed in Lemma 4. It is uniform provided that ω_K is bounded below by some positive constant. The same observation holds true for the separation estimate $\lambda_{K-1}(c) > \lambda_K(c)$, by (17)–(34)–(35)
2. The bound $\lambda_K(c) > 0$ will be analysed carefully in Lemma 8 below. □

3 Monotonicity of the Spatial Density

An important issue related to the existence of travelling waves for the coupled problem (4) is the monotonicity of the spatial density ρ. In this section, we establish the following result, without any restriction on the parameters, except (2).

Theorem 2 *Let $c \in \mathscr{C}$. Let f be the function defined in Theorem 1. The spatial density $\rho(z) = \sum_{k\in\mathscr{K}} \omega_k f(z, v_k)$ changes monotonicity at $z = 0$: it is increasing for $z < 0$, and decreasing for $z > 0$.*

Before presenting the proof of this statement, let us comment on the strategy. We present here the adaptation of the argument proposed in [7], but in the case of finite velocities. The version developed here is much simpler because many regularity issues can be overcome. However, we insist on the fact that [7] does not readily contain the finite velocity case.

- The first observation is that the monotonicity claimed in Theorem 2 is unlikely to be a straightforward consequence of the eigenmode decompositions (13), as all terms are changing signs.[3] More strikingly, the functions $z \mapsto f(z, v_k)$ does not have the required monotonicity for fixed v_k, see Fig. 3. Only the velocity average $\rho(z) = \sum \omega_k f(z, v_k)$ possesses the appropriate monotonicity.
- As an alternative, we establish that the spatial density ρ cannot change monotonicity as the weights (ω_k) vary continuously. This procedure is initialized with the case of two velocities only (all weights ω_k are set to zero except the extremal

[3] Moreover, we have no argument so far to determine the signs of the coefficients (a_j, b_j), if they have any.

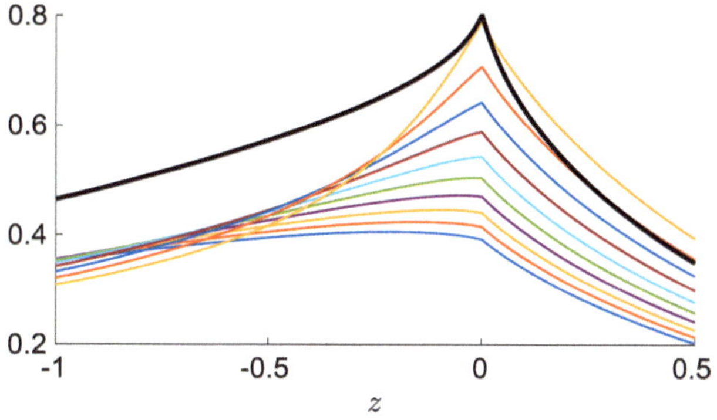

Fig. 3 Overshoot phenomena. The spatial density is plotted in *bold (black) line*. It changes monotonicity at $z = 0$, as claimed in Theorem 2. The various functions $z \mapsto f(z, v_k)$ are plotted in *coloured lines* for each $v_k < 0$. For the most negative velocities, the maximum of the function is reached at some negative z. Here, $\chi_N = 0.2$, $\chi_S = 0.48$, and $c = 0.25$

ones ω_K). Monotonicity is obvious in the latter case since the solution is explicit as the concatenation of decaying exponential functions on each side (13).

- The key argument, that we called enhancement of monotonicity, resembles a maximum principle. Essentially, we prove that, if monotonicity of ρ is appropriate, but in the large (non decreasing for $z < 0$, and non increasing for $z > 0$), then monotonicity is necessarily strict, as stated in Theorem 2. The proof goes through a refined description of the velocity distribution for any z. This central argument (Lemma 7) is complemented with various compactness estimates, and some connectedness argument.

Proof Let denote $\mathscr{K}^+ = \{k : v_k > c\}$ the set of positive relative velocities, and $\mathscr{K}^- = \{k : v_k < c\}$ the set of negative relative velocities. We introduce

$$\begin{cases} \rho^+(z) = \sum_{k\in\mathscr{K}^+} \omega_k f(z, v_k)\,, \\ \rho^-(z) = \sum_{k\in\mathscr{K}^-} \omega_k f(z, v_k)\,, \end{cases}$$

as being the contributions of positive and negative relative velocities to the spatial density, respectively. Finally, we recall the definition of the other macroscopic quantity of interest, which appears in the kinetic equation (4):

$$I(z) = \sum_{k\in\mathscr{K}} \omega_k T(z, v_k - c) f(z, v_k)\,. \tag{37}$$

We make two key observations:

- Firstly, we have the following elementary reconstruction

$$I(z) = \begin{cases} T_-^- \rho^-(z) + T_-^+ \rho^+(z) & \text{for } z < 0\,, \\ T_+^- \rho^-(z) + T_+^+ \rho^+(z) & \text{for } z > 0\,. \end{cases}$$

  Identical monotonicity of both ρ^+ and ρ^- implies the same monotonicity for I.
- Secondly, the Duhamel formulation along characteristic lines (23)–(24) enables to reconstruct the kinetic density f from the spatial density I. We deduce the following important information from (23) for negative relative velocities: **If I is non increasing for $z > 0$, and not constant,[4] then $z \mapsto f(z, v_k)$ is decreasing for $z > 0$ and $v_k < c$.**

As byproduct, we deduce that $f(z, v_k)$ is increasing with respect to velocity on $\mathcal{K}^-$:

$$(\forall (v_i, v_j) \in \mathcal{K}^- \times \mathcal{K}^-) \quad (v_i < v_j) \Rightarrow f(z, v_i) < f(z, v_j) < \frac{I(z)}{T_+^-}\,. \tag{38}$$

Similar result holds true with opposite signs on $\{z < 0\} \times \mathcal{K}^+$:

$$(\forall (v_i, v_j) \in \mathcal{K}^+ \times \mathcal{K}^+) \quad (v_i < v_j) \Rightarrow \frac{I(z)}{T_-^+} > f(z, v_i) > f(z, v_j)\,. \tag{39}$$

The next Lemma describes the shape of the velocity profiles on $\mathcal{K}^+$ for $z > 0$.

Lemma 6 (Qualitative Behaviour of the Stationary Density) *Assume that I is non increasing on $\{z > 0\}$, and non decreasing on $\{z < 0\}$. Let $z > 0$. For all $v_i > c$, we face the following alternative: either $T_+^+ f(z, v_i) \geq I(z)$, or for all $v_j > v_i$, $f(z, v_j) < f(z, v_i)$. Alternatively speaking, on the right hand side ($z > 0$), f is decreasing with respect to velocity on the subset $\mathcal{K}^+ \cap \{T_+^+ f < I\}$.*

Proof We introduce the notation $\mathfrak{f} = Tf$. From (24), we deduce

$$\begin{aligned} \mathfrak{f}(z+h, v_k) - I(z+h) &= (\mathfrak{f}(z, v_k) - I(z)) \exp\left(-\frac{hT_+^+}{v_k - c}\right) \\ &\quad + (I(z) - I(z+h)) \exp\left(-\frac{hT_+^+}{v_k - c}\right) \\ &\quad + \int_0^{\frac{h}{v_k - c}} (I(z+h-s(v_k-c)) - I(z+h)) \exp\left(-sT_+^+\right)\,. \end{aligned} \tag{40}$$

[4] This is clearly the case due to integrability over $\mathbb{R}$.

As a consequence, if $f(z, v_k) > I(z)$, then for all $h > 0$, $f(z+h, v_k) > I(z+h)$, because the last two contributions in (40) are non negative by assumption. On the contrary, if $f(z, v_k) < I(z)$, then for all $y \in (0, z)$, $f(z-y, v_k) < I(z-y)$.

To characterize monotonicity with respect to velocity, we get back to the kinetic transport formulation (4). Let $v_i > c$ such that $f(z, v_i) < I(z)$. Let $v_j > v_i$, then,

$$(v_j - c)\partial_z \left(f(z, v_j) - f(z, v_i)\right) + (v_j - v_i)\partial_z f(z, v_i) = -T_+^+ \left(f(z, v_j) - f(z, v_i)\right) .$$

Denoting by $g(z) = \frac{f(z,v_j)-f(z,v_i)}{v_j-v_i}$, we have:

$$(v_j - c)\frac{dg}{dz}(z) + T_+^+ g(z) = -\partial_z f(z, v_i) = \frac{1}{v_i - c}\left(f(z, v_i) - I(z)\right) .$$

We deduce,

$$g(z) = g(0)\exp\left(-\frac{zT_+^+}{v_j - c}\right)$$
$$+ \frac{1}{v_i - c}\int_0^{\frac{z}{v_j - c}} \left[f(z - s(v_j - c), v_i) - I(z - s(v_j - c))\right]\exp\left(-sT_+^+\right) ds .$$

Both contributions in the right hand side are negative. Indeed, we have $g(0) < 0$ by (39), and $f(z-y, v_i) < I(z-y)$ for $y \in (0, z)$ since $f(z, v_i) < I(z)$. In conclusion,

$$\left(f(z, v_i) < I(z)\right) \quad \Rightarrow \quad (\forall v_j > v_i) \quad f(z, v_j) < f(z, v_i) . \qquad \square$$

The properties stated in Lemma 6 enable to decipher the compensations in the velocity average $\rho(z) = \sum_{k\in\mathscr{K}} \omega_k f(z, v_k)$, yielding appropriate monotonicity for ρ.

Lemma 7 (Enhancement of Monotonicity) *Assume that I is non increasing on $\{z > 0\}$, and non decreasing on $\{z < 0\}$. Then both ρ^+ and ρ^- are decreasing on $\{z > 0\}$, and both ρ^+ and ρ^- are increasing on $\{z < 0\}$.*

Proof We compute first the derivative of ρ^- on $\{z > 0\}$:

$$\frac{d\rho^-}{dz}(z) = \sum_{k\in\mathscr{K}-} \omega_k \partial_z f(z, v_k) = \sum_{k\in\mathscr{K}-} \omega_k \frac{I(z) - T_+^- f(z, v_k)}{v_k - c} .$$

We deduce from (38) that each term in the sum is negative, because $v_k < c$ there. On the other hand, the derivative of ρ^+ reads as follows,

$$\frac{d\rho^+}{dz}(z) = \sum_{k\in\mathscr{K}+} \omega_k \partial_z f(z, v_k) = \sum_{k\in\mathscr{K}+} \omega_k \frac{I(z) - T_+^+ f(z, v_k)}{v_k - c} .$$

The key observation is that we can omit the decreasing weights $(v_k - c)^{-1}$ in the last sum:

$$\frac{d\rho^+}{dz}(z) \leq \sum_{k\in\mathscr{K}^+} \omega_k \left(I(z) - T_+^+ f(z, v_k)\right) = - \sum_{k\in\mathscr{K}^-} \omega_k \left(I(z) - T_+^- f(z, v_k)\right) < 0\,.$$

The last identity is a consequence of the very definition of I (37). In order to establish the inequality, we notice that the cumulative sum

$$H_k(z) = \sum_{j=J+1}^{k} \omega_j \left(I(z) - T_+^+ f(z, v_j)\right)\,,$$

satisfies the following properties:

$$H_J(z) = 0\,, \quad \text{and} \quad (\forall k > J) \quad H_k(z) < 0\,.$$

Indeed, for fixed $z > 0$, the sequence $\left(\omega_j \left(I(z) - T_+^+ f(z, v_j)\right)\right)_{j\in\mathscr{K}^+}$ has the following pattern: the terms are first negative, then possibly positive. The reason is that the sequence is increasing as soon as it becomes positive (Lemma 6). Hence, the sequence $(H_k(z))_{k\in\mathscr{K}^+}$ has the following pattern: it is first decreasing, then possibly increasing. As the overall sum $H_K(z)$ is negative by the very definition of I, we deduce that each intermediate cumulative summation gives a negative value: for all $k \in \mathscr{K}^+$, $H_k(z) < 0$. Consequently, we have after summation by parts,

$$\begin{aligned}\frac{d\rho^+}{dz}(z) &= \sum_{k=J+1}^{K} \frac{1}{v_k - c} (H_k - H_{k-1}) \\ &= \sum_{k=J+1}^{K-1} \left(\frac{1}{v_k - c} - \frac{1}{v_{k+1} - c}\right) H_k(z) + \frac{H_K(z)}{v_K - c} < 0\,. \qquad \square\end{aligned}$$

Lemma 7 is very useful to prove the monotonicity result stated in Theorem 2. To this end, we make the weights (ω_k) vary continuously from any initial configuration $\boldsymbol{\omega}^0 = \omega = (\omega_k)$ to the final state $\boldsymbol{\omega}^1 = \left(\frac{1}{2}, 0, \ldots, 0, \frac{1}{2}\right)$. For $s \in [0, 1]$, let define f^s the solution of (4) associated with the weight $\boldsymbol{\omega}^s = (1 - s)\boldsymbol{\omega}^0 + s\boldsymbol{\omega}^1$, having normalized mass (3).

Standard arguments enable to prove that the map $s \mapsto f^s$ is continuous for the topology of uniform convergence: Firstly, the solution of (4) is unique for a given set of weights. Secondly, the function is uniformly bounded and Lipschitz continuous. Lastly, it is uniformly small outside a compact interval $[-L, L]$, see Sect. 2.

These statements require some justifications, based on the results established in Sect. 2. As discussed previously, we are able to prove in a quantitative way the asymptotic behaviour (29), provided that we can bound b_K and $\lambda_K(c)$ from below, as well as the spectral gap $\lambda_{K-1}(c) - \lambda_K(c) > 0$. The bounds on b_K and $\lambda_{K-1}(c) - \lambda_K(c)$

rely on the non degeneracy of the extremal weight ω_K. This is guaranteed uniformly along the sequence of weights ω^s, as we have, by definition,

$$(\forall s \in [0,1]) \quad \omega_K^s \geq \min\left(\omega_K, \frac{1}{2}\right).$$

The bound on $\lambda_K(c)$ requires some additional argument, related to the confinement property. Indeed, the confinement by biased velocity-jump processes is equivalent to the inequality $c < c^\star$. However, c is fixed here, but $c^\star$ implicitly depends on the weights ω^s, which are not constant. Thus, it is mandatory to guarantee that confinement holds true along the sequence of weights (ω^s). This is the purpose of the next lemma. Let us emphasize that we use here the condition $\chi_N \leq \chi_S$ (2). We believe this restriction is not needed to ensure the final result. However, it simplifies the proof, as the sequence of weights must be redefined in a delicate way if $\chi_N > \chi_S$. In particular, the final state should not charge the extremal velocities $v_{\pm K}$, but rather some intermediate weights.

Lemma 8 (Uniform Confinement Along the Sequence) *Under condition* (2)*, the eigenvalue $\lambda_K(c)$ is uniformly bounded from below, uniformly for s in* $[0,1]$.

Proof As s belongs to a compact interval, it is sufficient to establish that $\lambda_K(c)$ remains positive for all s. The confinement on the right hand side is guaranteed for all s, provided that the mean algebraic run length is negative (18):

$$\sum_{k \in \mathcal{K}} \omega_k^s \frac{v_k - c}{T_+(v_k - c)} < 0.$$

As this expression is linear with respect to s, and negative at $s = 0$ by assumption, it is sufficient to check that the final value at $s = 1$ is smaller than the initial value at $s = 0$. Hence, we are reduced to establish the following inequality:

$$\frac{1}{2}\left(\frac{v_K - c}{T_+^+} + \frac{v_{-K} - c}{T_+^-}\right) \leq \sum_{k \in \mathcal{K}} \omega_k \frac{v_k - c}{T_+(v_k - c)} \tag{41}$$

We may recombine this inequality by factoring out $1/T_+^+$, on the one hand, and $1/T_+^-$, on the other hand:

$$\frac{1}{T_+^+}\left(\frac{1}{2}(v_K - c) - \sum_{k>J} \omega_k(v_k - c)\right) \leq \frac{1}{T_+^-}\left(-\frac{1}{2}(v_{-K} - c) + \sum_{k \leq J} \omega_k(v_k - c)\right).$$

We claim that both factors are equal and non negative:

$$\frac{1}{2}(v_K - c) - \sum_{k>J} \omega_k(v_k - c) = -\frac{1}{2}(v_{-K} - c) + \sum_{k \leq J} \omega_k(v_k - c) \geq 0.$$

This is an immediate consequence of $\sum \omega_k = 1$, and $\sum \omega_k v_k = 0$. The positive sign can be viewed easily on the left hand side: $\sum_{k>J} \omega_k(v_k - c) \leq (v_K - c)\sum_{k>J} \omega_k \leq (v_K - c)/2$. We conclude that (41) holds true, since $T_+^- < T_+^+$ by (2) and (7).

The same arguments lead to the opposite inequality on the left hand side, in order to ensure uniform confinement there as well:

$$(\forall s) \quad \sum_{k \in \mathscr{K}} \omega_k^s \frac{v_k - c}{T_-(v_k - c)} > 0 . \qquad \square$$

We are now in position to conclude the proof of Theorem 2. Lemma 5 is useful to compactify the space interval, since the appropriate monotonicity is guaranteed for $|z| > L$, uniformly with respect to $s \in [0, 1]$. Let $\mathscr{S}$ be the set of values s for which I^s has the appropriate monotonicity:

$$\mathscr{S} = \left\{ s \in [0, 1] \mid (\forall z \neq 0) \ (\text{sign } z) \frac{dI^s}{dz}(z) \leq 0 \right\} . \tag{42}$$

It is immediate to see that $\mathscr{S}$ contains the extremal value $s = 1$, simply because the eigenmode decompositions (13) are reduced to a single element on each side. Alternatively speaking, the solution is an exponential function on each side, so it is monotonic. Confinement (Lemma 8) guarantees that the exponential functions decay on both sides, so it has the appropriate monotonicity (42).

Theorem 2 is equivalent to say that the value $s = 0$ belongs to $\mathscr{S}$, also. We argue by connectedness.

- Firstly, $\mathscr{S}$ is open. Let $s_0 \in \mathscr{S}$. Lemma 7 says that monotonicity is strict on both sides $\{z < 0\}$, and $\{z > 0\}$. We claim that there exist $\varepsilon_0 > 0$, and a neighbourhood $\mathscr{V}_0$ of s_0 in $\mathscr{S}$, such that

$$(\forall s \in \mathscr{V}_0) \ (\forall z \in [-L, L] \setminus \{0\}) \quad (\text{sign } z) \frac{dI^s}{dz}(z) < -\varepsilon_0 .$$

  We argue by contradiction: otherwise, there would exist a sequence $s_n \to s_0$, and a sequence z_n with constant sign, converging towards some $z_0 \in [-L, L]$, including value 0, such that

$$\lim_{n \to +\infty} (\text{sign } z_n) \frac{dI^{s_n}}{dz}(z_n) = 0 .$$

  We can pass to the limit in the sequence of derivative functions $\frac{dI^{s_n}}{dz}$, uniformly over $[-L, 0]$ or $[0, L]$.[5] This yields a contradiction with the strict monotonicity at $s = s_0$.

---

[5]Notice that the value $z = 0$ is not an issue here. In fact, Lemma 7 includes the values $z = 0^+$ and $z = 0^-$ which must be distinguished from each other because I is not continuous at $z = 0$.

- Secondly, $\mathscr{S}$ is closed. This is an easy consequence of the continuity of the sequence $s \mapsto f^s$, for the topology of uniform convergence, which is compatible with monotonicity properties. □

4 Coupling with Reaction-Diffusion Equations: Three Cases

As soon as the existence of a function f is established for a given c, as in Theorem 1, with appropriate monotonicity properties as in Theorem 2, **the coupling with the reaction-diffusion equations through ρ in the travelling wave problem** (4) **is essentially not sensitive to the topology of the velocity space** (discrete as in here, or continuous as in [7]). Therefore, we do not repeat the analysis performed in [7], but we recall two useful propositions.

Proposition 1 *Assume that the function $\rho \in L^1$ is locally Lipschitz continuous on $\mathbb{R}^*$, and that it is increasing for $z < 0$, and decreasing for $z > 0$. Let S be the unique solution of the following elliptic problem,*

$$(\forall z \in \mathbb{R}) \quad -c\partial_z S(z) - D_S \partial_z^2 S(z) + \alpha S(z) = \beta\rho(z) \,.$$

Then S is unimodal, meaning that $\partial_z S$ vanishes once, and only once.

Proposition 2 *Assume that $c > 0$, and that ρ is exponentially bounded on both sides, namely*

$$\begin{cases} (\forall z < 0) & \rho(z) \leq C \exp(\lambda_- z) \\ (\forall z > 0) & \rho(z) \leq C \exp(-\lambda_+ z) \end{cases}$$

Then, there exist two constants N_-, N_+, ordered as $0 < N_- < N_+$, and a solution N of the following elliptic problem

$$-c\partial_z N(z) - D_N \partial_z^2 N(z) = -\gamma\rho(z)N(z) \,,$$

such that

$$\begin{cases} \lim_{z \to -\infty} N(z) = N_- \\ \lim_{z \to +\infty} N(z) = N_+ \end{cases}$$

Moreover, for all z in $\mathbb{R}$, $\partial_z N(z) > 0$.

We deduce from these two propositions that the framework emphasized in the Introduction is consistent, except that in point (i) it is required that the maximum of S is located at $z = 0$. The latter requirement is not true in general. In fact, it is intuitively clear that the location of the maximum depends on the value of c, which is the main unknown of the problem (4).

The last step of the construction of travelling waves consists in varying c to meet this last requirement about the location of the maximum at the origin. This is equivalent to say that the function Υ (8) has a root $\mathbf{c}$ in the interval of admissible velocities $(c_\star, c^\star)\cap(0, c^\star)$.[6] In the present work, we have $(c_\star, c^\star)\cap(0, c^\star) = (0, c^\star)$, because $c_\star \leq 0$ under condition (2).

Surprisingly enough, neither the existence nor the uniqueness of a root $\mathbf{c}$ such that $\Upsilon(\mathbf{c}) = 0$ do hold for any set of parameters (this is the main difference between the kinetic problem, and its macroscopic diffusive limit, as in [7, 29]).

Here, we present three examples of possible shapes for Υ, based on accurate numerical simulations. It is important to notice that the function Υ is not smooth in the case of discrete velocities (as opposed to the continuous velocity case, see [7]). In fact, it has jump discontinuities located on the set of velocities $(v_k)_{k\in\mathcal{K}}$. A thorough analysis of the sign and the size of these jump discontinuities was performed in [7], questing for counter-examples. Here, we bypass this analysis, and we present directly the counter-examples.

4.1 *Existence and Uniqueness of the Wave Speed*

Here, we consider a first set of velocities and weights for which there is numerical evidence that Υ has a unique admissible root $\mathbf{c}$, as in Fig. 4. The velocities are chosen as follows:

$$v_1 = 0.0848\,,\, v_2 = 0.2519\,,\, v_3 = 0.4118\,,\, v_4 = 0.5598\,,\, v_5 = 0.6917\,,$$
$$v_6 = 0.8037\,,\, v_7 = 0.8926\,,\, v_8 = 0.9558\,,\, v_9 = 0.9916\,.$$

[6] The restriction $c > c_\star$ is due to confinement on the right hand side, whereas the condition $c > 0$ is due to the arbitrary choice of the direction of propagation of the wave (here, left to right), which influences itself the monotonicity of N (here, increasing). It is an arbitrary choice, of course, since the problem is symmetric.

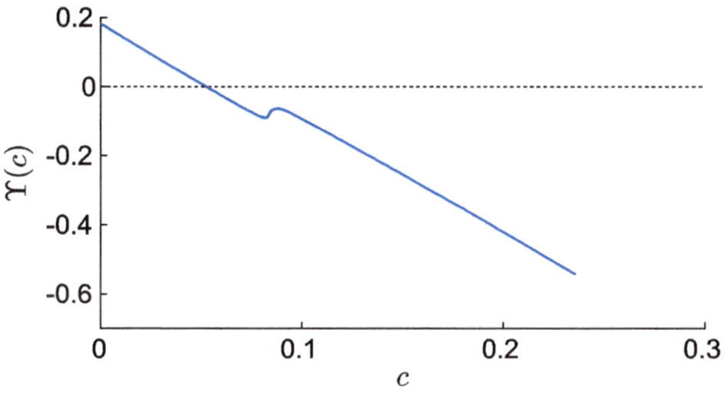

Fig. 4 Plot of the function Υ over $(0, c^\star)$ for the set of parameters described in Sect. 4.1. We observe that the function is not globally monotonic. It is seemingly decreasing on each side of the positive jump discontinuity. However, this jump discontinuity is not large enough to imply the co-existence of two wave speeds. In this case, we observe the existence of a unique wave speed

The weights are chosen as follows:

$$w_1 = 0.0846\,, w_2 = 0.0822\,, w_3 = 0.0774\,, w_4 = 0.0703\,, w_5 = 0.0613\,,$$
$$w_6 = 0.0505\,, w_7 = 0.0382\,, w_8 = 0.0249\,, w_9 = 0.0108\,.$$

We set $\omega_0 = 0$ for numerical purposes. However, this does not affect the results of Sects. 2 and 3. The other parameters are:

$$\chi_S = 0.3\,, \chi_N = 0.15\,, D_S = 0.5\,, D_N = 1\,, \alpha = 0.5\,, \beta = 1\,, \gamma = 1\,.$$

4.2 Non Uniqueness of the Wave Speed

The second example consists in a set of velocities and weights, for which there is seemingly two possible wave speeds, as the function Υ crosses at least twice the zero axis on the set of admissible velocities, see Fig. 5. There, we observe two crossing downwards, corresponding to two admissible wave speed. The vertical crossing upwards is not admissible as it corresponds to a jump discontinuity of Υ located at one of the discrete velocities $(v_k)_{k\in\mathcal{K}}$. The parameters are the same as in Sect. 4.1, except that the values of χ_S, χ_N and α are replaced with:

$$\chi_S = 0.5\,, \chi_N = 0.45\,, \alpha = 10\,.$$

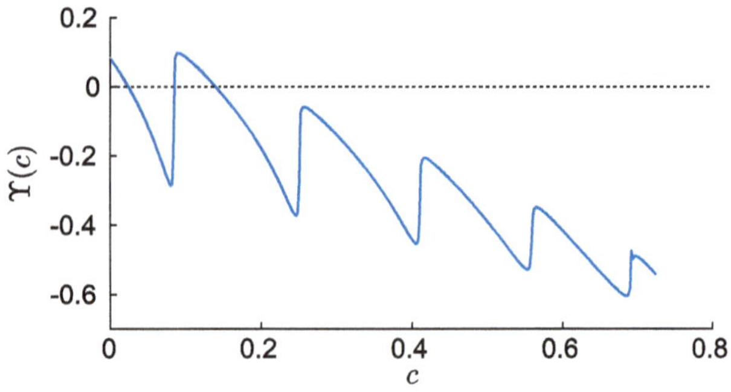

Fig. 5 Plot of the function Υ over $(0, c^\star)$ for the set of parameters described in Sect. 4.2. We observe that the function Υ possesses two admissible roots, as it crosses the horizontal axis downwards. Note that the crossing upwards is not admissible as it corresponds to a jump discontinuity of Υ

Beyond the static analysis dealing exclusively with the construction of travelling wave solutions, we ran numerical simulations of the Cauchy problem (1). The well-balanced upwind numerical scheme is designed as follows:

- The kinetic equation involves a formulation involving **a scattering S-matrix at each cell's interface**, as presented in [21] for slightly different kinetic models. Such a S-matrix is retrieved thanks to a Case's mode decomposition (13) associated with $c = 0$, see [9] for details, [20, Chap. 10] and [13, 19] as well.
- Reaction-diffusion equations are treated by means of a $\mathscr{L}$-spline interpolation leading to accurate numerical fluxes where all the terms (diffusive, reactive, drift) can be treated as a whole, allowing for the preservation of a delicate balance between each other: see especially [22].

Results are shown in Fig. 6. Beginning with an initial data which is concentrated on the left-hand-side, with arbitrary shape, we observe the formation of a wave moving at constant speed over a reasonably long time span (Fig. 6). The speed of propagation corresponds to the fastest of the two roots of Υ, as seen on Fig. 5.

A similar (but simpler) example of non uniqueness with only four velocities is extensively analysed in [9]. We conjecture that the two co-existing travelling waves are stable. Bistability is clearly shown using numerical simulations of the Cauchy problem with various initial data.

4.3 Non Existence of the Wave Speed

The third example consists in a set of velocities and weights, for which there is seemingly no admissible wave speed, as in Fig. 7. The velocities are chosen as follows:

$$v_1 = 0.015,\ v_2 = 0.03\,, v_3 = 0.05\,, v_4 = 0.08\,, v_5 = 0.12\,,$$
$$v_6 = 0.16\,, v_7 = 0.25\,, v_8 = 0.5\,, v_9 = 1\,.$$

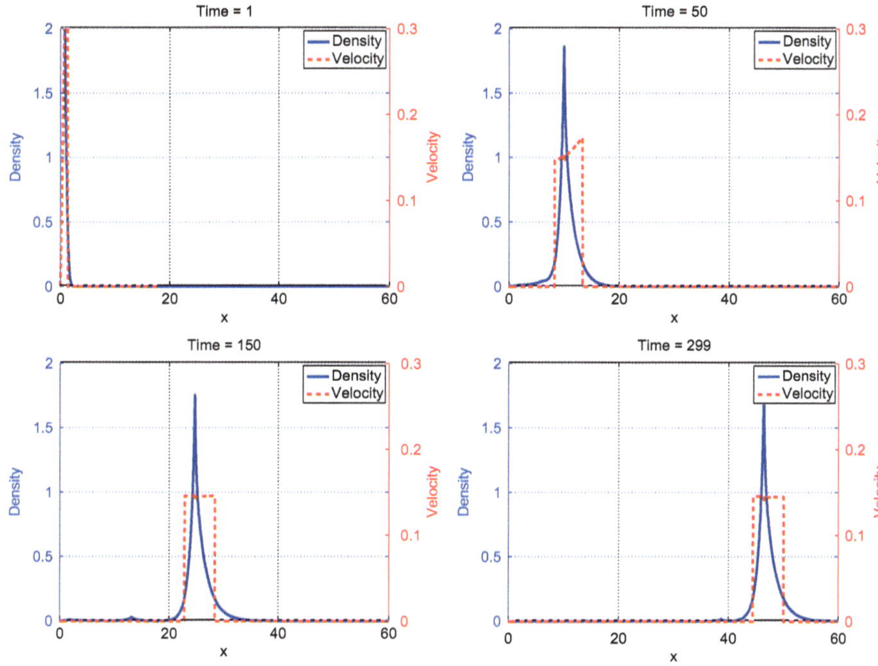

Fig. 6 Numerical simulations of the Cauchy problem in the case of multiple existence of travelling waves (second example in Sect. 4.2). The formation of a stable wave travelling at constant speed is clearly observed. It corresponds to the fastest wave: $\mathbf{c} \approx 0.15$ (compare with Fig. 5)

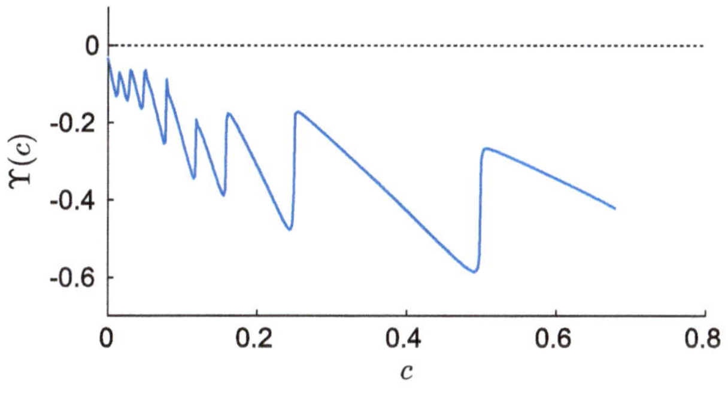

Fig. 7 Plot of the function Υ over $(0, c^{\star})$ for the set of parameters described in Sect. 4.3. We observe that the function is negative for all admissible c. This means that the maximum of the concentration S is always located on the left side of the origin. Consequently, the construction of travelling wave cannot be achieved in this case

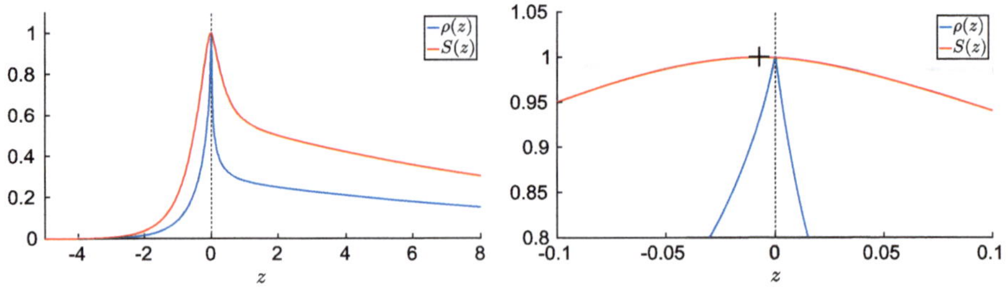

Fig. 8 Illustration of the counter-intuitive phenomenon occurring in the third example (Sect. 4.3). For $c = 0$, the cell density ρ is tilted to the right side for large $|z|$, because the positive gradient of nutrient yields net biased motion in the right direction (compare $\rho(z)$ and $\rho(-z)$ for large $z \gg 1$). However, it is tilted to the left side for small $|z| \ll 1$. As a consequence, the chemical concentration S reaches its maximum at a negative value, for a suitable choice of the reaction-diffusion parameters (see the zoom on the right). This prevents the existence of a travelling wave

The weights are uniform:

$$(\forall k \in [-K, K] \setminus \{0\}) \quad \omega_k = 1/18\,, \quad \omega_0 = 0\,.$$

The other parameters are as in Sect. 4.2.

Let us emphasize that this is a counter-intuitive result. Indeed, it is highly related to the fact that $\Upsilon(0) < 0$. So, let us focus on the case $c = 0$. We clearly have $\lambda_+(0) < \lambda_-(0)$. This is a way to express the net biased motion of cells to the right side, due to the contribution of the positive gradient of nutrient $\partial_z N > 0$ in the tumbling rate (5), together with $\chi_N > 0$. We could intuitively deduce that the spatial density ρ is globally tilted to the right side (see Fig. 8). But, this is not true. It may happen that, for small $0 < z \ll 1$, $\rho(z) < \rho(-z)$, meaning that the spatial density is locally tilted to the left side (see the zoom in Fig. 8). Then, by choosing appropriately the reaction-diffusion parameters α, D_S, it is possible to transfer this local asymmetry to ensure $\Upsilon(0) < 0$, meaning that the maximum of S is located on the left side.

Beyond this negative result, it is interesting to run numerical simulations of the Cauchy problem (1). Indeed, we guess that the net biased motion to the right side makes the wave propagating in a way that is not compatible with the formation of a stable travelling wave. Results are shown in Fig. 9. We observe the inclination to form a wave moving to the right after short time, as expected intuitively. However, the cell density splits quickly into smaller components, as opposed to Fig. 6, where it is maintained over the duration of the numerical test. This peculiar behaviour agrees with the non existence of a travelling wave.

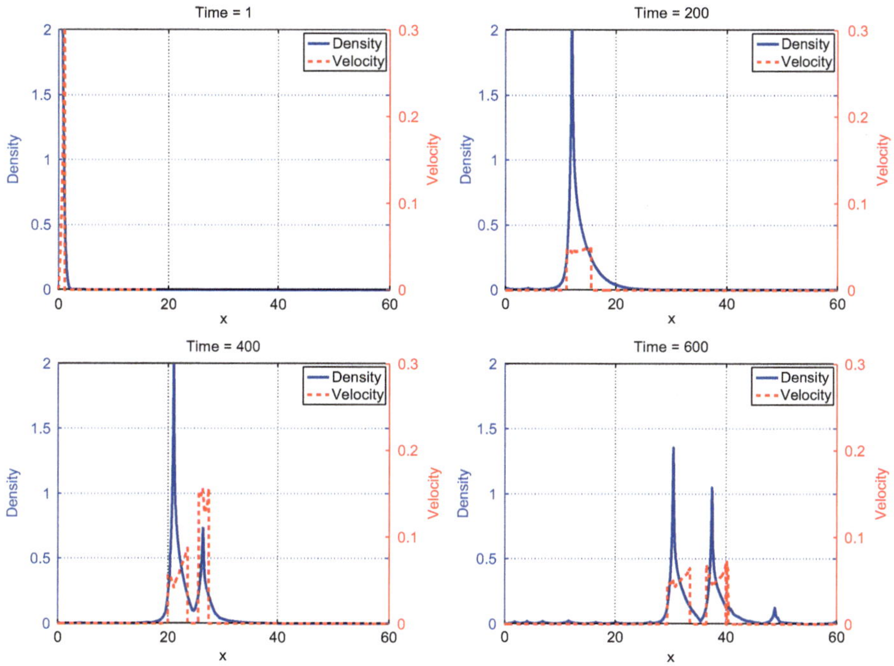

Fig. 9 Numerical simulations of the Cauchy problem in the case of non existence of a travelling wave (third example in Sect. 4.3). As opposed to Fig. 6, the cell density quickly splits into smaller units. However, we still observe global propagation of the cell population to the right side

Acknowledgements This project has received funding from the European Research Council (ERC) under the European Union's Horizon 2020 research and innovation programme (grant agreement No 639638). M.T. has benefited from the PICS Project CNR-CNRS 2015-2017 *Modèles mathématiques et simulations numériques pour le mouvement de cellules*.

References

1. Aamodt, R.E., Case, K.M.: Useful identities for half-space problems in linear transport theory. Ann. Phys. **21**, 284–301 (1963)
2. Adler, J.: Chemotaxis in bacteria. Science **153**(3737), 708–716 (1966)
3. Almeida, L., Emako, C., Vauchelet, N.: Existence and diffusive limit of a two-species kinetic model of chemotaxis (2014). arXiv:1404.4769
4. Alt, W.: Biased random walk models for chemotaxis and related diffusion approximations. J. Math. Biol. **9**(2), 147–177 (1980)
5. Bardos, C., Santos, R., Sentis, R.L.: Diffusion approximation and computation of the critical size. Trans. Am. Math. Soc. **284**(2), 617–649 (1984)
6. Berg, H.C.: E. coli in Motion. Springer, New York (2004)
7. Calvez, V.: Chemotactic waves of bacteria at the mesoscale (2016). arXiv:1607.00429

8. Calvez, V., Raoul, G., Schmeiser, C.: Confinement by biased velocity jumps: aggregation of Escherichia coli. Kinet. Relat. Models **8**(4), 651–666 (2015)
9. Calvez, V., Gosse, L., Twarogowska, M.: Traveling chemotactic aggregates at mesoscopic scale and bi-stability (2016, Submitted)
10. Case, K.M.: Elementary solutions of the transport equation and their applications. Ann. Phys. **9**, 1–23 (1960)
11. Chalub, F.A.C.C., Dolak-Struss, Y., Markowich, P., Oelz, D., Schmeiser, C., Soreff, A.: Model hierarchies for cell aggregation by chemotaxis. Math. Models Methods Appl. Sci. **16**, 1173–1197 (2006)
12. Dolak, Y., Schmeiser, C.: Kinetic models for chemotaxis: hydrodynamic limits and spatio-temporal mechanisms. J. Math. Biol. **5**(6), 595–615 (2005)
13. Emako, C., Tang, M.: Well-balanced and asymptotic preserving schemes for kinetic models (2016). arXiv:1603.03171
14. Emako, C., Gayrard, C., Buguin, A., de Almeida, L.N., Vauchelet, N.: Traveling pulses for a two-species chemotaxis model. PLOS Comput. Biol. **12**(4), e1004843 (2016)
15. Erban, R., Othmer, H.G.: From signal transduction to spatial pattern formation in *E. coli*: a paradigm for multiscale modeling in biology. Multiscale Model. Simul. **3**(2), 362–394 (2005)
16. Filbet, F., Yang, C.: An inverse Lax–Wendroff method for boundary conditions applied to Boltzmann type models. J. Comput. Phys. **245**, 43–61 (2013)
17. Filbet, F., Yang, C.: Numerical simulations of kinetic models for chemotaxis. SIAM J. Sci. Comput. **36**(3), B348–B366 (2014)
18. Franz, B., Xue, C., Painter, K.J., Erban, R.: Travelling waves in hybrid chemotaxis models. Bull. Math. Biol. **76**(2), 377–400 (2013)
19. Gosse, L.: A well-balanced scheme for kinetic models of chemotaxis derived from one-dimensional local forward-backward problems. Math. Biosci. **242**, 117–128 (2013)
20. Gosse, L.: Computing Qualitatively Correct Approximations of Balance Laws. SIMAI Springer Series, vol. 2. Springer Milan, Milano (2013)
21. Gosse, L.: Redheffer products and numerical approximation of currents in one-dimensional semiconductor kinetic models. SIAM Multiscale Model. Simul. **12**, 1533–1560 (2014)
22. Gosse, L.: Viscous equations treated with $\mathscr{L}$-splines and Steklov-Poincaré operator in two dimensions. In: Gosse, L., Natalini, R. (eds.) Innovative Algorithms and Analysis. Springer, Cham (2016). doi:10.1007/978-3-319-49262-9
23. Keller, E.F., Segel, L.A.: Traveling bands of chemotactic bacteria: a theoretical analysis. J. Theor. Biol. **30**(2), 235–248 (1971)
24. Mischler, S., Weng, Q.: On a linear runs and tumbles equation (2016). arXiv:1602.03474
25. Othmer, H.G., Dunbar, S.R., Alt, W.: Models of dispersal in biological systems. J. Math. Biol. **26**(3), 263–298 (1988)
26. Perthame, B., Tang, M., Vauchelet, N.: Derivation of the bacterial run-and-tumble kinetic equation from a model with biochemical pathway (2015). arXiv:1503.03979
27. Rousset, M., Samaey, G.: Simulating individual-based models of bacterial chemotaxis with asymptotic variance reduction. Math. Models Methods Appl. Sci. **23**(12), 2155–2191 (2013)
28. Salman, H., Zilman, A., Loverdo, C., Jeffroy, M., Libchaber, A.: Solitary modes of bacterial culture in a temperature gradient. Phys. Rev. Lett. **97**(11), 118101 (2006)
29. Saragosti, J., Calvez, V., Bournaveas, N., Buguin, A., Silberzan, P., Perthame, B.: Mathematical description of bacterial traveling pulses. PLoS Comput. Biol. **6**(8), e1000890 (2010)
30. Saragosti, J., Calvez, V., Bournaveas, N., Perthame, B., Buguin, A., Silberzan, P.: Directional persistence of chemotactic bacteria in a traveling concentration wave. PNAS **108**(39), 16235–16240 (2011)
31. Stroock, D.W.: Some stochastic processes which arise from a model of the motion of a bacterium. Z. Wahrscheinlichkeitstheorie verw Gebiete **28**(4), 305–315 (1974)

32. Tindall, M.J., Maini, P.K., Porter, S.L., Armitage, J.P.: Overview of mathematical approaches used to model bacterial chemotaxis II: bacterial populations. Bull. Math. Biol. **70**(6), 1570–1607 (2008)
33. Xue, C.: Macroscopic equations for bacterial chemotaxis: integration of detailed biochemistry of cell signaling. J. Math. Biol. **70**(1), 1–44 (2015)
34. Xue, C., Hwang, H.J., Painter, K.J., Erban, R.: Travelling waves in hyperbolic chemotaxis equations. Bull. Math. Biol. **73**(8), 1695–1733 (2010)
35. Yasuda, S.: A Monte Carlo simulation for kinetic chemotaxis models: an application to the traveling population wave (2015). arXiv:1503.08099

A Numerical Glimpse at Some Non-standard Solutions to Compressible Euler Equations

Elisabetta Chiodaroli and Laurent Gosse

Abstract In this note we review and recast some recent results on the existence of non-standard solutions to the compressible Euler equations as to make possible a preliminary numerical investigation. In particular, we are interested in studying numerically the forward in time evolution of some Lipschitz initial data which allow for non-standard solutions ("colliding data"). Numerical results indicate appearance of oscillations after the first break-up time along with a qualitative behavior seemingly compatible with relevant properties of non-standard solutions.

1 Introduction

Throughout this text, the focus is on the Cauchy problem for the compressible isentropic Euler system of gas dynamics in two space dimensions, namely

$$\begin{cases} \partial_t \rho + \mathrm{div}_{\mathbf{x}}(\rho \mathbf{v}) \; = \; 0 \\ \partial_t(\rho \mathbf{v}) + \mathrm{div}_{\mathbf{x}}\,(\rho \mathbf{v} \otimes \mathbf{v}) + \nabla_{\mathbf{x}}[p(\rho)] \; = \; 0 \\ \partial_t(\rho E) + \mathrm{div}_{\mathbf{x}}([\rho E + p(\rho)]\,\mathbf{v}) \leq 0 \\ \rho(\cdot, 0) \; = \; \rho^0, \; \mathbf{v}(\cdot, 0) \; = \; \mathbf{v}^0 := (u^0, v^0)\,, \end{cases} \tag{1}$$

for a certain pressure law $p(\rho) = \kappa \rho^\gamma$ with constants $\kappa > 0$, $\gamma \geq 1$ determining the total energy $E = e(\rho) + |\mathbf{v}|^2/2$, $\rho^2 e'(\rho) = p(\rho)$. Solutions $\mathbf{U} = (\rho, \mathbf{v})$ satisfying the inequality in the distributional sense are usually called "entropy solutions".

E. Chiodaroli (✉)
EPFL Lausanne, Station 8, 1015 Lausanne, Switzerland
e-mail: elisabetta.chiodaroli@epfl.ch

L. Gosse
IAC, CNR, via dei Taurini 19, 00185 Roma, Italy
e-mail: l.gosse@ba.iac.cnr.it

L. Gosse, R. Natalini (eds.), *Innovative Algorithms and Analysis*,
Springer INdAM Series 16, DOI 10.1007/978-3-319-49262-9_4

1.1 Theoretical Aspects and General Non-uniqueness

The Euler system (1) dates back to 1757, yet a complete understanding of the underlying phenomena is still missing. In particular, in more than one space dimension, a good theory for unique continuation of solutions after the formation of shocks is not available. Starting with the groundbreaking work of De Lellis and Székelyhidi [16, 17], it became possible to construct counterexamples to uniqueness of weak solutions for (1), with suitable initial data, even under the entropy condition (see [7, 10, 38]). This failure of uniqueness was previously suggested by Elling in [19]; indeed, quoting [21] "*this is due to the inclusion of vorticity (rotational flow) in an inviscid flow model. Uniqueness does not and should not hold since inviscid models neglect vorticity on small viscous scales that may propagate to have a large-scale effect. Only models with sufficient viscosity or other smoothing terms to establish a "bottom" to the hierarchy of scales would have a chance for uniqueness. Alternatively, inviscid models could be valid/reliable in cases where the effects of vorticity are negligible*". The method of convex integration developed by De Lellis and Székelyhidi yields a multiplicity of entropy solutions, see [7–10, 17], which exhibit a highly oscillatory behavior. We call them *non-standard solutions*. This series of results on non-standard solutions motivates some natural issues:

- if the entropy inequality does not yield uniqueness, are there alternative selection/admissibility criteria? Or should we look for other notions of solution?
- Moreover, is there any way these non-standard solutions can be visualized (even approximately) with existing means?

While the first two questions were (partially) tackled in [8, 11], respectively, in this note we plan to address the last question. We are particularly interested in the non-standard solutions constructed in [8, 9], where both Lipschitz and Riemann initial data for (1) were considered. A building block in these papers was a special class of Riemann problems for the system (1) in two-space dimensions: denoting the space variable as $\mathbf{x} = (x, y) \in \mathbb{R}^2$, an y-translation invariant data reads

$$\mathbf{U}^0(\mathbf{x}) = (\rho^0(\mathbf{x}), \mathbf{v}^0(\mathbf{x})) := \begin{cases} (\rho_-, \mathbf{v}_-) & \text{if } x < 0 \\ (\rho_+, \mathbf{v}_+) & \text{if } x > 0, \end{cases} \tag{2}$$

where $\rho_\pm, \mathbf{v}_\pm$ are constant states for density and velocity. In [8, 9] it was shown that for some classes of Riemann data like (2), that is, for some choices of constants $0 < \rho_\pm, \mathbf{v}_\pm$, the compressible Euler system (1) admits many non-standard solutions, satisfying also the entropy inequality. More precisely, follows from [8, 9]:

Theorem 1 *Let* $p(\rho) = \rho^\gamma$ *with* $\gamma \geq 1$. *For some Riemann data as in* (2) *there exist infinitely many bounded entropy solutions to* (1) *on* $\mathbb{R}^2 \times [0, \infty)$.

We refer to such Riemann data allowing for infinitely many forward in time non-standard solutions as to *wild Riemann data*. Moreover, in [9] it was proven that it is possible to choose wild Riemann data in such a way that they are generated (backward in time) by a compression wave: this allowed to conclude non-uniqueness even starting from smooth data (Lipschitz data is the compression wave at time $t = -1$). To get uniqueness in Theorem 1, moreover,

- admissible solutions must be self-similar, i.e. of the form

$$(\rho, \mathbf{v})(t, \mathbf{x}) = \left(r\left(\frac{\mathbf{x}}{t}\right), \mathbf{w}\left(\frac{\mathbf{x}}{t}\right) \right),$$

- and of locally bounded variation (standard theory of 1D Riemann problems).

Indeed, non-unique entropy solutions in Theorem 1 are truly two-dimensional, hence depend on both space variables: there is loss of translational invariance in y.

1.2 Numerical Trick Based on Rotational Invariance

The essence of counter-examples like the ones in [8] is that, starting from y-translation invariant initial data (2), it's possible to build entropy solutions to (1) for which that translational invariance is broken beyond shock onset. Clearly, as most numerical schemes rely on dimensional splitting, that is, solve sequentially first a subsystem involving partial derivatives in x, and then in y, such phenomena seems not to be observable in practice. However, one may draw advantage from a distinctive feature of (1), rotational invariance: consider any 2D rotation matrix,

$$R_\theta = \begin{pmatrix} \cos\theta & \sin\theta \\ -\sin\theta & \cos\theta \end{pmatrix}, \qquad \theta \in (0, \frac{\pi}{2}),$$

then a standard fact is that, for $\tilde{\mathbf{x}} = R_\theta(\mathbf{x})$, given a solution $\mathbf{U}(t, \mathbf{x})$ to (1),

$$\tilde{\mathbf{U}}(t, \tilde{\mathbf{x}}) = \begin{pmatrix} 1 & 0 & 0 \\ 0 & \cos\theta & \sin\theta \\ 0 & -\sin\theta & \cos\theta \end{pmatrix} \mathbf{U}(t, \mathbf{x}) = \begin{pmatrix} 1 & 0 & 0 \\ 0 & \cos\theta & \sin\theta \\ 0 & -\sin\theta & \cos\theta \end{pmatrix} \mathbf{U}(t, R_{-\theta}(\tilde{\mathbf{x}}))$$

is again a solution of (1).

Definition 1 Given $\mathbf{U}^0(\mathbf{x})$ an y-translation Riemann data like (2) and

$$\mathbf{V}^0(\tilde{\mathbf{x}}) = \begin{pmatrix} 1 & 0 & 0 \\ 0 & \cos\theta & \sin\theta \\ 0 & -\sin\theta & \cos\theta \end{pmatrix} \mathbf{U}^0(\mathbf{x})$$

"rotated data" for (1), a "back-rotated solution" is a vector $\mathbf{U}(t,\mathbf{x})$ obtained by,

$$\forall t > 0, \qquad \mathbf{U}(t,\mathbf{x}) = \begin{pmatrix} 1 & 0 & 0 \\ 0 & \cos\theta & -\sin\theta \\ 0 & \sin\theta & \cos\theta \end{pmatrix} \mathbf{V}(t, R_\theta(\mathbf{x})),$$

where $\mathbf{V}(t,\mathbf{x})$ is a solution of (1) emerging from Cauchy data $\mathbf{V}^0(\mathbf{x})$.

Yet, as long as we deal with self-similar, unique solutions emerging from (1), such a notion seems useless. However, we plan to consider "back-rotated approximate solutions" generated by standard, dimensional splitting, numerical schemes for (1). According to former studies, see [12, 33], so-called "grid effects" in both x, y directions should pollute the 1D Riemann fan, so that, given $\theta \in (0, \frac{\pi}{2})$, the numerical back-rotated $\mathbf{U}(t,\mathbf{x})$ can be expected to be quite different compared to the direct 1D simulation corresponding to $\theta = 0$. Indeed, artificial vorticity is likely to appear. Actually, the vorticity of the self-similar solution is always null: initially (for our choice of data), we have

$$\omega(t=0,\cdot,\cdot) = \partial_y u^0 - \partial_x v^0, \quad u^0(x,y) \text{ is a function of } x, \quad v^0(x,y) \equiv 0,$$

and later, for a self-similar solution $(\rho,\mathbf{v})(t,\mathbf{x}) = (\rho,\mathbf{v})(x/t)$ with $v(t,x) \equiv 0$ it still holds that $\omega(t,x,y) \equiv 0$. Moreover, if we denote the velocity vector $\mathbf{v} = (u, v)$ then, by rewriting $\omega(t,\mathbf{x}) = (\nabla^\perp)\cdot\mathbf{v}(t,\mathbf{x})$, it comes, for $\tilde{\mathbf{x}} = R_\theta(\mathbf{x})$,

$$\omega(t,\tilde{\mathbf{x}}) = \langle \nabla^\perp, \mathbf{v}(t,\tilde{\mathbf{x}})\rangle = \left\langle R_{-\theta}\tilde{\nabla}^\perp, \mathbf{v}(t,\tilde{\mathbf{x}})\right\rangle, \qquad \tilde{\nabla}^\perp = (\partial_{\tilde{y}}, -\partial_{\tilde{x}}),$$

so that, by considering a "rotated velocity vector", $\tilde{\mathbf{v}} = R_\theta \mathbf{v}$ as well,

$$\tilde{\omega}(t,\tilde{\mathbf{x}}) = \left\langle \tilde{\nabla}^\perp, \tilde{\mathbf{v}}(t,\tilde{\mathbf{x}})\right\rangle = \left\langle \tilde{\nabla}^\perp, R_\theta\mathbf{v}(t,\mathbf{x})\right\rangle = \left\langle R_{-\theta}\tilde{\nabla}^\perp, \mathbf{v}(t,\mathbf{x})\right\rangle = \omega(t,\mathbf{x}),$$

thereby showing that the vorticity is invariant under rotation.

1.3 Non-standard "Equivalent Solutions"?

Given a conservative numerical scheme for a scalar law with flux f, one usually speaks about its "equivalent equation" (or "modified" [36]) after putting it into the so-called viscous form, i.e. emphasizing its relations with viscous perturbations of the original equation. Being F a consistent numerical flux such that $F(u,u) = f(u)$, the exact one, the numerical viscosity coefficient $Q(u,v)$ is defined through,

$$\forall u, v, \qquad Q(u,v)(v-u) = \frac{1}{\Delta x}\left(\frac{f(u)+f(v)}{2} - F(u,v)\right),$$

so that any conservative numerical discretization,

$$\frac{u_j^{n+1} - u_j^n}{\Delta t} + \frac{F(u_j^n, u_{j+1}^n) - F(u_{j-1}^n, u_j^n)}{\Delta x} = 0$$

is a "viscous" stabilization of the centered, formally second-order, one,

$$\frac{u_j^{n+1} - u_j^n}{\Delta t} + \frac{f(u_{j+1}^n) - f(u_{j-1}^n)}{2\Delta x} = Q(u_j^n, u_{j+1}^n)(u_{j+1}^n - u_j^n) - Q(u_{j-1}^n, u_j^n)(u_j^n - u_{j-1}^n).$$

In order to gain insight of the fine properties of the numerical approximations u_j^n, it is now tempting to study its "equivalent (viscous) equation" which reads accordingly,

$$\partial_t \tilde{u} + \partial_x f(\tilde{u}) = \Delta x\, \partial_x\big(Q(\tilde{u}, \partial_x \tilde{u})\partial_x \tilde{u}\big), \qquad \Delta x \ll 1,$$

the algebraic complexity of $F(\cdot, \cdot)$ being transferred inside the artificial viscosity coefficient Q. Beside, the solution $\tilde{u}$ is called the "equivalent solution". The idea we propose hereafter mimics this rationale, in the sense that it appears that oscillating non-standard solutions may realize possible "equivalent solutions" to numerical approximations of (1) obtained by dimensional-splitting schemes. Although applying rotations in order to study numerical discrepancies is a standard procedure, see [12, 15, 33] (also [23, 28]), what we propose here is a bit different:

- starting from an y-translational invariant data like (2), a rotation matrix is applied in order to force the dimensional-split scheme to really "work in 2D";
- when the final time is reached, we "back-rotate" (in the sense of Definition 1) the numerical solution, so that performing a comparison with the self-similar solution of the 1D Riemann problem induced by (2) becomes easy;
- an examination of both the oscillations present inside the approximate velocity vector and the (slightly lower) values of the density ρ allows to think that a non-standard solution might be a good candidate for an "equivalent solution" of that type of widely-used numerical algorithms (see also [19, 20, 35]).
- by performing fine-grid computations with a (formally) second-order accurate scheme proposed in [3], we discovered on the same benchmark an intriguing phenomenon: for a rotation angle $\theta > 0$, spurious vortices can develop in the middle of the stagnation zone flanked by the two moving shocks, see Fig. 8. This agrees with heuristic ideas of Roe, [34, pp. 346–347]: "*Inaccurate recognition of the physical processes present destroys what I take to be the foundation of upwinding, and cause a shear to be misrepresented as the sum of two acoustic waves. The discrepancy, due solely to the dimensional-splitting, is dispersive and non-dissipative. We may deduce that as the components of an operator split method are improved, the results will get worse!*".

Hereafter, we summarize recent results on the existence of non-standard solutions and present a preliminary numerical investigation on the time evolution of some "wild data". Of salient interest are Lipschitz initial data which can be shown to allow for non-standard solutions to develop. Numerical results seem to signal the onset of oscillations (in the y direction) after break-up time, along with a qualitative behavior in accordance with the main properties of exact non-standard solutions.

2 Subsolutions and Riemann Data Generating Two Shocks

2.1 *Self-similar Solutions of 1D Riemann Problems*

We review here some basic theory for one-dimensional solutions of the Riemann problem for (1)–(2), i.e. functions $\rho(x,t)$ and $\mathbf{v}(x,t) = (u,v)(x,t)$ solving the one-dimensional compressible isentropic Euler system, which reads as follows

$$\begin{cases} \partial_t \rho + \partial_x(\rho u) = 0 \\ \partial_t(\rho u) + \partial_x\left(\rho u^2 + p(\rho)\right) = 0 \\ \partial_t(\rho v) + \partial_x(\rho u v) = 0 \end{cases} \tag{3}$$

with Riemann data (2). System (3) is strictly hyperbolic on the part of the state space where $\rho > 0$ (see [14]) since it has three real distinct *characteristic velocities*

$$\lambda_1 = u - \sqrt{p'(\rho)}, \qquad \lambda_2 = u, \qquad \lambda_3 = u + \sqrt{p'(\rho)} \tag{4}$$

with associated right eigenvectors

$$\mathbf{R}_1 = \begin{pmatrix} 1 \\ v \\ u - \sqrt{p'(\rho)} \end{pmatrix}, \quad \mathbf{R}_2 = \begin{pmatrix} 0 \\ 0 \\ 1 \end{pmatrix}, \quad \mathbf{R}_3 = \begin{pmatrix} 1 \\ v \\ u + \sqrt{p'(\rho)} \end{pmatrix}. \tag{5}$$

Recall that the λ_i are called the *i-characteristic speeds* of the system (3). The Riemann functions associated with (3)

$$w_3 = u + \int_0^\rho \frac{\sqrt{p'(\tau)}}{\tau} d\tau, \qquad w_2 = v, \qquad w_1 = u - \int_0^\rho \frac{\sqrt{p'(\tau)}}{\tau} d\tau \tag{6}$$

are, respectively, (1- and 2-), (1- and 3-), (2- and 3-) Riemann invariants, and

$$w_3 = u + \frac{\rho}{2} = \lambda_3, \quad w_1 = u - \frac{\rho}{2} = \lambda_1, \quad \text{when} \quad p(\rho) = \frac{\rho^3}{12}.$$

The 2-characteristic family of the system (3) is linearly degenerate, i.e. $D\lambda_2 \cdot \mathbf{R}_2 = 0$, whereas the 1- and 3-characteristic families are genuinely nonlinear. Moreover, the state variable v appears only in the third equation of (3) and thus the system decouples for smooth solutions. In particular, if Riemann initial data (2) satisfy

$v_- = v_+$, then the second component of the velocity vector of the self-similar solution must remain equal to this constant: $v(x,t) \equiv v_- = v_+$ (see [9, Sect. 8]). All Riemann data studied hereafter satisfy $v_- = v_+ = 0$ in order to focus onto a reduced system containing only the first component of the velocity vector,

$$\begin{cases} \partial_t \rho + \partial_x(\rho u) = 0 \\ \partial_t(\rho u) + \partial_x\left(\rho u^2 + p(\rho)\right) = 0 \end{cases} \tag{7}$$

with initial data

$$(\rho^0(x), u^0(x)) := \begin{cases} (\rho_-, u_-) & \text{if } x < 0 \\ (\rho_+, u_+) & \text{if } x > 0. \end{cases} \tag{8}$$

As it is well-known, the system (7) admits self-similar solutions, i.e. solutions of the form $(\rho, u)(t,x) = (\rho, u)(x/t)$, given by a gluing of shocks and rarefaction waves. Solving Rankine–Hugoniot conditions and using the characterization of rarefaction waves by Riemann invariants yields an explicit description of the structure of these self-similar solutions to (7). We recall from e.g., [8] such a characterization: first, a result for Riemann data connected by simple waves.

Lemma 1 *Let $p(\rho) = \rho^\gamma$ with $\gamma \geq 1$ and let $\rho_\pm > 0$. The states (ρ_-, u_-) on the left and (ρ_+, u_+) on the right are connected with*

- *admissible* 1*-shock if and only if*

$$\rho_+ > \rho_- \qquad u_- = u_+ + \sqrt{\frac{(\rho_+ - \rho_-)(p(\rho_+) - p(\rho_-))}{\rho_+\rho_-}} \tag{9}$$

the speed of the shock is $$s = u_+ - \sqrt{\frac{\rho_-(p(\rho_+) - p(\rho_-))}{\rho_+(\rho_+ - \rho_-)}} \tag{10}$$

- *admissible* 3*-shock if and only if*

$$\rho_+ < \rho_- \qquad u_- = u_+ + \sqrt{\frac{(\rho_+ - \rho_-)(p(\rho_+) - p(\rho_-))}{\rho_+\rho_-}} \tag{11}$$

the speed of the shock is $$s = u_+ + \sqrt{\frac{\rho_-(p(\rho_+) - p(\rho_-))}{\rho_+(\rho_+ - \rho_-)}} \tag{12}$$

- 1*-rarefaction wave if and only if*

$$\rho_+ < \rho_- u_- = u_+ - \int_{\rho_+}^{\rho_-} \frac{\sqrt{p'(\tau)}}{\tau} \mathrm{d}\tau \tag{13}$$

- *3-rarefaction wave if and only if*

$$\rho_+ > \rho_-, \qquad u_- = u_+ - \int_{\rho_-}^{\rho_+} \frac{\sqrt{p'(\tau)}}{\tau} \mathrm{d}\tau. \tag{14}$$

2.2 Loss of Translational-Invariance and Non-uniqueness

Using Lemma 1 a subset of entropy-admissible, self-similar solutions to the 1D Riemann problem (7)–(8) is reported, following [8].

Lemma 2 *Let $\rho_\pm, u_\pm$ be constants, $\rho_\pm > 0$, and let $p(\rho) = \rho^\gamma$ with $\gamma \geq 1$. Assume for simplicity that (ρ_-, u_-) and (ρ_+, u_+) do not lie on any simple i-wave curve (otherwise the form of the self-similar solution is given directly by Lemma 1).*

(1) If

$$u_+ - u_- \geq \int_0^{\rho_-} \frac{\sqrt{p'(\tau)}}{\tau} \mathrm{d}\tau + \int_0^{\rho_+} \frac{\sqrt{p'(\tau)}}{\tau} \mathrm{d}\tau, \tag{15}$$

then there exists a unique self-similar solution to the Riemann problem (7)–(8) consisting of a 1-rarefaction wave and a 3-rarefaction wave. The intermediate state is vacuum, i.e. $\rho_m = 0$.

(2) If

$$u_+ - u_- < -\sqrt{\frac{(\rho_+ - \rho_-)(p(\rho_+) - p(\rho_-))}{\rho_+ \rho_-}} \tag{16}$$

then there exists a unique self-similar solution to the Riemann problem (7)–(8) consisting of an admissible 1-shock and an admissible 3-shock. The intermediate state (ρ_m, u_m) is the solution of the system of equations

$$u_+ - u_- = -\sqrt{\frac{(\rho_m - \rho_-)(p(\rho_m) - p(\rho_-))}{\rho_m \rho_-}} - \sqrt{\frac{(\rho_m - \rho_+)(p(\rho_m) - p(\rho_+))}{\rho_m \rho_+}} \tag{17}$$

$$u_m = u_- - \sqrt{\frac{(\rho_m - \rho_-)(p(\rho_m) - p(\rho_-))}{\rho_m \rho_-}}. \tag{18}$$

Particularly relevant for us is the following result proved in [8].

Theorem 2 *Let $p(\rho) = \rho^\gamma$ with $\gamma \geq 1$. For every Riemann data (2) such that the (forward-in-time) self-similar solution to the Riemann problem (1)–(2) consists of* *an admissible* 1*-shock and an admissible* 3*-shock, i.e.* $v_- = v_+$ *and*

$$u_+ - u_- < -\sqrt{\frac{(\rho_- - \rho_+)(p(\rho_-) - p(\rho_+))}{\rho_-\rho_+}}, \tag{19}$$

there exist ***infinitely many entropy solutions*** *to (1)–(2).*

This theorem ensures that, even for $v_- = v_+ = 0$, every choice of constants $\rho_\pm, u_\pm$ satisfying (19) produces forward-in-time non-standard solutions. We remark once again that non-standard solutions are genuinely two-dimensional in that they depend on both space variables (x, y) though initial data do not. A natural question is: which behavior do approximate solutions display when starting out from Riemann data as in Theorem 2? For numerical approximation we will opt for the choice $\rho_\pm = 1$ and $u_\pm = \mp\frac{1}{2}$ ($v_- = v_+ = 0$) which is compatible with the hypothesis of Theorem 2; the reason for this choice will become clear in the next sections.

2.3 *Notion of Subsolution Emerging in Riemann Fans*

The proof of Theorem 2 hinges crucially on the following Proposition which is a main ingredient of both [8, 9] (we report here a simplified version which does not care about entropy conditions as we are not interested primarily in this aspect).

Proposition 1 *Let p be any* C^1 *function and* $(\rho_\pm, \mathbf{v}_\pm)$ *be such that there exists at least one fan subsolution of (1) with initial data (2). Then there are infinitely many bounded admissible solutions* $(\rho, \mathbf{v})$ *to (1), (2).*

Proposition 1 reduces the construction of non-standard solutions to the existence of a suitable subsolution to the Euler system (1) with appropriate initial data and relies on previous results from [16, 17]. The notion of subsolution for this framework has been introduced for the first time in [9] and then used again in [8]. We recall here its definition for the sake of completeness.

Definition 2 (Fan Partition) A *fan partition* of $\mathbb{R}^2 \times (0, \infty)$ consists of three open sets P_-, P_1, P_+ of the following form

$$P_- = \{(x, t) : t > 0 \quad \text{and} \quad x < \nu_- t\} \tag{20}$$

$$P_1 = \{(x, t) : t > 0 \quad \text{and} \quad \nu_- t < x < \nu_+ t\} \tag{21}$$

$$P_+ = \{(x, t) : t > 0 \quad \text{and} \quad x > \nu_+ t\}, \tag{22}$$

where $\nu_- < \nu_+$ is an arbitrary couple of real numbers.

We denote by $\mathscr{S}_0^{2\times 2}$ the set of all symmetric 2×2 matrices with zero trace.

Definition 3 (Fan Subsolution) A *fan subsolution* to compressible Euler equations (1) with data (2) is a triple $(\overline{\rho}, \overline{\mathbf{v}}, \overline{\mathbf{u}}) : \mathbb{R}^2 \times (0, \infty) \rightarrow (\mathbb{R}^+, \mathbb{R}^2, \mathcal{S}_0^{2\times 2})$ of piecewise constant functions satisfying the following requirements.

(i) There is a fan partition P_-, P_1, P_+ of $\mathbb{R}^2 \times (0, \infty)$ such that

$$(\overline{\rho}, \overline{\mathbf{v}}, \overline{\mathbf{u}}) = (\rho_-, \mathbf{v}_-, \mathbf{u}_-)\mathbf{1}_{P_-} + (\rho_1, \mathbf{v}_1, \mathbf{u}_1)\mathbf{1}_{P_1} + (\rho_+, \mathbf{v}_+, \mathbf{u}_+)\mathbf{1}_{P_+}$$

where $(\rho_1, \mathbf{v}_1, \mathbf{u}_1) \in \mathbb{R}^+ \times \mathbb{R}^2 \times \mathcal{S}_0^{2\times 2}$ are constants with

$$\rho_1 > 0, \qquad \mathbf{u}_\pm = \mathbf{v}_\pm \otimes \mathbf{v}_\pm - \frac{1}{2}|\mathbf{v}_\pm|^2 \mathrm{Id};$$

(ii) There exists a positive constant C such that

$$\mathbf{v}_1 \otimes \mathbf{v}_1 - \mathbf{u}_1 < \frac{C}{2}\mathrm{Id}\,; \tag{23}$$

(iii) The triple $(\overline{\rho}, \overline{\mathbf{v}}, \overline{\mathbf{u}})$ solves the following system in the sense of distributions:

$$\partial_t \overline{\rho} + \mathrm{div}_{\mathbf{x}}(\overline{\rho}\,\overline{\mathbf{v}}) = 0 \tag{24}$$

$$\partial_t(\overline{\rho}\,\overline{\mathbf{v}}) + \mathrm{div}_{\mathbf{x}}\,(\overline{\rho}\,\overline{\mathbf{u}}) + \nabla_{\mathbf{x}}\left(p(\overline{\rho}) + \frac{1}{2}\left(C\rho_1\mathbf{1}_{P_1} + \overline{\rho}|\overline{\mathbf{v}}|^2\mathbf{1}_{P_+\cup P_-}\right)\right) = 0. \tag{25}$$

A remarkable feature of fan subsolutions is that, due to their "fan" structure, their existence can be checked by solving a set of algebraic equalities (coming from Rankine–Hugoniot conditions) and inequalities. This makes easier the possibility of investigating whether or not given Riemann data allow for the existence of a fan subsolution. Using the notation of [8], in Sect. 4.2, we shall show that for $\rho_\pm = 1$ and $u_\pm = \mp\frac{1}{2}$ $(v_- = v_+ = 0)$ such a subsolution exists for a pressure $p(\rho) = \frac{\rho^3}{12}$.

3 Riemann Data Generated by Compression Wave

3.1 Theoretical Approach

We now look at the non-standard solutions constructed in [9] as we are interested in non-standard solutions starting out from Lipschitz initial data as well. Moreover our aim is to combine the ideas from [9] with the ones from [8] to prepare a numerical investigation. As we disclosed in the Introduction, such solutions are obtained via a suitable Riemann problem. From Lemma 1 we know that if

$$\rho_+ < \rho_- \tag{26}$$

$$u_- = u_+ - \int_{\rho_+}^{\rho_-} \frac{\sqrt{p'(\tau)}}{\tau} d\tau \tag{27}$$

then the self-similar solution of (1)–(2) consist of a simple 1-*rarefaction wave*, i.e. a solution depending only on t and x which is locally Lipschitz for positive t and constant on lines emanating from the origin (see [14, Sect. 7.6] for the precise definition). Reversing their order (i.e. exchanging $+$ and $-$) the very same constants allow for a *compression wave* solution on $\mathbb{R}^2 \times]-\infty, 0[$ converging, for $t \uparrow 0$, to the jump discontinuity of (2). When this is the case we will then say that the data (2) are *generated by a classical compression wave*. This idea combined with the subsolution technique explained in the previous section is at the core of the construction in [9]. The main theorem of [9] is the following:

Theorem 3 *Assume $p(\rho) = \rho^2$. Then there are Riemann data as in (2) for which there are infinitely many bounded admissible solutions $(\rho, \mathbf{v})$ of (1) on $\mathbb{R}^2 \times [0, \infty[$ with* $\inf \rho > 0$. *Moreover, these data are generated by classical compression waves.*

As an obvious corollary of Theorem 3 we have:

Corollary 1 *Assume $p(\rho) = \rho^2$. There are Lipschitz initial data $(\rho^0, \mathbf{v}^0)$ for which there are infinitely many bounded admissible solutions $(\rho, \mathbf{v})$ of (1) on $\mathbb{R}^2 \times [0, \infty[$ with* $\inf \rho > 0$. *These solutions are all locally Lipschitz on a finite interval on which they all coincide with the unique classical solution.*

For the proofs we refer to [9]. Once more the existence of non-standard solutions $(\rho, \mathbf{v})$ forward-in-time starting from some Riemann data (2) is ensured by the existence of a fan subsolution with same Riemann data as Proposition 1 states. The choice of data can be made such that they are also generated by a compression wave; thus, by Proposition 1 infinitely many non-standard solutions starting out from Lipschitz data (chosen as the compression wave at time $t = -1$) can be proven.

Riemann data in Theorem 3, and considered in [9], don't satisfy (19). In particular, the classical self-similar evolution of the Lipschitz data in Corollary 1, yields a compression wave which breaks into a discontinuity that further evolves into a composite wave consisting in a 1-shock and 3-rarefaction wave. Constructions in [8, 9] suggest to investigate a problem for which compression waves could be explicitly computed in order to study Riemann data, still generated by a compression wave, but ensuring the existence of a fan subsolution as in Theorem 2, thus combining the strategies of [8, 11].

3.2 Decoupling of Euler System with $\gamma = 3$

For numerical investigations the choice of pressure $p(\rho) = \frac{\rho^3}{12}$ turned out to be convenient as it allows to decouple the system (7) into two Burgers equations (see [6]). Indeed, let us now consider, for $x \in \mathbb{R}, t > 0$, the elementary 1D system,

$$\partial_t \rho + \partial_x(\rho u) = 0, \qquad \partial_t(\rho u) + \partial_x(\rho u^2 + \frac{\rho^3}{12}) = 0. \tag{28}$$

In order to cope with the notation in [6, pp. 171–172] (where these systems were studied and then used to produce numerical geometric optics results in [25]),

$$a(t,x) = u + \frac{\rho}{2} \geq u - \frac{\rho}{2} = b(t,x).$$

Remarkably, inside areas where the solution is smooth, (28) decouple into,

$$\partial_t a + \lambda^+(a,b)\partial_x a = 0, \qquad \partial_t b + \lambda^-(a,b)\partial_x b = 0,$$

and since $\lambda^+ = a$, and $\lambda^- = b$, there remain just two Burgers equations,

$$\boxed{\partial_t a + \frac{1}{2}\partial_x(a^2) = 0, \qquad \partial_t b + \frac{1}{2}\partial_x(b^2) = 0.}$$

4 Explicit Initial Data Leading to Subsolutions

4.1 Lipschitz y-Independent Compression Wave

Coupling in (28) occurs only because of Rankine–Hugoniot conditions inside a shock wave. By following ideas from [25, p. 172], where the same system (28) was set (for different purposes!), the following data in $t = 0$ are obtained:

$$a(t=0,x) = \left(\chi_{x<-1} - x\chi_{x\geq -1}\right)\chi_{x<0}, \quad b(t=0,x) = a(t=0,x-1) - 1, \tag{29}$$

with χ_A standing for the indicator of the set A. By using the formulas,

$$\rho = a - b \geq 0, \qquad \rho u = \frac{1}{2}(a^2 - b^2), \qquad u = \frac{a+b}{2},$$

one recovers corresponding initial data in "macroscopic variables": see Fig. 2. The main drawback is that $\rho(t = 0, \cdot)$ goes very close to vacuum in $x = 0$. There is breakup at time $t = 1$, and emerging dynamics are displayed in the plane x, t on Fig. 1 in variables ρ, u: Final data, at time $t = 1.95$, are displayed on

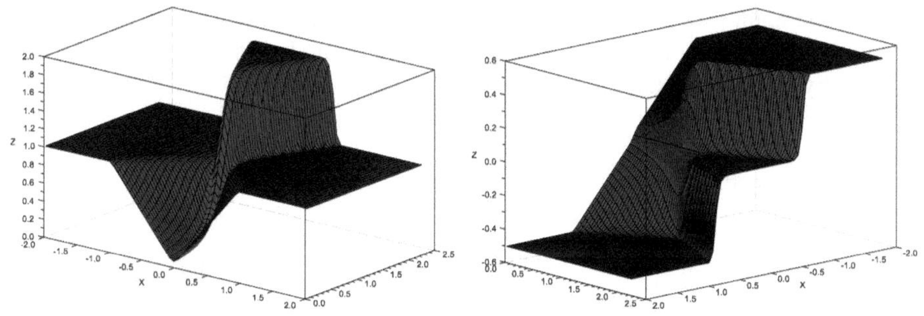

Fig. 1 Time evolution before and after breakup for ρ (*left*) and u (*right*)

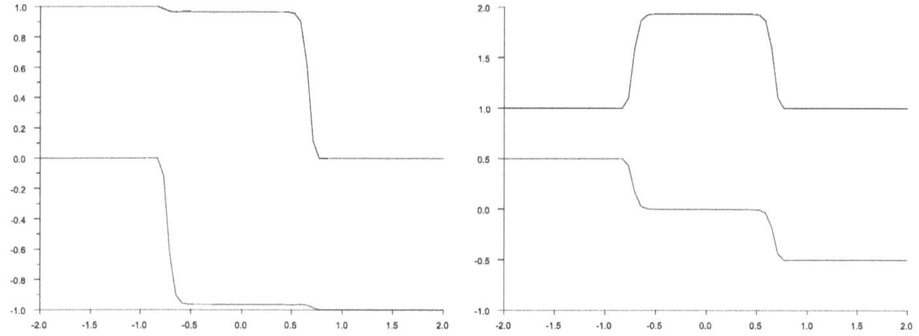

Fig. 2 Final results at time $t = 2$ for a, b (*left*), and ρ, u (*right*)

Fig. 2, in both a, b and ρ, u variables: two shocks separated by a constant state. Hereafter, the existence of non-standard solutions associated with the preceding $\rho(t = 0, \cdot, \cdot), u(t = 0, \cdot, \cdot)$ is established by means of a simple variation of the arguments given in [8]. A simple criterion is singled out in order to (at least, try to) look after them numerically, through a specific benchmark:

- the value of $1 \leq \rho_m$ in the triangular zone (see Fig. 1, left) is slightly lower for a non-standard solution compared to the usual self-similar one;
- non-standard vertical velocity $v(t, \cdot, \cdot)$, $t > 1$ beyond breakup, isn't null (along with the vorticity). Horizontal velocity $u(t, \cdot, \cdot)$ oscillates around 0.

4.2 *Existence Criterion for a Subsolution*

In order to establish the existence of non-standard solutions for our situation, the following Riemann problem for the simple 1D system (28) is relevant,

$$\rho_{\pm} = 1, \qquad u_{\pm} = \mp\frac{1}{2}, \qquad v_{\pm} = 0. \tag{30}$$

Such a Riemann problem yields a self-similar solution involving 1-shock and a 3-shock. The middle state $\rho_m > 1$ has to be determined, and the middle velocity $u_m = 0$ by symmetry. Rankine–Hugoniot relations give,

$$u_- = u_m + \sqrt{\frac{(\rho_m - \rho_-)(p(\rho_m) - p(\rho_-))}{\rho_m \rho_-}} \Rightarrow \frac{1}{4} = \frac{(\rho_m - 1)(\rho_m^3 - 1)}{12\rho_m}.$$

So, the middle state $0 \leq \rho_m \approx 1.93251$ is a root[1] of the polynomial,

$$P(X) = X^4 - X^3 - 4X + 1 = 0, \qquad X_+ \simeq 1.93251\,, X_- = 0.24716.$$

The kinetic Riemann solver used to produce Fig. 1, with $2^6 = 64$ grid points in x, gives a numerical value of $\rho_m^{\Delta x} = 1.9294259$, so a scheme's deviation of

$$1.93251 - 1.9294259 = 0.0030841, \qquad \to 0.16\%.$$

Yet, applying the formulas (4.51)–(4.54), written in [8, p. 1038], it comes:

$$R = 0, \quad \beta = 0, \quad \nu_\pm = \pm \frac{|u_+ - u_-|}{2(\tilde{\rho} - 1)}, \quad \epsilon_1 = \frac{|u_+ - u_-|^2}{4(\tilde{\rho} - 1)} - \frac{p(\tilde{\rho}) - p(1)}{\tilde{\rho}}.$$

In order to get non-standard solutions with u oscillating around zero ($\to \beta = 0$), we need to find ϵ_1 and ϵ_2 (see [11] for notation) such that

$$0 < \epsilon_1 \ll 1, \qquad \epsilon_2 > 0 \text{ as given in formulas (4.82)–(4.83) in [8].}$$

It is easy to verify (as in [8, Lemma 4.5]), that, it is possible to define a function,

$$\boxed{\mathbb{R}_*^+ \ni \rho \mapsto \epsilon_1(\rho), \qquad \epsilon_1(\rho_m) = 0, \qquad \frac{d\epsilon_1}{d\rho} < 0.}$$

Indeed, as $|u_+ - u_-| = 1$, our self-similar value $\rho_m \approx 1.93251$ solves,

$$\frac{(\rho_m - 1)(p(\rho_m) - 1)}{\rho_m} = \frac{1}{4}, \Leftrightarrow \frac{|u_+ - u_-|}{4(\rho_m - 1)} - \frac{p(\rho_m) - 1}{\rho_m} = \epsilon_1(\rho_m) = 0.$$

For the derivative,

$$\frac{d\epsilon_1}{d\rho}(\rho) = -\frac{1}{4(\rho - 1)^2} - \frac{\rho^3/6 + 1}{\rho^2} < 0.$$

[1]https://www.symbolab.com/solver/polynomial-equation-calculator/.

By smoothness of both $\rho \mapsto \epsilon_1(\rho)$ and $\rho \mapsto \epsilon_2(\rho)$, there exists a small neighborhood of (the self-similar value) $\rho_m \approx 1.93251$, say (ρ_ℓ, ρ_m) such that,

$$\boxed{\forall \rho \in (\rho_\ell, \rho_m), \qquad 0 < \epsilon_1(\rho) \ll 1, \quad \epsilon_2(\rho) > 0.} \tag{31}$$

Theorem 4 *Compressive, Lipschitz initial data (29), leading to the Riemann problem (30) for (28), give rise to an infinity of non-standard L^∞ solutions of the 2D system (1): these "wild solutions" are such that, inside "the P_1 zone",*

- *shocks flanking the "P_1 zone" move at speeds $v_\pm = \frac{\pm 1}{2(\tilde{\rho}-1)}$,*
- *in between, $\rho(t, x, y) = \tilde{\rho} < \rho_m \approx 1.93$ satisfies (31),*
- *and $u(t, x, y)$ oscillates around zero (because $\beta = 0$),*
- *both $v(t, x, y) \neq 0$ and the vorticity $\omega(t, x, y) = u_y - v_x \neq 0$.*

This is an immediate consequence of Theorem 1 stated in [8, p. 1022].

Initial data (29) are usually referred to as to "colliding flow data", or "colliding winds" leading to so-called "thin shell instabilities" in astrophysics [5]) (in contrast to shear flow data, see [1]). Such data are notoriously difficult to handle numerically, as explained in e.g. [30, §7.7]; it generates "overheating" problems, too, when pressure depends on temperature, [37]. Finally, Morton and Roe conjectured that *anomalous Euler solutions result from the systematic creation of vorticity due to truncation error*, in [31, p. 189].

5 Preliminary Numerical Results on Coarse Grid

All the forthcoming numerical results are obtained with a uniform 2D Cartesian grid with square computational cells of width $\Delta x = \Delta y = h > 0$. The standard CFL restriction imposes that the time-step satisfies: $\sup |2u \pm \rho| \Delta t \leq h$. Neither the numerical values before shock onset nor the macroscopic density ρ will be displayed hereafter because they are devoid of any particular phenomenon.

5.1 Snapshot of Dimensional-Split Kinetic Flux-Splitting

Denoting $\mathbf{U}^n_{i,j} \simeq \mathbf{U}(t^n, x_i, y_j)$ the piecewise-constant approximation of a solution of (1) at $t^n = n\Delta t$, with $x_i = i\Delta x$, $y_j = j\Delta y$, $n, i, j \in \mathbb{N} \times \mathbb{Z}^2$, the standard alternating-directions, kinetic flux-splitting, marching scheme reads, (see e.g. [13, 27])

$$\mathbf{U}^{n}_{i,j} = \mathbf{U}^{n}_{i,j} - \frac{\Delta t}{h}\left[F^{+}(\mathbf{U}^{n}_{i,j}) + F^{-}(\mathbf{U}^{n}_{i+1,j}) - F^{+}(\mathbf{U}^{n}_{i-1,j}) - F^{-}(\mathbf{U}^{n}_{i,j})\right]$$
$$-\frac{\Delta t}{h}\left[G^{+}(\mathbf{U}^{n}_{i,j}) + G^{-}(\mathbf{U}^{n}_{i,j+1}) - G^{+}(\mathbf{U}^{n}_{i,j-1}) - G^{-}(\mathbf{U}^{n}_{i,j})\right].$$

Split-flux functions in x and y read $F^{\pm}$ and $G^{\pm}$, respectively. Thanks to the particular structure inherited from the pressure law $\gamma = 3$, they are essentially given by [26, Eqs. (6.30)–(6.31)]. More precisely, let $a = u + \frac{\rho}{2}$, $b = u - \frac{\rho}{2}$, and define,

$$M_2(a,b) = (a-b, \tfrac{a^2-b^2}{2}) = (\rho, \rho u),$$
$$F_2(a,b) = (\tfrac{a^2-b^2}{2}, \tfrac{a^3-b^3}{3}) = (\rho u, \rho u^2 + \tfrac{\rho^3}{12}),$$

the exact flux function of (1) in the x-direction reads,

$$F(\rho,u,v) = (\rho u, \rho u^2 + \frac{\rho^3}{12}, \rho u v) = (F_2(a,b),\ F_{2,1}(a,b)v).$$

Analogously, let $c = v + \frac{\rho}{2}$, $d = v - \frac{\rho}{2}$, the flux function in y reads,

$$G(\rho,u,v) = (\rho v, \rho u v, \rho v^2 + \frac{\rho^3}{12}) = (F_{2,1}(c,d),\ F_{2,1}(c,d)u, F_{2,2}(c,d)).$$

Denoting $a^{+} = \max(0,a)$ and $a^{-} = \min(0,a)$, the expression of flux functions is

$$F^{\pm}(\mathbf{U}) = \left(F_2(a^{\pm},b^{\pm}),\ F_{2,1}(a^{\pm},b^{\pm})v\right),$$
$$G^{\pm}(\mathbf{U}) = \left(F_{2,1}(c^{\pm},d^{\pm}),\ F_{2,1}(c^{\pm},d^{\pm})u, F_{2,2}(c^{\pm},d^{\pm})\right).$$

The treatment of simply transported velocities is directly inspired by standard results for multi-component flows [29]. Boundary conditions are purely incoming on both left and right sides of the computational domain since $u^{\pm} = \mp\frac{\rho^{\pm}\equiv 1}{2}$, so that they can be considered supersonic. Boundary data in the y direction are more delicate, because they are partly incoming, partly exiting. More precisely, being $\theta > 0$ the rotation angle, we interpolate non-conservative variables ρ, u, v like, e.g.,

$$\tilde{\rho}^{n}_{i,j=1} = (1-\tan(\theta))\rho^{n}_{i,1} + \tan(\theta)\rho^{n}_{i+1,1}.$$

Similarly, a formal second-order accuracy in space can be reached by setting the usual slope-limiting strategy; we chose to apply it again to non-conservative variables (because before shock onset, these variables are piecewise linear). Beyond shock onset, there are only constant states so that reconstructions have a marginal role. Concerning the time integrator, a standard "predictor-corrector" was used.

5.2 Transient Loss of Translational Invariance

A distinctive feature of non-standard solutions is their loss of translational invariance beyond shock onset, even starting from very elementary Riemann data like (2). On Fig. 3, we display at two consecutive instants both the components u, v of the velocity vector and the centered vorticity. Especially, both a contrast-enhanced colormap and a hard thresholding were used,

$$\min(0.003, \max(-0.003, [u(t^n, x_i, y_j), \text{ or } v(t^n, x_i, y_j)])), \quad t^n = 1.9 \text{ or } 2.6,$$

so that small fluctuations (of the order of 1%) are easily visualized. It appears that translational invariance are mostly kept on horizontal velocity, except very close to

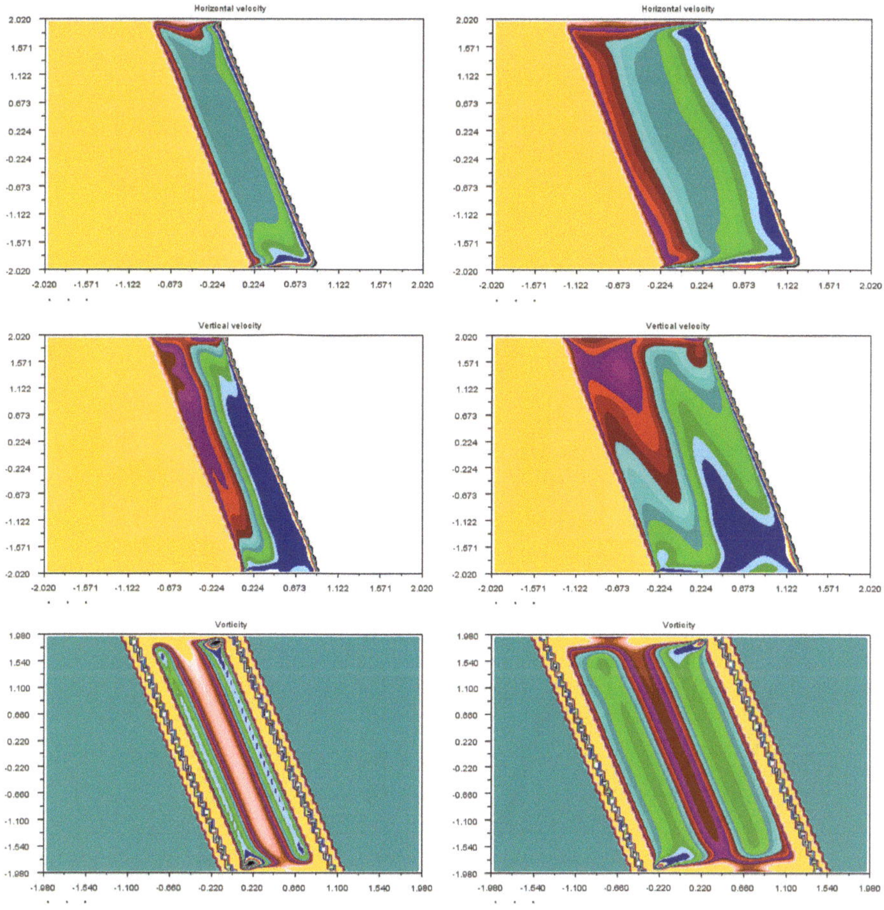

Fig. 3 Horizontal, vertical velocity (*top*, *middle*), vorticity (*down*) at $t = 1.9$ (*left*), $t = 2.6$ (*right*)

each of the vertical boundaries, and, more surprisingly, on the centered vorticity. Oppositely, the vertical component of the velocity completely lost its translational invariance. One may think it's a consequence of a bad boundary treatment; however, overall situation is satisfying during the whole pre-shock phase.

5.3 *Back-Rotated Solution and Oscillating Velocities*

To complete the former picture, we display on both the first rows of Fig. 4 the contours of the velocity components of the "back-rotated solution" (according to Definition 1), along with several "cuts" along the y direction. Cutting along the x-direction reveals only a smooth behavior. Oppositely, small oscillations appear along the y-direction in the whole area flanked by the two transonic shocks. The situation is a bit different for horizontal and vertical velocities: for the first one, the amplitude of variations is $\pm 5 \times 10^{-4}$, so 0.1% only. Instead, for the second, amplitude (of shear-like motion) goes slightly over 10^{-3} instead of being uniformly null. Examining the arrows of the velocity field reveals that the origin of such oscillations is related to the numerical scheme's preference to confine motion along the grid's axes inside the zone bordered by shocks. Put in another way: **when velocities go below some**

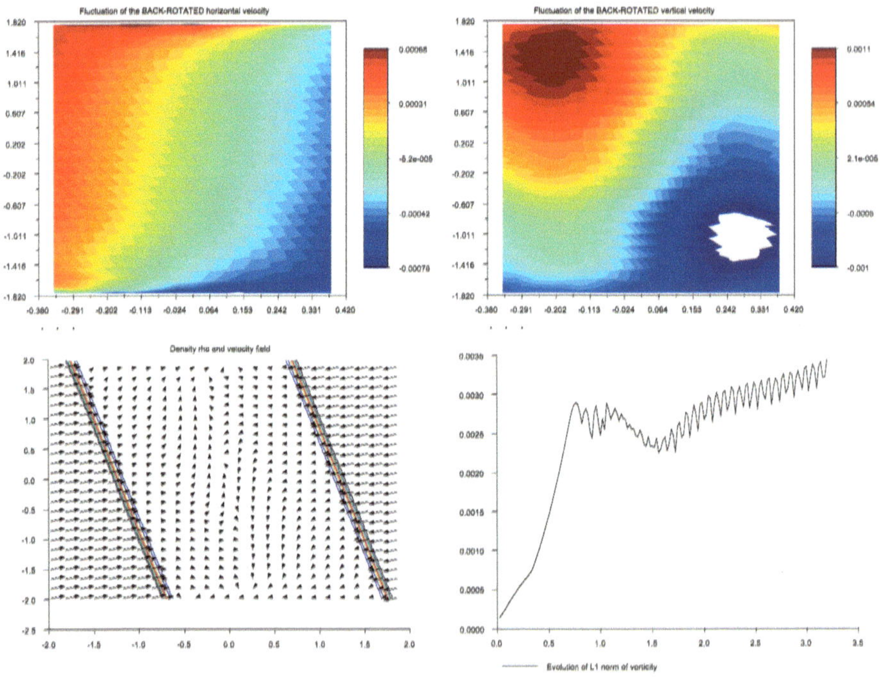

Fig. 4 Horizontal, vertical velocity (*top*), vorticity (*bottom*) for an angle $\theta = \frac{\pi}{12}$

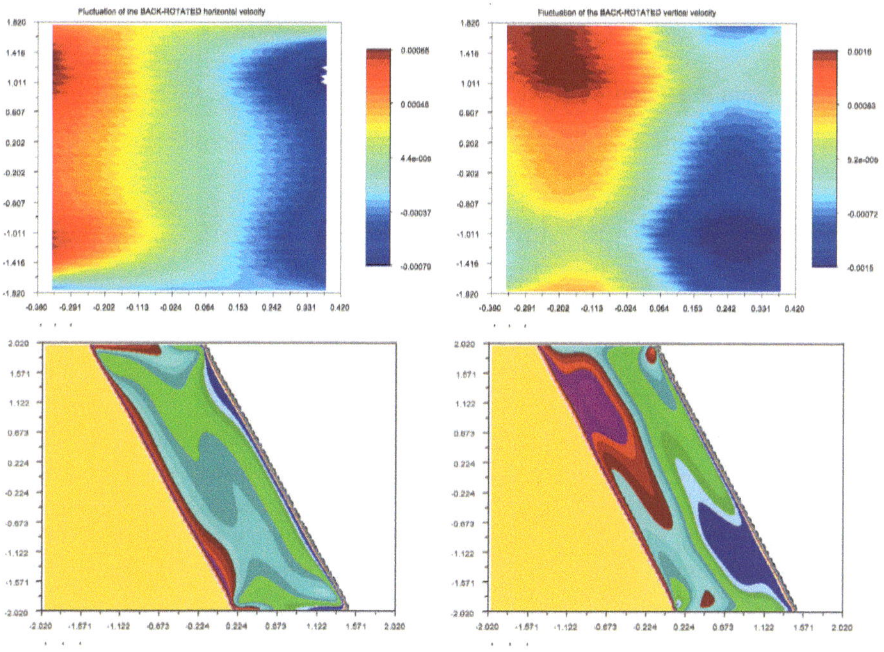

Fig. 5 Horizontal, vertical velocity (*top*), fluctuations (*bottom*) for an angle $\theta = \frac{\pi}{8}$

value, they are preferably directed according to the computational grid: this partly explains time-growth of numerical vorticity (bottom, right of Fig. 4).

Yet, increasing the value of the rotation angle θ from $\frac{\pi}{12}$ to $\frac{\pi}{8}$, the results of Fig. 5 are obtained. Overall, the situation is kept roughly the same. Even the amplitudes of oscillations are comparable. Examining the analogue of Fig. 3 reveals that for such value of the rotation angle, the dimensional-split scheme is less able to keep translational invariance in the horizontal component of the velocity vector. However, the situation on the vertical component appears to be slightly better.

6 Riemann Data Yielding Vortices in the Stagnation Zone

An inspection of Figs. 3 and 5 suggests that an approximately vertical shear flow develops inside the stagnation area flanked on both sides by the moving oblique shocks. In [32–34], it is thoroughly explained how standard upwind schemes usually fail at recognizing that type of oblique flows and inevitably create artificial acoustic waves in order to satisfy global conservation requirements.

6.1 Reformulation in Terms of a Commutator

In complement to Definition 1, we state:

Definition 4 For any $\theta \in (0, \pi/2)$, let the operator $\mathscr{R}_\theta : \mathbb{R}^3 \to \mathbb{R}^3$ stand for,

$$\forall \mathbf{x} \in \mathbb{R}^2, \qquad \mathscr{R}_\theta[\mathbf{U}](\mathbf{x}) = \begin{pmatrix} 1 & 0 \\ 0 & R_\theta \end{pmatrix} \mathbf{U}(R_\theta(\mathbf{x})), \qquad R_\theta = \begin{pmatrix} \cos\theta & \sin\theta \\ -\sin\theta & \cos\theta \end{pmatrix},$$

along with a commutator, denoted $\mathscr{C}_\theta$, built on a "solution operator S_t",

$$\boxed{\forall t > 0, \qquad \mathscr{C}_\theta[\mathbf{U}_0](t, \cdot) = [S_t, \mathscr{R}_\theta](\mathbf{U}_0) = S_t \circ \mathscr{R}_\theta(\mathbf{U}_0) - \mathscr{R}_\theta \circ S_t(\mathbf{U}_0).}$$

An interesting feature of $\mathscr{C}_\theta$ lies in the fact that it can be set up, even for an approximate, numerical, solution operator; in particular, we shall set it up with the "kinetic dimensional splitting" S_t^{KDS} formerly presented in Sect. 5.1. Clearly, such a numerical scheme produces the unique self-similar solution when $\theta = 0$, so that the initial discontinuity of (2) is aligned with the grid. Hence, for $\Delta x = \Delta y > 0$ and $\theta \in (0, \pi/2)$, the commutator $[S_t^{KDS}, \mathscr{R}_\theta]$ allows to visualize the deviation from the self-similar solution induced by the dimensional-splitting mechanism.

Lemma 3 *Let $\mathbf{U}_0(\mathbf{x})$ be a (possibly wild) initial Riemann data like (2), then if $\mathbf{U}(t, \cdot) := S_t(\mathbf{U}_0)$ is the unique, exact self-similar solution of (1)–(2), then,*

$$\forall t > 0, \qquad \mathscr{C}_\theta[\mathbf{U}_0](t, \cdot) = 0.$$

Accordingly, our numerical strategy is yet to:

- prescribe Riemann data like (30) for (1) with $\gamma = 3$ pressure;
- choose an angle $\theta \in (0, \pi/2)$ sufficiently big to create a noticeable effect;
- compute the commutator $[S_t^{KDS}, \mathscr{R}_\theta]$ for $t > 0$ for waves to have propagated;
- so that it may reveal quantitatively the deviation of the numerical approximation obtained with $\theta > 0$ from the self-similar one, which corresponds to $\theta = 0$.

6.2 Practical Visualization of That Commutator

Given a Cartesian computational mesh, characterized by $\Delta x = \Delta y$, along with a time-step Δt satisfying the usual CFL stability restriction, one can build a whole family of compressive Riemann initial data, of the type (30), and leading to non-standard solutions as well. More precisely, (30) asks for a mild restriction on S_t^{KDS},

$$\max(|a|, |b|) = \max(|u| \pm \frac{\rho}{2}) \simeq 1, \qquad \text{so that } \Delta t \leq \Delta x.$$

On exactly the same grid, one can work out the following Riemann data, too,

$$\rho_\pm = \rho_0 \in (0,2), \qquad u_\pm = \mp(1 - \frac{\rho_0}{2}), \qquad v_\pm = 0.$$

This gives flexibility in choosing a "big mass moving slowly", or the opposite; (30) is recovered from fixing $\rho_0 = 1$. Instead of numerically visualizing the aforementioned $\mathscr{C}_\theta$, it was preferred to display results aligned with the grid's axes, so

$$\tilde{\mathscr{C}}_\theta[\mathrm{U}_0](t,\cdot) = \mathscr{R}_\theta^{-1} \circ S_t^{KDS} \circ \mathscr{R}_\theta(\mathrm{U}_0) - S_t^{KDS}(\mathrm{U}_0).$$

Figure 6 displays, on its left side, the stagnation zone (flanked by two Lax shocks) of the back-rotated numerical solution generated by a (formal) second-order extension (in both space and time) of the kinetic dimensional-splitting "solution operator" S_t^{KDS}, at time $t \simeq 0.76$, for a griding of $x, y \in (-2,2)^2$ given by $\Delta x = \Delta y = 10^{-2}$. **Such a numerical process obviously preserves a self-similar solution as long as its waves are perfectly aligned with the grid's axes**; on the contrary, the operator $\mathscr{R}_\theta$ with $\theta = \frac{\pi}{8}$ is meant to reveal the effects of sensibly rotating initial data. These effects, which are roughly grid-size independent (see below), are:

- a fingering process on horizontal velocity, which breaks translational invariance with respect to the y variable. Such a loss of invariance is stronger compared to what was observed, when starting from Lipschitz compression waves, in Sect. 5.3;
- an even more complex redistribution of vertical velocity, involving an oscillating pattern with respect to both x and y variables;
- surprisingly, one would expect such a disordered agitation to induce a chaotic vorticity. On the contrary, it appears that, from the very first iterations of the numerical process, the centered approximation of vorticity presents many "patches", which, as time goes on, interact and merge with each other, into a quite stable pattern. To each patch corresponds a local vortex.

While being rather spectacular, these fluctuations are of small amplitude, and this is necessary in order for the whole numerical approximation to keep on having two stable shocks propagating at a velocity slightly higher than in the self-similar solution. This portrait doesn't vary notably when changing the value of $\rho_0 \in (0,2)$ and the grid-size Δx; note that the time-step Δt is fixed so as to reduce numerical dissipation at its minimum (see [35] for interesting results in this direction). Yet, on the right of Fig. 6, are displayed "cuts" along the y-axis of each quantity, u, v, ω (horizontal and vertical velocities, vorticity), that is to say, we superimposed

$$u(t = 0.76, x_i, \cdot), \qquad v(t = 0.76, x_i, \cdot), \qquad \omega(t = 0.76, x_i, \cdot),$$

for several values of i such that x_i roughly remains in the stagnation area of the self-similar Riemann fan. Deriving these figures was the main reason for having worked with $\tilde{\mathscr{C}}_\theta$, instead of $\mathscr{C}_\theta$. Clearly, a rather regular pattern of oscillations is

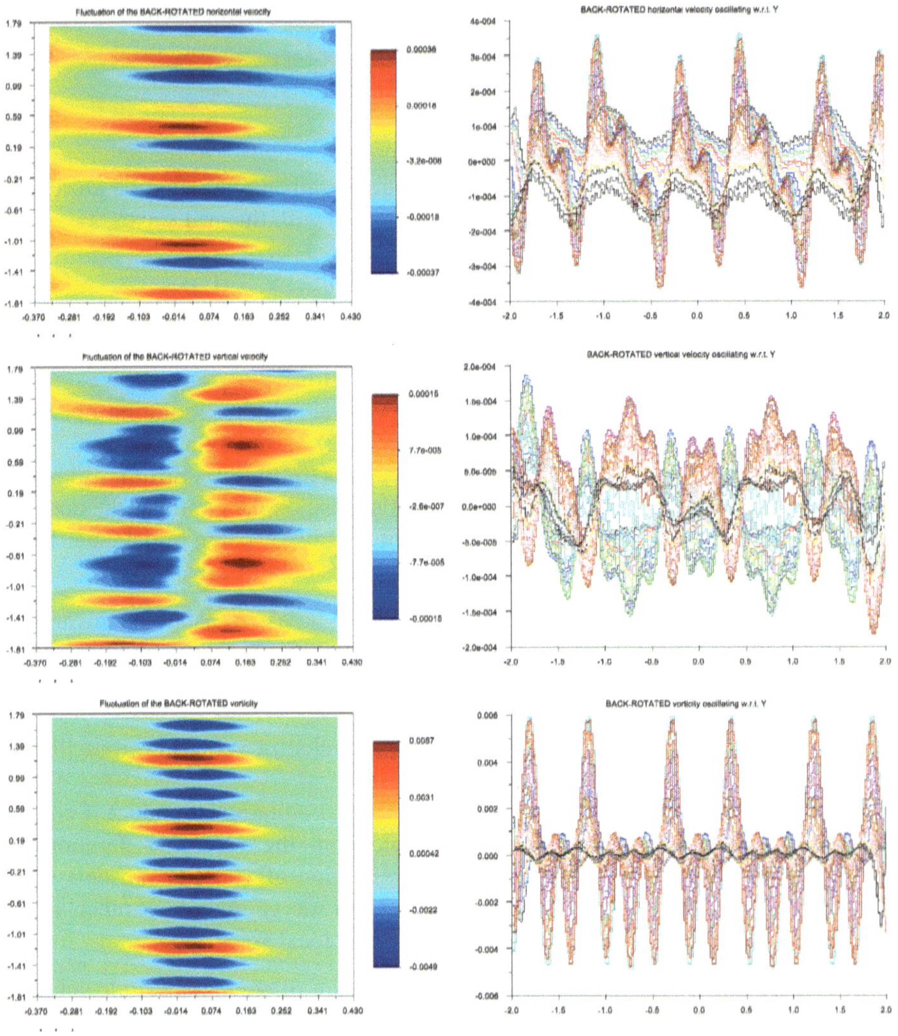

Fig. 6 Horizontal, vertical velocity and vorticity (*top to down, left*) in the stagnation area, corresponding fluctuations (*right*), at time $t \simeq 0.76$, for a rotation angle $\theta = \frac{\pi}{8}$ and grid 200×200

revealed, on both horizontal and vertical velocities, but also on the more sensitive vorticity. Such observations suggest that **the commutator $\tilde{\mathscr{C}}_\theta$ isn't made of numerical disordered noise; instead it appears to contain information encoded in the frequency of oscillations**. This frequency is related to the grid's fineness, as illustrated on Fig. 7: **the finer the grid, the higher the oscillations frequency**. Similar illustrations are obtained for both horizontal and vertical velocities, with increasing frequencies, too, invalidating strong convergence (in accordance with [24]).

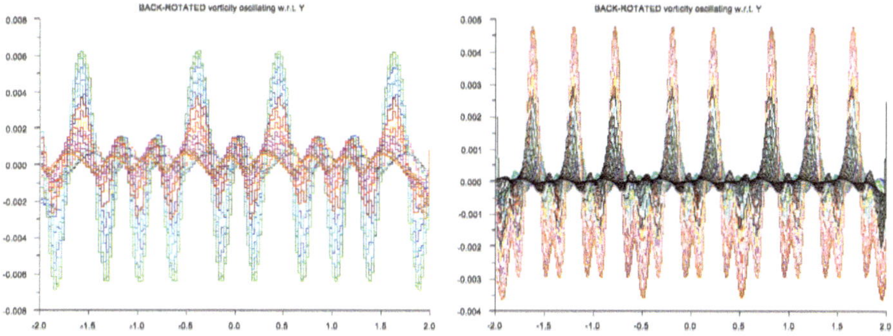

Fig. 7 Vorticity oscillations for $\theta = \frac{\pi}{8}$ and grids 150 × 150 (*left*), 300 × 300 (*right*)

Accordingly, we display supplementary numerical results, obtained on finer computational grids, and using an accurate BGK scheme, proposed in [3]. In order to emphasize phenomena occurring inside the stagnation area, the velocity vector is multiplied by 10^{10} inside it, and zeroed outside. The test again consists in simulating only the Riemann data (30) for (1) with a rotation angle $\theta = \frac{\pi}{12}$. On Fig. 8, artificial numerical vortices are revealed after breakup time with rather fine grids; besides, refining brings smaller, but more numerous vortices. Hence, vortices are not specific to the algebra of the kinetic flux-splitting scheme presented in Sect. 5.1.

6.3 An Instability Due to the Grid's Cutting Frequency

This section is meant to propose an explanation for the phenomenon of "vortex generation", displayed in the preceding section, exclusively at the numerical level; such an explanation is complementary to the early studies of Roe [33, 34] (i.e. much before the publication of [7, 16]). To this end, it is useful to compare the approximations generated by S_t^{KDS} with Riemann data like (30) at the first iteration, $t = \Delta t = \Delta x$, with the ones obtained out of smooth compressive data (like in Sect. 5.3), **leading to the exact same Riemann problem**, at a time $t \simeq 1 + \Delta t$: see Fig. 9. By inspecting Fig. 9, which involves only 50 × 50 grid points,

- one notices that, for $\theta > 0$, the approximation emerging from Riemann data involving a constant $\rho(t = 0, \cdot, \cdot) \equiv \rho_0$ displays a very spiky pattern, already signaling that it doesn't really match the self-similar solution's structure;
- oppositely, the numerical solution resulting from the compression wave shows a "W-profile", typical of Gibbs phenomena for spectral approximations in presence of excessively high frequencies. Scrutinizing wave interactions around time $t \simeq 1$, which should normally bring out a Riemann problem like (30), one sees that artificial dissipation prevents the constant density ρ_0 to appear (see also [22]).

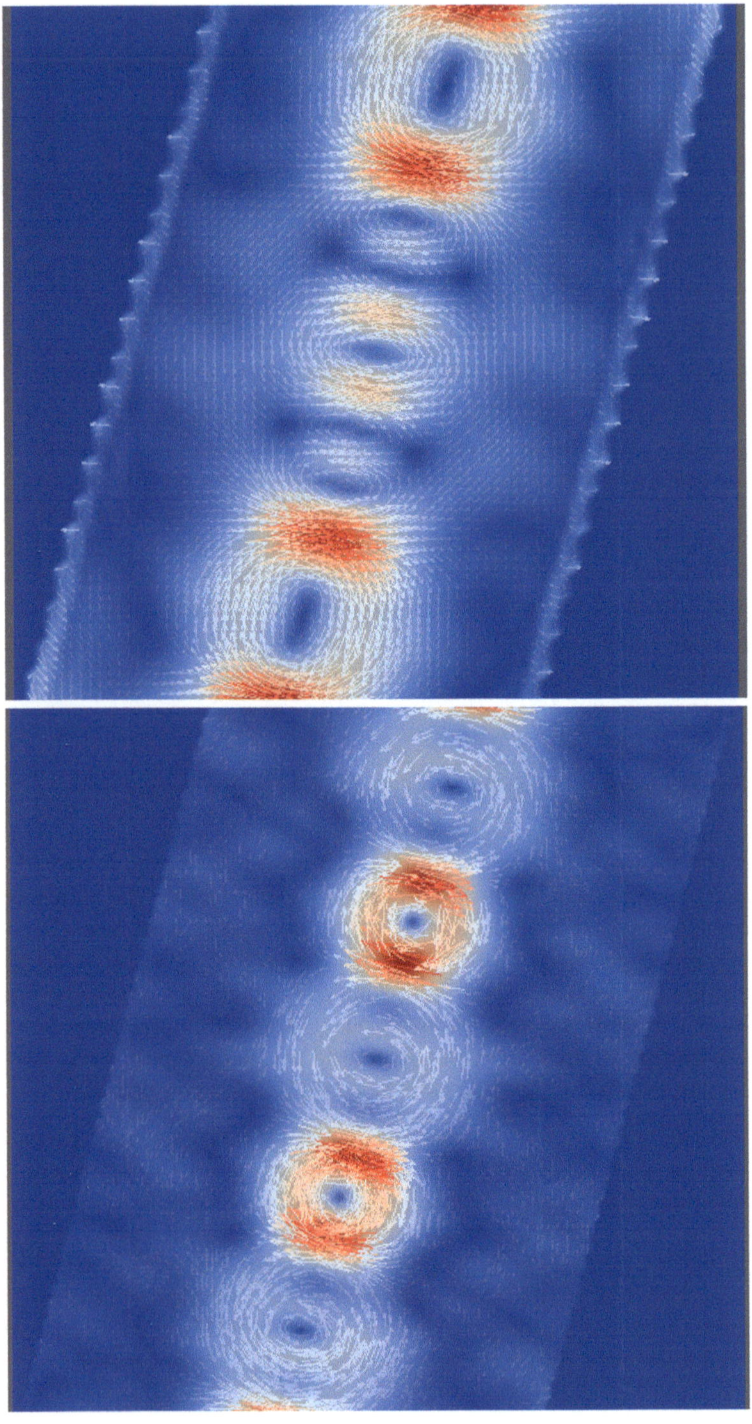

Fig. 8 Velocity fields obtained from [3] with grid 150 × 150 (*top*), 300 × 300 (*bottom*)

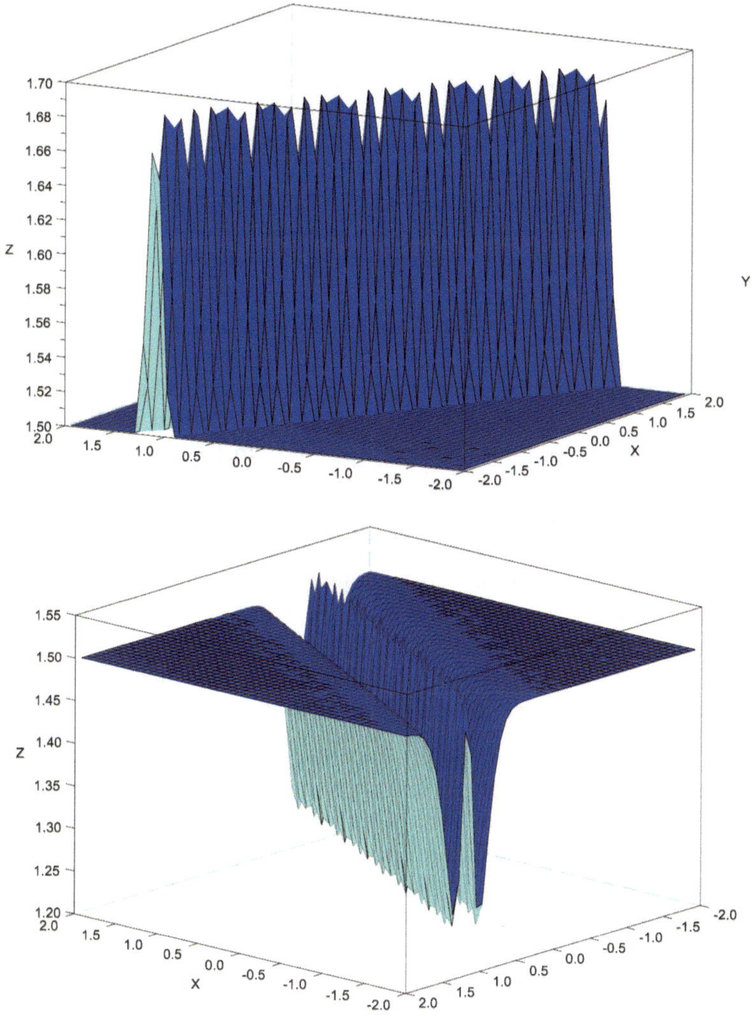

Fig. 9 First iteration of $\rho(\Delta t, \cdot, \cdot)$, with Riemann data (*top*) vs. Lipschitz data (*bottom*)

Of course, both the grid's fineness and the angle θ are kept identical. The self-similar solution to Riemann data (30) has two shocks moving at velocities,

$$\pm\sigma = \frac{\rho_0 u_+}{\rho_m - \rho_0} \simeq \frac{0.5}{0.93} \simeq 0.53.$$

When the rotation of angle $\theta = \frac{\pi}{8}$ is applied, it breaks into two components,

$$\sigma(\cos\theta, \sin\theta) \simeq 0.53(0.924, 0.383) \simeq (0.488, 0.2).$$

At the first iteration, the self-similar solution ρ is an indicator function,

$$\forall x, y, \qquad \rho(t = \Delta t, x, y) = \rho_0 + (\rho_m - \rho_0)\chi_{R_{-\theta}(x,y)\in(-\sigma\Delta t, \sigma\Delta t)}.$$

In particular, when sampled onto the grid, it is supported in the domain,

$$(-\sigma\Delta t\ \cos\theta, \sigma\Delta t\ \cos\theta) \times (-\sigma\Delta t\ \sin\theta, \sigma\Delta t\ \sin\theta).$$

However, such a domain is strictly smaller than any computational cell,

$$\sigma\Delta t \cos\theta < \frac{\Delta x}{2}, \qquad \text{so that } (-\sigma\Delta t\ \cos\theta, \sigma\Delta t\ \cos\theta) \subset (-\frac{\Delta x}{2}, \frac{\Delta x}{2}),$$

and this inclusion is strict (it's even worse along the vertical axis). Consequently, the self-similar ρ isn't representable on the grid at the first iteration, hence the dimensional-split scheme "does something" so as to locally preserve total mass (conservation property) and remain consistent with a weak solution (Lax-Wendroff property). As the time-evolution of S_t^{KDS} reveals that ρ ceases to be sharply-varying, we believe that the scheme adjusts the 2D velocities in order to compensate for the pressure gradients created at the very first iteration: this creates the small vortices. At this level, we pause in order to make it clear that refining the grid is useless: indeed, dividing $\Delta x = \Delta y$ by some factor bigger than one implies, by the CFL restriction, that the time-step Δt must be divided, too. And since the support of the indicator function depends on Δt, the issue remains.

Remark 1 One may think that vortices may be generated because of our pointwise application of rotation matrices and that applying some sort of averaging process at time $t = 0$ would tame the instability. More precisely, when any computational cell is cut by the discontinuity line separating U_{left} from U_{right}, its value can be chosen so as to be an average of those left and right states, according to the ratio of the area of the cell located at each side of the jump line. This somehow amounts to applying a bit of artificial viscosity at time $t = 0$, prior to starting the marching scheme. Corresponding numerical results, shown in Fig. 10, suggest that such a regularization induces marginal changes, suggesting that the vorticity generation process is robust, as it appears to be stable to small perturbations.

This agrees with some of Roe's ideas in [33] and Elling's ones in [20, 21]. Now, one may think that artificial viscosity is about to smooth out everything after a few iterations, but it's not how it goes. Small vortices interact with each other: some merge into bigger ones, other are canceled, but in any case, a structure looking like Fig. 8 appears in the stagnation area while ρ goes back to look like being piecewise-constant. After many iterations, the vortex structure is well described by Fig. 6, and seems more compatible with a non-standard solution rather than with a self-similar one. Back-rotating such vortex structures may yield plausible "numerical non-standard solutions", which details depend on the internal algebra of each type of algorithm: this would be compatible with the fact that there exists an infinity of them. Beside, it was observed [35] that artificial viscosity theory, mostly developed

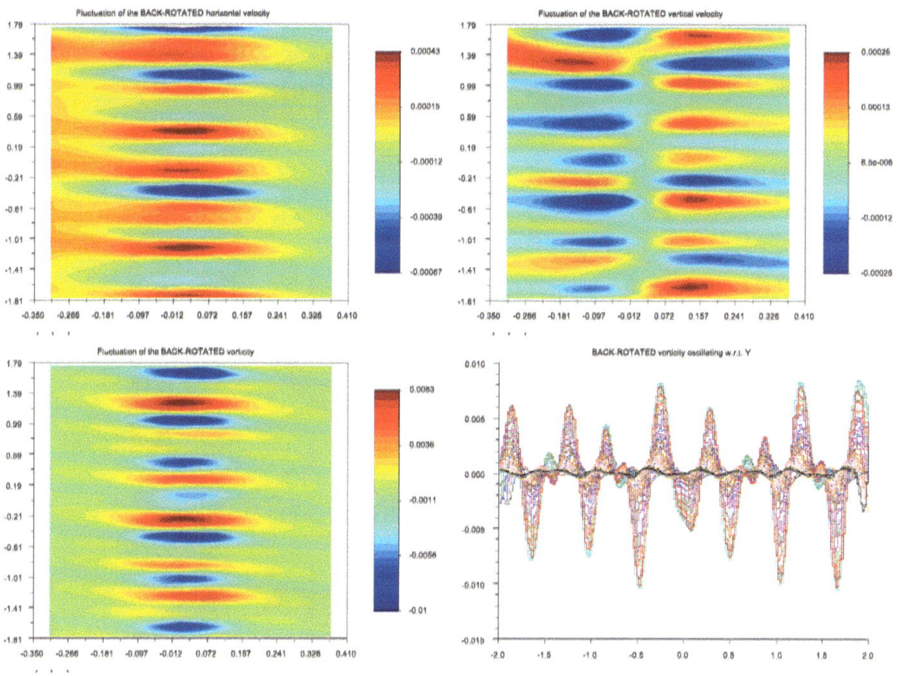

Fig. 10 Velocity (*top*) and vorticity for data like Fig. 6, but with initial averaging (200 × 200)

for L^1-contracting 1D scalar laws, may be insufficient to ensure strong stabilizing effects in the context of 1D systems (for which well-posedness was established in [4]). Yet, as system (1) is ill-posed in L^∞, it seems plausible that Lax-Wendroff properties might not be enough to prevent a dimensional-split scheme [13, 27] from approximating one (among many) non-standard, entropy-admissible solution.

7 Conclusions and Future Work

The present note presented an explicit form of Lipschitz initial data, yielding a composite compression wave (involving both characteristic families) bringing out the Riemann problem (30) for (28) for which, thanks to the special pressure law $p(\rho) = \rho^3/12$, an existence theorem for non-standard solutions can be proved (relying on [8]). As the initial data is explicit, one can set up a dimensional-splitting kinetic numerical scheme for (1) and march it forward in time, beyond breakup, especially after rotating it with R_θ in order to force the scheme to "work in 2D". Non-standard solutions are known to have specific properties (slightly faster shocks, density ρ below the self-similar value ρ_m), which can be checked rather easily on "back-rotated" numerical approximations. Besides, an oscillating

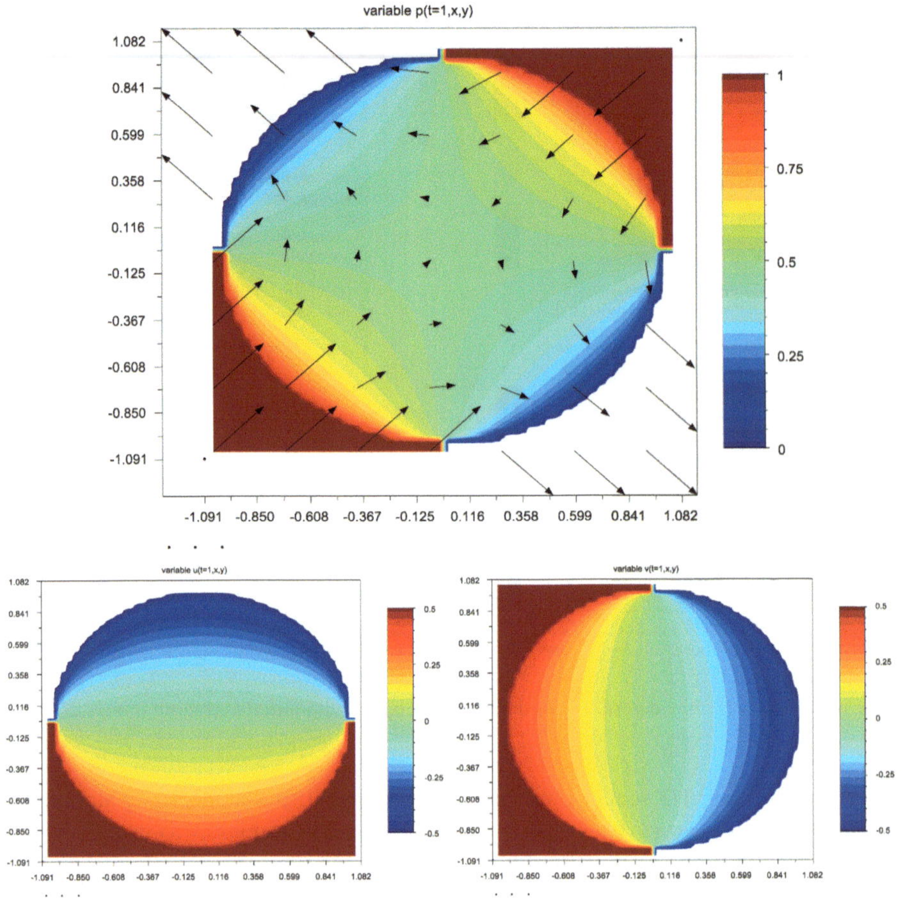

Fig. 11 Exact solution of a "Double Sod" Riemann problem inspired by Aregba-Driollet and Natalini [3, Fig. 6.10]

profile in both u and v, consistent with a slightly lower value of ρ compared to the self-similar ρ_m is observed for several values of the rotation angle θ, too. Such observations are consistent with former ones, displayed in both [33, Sect. VII] and [12, §3] and achieved by means of the dimensional-split Glimm scheme (or Random-Choice Method). Besides, from a numerical perspective, it may also be interesting to devise new "genuinely multi-dimensional" algorithms, even for simpler, linear Euler systems. As an illustrative example, we display on Fig. 11 the exact solution of the 2D Riemann problem written in [2, §6.2.1] for the "double Sod test" considered in [3, Fig. 6.10]. Despite all the nonlinearities were wiped off, the initial pressure jumps are small enough so that the linearized solution appears to be meaningful. Another strategy is to take a step back and focus on "compressible potential flows", [21]; corresponding systems are indeed a 2D generalization of 1D hyperbolic systems belonging to a so-called "$\mathscr{K}$ class", as emphasized by DiPerna in [18, §5].

Acknowledgements Both the authors are happy to thank Denise Aregba-Driollet (Bordeaux) and Roger Käppeli (Zürich) who gently accepted to perform fine-grid computations, partly reported in §6, by using (formally) second-order extensions of 2D BGK and HLL schemes. Eleuterio Toro (Trento) suggested to set up the averaging process reported in Remark 1.

References

1. Acharya, A., Chen, G.Q., Slemrod, M., Wang, D.: Fluids, Elasticity, Geometry, and the Existence of Wrinkled Solutions (2016). arXiv: 1605.03058
2. Amadori, D., Gosse, L.: Error Estimates for Well-Balanced Schemes on Simple Balance Laws: One-Dimensional Position-Dependent Models. BCAM SpringerBriefs in Mathematics. Springer, New York (2015).
3. Aregba-Driollet, D., Natalini, R.: Discrete kinetic schemes for multi-dimensional systems of conservation laws. SIAM J. Numer. Anal. **37**, 1973–2004 (2000)
4. Bianchini, S., Bressan, A.: Vanishing viscosity solutions of nonlinear hyperbolic systems. Ann. Math. **161**, 223–342 (2005)
5. Blondin, J.M., Marks, B.S.: Evolution of cold shock-bounded slabs. New Astronomy **1**, 235–244 (1996)
6. Brenier, Y., Corrias, L.: A kinetic formulation for multi-branch entropy solutions of scalar conservation laws. Ann. Inst. Henri Poincare Nonlinear Anal. **15**, 169–190 (1998)
7. Chiodaroli, E.: A counterexample to well-posedness of entropy solutions to the compressible Euler system. J. Hyperbolic Differ. Equ. **11**, 493–519 (2014)
8. Chiodaroli, E., Kreml, O.: On the energy dissipation rate of solutions to the compressible isentropic Euler system. Arch. Ration. Mech. Anal. **214**, 1019–1049 (2014)
9. Chiodaroli, E., De Lellis, C., Kreml, O.: Global ill-posedness of the isentropic system of gas dynamics. Commun. Pure Appl. Math. **68**, 1157–1190 (2015)
10. Chiodaroli, E., Feireisl, E., Kreml, O.: On the weak solutions to the equations of a compressible heat conducting gas. Ann. Inst. H. Poincaré Anal. Non Linéaire **32**, 225–243 (2015)
11. Chiodaroli, E., Feireisl, E., Kreml, O., Wiedemann, E.: $\mathscr{A}$-free rigidity and applications to the compressible Euler system (2015). arXiv:1511.03114. https://arxiv.org/abs/1511.03114
12. Colella, P.: Glimm's method for gas dynamics. SIAM J. Sci. Stat. Comput. **3**, 76–110 (1982)
13. Crandall, M., Majda, A.: The method of fractional steps for conservation laws. Numer. Math. **34**, 285 (1980)
14. Dafermos, C.M.: Hyperbolic conservation laws in continuum physics. Grundleheren der Mathematischen Wissenschaften [Fundamental Principles of Mathematical Sciences], vol. 325, 3rd edn. Springer, Berlin (2010)
15. Davis, S.: A rotationally biased upwind difference scheme for the Euler equations. J. Comput. Phys. **56**, 65–92 (1984)
16. De Lellis, C., Székelyhidi, L.J.: The Euler equations as a differential inclusion. Ann. Math. **170**, 1417–1436 (2009)
17. De Lellis, C., Székelyhidi, L.J.: On admissibility criteria for weak solutions of the Euler equations. Arch. Ration. Mech. Anal. **195**, 225–260 (2010)
18. DiPerna, R.: Global solutions to a class of nonlinear hyperbolic systems of equations. Commun. Pure Appl. Math. **26**, 1–28 (1973)
19. Elling, V.: A possible counterexample to well posedness of entropy solutions and to Godunov scheme convergence. Math. Comput. **75**, 1721–1733 (2006)
20. Elling, V.: The carbuncle phenomenon is incurable. Acta Math. Sci. **29B**(6), 1647–1656 (2009)

21. Elling, V.: Relative entropy and compressible potential flow. Acta Math. Sci. **35B**(4), 763–776 (2015)
22. Engquist, B., Sjögreen, B.: The convergence rate of finite difference schemes in the presence of shocks. SIAM J. Numer. Anal. **35**, 2464–2485 (1998)
23. Fjordholm, U.S., Mishra, S.: Vorticity preserving finite volume schemes for the shallow water equations. SIAM J. Sci. Comput. **33**, 588–611 (2011).
24. Fjordholm, U.S., Mishra S., Tadmor, E.: On the computation of measure-valued solutions. Acta Numer. **25** 567–679 (2016)
25. Gosse, L.: Using K-branch entropy solutions for multivalued geometric optics computations. J. Comput. Phys. **180**, 155–182 (2002)
26. Gosse, L.: Computing Qualitatively Correct Approximations of Balance Laws. Exponential-Fit, Well-Balanced and Asymptotic-Preserving. SIMAI Springer Series. Springer, Milan (2013). ISBN: 978-88-470-2891-3
27. Holden, H., Karlsen, K.H., Lie, K.-A., Risebro, N.H.: Splitting Methods for Partial Differential Equations with Rough Solutions. Analysis and MATLAB Programs. Series of Lectures in Mathematics. European Mathematical Society (EMS), Zürich (2010)
28. Jeltsch, R., Torrilhon, M.: On curl-preserving finite volume discretizations for shallow water equations. BIT Numer. Math. **46**, S35–S53 (2006)
29. Larroutourou, B.: How to preserve the mass fractions positivity when computing multi-component flows. J. Comput. Phys. **95**, 59–84 (1991)
30. LeVeque, R.J.: Nonlinear conservation laws and finite-volume methods. In: Computational Methods for Astrophysical Fluid Flow. The series Saas-Fee Advanced Courses, vol. 27, pp. 1–159. Springer (1998)
31. Morton, K.W., Roe, P.L.: Vorticity-preserving lax-Wendroff type schemes for the system wave equation. SIAM J. Sci. Comput. **23**, 170–192 (2001)
32. Roe, P.L.: Discrete models for the numerical analysis of time- dependent multidimensional gas dynamics. J. Comput. Phys. **63**, 458–476 (1986)
33. Roe, P.L.: Discontinuous solutions to hyperbolic systems under operator splitting. Numer. Methods Partial Differ. Equ. **7**, 277–297 (1991)
34. Roe, P.L.: Beyond Riemann problem. In: Algorithmic Trends in Computational Fluid Dynamics. Part of the Series ICASE/NASA LaRC Series, pp. 341–367. Springer, New York (1993)
35. Wang, Y., Li, J.: Numerical defects of the HLL scheme and dissipation matrices for the Euler equations. SIAM J. Numer. Anal. **52**, 207–219 (2014)
36. Warming, R.F., Hyett, B.J.: The modified equation approach to the stability and accuracy analysis of finite difference methods. J. Comput. Phys. **14**, 159–179 (1974)
37. Xu, K., Hu, J.: Projection dynamics in Godunov-type schemes. J. Comput. Phys. **42**, 412–427 (1998)
38. Zheng, Y.: Systems of Conservation Laws: Two-Dimensional Riemann Problems. Progress in Nonlinear Differential Equations and Their Applications, vol. 38. Birkhauser, Basel (2001)

On Hyperbolic Balance Laws and Applications

Cleopatra Christoforou

Abstract An overview of the current state of the theory of general strictly hyperbolic systems of balance laws in one space dimension is documented in this article. Results on global existence, stability and uniqueness of entropy weak solutions are stated and properties such as the decay of positive waves and the rate of convergence of viscous approximations are presented. The article concludes with an application on the existence of non-smooth isometric immersions into $\mathbb{R}^3$.

1 Introduction

This article serves as an exposition of the current state of the theory of general strictly hyperbolic systems of balance laws in one-space dimension. Systems of balance laws in one space dimension are systems of the form

$$\partial_t U + \partial_x F(U, x, t) + G(U, x, t) = 0, \quad x \in \mathbb{R}, \quad t > 0 \tag{1}$$

where $U = U(x, t)$ is the unknown function from $\mathbb{R} \times [0, \infty)$ to $\mathbb{R}^n$. Also, the flux F and the source G are given smooth, nonlinear functions from $\mathbb{R}^n \times \mathbb{R} \times \mathbb{R}_+$ to $\mathbb{R}^n$. Such equations arise in Continuum Physics and are called *conservation laws* for $G \equiv 0$ and balance laws for $G \neq 0$ because the conservation of mass, momentum and energy are governed by equations of this type. In this context, $U(x, t) \in \mathbb{R}^n$ determines the state of the medium at the point (x, t) and $F(U, x, t)$ is the flux of U. System (1) is strictly hyperbolic, that is, the Jacobian matrix $A(U, x, t) = D_U F(U, x, t)$ has n real and distinct eigenvalues

$$\lambda_1(U, x, t) < \lambda_2(U, x, t) < \ldots < \lambda_n(U, x, t), \tag{2}$$

C. Christoforou (✉)
Department of Mathematics and Statistics, University of Cyprus, Nicosia 1678, Cyprus
e-mail: Christoforou.Cleopatra@ucy.ac.cy

L. Gosse, R. Natalini (eds.), *Innovative Algorithms and Analysis*,
Springer INdAM Series 16, DOI 10.1007/978-3-319-49262-9_5

known as characteristic speeds and thereby, n linearly independent right eigenvectors $r_1(U,x,t), r_2(U,x,t), \ldots, r_n(U,x,t)$. The system is supplemented by initial data

$$U(x,0) = U_0(x) \qquad x \in \mathbb{R} \tag{3}$$

and the object of our study is the Cauchy problem (1)–(3). Throughout this article, we consider general systems that are strictly hyperbolic and discuss the state of the theory in the general setting that applies to plenty physical models. We note that F and G may depend explicitly on (x,t) and an emphasis will be given when this inhomegeneity will be crucial.

The theory of hyperbolic conservation laws is confronted with two major challenges: First, classical solutions, starting out from smooth and small initial values, develop discontinuities in finite time that propagate as shocks. Hence, inevitably, the theory must deal with weak solutions. This difficulty is compounded further by the fact that in the context of weak solutions, uniqueness is lost, thus raising the need to devise proper criteria that will weed out unstable, physically irrelevant solutions and hopefully single out a unique admissible solution. Continuum Physics naturally induces such admissibility criteria through the Second Law of thermodynamics; admissible solutions should satisfy "entropy inequalities". Equivalently, in order to single out the thermodynamically admissible solution, hyperbolic systems may be viewed as the "vanishing viscosity" limit of a family of systems with diffusion, which has negligible effect on smooth solutions but reacts stiffly in the presence of shocks. Exposition of the current state of the theory can be found in Dafermos [23].

The Cauchy problem has been studied in great detail for the case of conservation laws, i.e. $G \equiv 0$. Assume that the initial data have small total variation over $\mathbb{R}$, then global admissible solution U can be constructed in the space of bounded variation (BV). Over the years three different techniques have been developed for constructing BV solutions global in time namely the random choice method of Glimm [31], the front tracking algorithm of Bressan et al. [7] and Holden and Risebro [35] and the vanishing viscosity method of Bianchini and Bressan [5]. It should be noted that the smallness condition of the total variation is an essential limitation in applying these methods. There is another method that is functional analytic in character and provides solutions in L^p space, that is the method of compensated compactness; cf. [44]. We remark however that discontinuous solutions in one-space dimension can be at most BV. An exposition of the current state of the theory of conservation laws can be found in the manuscripts [7, 23, 35, 44].

The presence of a nonzero source G plays a crucial role on the well-posedness of weak solutions to strictly hyperbolic balance laws even if the initial data are sufficiently small. The existing results on this topic are presented in detail in Sect. 2. Two major properties of solutions to such systems are mentioned in Sect. 3. Specifically, these are the decay of positive waves and the rate of convergence of vanishing viscosity approximations. We conclude this article with an application of the theory of balance laws to the isometric immersion problem in Differential Geometry. The connection between the two disciplines of Continuum Physics and Differential Geometry is an interesting discovery that has been identified and explored in the recent years. This connection and the application of the theory of balance laws of Sect. 2 are summarized in Sect. 4. It should be mentioned that although the extension of the theory of conservation laws (with $G \equiv 0$) to balance laws ($G \neq 0$) may naively be considered not substantial, it has important consequences and value in such an unexpected area of application which is the isometric immersion problem of negatively curved 2-d Riemannian manifolds.

2 Well-Posedness of Global Weak Solutions

For systems of balance laws, a nonzero source G may enhance stability or induce instability. In other words, for general source term G, blow-up of solutions in finite time is expected even when the initial data is of small variation. In fact, the presence of the production term G results to the amplification in time of even small oscillations in the solution. Because of this feature of hyperbolic balance laws, one does not expect in general long term stability in BV. Local in time existence of BV solutions was first established by Dafermos and Hsiao [24], using a modification of the random choice method of Glimm [31].

However, one may hope for global existence if a special mechanism induced by the source that would not allow the total variation of approximate solutions to amplify in time. Different mechanisms have been studied and incorporated in the methods of random choice, front tracking and vanishing viscosity. Indeed, these three methods have been successfully modified and applied to strictly hyperbolic systems of balance laws for initial data close to a constant equilibrium solution U^* under appropriate conditions on F and G. First, Dafermos and Hsiao in [24] introduced a suitable *dissipation assumption* on G to achieve global existence by the modified random choice method. This *dissipation hypothesis* introduced in [24] describes a special structure on the dependence of G on the state vector U and induces a special mechanism, which controls the total variation of the solution for all times. Under this *dissipation hypothesis* on G and existence of a constant equilibrium solution, system (1) has been studied independently using the three aforementioned construction methods by modifying them appropriately in the presence of the source G. Roughly speaking, the operator splitting is incorporated in an appropriate setting for each construction method. More precisely, under the *dissipation hypothesis* on G, global existence to (1) has been established by Dafermos and Hsiao [24] using the random choice method, by Amadori and Guerra [2, 3] using the front-tracking algorithm and by Christoforou [12, 13] using the vanishing viscosity method. In [24], system (1) has been treated in the general case that F and G depend explicitly on (x,t) imposing uniform bounds on F and G and its derivatives with respect to (x,t). However, in [2, 3, 12, 13], the case that both F and G depend only on the state U has been considered. Also, in [12, 13], strictly hyperbolic systems that are not necessarily of conservative form have been included. However in [2, 3, 24], the systems under consideration should be conservative and also the characteristic fields should be either genuine nonlinear or linearly degenerate. It should be added that except of existence also the stability of solutions with respect to initial data in L^1 is established in [3, 12].

In summary, if $F = F(U)$ and $G = G(U)$, the results of [2, 3, 12, 13, 24], for data U_0 with sufficiently small total variation, yield the same entropy BV weak solution $U(x,t)$ to (1) defined on $\mathbb{R} \times [0, +\infty)$ that has uniformly bounded total variation for all times, it is stable in L^1 with respect to initial data and time and as $t \to \infty$, $U(x,t)$ tends to the equilibrium state U^* exponentially fast. More precisely, the following bounds hold:

$$
\begin{aligned}
TV_x U(\cdot,t) &\le Ce^{-\mu t} TV_x U_0(\cdot) \qquad && t > 0 \\
\|U(t) - V(t)\|_{L^1} &\le Ce^{-\mu t} \|U_0 - V_0\|_{L^1} \qquad && t > 0 \\
\|U(t) - U(t')\|_{L^1} &\le Ce^{-\mu t} |t - t'|, \qquad && t < t' ,
\end{aligned}
$$

for some constants $\mu > 0$ and $C > 0$. It should be noted that if F and G depend explicitly on (x,t), then additional terms appear on the right-hand side of the above bounds. Also, the above bounds imply that U is a function of bounded variation in both variables (x,t). Dafermos revisited this problem in [22] and relaxed the dissipation hypothesis introduced in [24] to the so-called "*weak dissipation*" by assuming apriori that the Cauchy data is stable in L^1, i.e. for every $t \in [0,T]$

$$
\int_{-\infty}^{\infty} |U(x,t) - U^*|\, dx < \beta \int_{-\infty}^{\infty} |U_0(x) - U^*|\, dx
$$

for some constant $\beta > 0$ using again the modified random choice method.

Global existence of entropy weak solutions to (1)–(3) using the modification of Glimm's scheme as constructed in [24], but under different hypotheses is established also in [15]. Actually in [15], there are two main characteristics that are not present in the aforementioned articles: (i) existence of a constant equilibrium solution to (1) is not assumed and (ii) global existence is established without any conditions on the structure of the source G on U but only through the rapid decay of inhomogeneity of the flux $F(\cdot,x,t)$ and the source $G(\cdot,x,t)$ with respect to (x,t). The analysis involves mainly ideas and techniques that are present in the *local existence* result of [24], but implemented under sufficiently rapid decay in the inhomogeneity. This result is actually stated in Dafermos's book without proof, cf. [23, Chap. 13.9] and in [15], the proof in detail is provided. For the sake of completeness, we remark that the existence of solutions is also established for other classes of systems of balance laws under different assumptions, for example [4, 38, 39] when $F = F(U)$ and $G = G(U,x)$.

In this following, we will describe two different mechanisms that we mentioned above and present the corresponding existence results of entropy weak solutions to strictly hyperbolic systems of balance laws. These mechanisms are the *rapid decay of the inhomogeneity* in Sect. 2.1 as proven in [15] and the *dissipation hypothesis* of Dafermos and Hsiao [24] in Sect. 2.2 via the vanishing viscosity method of Christoforou [12, 13].

2.1 Rapidly Decaying Inhomogeneity

We assume that system (1) is not only strictly hyperbolic, i.e. the Jacobian matrix $D_U F(U,x,t)$ has n real distinct eigenvalues as mentioned in (2), which remain uniformly separated, i.e.

$$|\lambda_i(U,x,t) - \lambda_j(U,x,t)| > A^{-1}, \qquad i \neq j, \tag{4}$$

for some positive constant A, but also that the characteristic speeds are uniformly away zero, i.e.

$$|\lambda_i(U,x,t)| > A^{-1}, \qquad i = 1,\dots,n, \tag{5}$$

for all $U \in \mathscr{B}$, $x \in \mathbb{R}$ and $t \geq 0$, where $\mathscr{B}$ denotes a ball in $\mathbb{R}^n$. Moreover, we assume that the flux and the source satisfy the uniform bounds

$$|D_U F(U,x,t)|,\ |D_U^2 F(U,x,t)|,\ |D_U G(U,x,t)| \leq A \tag{6}$$

for $U \in \mathscr{B}$, $x \in \mathbb{R}$ and $t \geq 0$. Further, there is a positive function $\varphi \in W^{1,1}(\mathbb{R})$ with

$$\int_{-\infty}^{\infty} \varphi(x)\, dx \leq \omega \tag{7}$$

and a bounded function $\psi \in L^1(0,\infty)$, for which

$$|F_x(U,x,t)|,\ |G(U,x,t)| \leq \omega\psi(t), \tag{8}$$

$$|D_U F_U, x,t)|,\ |D_U F_t(U,x,t)|,\ |D_U G(U,x,t) \leq \omega\psi(t), \tag{9}$$

$$|F(U,x,t)|,\ |D_U F(U,x,t)|,\ |F_{tx}(U,x,t)|,\ |G_U,x,t)| \leq \varphi(x)\,\psi(t), \tag{10}$$

for every $U \in \mathscr{B}$, $x \in \mathbb{R}$ and $t \geq 0$. Assumptions (7)–(10) correspond to the mechanism of rapid decay of inhomogeneity. Thus it is crucial that F and G depend explicitly on (x,t) for these to hold. In [15], global existence of entropy weak solutions to the Cauchy problem is established for initial data of small bounded variation using the approximate sequence $\{U_h\}$ constructed via the random choice method in conjunction with the operator splitting. The main result of [15] is

Theorem 1 (Christoforou [15]) *Under the assumptions (4)–(7) on the flux F and the source G terms, there are positive constants ω_0 and δ_0, depending only on A and the radius R of the ball $\mathscr{B}_R$, such that when (8)–(10) are satisfied with $\omega < \omega_0$ and $TVU_0 < \delta_0$, then if $U_0 \in \mathscr{B}_{R/2} \subset \mathscr{B}_R$,*

1. *there exists $h_0 > 0$ small enough for which the family of approximate solutions $\{U_h(x,t)\}_h$, with h being $0 < h < h_0$, is globally defined with $(x,t) \in \mathbb{R} \times [0,\infty)$ for each sequence $\{a_s\}$;*

2. *there is a converging subsequence* $\{U_{h_k}\}$, *with* $h_k \to 0+$ *as* $k \to \infty$, *such that* U_{h_k} *converges in* L^1_{loc} *to a function* U, *for almost every sequence* $\{a_s\}$, *which is the entropy admissible weak solution to the Cauchy problem (1)–(3).*

Further, for each $t > 0$, $U(t)$ *is a function of bounded variation on* $(-\infty, \infty)$ *and*

$$TV_{(-\infty,\infty)}\{U(t)\} \leq C_1 e^{\sigma}(TV\{U_0\} + \omega), \tag{11}$$

$$\|U(t)\|_{L^\infty} \leq \|U_0\|_{L^\infty} + C_2 e^{\sigma}(TV\{U_0\} + \omega). \tag{12}$$

where C_1 *and* C_2 *are constants and* $\sigma = O(1)\omega\|\psi\|_{L^1[0,\infty)}$.

Hereafter, notation $O(1)$ stands for a term which is bounded by a constant. For a description of the construction of the sequence $\{U_h\}$, see [15, Sect. 2].

2.2 Dissipation and the Vanishing Viscosity Method

The vanishing viscosity method for systems of conservation laws had been a major open problem for a long time. It was solved in a spectacular way by Bianchini and Bressan [5]. Christoforou has extended their fundamental work to systems of balance laws in [12] under the dissipation assumptions on G introduced by Dafermos and Hsiao [24]. These conditions on the source G assumed in [3, 12, 24] is as follows: Consider a constant equilibrium solution U^* to (1), i.e. $G(u^*, x, t) = 0$. If we linearize the hyperbolic system (1) about U^* and then decompose the solution U along the right eigenvectors of $D_U F(U^*, x, t)$, the resulting linear system is

$$v_{i,t} + \lambda_i(U^*, x, t)v_{i,x} + \sum_{j=1}^{n} B_{ij}(U^*, x, t)v_j = 0\ ,$$

where $B_{ij}(U^*, x, t)$ are the entries of the $n \times n$ matrix

$$B(U, x, t) = [r_1(U, x, t), \dots, r_n(U, x, t)]^{-1} D_U G(U, x, t)[r_1(U, x, t), \dots, r_n(U, x, t)] \tag{13}$$

evaluated at $U = U^*$. The hypothesis in [3, 12, 24] states that the matrix $B(U^*, x, t)$ is *strictly column diagonally dominant*, i.e. there is a positive constant $\tilde{\beta}$ such that

$$\boxed{B_{ii}(U^*, x, t) - \sum_{j \neq i} |B_{ji}(U^*, x, t)| > \tilde{\beta} > 0,} \tag{14}$$

for all $i = 1, \dots, n$, $x \in \mathbb{R}$, and $t > 0$. This hypothesis depends on the choice of the right eigenvectors r_i. In fact, for practical use, one can find an equivalent condition in [2] which is independent of the choice of the eigenvectors.

Now we will present the existence and stability results of Christoforou [12] under the dissipation hypothesis (14) and note that in [12], both F and G depend only on U. Hence the Cauchy problem (1), (3) reduces to

$$U_t + A(U)\, U_x + G(U) = 0 \tag{15}$$

with data

$$U(x, 0) = U_0(x), \tag{16}$$

with $A = DF(U)$. Global weak solutions to (15) were constructed via the vanishing viscosity method in[12], namely, as the $\varepsilon \downarrow 0+$ limit of a family $\{U^\varepsilon\}$ of functions that satisfy the parabolic system

$$U_t^\varepsilon + A(U^\varepsilon)U_x^\varepsilon + G(U^\varepsilon) = \varepsilon U_{xx}^\varepsilon. \tag{17}$$

Theorem 2 (Christoforou [12]) *Consider the Cauchy problem*

$$U_t^\varepsilon + A(U^\varepsilon)\, U_x^\varepsilon + G(U^\varepsilon) = \varepsilon U_{xx}^\varepsilon \tag{18}$$

$$u^\varepsilon(0, x) = u_0(x). \tag{19}$$

Assume that the matrices $A(U) = DF(U)$ have real distinct eigenvalues $\lambda_1(U) < \lambda_2(U) < \ldots < \lambda_n(U)$ and thereby n linearly independent eigenvectors $r_1(U)$, $r_2(U)$, $\ldots, r_n(U)$. Under the assumption that the matrix $B(U^*)$ defined by (13) is strictly diagonally dominant, there exists a constant $\delta_0 > 0$ such that if $U_0 - U^* \in L^1$ and

$$TV\{U_0\} < \delta_0, \tag{20}$$

then for each $\varepsilon > 0$ the Cauchy problem (18)–(19) has a unique solution U^ε, defined for all $t \geq 0$. Moreover,

$$TV\{U^\varepsilon(t, \cdot)\} \leq C\, e^{-\mu t} TV\{U_0\}, \tag{21}$$

$$\|U^\varepsilon(t) - U^\varepsilon(s)\|_{L^1} \leq L' \left(|t - s| + \sqrt{\varepsilon}|\sqrt{t} - \sqrt{s}|\right) e^{-\mu s}, \qquad for \quad t > s, \tag{22}$$

where μ is a positive constant that depends on $B(U^*)$. Furthermore, if V^ε is another solution of (18) with initial data V_0, then

$$\boxed{\|U^\varepsilon(t) - V^\varepsilon(t)\|_{L^1} \leq L\, e^{-\mu t} \|U_0 - V_0\|_{L^1}.} \tag{23}$$

Finally, as $\varepsilon \to 0$, U^ε converges in L^1_{loc} to a function U, which is the admissible weak solution U of (15)–(16), when the system is in conservation form, $A = DF$.

In the conservative case, $A = DF$, every vanishing viscosity limit is an admissible weak solution to $U_t + (F(U))_x + G(U) = 0$ and the stability estimate (23) implies uniqueness within the family of solutions obtained via the vanishing viscosity method (cf. Theorem 2). In [13], uniqueness of solutions to (15) is established by means of local integral estimates. The aim is twofold: namely treating the general case in which the system is not necessarily conservative as well as establishing uniqueness within a broader class of solutions.

Theorem 3 (Christoforou [13]) *Suppose that the hypotheses of Theorem 2 hold. Let $\mathcal{P} : \mathcal{D} \times [0,\infty) \mapsto \mathcal{D}$ be the semigroup of vanishing viscosity solutions constructed as limit of the vanishing viscosity approximations via (18). Then* ***every trajectory*** *$U(t) = \mathcal{P}_t U(0)$, $U(\cdot,t) \in \mathcal{D}$* ***satisfies the following conditions****:*

i. *At every point (ξ,τ), for every $\beta' > 0$ one has*

$$\boxed{\lim_{h\to 0+} \frac{1}{h}\int_{\xi-\beta' h}^{\xi+\beta' h} |U(x,\tau+h) - U^{\sharp}_{(u;\xi,\tau)}(x-\xi,h)|\,dx = 0,} \tag{24}$$

where $U^{\sharp}_{(u;\xi,\tau)}$ is solution of the homogeneous system, (i.e (15) with $G = 0$), with Riemann data given by the left and right states of $U(x,\tau)$ as $x \to \xi\pm$.

ii. *There exist constants C, $\beta > 0$ such that for every $\tau \geq 0$ and $\xi \in (a,b)$,*

$$\boxed{\limsup_{h\to 0+} \int_{a+\beta h}^{b-\beta h} \left|U(x,\tau+h) - U^{\flat}_{(u;\xi,\tau)}(x,\tau+h)\right| \frac{dx}{h} \leq C\left(TV_{(a,b)}\{U(\tau)\} + b - a\right)^2} \tag{25}$$

where $U^{\flat}_{(u;\xi,\tau)}$ is solution of the system (15) linearized about (ξ,τ) (i.e. $w_t + \hat{A}w_x + \hat{G} = 0$ with $\hat{A} = A(U(\xi,\tau))$ and $\hat{G} = G(U(\xi,\tau))$) and data $U(x.\tau)$.

Conversely, let $U : [0,T] \mapsto \mathcal{D}$ be Lipschitz continuous map with values in $L^1(\mathbb{R},\mathbb{R}^n)$ and assume that the conditions (i) and (ii) hold at almost every time τ. Then $U(t)$ ***coincides with a (unique) trajectory of the semigroup*** *$\mathcal{P}$.*

Theorem 3 implies the convergence of $\mathcal{P}^{\varepsilon}_t U_0$ as $\varepsilon \downarrow 0+$ (as a whole sequence and not in the context of a subsequence $\{\varepsilon_m\}$) to a unique limit $\mathcal{P} U_0$. Furthermore, it characterizes the trajectories of the semigroup by means of local integral estimates (24) and (25). Accordingly, if a *viscosity solution* is a Lipschitz function that satisfies (24) and (25), then the limit $\mathcal{P}_t U_0$ is a *viscosity solution* and every *viscosity solution* can be constructed as a limit of vanishing viscosity approximations. Roughly speaking, in view of the above result, any solution U to the hyperbolic system (15) can be approximated by the self-similar solution of a Riemann problem to $U_t + A(U)U_x = 0$ in a neighborhood of (τ,ξ). Also, it relates U to the solution of the corresponding linear hyperbolic system $U_t + \hat{A}U_x + \hat{G} = 0$ with constant coefficients in terms of the total variation over the interval (a,b) and the length $b-a$ of the interval. See also [4] for a related work on the front tracking approximation.

3 Properties of Solutions

Historically, decay rates were used to provide *quantitative* and *qualitative* information on the solutions to hyperbolic conservation laws. Quantitative results include the establishment of convergence rates for approximating procedures and numerical schemes. Qualitative results include the establishment of results on uniqueness and regularity as well as the ability to visualize the waves and their evolution in time. In the scalar case, in particular, some celebrated results include the uniqueness of the solution to the one-dimensional conservation laws by Oleinik [43], local error estimates by Tadmor [45] and results on the structure of entropy solutions for multidimensional scalar conservation laws by De Lellis et al. [26].

Throughout this section, we consider the strictly hyperbolic system (15)–(16) and present the results of Christoforou and Trivisa [18, 20] on the spreading of rarefaction waves. Actually, we present two decay estimates on the positive waves for systems of hyperbolic and genuinely nonlinear balance laws satisfying the dissipative mechanism (14). These estimates are obtained by employing the continuity of Glimm-type functionals and the method of generalized characteristics [7, 23]. Having these, we estimate the convergence rate of the viscous approximations U^ε in L^1 as established in [19]. In addition to strict hyperbolicity, all characteristic fields are genuinely nonlinear, namely for each $i = 1, \dots, n$, the product of the gradient of the characteristic speed with the corresponding right eigenvector is nonzero. Without loss of generality, this condition can be expressed as

$$D\lambda_i(U)r_i(U) \equiv 1, \qquad i = 1, \dots, n. \tag{26}$$

3.1 *Decay of Positive Waves*

Estimates on positive waves for systems of balance laws (15) under the dissipation condition (14) are established using two different approaches. In Sect. 3.1.1, an Oleinik-type estimate is obtained by employing the lower semicontinuity properties of Glimm-type functionals and the method of generalized characteristics. In Sect. 3.1.2, a sharp decay estimate is established by introducing a partial ordering within the family of positive Radon measures, using symmetric rearrangements and a comparison with a solution of Burgers equation with impulsive sources.

3.1.1 Oleynik-Type of Decay

While linearly degenerate families do not gain any regularity during the time evolution, the genuinely nonlinear families show a regularizing effect due to the nonlinearity of the eigenvalue. In the context of the scalar hyperbolic conservation law $u_t + (f(u))_x = 0$ the celebrated result by Oleinik which yields the decay of positive waves has the form:

$$u_x^+ \leq \frac{1}{kt}, \tag{27}$$

where k is the genuinely nonlinearity constant, $k = \inf_u f''(u) > 0$. Here, u_x^+ denotes the positive part of the Radon measure u_x. For genuinely nonlinear systems of conservation laws, this decay estimate takes the form (cf. [7, 23]):

$$\mu_t^{i+}(B) \leq C\left\{\frac{\mathscr{L}^1(B)}{t-s} + \mathbf{Q}(s) - \mathbf{Q}(t)\right\}, \qquad 0 \leq s < t, \tag{28}$$

where μ_t^{i+} denotes the positive i-wave measure over a Borel set B, $\mathbf{Q}$ is the Glimm interaction potential and the constant C depends on the constant k of genuinely nonlinearity condition (the genuinely nonlinearity condition here requires that $D\lambda_i(U)r_i(U) \geq k > 0$). Estimate (28) takes into account the fact that new positive waves are created at a later time $t > 0$ in contrast to the scalar equation. Let $U(\cdot, t)$ be a function of bounded variation, as such its distributional derivative $D_x U$ is a Radon measure and fix an index i, $i = 1, \ldots, n$. Following [7], we define the wave measures μ^i of i-waves in U:

$$\mu^i = l_i(U) \cdot D_x U$$

on the sets of continuity of U and

$$\mu^i(\{x\}) = \sigma_i$$

at the points of jump of U. Here, σ_i denotes the strength of the i-wave in the Riemann solution with data the left $U_l = U(x-)$ and right $U_r = U(x+)$ states. If we denote by $U_l = U_0, U_1, \ldots, U_n = U_r$ the states in the Riemann solution, and take into consideration the fact that all characteristic fields are genuine nonlinear, we can write the strength of the i-wave to be

$$\sigma_i = \lambda_i(U_i) - \lambda_i(U_{i-1}) \ .$$

Taking into account that

$$\sigma_i = l_i(U_r) \cdot (U_r - U_l) + O(1) \cdot |U_r - U_l|^2,$$

one can write the measure μ^i as

$$\mu^i = l_i \cdot D_x U, \tag{29}$$

for an appropriate vector $l_i = l_i(x)$. We denote by μ^{i+} and μ^{i-} the positive and negative parts of μ^i and by $|A|$ the Lebesgue measure of a set A.

Estimate (28) was employed by Bianchini and Caravenna [6] for the establishment of a regularity result stating that for hyperbolic conservation laws, the function $U(t)$ is a function of *special bounded variation* (SBV), in the sense that, its distributional derivative U_x is a measure with no cantorian part. The result presented here is the analogue of (28) in the context of hyperbolic balance laws.

Theorem 4 (Christoforou and Trivisa [20]) *Consider a system of hyperbolic balance laws in one space dimension (15) that is strictly hyperbolic, genuinely nonlinear and satisfying the dissipation hypothesis (14), and entropy solutions with range in a small neighborhood of the constant state U^*. Then, there exist positive constants C and β such that, for every solution U with tame variation and every interval $I \subset \mathbb{R}$, the following estimate holds:*

$$\boxed{\mu_T^{i+}(I) \le C\left(meas(I)\,\frac{\beta}{(e^{\beta T}-1)} + (\mathbf{Q}(0) - \mathbf{Q}(T))\right).} \tag{30}$$

Roughly speaking, this theorem yields that the rarefaction waves in any i-family within an interval $[0, T]$ are of two types:

1. Waves already present at time $t = 0$ and which propagated up to time $t = T$: these waves had enough time to decay and, on any interval $I \subset \mathbb{R}$, are like

$$\mathcal{O}(1)\,\text{meas}(I)\frac{\beta}{e^{\beta T - 1}}.$$

2. Waves generated by interactions which took place during the time interval $[0, T]$: these waves are estimated by the decrease in the interaction potential:

$$\mathcal{O}(1)\left(\mathbf{Q}(0) - \mathbf{Q}(T)\right).$$

See also Goatin and Gosse [32] for a similar estimate for systems of balance laws satisfying a different mechanism from (14).

3.1.2 Sharp Decay Estimate

A sharper decay estimate is obtained by Christoforou and Trivisa in [18]. The sharp decay result presented by Bressan-Yang [8] for hyperbolic conservation laws and by Christoforou-Trivisa [18] for hyperbolic balance laws relates the positive measure that corresponds to the solution of the *system* of hyperbolic balance laws to the measure of the solution of the Burgers's-type *scalar* equation with impulsive source term and shows that the latter is in fact more singular. This comparison estimate is extremely meaningful because it enables us to obtain powerful results for systems of hyperbolic balance laws that were previously unattainable. One such result is the establishment of precise rate of convergence for vanishing viscosity approximations to hyperbolic balance laws [19] presented in Sect. 3.2 (see also [9], for $G \equiv 0$). Here we present the estimate in the context of hyperbolic balance laws satisfying the *dissipation condition* (14) sharper than (30). The result is obtained (cf. Christoforou-Trivisa [18]) by introducing a partial ordering within the family of positive Radon measures, using symmetric rearrangements and a comparison with a solution of Burgers's equation with impulsive sources as well as lower semicontinuity properties of continuous Glimm-type functionals. Now, we define the partial ordering between two Radon measures as introduced in [8].

Definition 1 Let μ, μ' be two positive Radon measures: $\mu \preceq \mu'$ if and only if

$$\forall s > 0, \qquad \sup_{|A| \le s} \mu(A) \le \sup_{|B| \le s} \mu'(B). \tag{31}$$

The sharp decay rate for system (15) is:

Theorem 5 (Christoforou and Trivisa [18]) *Consider system of balance laws (15) and assume that the system is strictly hyperbolic, each characteristic field is genuinely nonlinear and the dissipation hypothesis (14) holds. Let $w(t,x)$ solve the Cauchy problem for the scalar Burgers equation with impulsive source term,*

$$\partial_t w + \partial_x \left(\frac{w^2}{2} \right) = -\beta w - ksgn(x) \cdot \frac{d}{dt} Q(u(t)), \tag{32}$$

$$w(0,x) = sgn(x) \cdot \sup_{meas(A) < 2|x|} \frac{\mu_0^{i+}}{2} \tag{33}$$

with k some positive constant. Then, for every solution $U(t,x)$ of the Cauchy problem for (15) with small total variation, and for every $t \ge 0$, and $i = 1, \dots, n$,

$$\boxed{\mu_t^{i+} \preceq D_x w(t).} \tag{34}$$

This result relates the measure that corresponds to the solution of the system of hyperbolic balance laws to the measure that corresponds to the solution of the Burgers's equation with impulsive source term and shows that the latter is in fact more singular. The proof can be found in [18]. As a consequence, a result on the rate of convergence for vanishing viscosity approximations is established.

3.2 Convergence Rate of Vanishing Viscosity Approximations

The rate of convergence of the vanishing viscosity approximations is established by first estimating the $\mathbf{L}^1$ distance between an exact BV solution U to the balance law and a viscous approximation U^ε and letting the viscosity coefficient $\varepsilon \to 0$. Error estimates are derived by introducing new Lyapunov functionals in order to measure the interaction of shock waves, while the rarefaction waves are treated by employing the sharp decay estimate (34). Consider the viscous approximations satisfying the parabolic system (17). Given initial data (16) with small total variation and close to U^* in $\mathbf{L}^1$, the solutions $\{U^\varepsilon\}$ of (17), (16) exist for all $t \geq 0$, have uniformly small total variation and converge to a unique solution of problem (15)–(16) as $\varepsilon \to 0$ as described in Sect. 2.2. Under the dissipation hypothesis (14) on the source term $G(U)$, we obtain results on the rate of convergence of the vanishing viscosity approximations by estimating the $\mathbf{L}^1$ distance

$$\|U^\varepsilon(t) - U(t)\|_{L^1}$$

and letting the viscosity coefficient $\varepsilon \to 0$.

Theorem 6 (Christoforou and Trivisa [19]) *Let system (15) be strictly hyperbolic and assume that each characteristic field is genuinely nonlinear and the dissipation hypothesis (14) holds. Then, given any initial data $U(\cdot, 0) = U_0$ with small total variation and $U_0 - U^* \in L^1$, there exists a positive constant β, such that for every $\tau > 0$, the solutions U and U^ε to the Cauchy problem for (15) and (17), respectively, satisfy the following estimate*

$$\begin{aligned} \|U^\varepsilon(\cdot,\tau) - U(\cdot,\tau)\|_{L^1} = \; & \mathcal{O}(1)\left(1 + \tfrac{1-e^{-\beta\tau}}{\beta} + e^{-\beta\tau}\right) \cdot \sqrt{\varepsilon}|\ln\varepsilon| TV\{U_0\} \\ & + \mathcal{O}(1)\varepsilon\|DG\|_\infty(\|U_0 - U^*\|_{L^1} + \sqrt{\varepsilon} TV\{U_0\}) \,. \end{aligned} \tag{35}$$

Bressan and Yang [9] considered the case of general BV solutions to hyperbolic systems of n conservation laws and established a convergence rate of order

$$\mathcal{O}(1)(1+\tau)\sqrt{\varepsilon}|\ln\varepsilon| TV\{U_0\}.$$

4 Isometric Immersions via Balance Laws

The isometric immersion problem for two-dimensional manifolds embedded into $\mathbb{R}^N$ is a fundamental and conceptually important problem. By generalizing the classical objects of curves and surfaces in $\mathbb{R}^3$, Riemann introduced Riemannian manifolds with a metric structure, and naturally, this gave rise to the issue whether a Riemannian manifold can be immersed into some Euclidean space $\mathbb{R}^N$ with its given induced metric. Of course this problem can also be generalized in dimensions more than two; cf. [37, 42, 43]. In the following subsections, we identify the connection of this problem to systems of balance laws and we expose the consequences of the application of the theory of Sect. 2 to isometric immersions. Its outcome reveals corrugated immersions, otherwise called "wrinkled surfaces".

4.1 The Problem

Let us describe this problem for the case of a two dimensional manifold into a three dimensional Euclidean space: Consider an open set $\Omega \subset \mathbb{R}^2$ and take a map $\mathbf{y} : \Omega \to \mathbb{R}^3$ corresponding to the surface $\mathbf{y}(\Omega) \subset \mathbb{R}^3$, i.e. its tangent plane at $\mathbf{y}(x_1, x_2)$ is spanned by the vectors $\{\partial_1 \mathbf{y}, \partial_2 \mathbf{y}\}$ with the unit normal vector to the surface $\mathbf{y}(\Omega)$ denoted by $\mathbf{n}$ and the distance measured along the surface by the formula

$$ds^2 = (\partial_1 \mathbf{y} \cdot \partial_1 \mathbf{y})(dx_1)^2 + 2(\partial_1 \mathbf{y} \cdot \partial_2 \mathbf{y}) dx_1\, dx_2 + (\partial_2 \mathbf{y} \cdot \partial_2 \mathbf{y})(dx_2)^2 \,. \tag{36}$$

Given (g_{ij}) $i,j = 1,2$ functions in Ω with $g_{12} = g_{21}$, the isometric immersion problem deals with the question of whether there exists a map $\mathbf{y} : \Omega \to \mathbb{R}^3$ so that

$$d\mathbf{y} \cdot d\mathbf{y} = g_{11}(dx_1)^2 + 2g_{12} dx_1 dx_2 + g_{22}(dx_2)^2 \,, \tag{37}$$

or equivalently,

$$\partial_1 \mathbf{y} \cdot \partial_1 \mathbf{y} = g_{11}, \qquad \partial_1 \mathbf{y} \cdot \partial_2 \mathbf{y} = g_{12}, \qquad \partial_2 \mathbf{y} \cdot \partial_2 \mathbf{y} = g_{22} \tag{38}$$

having the set $\{\partial_1 \mathbf{y}, \partial_2 \mathbf{y}\}$ as a linearly independent set in $\mathbb{R}^3$. In other words, the isometric immersion problem raises the issue of finding a surface $\mathbf{y}$ along which the distance is measured to be the prescribed metric g. From (38), one can see immediately that the isometric immersion problem is fully nonlinear in the three unknowns being the three components of the map $\mathbf{y}$. A two-dimensional manifold $\mathcal{M}$ parametrized by Ω with associated metric (g_{ij}) admits two fundamental forms;

- the first fundamental form I is

$$I \doteq g_{11}(dx_1)^2 + 2g_{12} dx_1 dx_2 + g_{22}(dx_2)^2 \tag{39}$$

- the second fundamental form II is

$$II \doteq h_{11}(dx_1)^2 + 2h_{12} dx_1\, dx_2 + h_{22}(dx_2)^2 \,. \tag{40}$$

If to the manifold $(\mathcal{M}, g)$ corresponds an isometric immersion, then one gets $II = -d\mathbf{n} \cdot d\mathbf{y}$ and the coefficients (h_{ij}) represent the orthogonality of $\mathbf{n}$ to the tangent plane. Standard notation from differential geometry is the following: $(E, F, G) = (g_{11}, g_{12}, g_{22})$ and $(L, M, N) = (h_{11}, h_{12}, h_{22})$. In view of the above analysis, the isometric immersion problem can be formulated as a special system of partial differential equations; the so-called Gauss-Codazzi system

$$\begin{aligned} \partial_1 M - \partial_2 L &= -\Gamma_{12}^{(1)} L + (\Gamma_{11}^{(1)} - \Gamma_{12}^{(2)}) M + \Gamma_{11}^{(2)} N \\ \partial_1 N - \partial_2 M &= -\Gamma_{22}^{(1)} L + (\Gamma_{21}^{(1)} - \Gamma_{22}^{(2)}) M - \Gamma_{21}^{(2)} N \end{aligned} \tag{41}$$

with the condition

$$LN - M^2 = K\,|g|\;. \tag{42}$$

Here $|g| \doteq det(g_{ij}) = g_{11}g_{22} - g_{12}^2$ and $K(x_1,x_2) \doteq \frac{R_{1212}}{|g|}$ is the Gauss curvature, where R_{ijkl} is the curvature tensor given by

$$R_{ijkl} = g_{lm}\left(\partial_k\Gamma_{ij}^{(m)} - \partial_j\Gamma_{ik}^{(m)} + \Gamma_{ij}^{(n)}\Gamma_{nk}^{(m)} - \Gamma_{ik}^{(n)}\Gamma_{nj}^{(m)}\right)\,, \tag{43}$$

and $\Gamma_{ij}^{(k)}$ is the Christoffel symbol given by

$$\Gamma_{ij}^{(k)} \doteq \frac{1}{2}g^{kl}\left(\partial_j g_{il} + \partial_i g_{jl} - \partial_l g_{ij}\right)\;. \tag{44}$$

Here, the indices $i, j, k, l = 1,2$, $(\partial_1,\partial_2) = (\partial_{x_1},\partial_{x_2})$ and the summation convention is used. Also, (g^{kl}) is the inverse of (g_{ij}). By rescaling the second fundamental form,

$$L^* = \frac{h_{11}}{\sqrt{|g|}},\qquad M^* = \frac{h_{12}}{\sqrt{|g|}},\qquad N^* = \frac{h_{22}}{\sqrt{|g|}}, \tag{45}$$

the Gauss-Codazzi system (41)–(42) is equivalent to

$$\begin{aligned}\partial_1 M^* - \partial_2 L^* &= \Gamma_{22}^{(2)}L^* - 2\Gamma_{12}^{(2)}M^* + \Gamma_{11}^{(2)}N^*\\ \partial_1 N^* - \partial_2 M^* &= -\Gamma_{22}^{(1)}L^* + 2\Gamma_{12}^{(1)}M^* - \Gamma_{11}^{(1)}N^*\end{aligned} \tag{46}$$

with the condition

$$L^*N^* - (M^*)^2 = K\;. \tag{47}$$

The Fundamental Theorem of Surfaces states that given forms I and II with (g_{ij}) being positive definite and smooth coefficients (g_{ij}) and (h_{ij}) that satisfy the Gauss-Codazzi system (41)–(42), or equivalently (45)–(47), then there exists a surface embedded into $\mathbb{R}^3$ with first and second fundamental forms I and II. Thanks to this, the isometric immersion problem is equivalent to solving the nonlinear system of PDEs (41)–(42), or equivalently (45)–(47), for a given metric (g_{ij}) and recover the solution (L,M,N) or equivalently (h_{ij}). Note that by Gauss's Theorema Egregium, the Gauss curvature K can be determined entirely by the given metric (g_{ij}). It should be added that there are several aspects and questions raised on this problem related to local or global issues.

Now the Gauss-Codazzi system is in general of mixed elliptic-hyperbolic type. Such problems are quite challenging since no general theory has been developed so far especially for systems that change type. The Gauss-Codazzi system changes type according to the sign of the Gauss curvature. The embedding of surfaces with positive Gauss curvature can be formulated as an elliptic boundary value problem and the isometric realization of such surfaces has been studied extensively; see Yau [46] and the references therein. On the other hand, for surfaces with negative Gauss curvature, the system is hyperbolic and the corresponding problem can be formulated as an initial or initial-boundary value problem. **Throughout this section, we consider only the global existence of negatively curved two-dimensional Riemannian manifolds into $\mathbb{R}^3$, i.e. $n = 2$ and $N = 3$ with $K < 0$.** A search for "corrugated immersions" that ask the data to be "rough" and not in C^1 is the issue pursued in the recent articles [10, 11, 14, 17]. In particular, we focus our discussion on discontinuous data for which the methods developed for balance laws have been successfully applied and we view the Gauss-Codazzi system as a system of two balance laws in the setting of Continuum Physics.

The Fundamental Theorem of Surfaces was extended when $(h_{ij}) \in L^\infty(\Omega)$ for given $(g_{ij}) \in C^{1,1}(\Omega)$ and then, the embedded surface is $C^{1,1}(\Omega)$. Thus, the isometric immersion problem reduces to solving the Gauss-Codazzi system (41)–(42), or (45)–(47), for $(h_{ij}) \in L^\infty(\Omega)$ with a given positive definite metric $(g_{ij}) \in C^{1,1}(\Omega)$ and then, immediately, we recover the immersion surface $\mathbf{y}(\Omega)$, which is $C^{1,1}$. We refer the reader to books [28, 33] for an exposition of the surface theory and to Mardare [40] for the extension of the aforementioned results to $(h_{ij}) \in L^\infty$. The recent applications [10, 11, 17] of the method of compensated compactness and [14] of BV techniques prove existence of global non-smooth isometric immersions of a two dimensional Riemannian manifold with negative Gauss curvature into three dimensional Euclidean space. In each paper, the authors view the system of Gauss-Codazzi as a hyperbolic system of two balance laws, solve an initial value problem using these techniques developed for the needs of the theory of Continuum Mechanics and construct discontinuous solutions. According to Mardare's theorem, there exists an immersion which lies in the Sobolev space $W^{2,\infty}_{loc}$ and hence will be locally in $C^{1,1}$, cf. Evans [30, Chap. 5]. This means that the immersion is smooth enough so that the Gauss curvature is well defined.

In short, Chen et al. [11] first established corrugated immersions using the method of compensated compactness and the case of the catenoidal shell of revolution is captured. Then this result has been extended for corrugated immersions having second derivatives of bounded variation by Christoforou in [14]. Next, Cao et al. [10] captured the case of the helicoid metric by introducing a different change of variables on the Gauss-Codazzi system and the method of compensated compactness. Last, Christoforou and Slemrod in [17] solve the isometric immersion problem in the non-smooth regime using the optimal so far slower decay rate of the Gauss curvature. This is the same decay rate in the work of Hong [36] on smooth immersions.

4.2 Corrugated Immersions with BV Techniques

In this subsection, the results of Christoforou [14] on corrugated immersions are presented. Specifically, the techniques developed in the theory of weak solutions of bounded variation in continuum physics are employed in [14] to deal with the isometric immersions. We begin by stating the *BV Framework* that ensures the convergence of approximate solutions of bounded variation (BV) to the Gauss-Codazzi system and yields the $C^{1,1}$ isometric realization of two-dimensional surfaces into $\mathbb{R}^3$, even if the underlying system of partial differential equations is of mixed-type. Consider a sequence of functions $(L^\varepsilon, M^\varepsilon, N^\varepsilon)$ defined on an open set $\Omega \subset \mathbb{R}^2$ that satisfies the following framework: **Framework (BV)**

(BV.1) $TV_{\mathscr{X}}(L^\varepsilon, M^\varepsilon, N^\varepsilon)$ and $\|(L^\varepsilon, M^\varepsilon, N^\varepsilon)\|_{L^1(\mathscr{X})}$ are uniformly bounded (independently of ε) on every open bounded set $\mathscr{X}$ with $\bar{\mathscr{X}} \subset \Omega$.

(BV.2) There exist $\delta_j^\varepsilon(1), j = 1, 2, 3$ with $\delta_j^\varepsilon(1) \to 0$ in the sense of distributions as $\varepsilon \to 0$ such that

$$\partial_1 M^\varepsilon - \partial_2 L^\varepsilon = \Gamma_{22}^{(2)} L^\varepsilon - 2\Gamma_{12}^{(2)} M^\varepsilon + \Gamma_{11}^{(2)} N^\varepsilon + \delta_1^\varepsilon(1) \tag{48}$$

$$\partial_1 N^\varepsilon - \partial_2 M^\varepsilon = -\Gamma_{22}^{(1)} L^\varepsilon + 2\Gamma_{12}^{(1)} M^\varepsilon - \Gamma_{11}^{(1)} N^\varepsilon + \delta_2^\varepsilon(1) \tag{49}$$

and

$$L^\varepsilon N^\varepsilon - (M^\varepsilon)^2 = K + \delta_3^\varepsilon(1).$$

Then the following theorem states the convergence of this sequence.

Theorem 7 ([14, Theorem 3.2]) *Let a sequence of functions $\{(L^\varepsilon, M^\varepsilon, N^\varepsilon)\}$ in $BV_{loc}(\Omega)$, $\Omega \subset \mathbb{R}^2$ for every $\varepsilon > 0$, satisfy Framework (BV). Then, there exists a subsequence $\{(L^{\varepsilon_k}, M^{\varepsilon_k}, N^{\varepsilon_k})\}$ that converges as $\varepsilon_k \to 0$ in $L^1_{loc}(\Omega)$ to $(L^*, M^*, N^*) \in BV_{loc}(\Omega)$ with $TV_{\mathscr{X}}(L, M, N)$ uniformly bounded. Moreover, the limit (L^*, M^*, N^*) is a BV weak solution to Gauss-Codazzi system (45)–(47) on Ω, which yields a $C^{1,1}(\Omega)$ isometric immersion of the corresponding two-dimensional Riemannian manifold into $\mathbb{R}^3$ with second fundamental form having coefficients $(h_{ij}) \in BV_{loc}(\Omega)$ for a given positive definite metric (g_{ij}) in $C^{1,1}(\Omega)$.*

To secure hyperbolicity, we restrict ourselves to negatively curved manifolds,

$$K = -\gamma(x_1, x_2)^2 \leq 0 \qquad (x_1, x_2) \in \Omega \tag{50}$$

Next, by scaling the variables

$$\tilde{L} = \frac{L^*}{\gamma}, \qquad \tilde{M} = \frac{M^*}{\gamma}, \qquad \tilde{N} = \frac{N^*}{\gamma}, \tag{51}$$

relation (47) becomes

$$\tilde{L}\tilde{N} - \tilde{M}^2 = -1 . \tag{52}$$

Then, "fluid variables" (u, v, ρ, p) of Chen et al. [11] in this geometry are as follows:

$$\tilde{L} = \rho v^2 + p, \qquad \tilde{M} = -\rho uv, \qquad \tilde{N} = \rho u^2 + p , \tag{53}$$

with $q^2 = u^2 + v^2$ and $p = -1/\rho$. Hence, the Gauss-Codazzi system (45)–(47) is equivalent to a constrained hyperbolic system of balance laws:

$$\partial_1 U + \partial_2 f(U) + \varphi(U, x_1, x_2) = 0, \qquad p^2 + p\rho q^2 = -1, \tag{54}$$

where

$$U = \begin{pmatrix} \rho uv \\ \rho u^2 + p \end{pmatrix}, \qquad f(U) = \begin{pmatrix} \rho v^2 + p \\ \rho uv \end{pmatrix}, \tag{55}$$

and

$$\varphi(U, x_1, x_2) = \begin{pmatrix} (\rho v^2 + p)\tilde{\Gamma}^{(2)}_{22} + 2\rho uv\tilde{\Gamma}^{(2)}_{12} + (\rho u^2 + p)\tilde{\Gamma}^{(2)}_{11} \\ (\rho v^2 + p)\tilde{\Gamma}^{(1)}_{22} + 2\rho uv\tilde{\Gamma}^{(1)}_{12} + (\rho u^2 + p)\tilde{\Gamma}^{(1)}_{11} \end{pmatrix} . \tag{56}$$

This system is in conservative form, the eigenvalues of the matrix $D_U f$ can easily be computed together with the corresponding eigenvectors and both characteristic fields are linearly degenerate. For convenience, we express u and v in polar coordinates (q, θ); i.e. $u = q\cos\vartheta$ and $v = q\sin\vartheta$. We assign data at $x_1 = 0$ via

$$q(0, x_2) = q_0(x_2), \qquad \vartheta(0, x_2) = \vartheta_0(x_2), \qquad x_2 \in \mathbb{R} \tag{57}$$

that yield the initial data $U(0, x_2) = U_0(x_2)$ of (54)–(56), where $U_0 \in \mathbb{R}^2$ can be expressed by q_0 and ϑ_0. Since in the theory of balance laws, cf. Sect. 2, we require a constant equilibrium configuration, we assume that U^* is a constant equilibrium, i.e. $\varphi(U^*, x_1, x_2) = 0$ for all $(x_1, x_2) \in [0, \infty) \times \mathbb{R}$. For simplicity, let the equilibrium U^* correspond to the coordinates $q = q^*$, $\vartheta = 0$. Then the identities

$$\Gamma^{(2)}_{11} = (q^{*2} - 1)(\Gamma^{(2)}_{22} + \frac{\gamma_2}{\gamma}), \qquad \Gamma^{(1)}_{11} + \frac{\gamma_1}{\gamma} = (q^{*2} - 1)\Gamma^{(1)}_{22} , \tag{58}$$

hold true for all (x_1, x_2). Following the theory of Sect. 2 to establish global existence, a special structure is needed in the following 2×2 matrix

$$B(x_1, x_2) \doteq \begin{bmatrix} 2\tilde{\Gamma}^{(2)}_{12} & (q^{*2} - 1)\tilde{\Gamma}^{(2)}_{22} + \tilde{\Gamma}^{(2)}_{11} \\ 2\tilde{\Gamma}^{(1)}_{12} & (q^{*2} - 1)\tilde{\Gamma}^{(1)}_{22} + \tilde{\Gamma}^{(1)}_{11} \end{bmatrix} . \tag{59}$$

which is defined via the relation $B \doteq R^{-1}D\varphi R$, where R is the matrix of the right eigenvectors. We state three different hypotheses on problem (54)–(56), (57) that yield global existence in $\Omega = [0, +\infty) \times \mathbb{R}$:

- **Hyp. 1** *Rapid decay of the inhomogeneity and the source as* $x_1, x_2 \to \infty$: There are functions $\chi \in L^1[0,\infty)$ and $\psi \in L^1(\mathbb{R})$ such that

$$|\varphi(U, x_1, x_2)| \leq \omega_0 \chi(x_1), \qquad |D\varphi(U, x_1, x_2)| \leq \omega_0 \chi(x_1)$$
$$|\partial_2 \varphi(U, x_1, x_2)| \leq \psi(x_2)\chi(x_1)$$

  hold true for every U in a small neighborhood of U^*, $x_1 \in [0,\infty)$ and $x_2 \in \mathbb{R}$.
- **Hyp. 2** *Dissipative source term:* The matrix $B(x_1, x_2)$ given by (59) is column diagonally dominant, i.e. there is a constant $\mu > 0$ such that for every $i = 1, 2$,

$$B_{ii}(x_1, x_2) - \sum_{j \neq i} |B_{ji}(x_1, x_2)| > \mu > 0,$$

  for $x_1 \in [0,\infty)$, $x_2 \in (-\infty, \infty)$. Also, there is a function $\psi \in L^1(\mathbb{R})$ such that

$$|\partial_2 \varphi(U, x_1, x_2)| \leq \psi(x_2), \quad x_2 \in \mathbb{R}, \tag{60}$$

  for every U in a small neighborhood of U^* and $x_1 \in [0, \infty)$.
- **Hyp. 3** *Weakly dissipative source and L^1 stable equilibrium:* The matrix $B(x_1, x_2)$ given by (59) has positive diagonal entries, i.e.

$$B_{ii}(x_1, x_2) > 0, \qquad i = 1, 2$$

  for every $x_1 \in [0,\infty)$ and $x_2 \in \mathbb{R}$, the equilibrium U^* is apriori L^1-stable, i.e. there is a constant $\beta > 0$ such that

$$\int_{-\infty}^{\infty} |U(x_1, x_2) - U^*|\, dx_2 \leq \beta \int_{-\infty}^{\infty} |U_0(x_2) - U^*|\, dx_2 \tag{61}$$

  for every $x_1 \in [0, \infty)$ and, also, there is a function $\psi \in L^1(\mathbb{R})$ such that

$$|\partial_2 \varphi(U, x_1, x_2)| \leq \psi(x_2), \quad x_2 \in \mathbb{R}, \tag{62}$$

  for every U in a small neighborhood of U^* and $x_1 \in [0, \infty)$.

Under one of the three above hypotheses, a global isometric immersion exists:

Theorem 8 (Global Existence [14, Theorem 5.2]) *Consider problem (54)–(56) with data (57), assume that (58) and one of the three aforementioned hypotheses holds. Then, for small enough positive numbers ω_0 and δ_0 such that*

$$\int_{-\infty}^{\infty} \psi(x_2)\, dx_2 \leq \omega_0, \quad \textit{and} \quad TV_{x_2} U_0 < \delta_0, \tag{63}$$

with $\|U_0 - U^*\|_{L^1}$ sufficiently small, there is an admissible weak solution $U(x_1, x_2)$ to (54)–(56) with data (57) defined in $\Omega = [0,\infty)\times(-\infty,\infty)$. For any $x_1 \in [0,\infty)$, the restriction $U(x_1,\cdot)$ is a function of bounded variation on $(-\infty,\infty)$ and

- *if Hyp. 1 holds, then*

$$TV_{x_2} U(x_1,\cdot) \leq C\,[TV_{x_2} U_0 + \omega_0] \tag{64}$$

- *if Hyp. 2 holds, then*

$$TV_{x_2} U(x_1,\cdot) \leq C\left[e^{-\tilde{\mu}x_1} TV_{x_2} U_0 + \omega_0\right] \tag{65}$$

- *if Hyp. 3 holds, then*

$$TV_{x_2} U(x_1,\cdot) \leq C\left[e^{-\tilde{\mu}x_1} TV_{x_2} U_0 + \tilde{\beta}\|U_0 - U^*\|_{L^1} + \omega_0\right] \tag{66}$$

for some $\tilde{\mu} > 0$, $\tilde{\beta} > 0$; C is a universal large positive constant. The isometric immersion problem (45)–(47) and (57) has a weak solution (L^*, M^*, N^*) in $BV(\Omega)$, yielding a $C^{1,1}(\Omega)$ immersion of the Riemannian manifold into $\mathbb{R}^3$.

Now Hypotheses 2 and 3 of Theorem 8 can be analyzed (and simplified) for a special class of so-called **conformal metrics**, $g = E(x_1)(dx_1^2 + dx_2^2)$, with

$$E(x_1,x_2) = G(x_1,x_2) \doteq E(x_1), \qquad G(x_1,x_2) = 0 \tag{67}$$

as stated in Theorem 5.3 of [14]. Let us only mention here that within this class, condition (58) becomes the ordinary differential equation

$$\frac{\gamma'(x_1)}{\gamma(x_1)} + q^{*2}\frac{E'}{2E} = 0, \quad \Longleftrightarrow \quad \frac{K'(x_1)}{K(x_1)} + q^{*2}\frac{E'(x_1)}{E(x_1)} = 0\,. \tag{68}$$

that yields

$$K(x_1) = -k_0 E(x_1)^{-q^{*2}}$$

with k_0 a positive constant. Note that the source φ now in system (54) is independent of x_2. Now we observe that the isometric immersion problem for equilibrium configurations of a catenoidal shell of revolution about the x_1-axis satisfies such conditions. Indeed, consider **the metric associated with the catenoid**,

$$E(x_1) = G(x_1) = \left(\cosh(cx_1)\right)^{\frac{2}{|q^*|^2-1}}, \qquad F = 0\,, \tag{69}$$

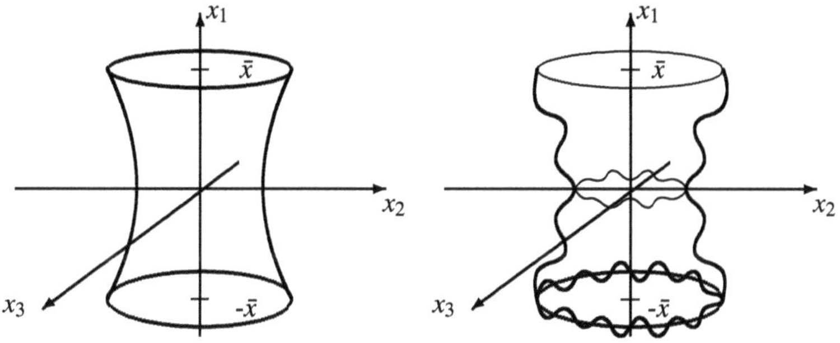

Fig. 1 A catenoid-type corrugated surface (on the *right*) having the same metric as the catenoid (on the *left*) in $\mathbb{R}^3$

where $c \neq 0, q^* > 1$ are constants. Assigning periodic data U_0 at $x = -\bar{x} < 0$ to be,

$$(q(-\bar{x}, x_2), \vartheta(-\bar{x}, x_2)) = (q_0(x_2), \vartheta_0(x_2)), \qquad x_2 \in \mathbb{R}\,. \tag{70}$$

to system (54)–(56), we have that if $TV_{x_2} U_0$ and $\|U_0 - U^*\|_{L^1}$ are sufficiently small and either (i) $q^* > \sqrt{2}$ or (ii) $q^* > 2/\sqrt{3}$ and the equilibrium $U^* = (0, (q^{*2} - 1)^{-1/2})$ is apriori L^1 stable, then there exists an admissible BV solution (L^*, M^*, N^*) to the Gauss-Codazzi system in $\Omega = [-\bar{x}, 0) \times \mathbb{R}$ with coefficients $(h_{ij}) \in BV(\Omega)$ and therefore, a $C^{1,1}(\Omega)$ isometric immersion into $\mathbb{R}^3$. This is the context of [14, Theorem 5.4]. Thus, the corrugated immersion with non-circular cross sections drawn in Fig. 1 is also captured here with $(h_{ij}) \in BV$. The above results are not only captured in BV space, but first in L^∞ in the work of Chen et al. [11].

4.3 Slowly Decaying Gauss Curvature

The analysis of isometric immersions of two dimensional Riemannian manifolds with negative Gauss curvature into $\mathbb{R}^3$ has a long history going back to the classical result of Hilbert of 1901 [34]. Specifically he proved that the hyperbolic half-plane,

$$x \in \mathbb{R},\ t > 0, \qquad g = \frac{dt^2 + dx^2}{t^2},$$

which metric is globally conformal, cannot be isometrically immersed into $\mathbb{R}^3$. In that case the Gauss curvature is a negative constant. Hilbert's result was generalized in 1963 by Efimov [29] to complete manifolds where the negative Gauss curvature has a slow rate of decay. Efimov's result [29] reads

Theorem 9 (Han and Hong [33, Theorem 10.0.1]) *Let $(\mathcal{M}, g)$ be a complete negative curved smooth Riemannian manifold with Gauss curvature K satisfying*

$$\forall p \in \mathcal{M}, \qquad \limsup_{p\to\infty} \left| D(\frac{1}{\sqrt{|K|}})(p) \right| < \frac{1}{2}\sqrt{\sqrt{41}-5},$$

D denoting the gradient. Then $(\mathcal{M}, g)$ admits no C^3 isometric immersion into $\mathbb{R}^3$.

If $(\mathcal{M}, g)$ is a geodesically complete, simply connected, smooth two-dimensional Riemannian manifold with non-positive Gauss curvature, then there exists a global geodesic coordinate system (x, t) in $\mathcal{M}$ in which the metric reads,

$$g = dt^2 + B^2(x,t)dx^2, \tag{71}$$

where B is a smooth function satisfying $B(x, 0) = 1$ and $\partial_t B(x, 0) = 0$ for $x \in \mathbb{R}$.

A direct substitution of (71) in (41) then yields that (L, M, N) satisfy the Gauss-Codazzi system in the form

$$\begin{aligned} \partial_t L - \partial_x M &= L\partial_t \ln B - M\partial_x \ln B + NB\partial_t B, \\ \partial_t M - \partial_x N &= -M\partial_t \ln B, \end{aligned} \tag{72}$$

with

$$LN - M^2 = KB^2, \tag{73}$$

where the Gauss curvature K is defined by the equation $\partial_{tt}B = -KB$ in terms of the metric. Now, we consider the case in which a negatively curved manifold has the metric in geodesic polar coordinates given by

$$\boxed{g = d\rho^2 + B^2(\rho)d\theta^2, \qquad \rho > 0,\ \theta \in [0, 2\pi],}$$

where B is defined by $\partial_{\rho\rho}B = -KB$ with $B(0) = 0$, $\partial_\rho B(0) = 1$, and the Gauss curvature $K = K(\rho)$ is a negative smooth even function satisfying

$$K = -a^2/(\rho^2 - 1),$$

if $|\rho| \geq 2$ for a positive constant a. We see that $\rho^2 |K(\rho)|$ decreases as ρ increases when $|\rho| \geq 2$ but the decay rate as measured by

$$D(1/\sqrt{|K|}) = \frac{1}{a}\partial_\rho\sqrt{(\rho^2-1)} = \frac{\rho}{a\sqrt{\rho^2-1}}$$

has the limit $1/a$ as $\rho \to \infty$. Hence in this case, **Efimov's theorem (Theorem 9) asserts there is no C^3 isometric immersion if a is sufficiently large**. For a long period, research aimed at proving nonexistence of isometric immersions in $\mathbb{R}^3$. Then, in 1980s, Yau proposed to find a sufficient condition for existence of isometric immersion of negatively curved manifolds in $\mathbb{R}^3$. Since K in the above example is $\mathcal{O}(\rho^{-2})$, it appears natural to try when K is $\mathcal{O}(\rho^{-2-\delta})$ as $\rho \to \infty$. Indeed, this is the result of Hong [36] in 1993, who proved that **geodesically complete negatively curved manifolds can be isometrically embedded in $\mathbb{R}^3$ if their negative Gauss curvature decays to zero at a certain rate at infinity**.

Theorem 10 (Han and Hong [33, Theorem 10.2.2]) *For a complete simply connected two dimensional Riemannian manifold $(\mathcal{M}, g)$ with negative Gauss curvature K with metric $g = dt^2 + B^2(x,t)dx^2$, assume for some constant $\delta > 0$*

(i) $t^{2+\delta}\,|K|$ is decreasing in $|t|$, $|t| > T$;
(ii) $\partial_x^i \ln|K|$, for $i = 1,2$ and $t\partial_t\partial_x \ln|K|$ are bounded;
(iii) K is periodic in x with period 2π.

Then $(\mathcal{M}, g)$ admits a smooth isometric immersion in $\mathbb{R}^3$.

The proof of Hong is a careful study of the hyperbolic system of two balance laws (the two Codazzi equations) and one closure relation (the Gauss equation) and requires two separate steps. The first step is to establish existence of smooth solutions to the balance laws for small, smooth data prescribed at a large enough time $t = T_1$. The reason for this part is that is only after large time that the decay of the Gauss curvature may be exploited to obtain the relevant C^1 a priori estimates. The second part of the proof is rather standard and simply asks for the initial data at $t = 0$ to be sufficiently small and smooth to enable us to get a solution up to $t = T_1$. Yet, a question is **whether one can capture the existence result of Hong with the slow decay rate of the Gauss curvature in the setting of non-smooth immersions**. This is the issue pursued by Christoforou and Slemrod in [17]; the main difference among [10, 11, 14, 17] is the Gauss curvature's decay rate, so that Hong's optimal (so far) rate $t^{-(2+\delta)}$ is achieved in [17] in a non-smooth regime

Theorem 11 (Christoforou and Slemrod [17, Theorem 6.2]) *Let $(\mathcal{M}, g)$ be a geodesically complete simply connected smooth two dimensional Riemannian manifold with non-positive Gauss curvature K and a metric of the form (71) defined on $\Omega = \mathbb{R} \times [T_1, T_2]$. Assume that $h \doteq B$ and $k^* = |K|$ are independent of x satisfying*

$$\partial_{tt}h = k^* h, \quad h(0) = 1, \quad \partial_t h(0) = 0. \tag{74}$$

and K is given by

$$K = -\frac{C}{(1+|t|)^{2+\frac{\delta}{2}}}, \quad C > 0 \tag{75}$$

for $t > T_1$. Then, for T_1 large enough, there exists $\mathbf{y} \in W^{2,\infty}_{loc}(\Omega)$ satisfying the embedding equations of Mardare's Theorem for any $T_2 > T_1$.

This theorem provides an isometric immersion on $\Omega = \mathbb{R}\times[T_1, T_2]$ for some $T_1 > 0$ large enough. To achieve a global result different from Hong's smooth case, one has to choose non-smooth data at $t = 0$ and get a solution up to $t = T_1$ in the non-classical sense. This is accomplished by using the local existence result of Dafermos and Hsiao [24], hence one can reach time $t = T_1$ from $t = 0$ starting at $t = 0$ with small BV data depending on the size of T_1 and the two pieces are glued together to provide a global immersion on the plane. See Theorem 6.3 in [17]. The isometric immersion constructed in [17] has no equilibrium configuration in contrast to the other three [10, 11, 14]. This is an important step in this research program since,

- in *Differential Geometry*, it is crucial to solve the isometric immersion problem with as general metric as possible. However, requiring an equilibrium configuration such as the catenoid or the helicoid as done in [10, 11, 14] produce constraints, which appear as identities on the metric g;
- in *Partial Differential Equations*, the importance is to solve an initial value problem with as general initial data as possible. This is not the case considered in [17] since constant initial data for the Riemann invariants could also be assumed to deal with as general metric as possible.

We refer the reader to the book [16] for a chapter on a survey of corrugated immersions using techniques on hyperbolic PDEs [10, 11, 14, 17] and specifically how the BV existence results presented in Sect. 2 are used in [14].

4.4 Applications of Corrugated Immersions

Let us also mention two of them, that are also described in [16]:

(i) Big data processing and deep learning are applied to fields like computer vision, automatic speech recognition, natural language processing, audio recognition and bio-informatics where they have been shown to produce state-of-the-art results on various tasks. The natural mathematical interpretation of this branch in machine learning would be geometrically via isometric immersions of lower-dimensional "surfaces" into higher dimensions. See [25] for a short description.
(ii) Another application of different nature is the phenomenon of fluid turbulence as presented in [21] which may be better understood through the connection of continuum mechanics with isometric immersions as the evolution of a non-smooth surface. See [1] for a connection with the theory of *wild solutions* of De Lellis and Szekelyhidi [27] to Euler equations.

A common example on *corrugated immersions* is that of paper folding: how a sheet of paper can be folded. It can be bended, folded, or crumpled but can be neither compressed nor stretched. Instead of considering surfaces with zero curvature, (i.e. corresponding to a sheet of paper), here we consider manifolds with negative Gauss curvature. Hence, establishing the isometric immersion problem would yield existence of corrugated surfaces with alternating grooves and ridges having negative curvature. Actually, the immersions established in [10, 11, 14, 17] correspond to discontinuous second derivatives and more precisely, $(h_{ij}) \in L^\infty$ or BV.

Acknowledgements The author would like to thank the organizers of *INdAM Workshop on Innovative Algorithms and Analysis* that took place in Rome from May 17th until 20th of 2016 for the invitation and the warm hospitality.

References

1. Acharya, A., Chen, G.Q., Slemrod, M., Wang, D.: Fluids, elasticity, geometry and the existence of wrinkled solutions. arXiv:1605.03058, preprint 2016
2. Amadori, D., Guerra, G.: Global weak solutions for systems of balance laws. Appl. Math. Lett. **12**(6), 123–127 (1999)
3. Amadori, D., Guerra, G.: Uniqueness and continuous dependence for systems of balance laws with dissipation. Nonlinear Anal. **49**, 987–1014 (2002)
4. Amadori, D., Gosse, L., Guerra, G.: Global BV entropy solutions and uniqueness for hyperbolic systems of balance laws. Arch. Ration. Mech. Anal. **162**(4), 327–366 (2002)
5. Bianchini, S., Bressan, A.: Vanishing viscosity solutions of nonlinear hyperbolic systems. Ann. Math. **161**, 223–342 (2005)
6. Bianchini, S., Caravenna, L.: SBV regularity for genuinely nonlinear, strictly hyperbolic systems of conservation laws in one space dimension. Commun. Math. Phys. **313**(1), 1–33 (2012)
7. Bressan, A.: Hyperbolic Systems of Conservation Laws: The One-Dimensional Cauchy Problem. Oxford University Press, Oxford (2000)
8. Bressan, A., Yang, T: A sharp decay estimate for positive nonlinear waves. SIAM J. Math. Anal. **36**(2), 659–677 (2004)
9. Bressan, A., Yang, T.: On the convergence rate of vanishing viscosity approximations. Commun. Pure Appl. Math. **57**, 1075–1109 (2004)
10. Cao, W., Huang, F., Wang, D.: Isometric immersion of surfaces with two classes of metrics and negative Gauss curvature. Arch. Ration. Mech. Anal. **218**(3), 1431–1457 (2015)
11. Chen, G.-Q., Slemrod, M., Wang, D.: Isometric immersions and compensated compactness. Commun. Math. Phys. **294**, 411–437 (2010)
12. Christoforou, C.: Hyperbolic systems of balance laws via vanishing viscosity. J. Differ. Equ. **221**(2), 470–541 (2006)
13. Christoforou, C.: Uniqueness and sharp estimates on solutions to hyperbolic systems with dissipative source. Commun. Partial Differ. Equ. **31**(12), 1825–1839 (2006)
14. Christoforou, C.: BV weak solutions to Gauss-Codazzi system for isometric immersions. J. Differ. Equ. **252**, 2845–2863 (2012)
15. Christoforou, C.: A remark on the Glimm scheme for inhomogeneous hyperbolic systems of balance laws. J. Hyperbolic Differ. Equ. **12**(4), 787–797 (2015)
16. Christoforou, C.: Isometric immersions via continuum mechanics. In: Partial Differential Equations: Ambitious Mathematics for Real-Life Applications SEMA-SIMAI Springer Series (submitted)
17. Christoforou, C., Slemrod, M.: Isometric immersions via compensated compactness for slowly decaying negative Gauss curvature and rough data. Z. Angew. Math. Phys. **66**(6), 3109–3122 (2015)
18. Christoforou, C., Trivisa, K.: Sharp decay estimates for hyperbolic balance laws. J. Differ. Equ. **247**(2), 401–423 (2009)
19. Christoforou, C., Trivisa, K.: Rate of convergence for vanishing viscosity approximations to hyperbolic balance laws. SIAM J. Math. Anal. **43**(5), 2307–2336 (2011)

20. Christoforou, C., Trivisa, K.: Decay of positive waves of hyperbolic balance laws. Acta Math. Sci. Ser. B Engl. Ed. **32**(1), 352–366 (2012)
21. Courant, R., Friedrichs, K.O.: Supersonic Flow and Shock Waves. Springer, New York (1948)
22. Dafermos, C.M.: Hyperbolic systems of balance laws with weak dissipation. J. Hyperbolic Differ. Equ. **3**(3), 505–527 (2006)
23. Dafermos, C.M.: Hyperbolic Conservation Laws in Continuum Physics. Grundlehren Math. Wissenschaften Series, vol. 325. Springer, Berlin (2010)
24. Dafermos, C.M., Hsiao, L.: Hyperbolic systems of balance laws with inhomogeneity and dissipation. Indiana Univ. Math. J. **31**, 471–491 (1982)
25. Daubechies, I.: Big Data's Mathematical Mysteries, Quantasized: Mathematics, Quanta Magazine, December 2015
26. De Lellis, C., Otto, F., Westdickenberg, M.: Structure of entropy solutions for multi-dimensional scalar conservation laws. Arch. Ration. Mech. Anal. **170**(2), 137–184 (2003)
27. De Lellis, C., Szekelyhidi, L.: The Euler equations as a differential inclusion. Ann. Math. **170**(2), 1417–1436 (2009)
28. do Carmo, M.P.: Riemannian Geometry (Transl. by F. Flaherty). Birkhauser, Boston, MA (1992)
29. Efimov, N.V.: The impossibility in Euclidean 3-space of a complete regular surface with a negative upper bound of the Gaussian curvature. Dokl. Akad. Nauk SSSR (N.S.) **150**, 1206–1209 (1963); Sov. Math. Dokl. **4**, 843–846 (1963)
30. Evans, L.C.: Partial Differential Equations. Graduate Studies in Mathematics, vol. 19, 2nd edn. American Mathematical Society, Providence (2010)
31. Glimm, J.: Solutions in the large for nonlinear hyperbolic systems of equations. Commun. Pure Appl. Math. **4**, 697–715 (1965)
32. Goatin, P., Gosse, L.: Decay of positive waves for $n \times n$ hyperbolic systems of balance laws. Proc. Am. Math. Soc. **132**(6), 1627–1637 (2004)
33. Han, Q., Hong, J.-X.: Isometric Embedding of Riemannian Manifolds in Euclidean Spaces. American Mathematical Society, Providence (2006)
34. Hilbert, D., Ueber Flächen von constanter Gaussscher Krümmung, (German) [On surfaces of constant Gaussian curvature] Trans. Am. Math. Soc. **2**(1), 87–99 (1901)
35. Holden, H., Risebro, N.H.: Front Tracking for Hyperbolic Conservation Laws. Springer, New York (2002)
36. Hong, J.X.: Realization in $\mathbb{R}^3$ of complete Riemannian manifolds with negative curvature. Commun. Anal. Geom. **1**, 487–514 (1993)
37. Janenko, N., Rozdestvenskii, B.L.: Systems of Quasilinear Equations and Their Applications to Gas Dynamics (Translations of Mathematical Monographs), vol. 55. American Mathematical Society, Providence (1983)
38. Liu, T.-P.: Quasilinear hyperbolic systems. Commun. Math. Phys. **68**(2), 141–172 (1979)
39. Liu, T.-P.: Nonlinear stability and instability of transonic flows through a nozzle. Commun. Math. Phys. **83**(2), 243–260 (1982)
40. Mardare, S.: The fundamental theorem of surface theory for surfaces with little regularity. J. Elast. **73**, 251–290 (2003)
41. Nash, J.: C^1 isometric imbeddings. Ann. Math. (2) **60**, 383–396 (1954)
42. Nash, J.: The imbedding problem for Riemannian manifolds. Ann. Math. (2) **63**, 20–63 (1956)
43. Oleinik, O.: Discontinuous solutions of nonlinear differential equations. Am. Math. Soc. Transl. Ser. **26**, 95–172 (1963)
44. Serre, D.: Systems of Conservation Laws, I, II. Cambridge University Press, Cambridge (1999)
45. Tadmor, E.: Local error estimates for discontinuous solutions of nonlinear hyperbolic equations. SIAM J. Numer. Anal. **28**(4), 891–906 (1991)
46. Yau, S.-T.: Review of geometry and analysis. In: Arnold, V., Atiyah, M., Lax, P., Mazur, B. (eds.) Mathematics: Frontiers and Perspectives, pp. 353–401, International Mathematics Union. American Mathematical Society, Providence (2000)

Viscous Equations Treated with $\mathscr{L}$-Splines and Steklov-Poincaré Operator in Two Dimensions

Laurent Gosse

Abstract Well-balanced schemes, nowadays well-known for 1D hyperbolic equations with source terms and systems of balance laws, are extended to strictly parabolic equations, first in 1D, then in 2D on Cartesian computational grids. The construction heavily relies on a particular type of piecewise-smooth interpolation of discrete data known as $\mathscr{L}$-splines. In 1D, several types of widely-used discretizations are recovered as particular cases, like the El-Mistikawy-Werle scheme or Scharfetter-Gummel's. Moreover, a distinctive feature of our 2D scheme is that dimensional-splitting never occurs within its derivation, so that all the multi-dimensional interactions are kept at the discrete level. This leads to improved accuracy, as illustrated on several types of drift-diffusion equations.

1 Introduction and Main Topics

We study numerical approximations of weak solutions of Cauchy problems for,

$$\partial_t u - \varepsilon \partial_{xx} u + p(x)\partial_x u + q(x)u = f(x), \quad x \in (-1,1), \quad t > 0, \tag{1}$$

where $\varepsilon > 0$ is allowed to become small with respect to the characteristic scale of the computational grid. Yet, initial and boundary data must be prescribed: for instance,

$$u(t=0,x) = u^0(x), \qquad u(.,x=\pm 1) = u_\pm, \text{ with } u^0(\pm 1) = u_\pm.$$

Standard assumptions on the coefficients read,

$$\varepsilon > 0, \quad p \in W^{1,\infty}(-1,1), \quad 0 \le q \in L^\infty(-1,1), \quad f \in L^\infty(-1,1). \tag{2}$$

L. Gosse (✉)
IAC, CNR, via dei Taurini, 19, 00185 Roma, Italy
e-mail: l.gosse@ba.iac.cnr.it

L. Gosse, R. Natalini (eds.), *Innovative Algorithms and Analysis*,
Springer INdAM Series 16, DOI 10.1007/978-3-319-49262-9_6

As (1) satisfies the maximum principle, all the data are chosen to be non-negative. Analogously, in the square domain $\Omega = (-1, 1)^2$, consider the Cauchy problem,

$$\partial_t u(t,x,y) - \varepsilon \Delta u + \alpha(x,y)u = f(x,y), \qquad u(t=0,\cdot,\cdot) = u^0(\cdot,\cdot) \geq 0, \tag{3}$$

where Δ stands for the Laplace operator in $\mathbb{R}^2$ and Dirichlet boundary data is prescribed on $\partial\Omega$ in accordance with initial data u^0. Parameters are assumed to be positive, $0 \leq f \in L^\infty(\Omega)$, $0 < \alpha \in L^\infty(\Omega)$, so as to ensure existence and uniqueness of a variational solution in $C^0(0,T;L^2(\Omega)) \cap L^2(0,T;H^1(\Omega))$. Lax-Milgram lemma and a classical bootstrap argument yield a stationary solution $u^\infty(x,y)$ in $W^{2,\infty}(\Omega)$.

1.1 An Overview of 1D "Piecewise-Steady" Schemes

A thorough examination of the literature reveals that, besides the seminal paper of Trefftz [66] (and follow-ups, e.g. [20, 41]), a systematic use of piecewise-exact curves of a (linear) differential operator within the algebraic derivation of a numerical scheme starts with Allen's article [5]. In the Soviet Union, Tikhonov and Samarskii developed so-called "homogeneous schemes", by converting through quadrature rules, the expression of the exact solution of 1D elliptic or parabolic PDE's given by the Green function into finite difference algorithms [64], see also [45] and [56, pp. 207–214]. The motivation was partly coming from couplings with hydrodynamic codes which numerical solutions bringing possibly discontinuous coefficients into the elliptic counterpart. The derivation of such robust schemes was also proceeding by setting up so-called "pattern functions" [45, p. 105], which are local solutions of the differential operator: in the case that differential operator is endowed with constant coefficients, pattern functions are expressed in closed form, see [56, pp. 208 and 212]. At this level, the connection with the methodology advocated by Scharfetter and Gummel in [57, p. 73] is obvious: the derivation of a numerical flux doesn't proceed by "standard difference approximations", but rather by integrating a differential equation at each interface. Assuming both drift and diffusion coefficients to be constant on each staggered computational cell, such a (steady-state) differential equation is solved by Bernoulli functions, granting what is called nowadays an "asymptotic-preserving" property in the vanishing viscosity limit. Along with others, Jerome (see [39, p. 19]) was able to reformulate this "box method" by means of "generalized B-splines", further called "$\mathscr{L}$-splines" and developed earlier in the context of data fitting: see [38, 40, 59, 60]. A first version of L-splines, for normal differential operators L^*L as proposed in [59], is endowed with a variational interpretation [19]; it is well-suited to numerical treatment of self-adjoint differential operators. However, a more general version was later studied in [3, 38] (theoretically) and then [4, 24, 33, 46, 50, 58] on the numerical side.

Two distinct families of schemes for PDE's based on $\mathscr{L}$-splines were developed:

- "Generalized Finite-Element methods", as advocated in [9], rely on generalized "hat functions" sometimes called "patch (or 'hill') functions", that are $\mathscr{L}$-splines with minimal support [60, p. 435] spanning either trial or test subspaces [2, 52, 53]. Using this type of $\mathscr{L}$-splines yield refined interpolation estimates, along with "nearly-optimal error estimates" (see also [17, 42]).
- "Uniformly accurate finite-difference methods", see e.g. [12, 13, 26, 27], which were essentially meant to solve boundary-value problems for the stationary equation of (1) with an accuracy depending only on the grid-size (and not on ε). Roughly speaking, the derivation of that sort of schemes proceeds by imposing a certain smoothness at each node of a computational grid where piecewise-exact curves connect with each other: [54, pp. 62–65] and [55].
- Extension to non-linear problems: beside [47, 49], this technique of $\mathscr{L}$-spline data interpolation was set up for time-dependent, possibly non-linear, PDE's. Perhaps, the most famous example was the existence result for strictly hyperbolic, non-resonant, systems of position-dependent balance laws [44]; among its follow-ups, [36, 62, 67], the Glimm-Sharp article [28] may be distinguished as it relates scattering processes with "zoom-out" asymptotic limits in systems of balance laws. This led to the idea of reducing artificial viscosity in Riemann-based schemes by "embodying" the source term inside numerical fluxes by means of a supplementary, static, so-called "standing wave" [37, 69], which further developed into what is nowadays referred to "well-balanced schemes" [29–31]. By setting up modern L^1 stability techniques [6], it's been possible to put on firm grounds the empirical idea that "piecewise-steady" $\mathscr{L}$-spline interpolation techniques were accurate for problems involving rough, sharply varying, coefficients.

Recent algorithms mix several points of view: for instance, Ultra-Weak Variational Formulation (UWVF) [18], or the Virtual Elements method [11]. See also [16, 34] along with new formulations of well-balanced methods for linear dissipative systems, [22] or Fokker-Planck models, [68]. Links between $\mathscr{L}$-splines and NURBS (Non-Uniform Rational B-Splines) are studied in [61].

1.2 Treating Two-Dimensional Equations Without Splitting

The most common manner of handling numerically equations set in several space dimensions is through "fractional steps algorithms", for instance the Alternating-Dimensions Implicit (ADI) method. Such methods may be endowed with various names, but they all rely on the simple idea which consists in decomposing a given model into components involving simpler physics over a relatively limited range of scales, and then seeking the solution of the entire system by mixing together numerical solutions of each individual component. An obvious application of that idea to the treatment of multi-dimensional problems is presented in [10]. Subsequent developments are presented in [43, 65] and references therein. However, a drawback of such a process lies in the fact that it discards, at the discrete level, any type of interactions possibly occurring between these "simpler components" with each other. It may henceforth be of interest to seek whether any feasible discretization can be derived, for instance in 2D, in order to approximate drift-diffusion equations (i.e. the ones considered originally in [57]) without splitting directions. A motivation would be twofold: first, to resolve boundary or internal layers possibly developing in the solution with a relatively coarse grid, and then to have better control over the large-time behavior of the numerical scheme compared to the continuous solution's. Following the nowadays usual canvas of well-balanced schemes [29], we derive a numerical discretization of (3) which originality is to involve 2D numerical fluxes where all the terms composing the (discrete) time-derivative would be lumped altogether and treated as a single object. Such methods were successfully applied in other areas of application, e.g. singular perturbations of 1D stationary diffusion equations [54]. Our numerical scheme makes heavy use of Bessel functions: previous papers already considered them as a powerful building block so as to ensure low truncation errors and strong stability properties, see especially [26] (rephrased as "tailored scheme" in [32]) and [7, 35]. Our approach here is notably different because we don't seek to fit the algebra of the scheme in such a way that it vanishes on grid values resulting from a sampling of distinguished stationary regimes (which is actually Gartland's procedure in [25]), but instead: first, a set of so-called "$\mathscr{L}$-splines" [59, 60] is derived, then Steklov-Poincaré operator [1, 8, 21, 51] is set to impose transmission conditions at the stencil's center. Monotonicity, stability and consistency properties are checked. Moreover, a well-known change of variable allows to draw advantage from algorithms developed for (3) in order to treat more delicate drift-diffusion problems: such discretizations can be shown to generalize in 2D the original Scharfetter-Gummel's algorithm (see e.g. [15] and recently, [23]).

2 Smooth Static Solutions and $\mathscr{L}$-Spline Approximations

A uniform Cartesian computational grid of the interval $(-1, 1)$ is defined as follows:

$$\Delta = \{x_1, x_2, \ldots, x_J\}, \quad x_0 = -1, \quad x_{J+1} = 1, \quad x_j = -1 + j\Delta x, \quad \Delta x = \frac{2}{J+1},$$

along with its tensorial product which yields a two-dimensional griding of $(-1, 1)^2$.

2.1 Continuous, Piecewise-Steady, 1D $\mathscr{L}$-Splines

A variational stationary profile $v(x)$ exists under assumptions (2):

Theorem 1 *Under the assumptions (2) and $u_\pm \geq 0$, if $p'(x)$ is such that,*

$$\forall x \in (-1,1), \qquad \frac{-p'(x)}{2} + q(x) \geq 0, \quad f(x) \geq 0, \tag{4}$$

there exists a unique $0 \leq v \in H^1(-1,1)$. Moreover, $v \in W^{2,\infty}(-1,1) \subset C^1(-1,1)$.

Theorem 1 gives uniqueness in the class of H^1, i.e. continuous, weak stationary solutions for any boundary data $u_\pm$. So, if $\mathscr{L}$ stands for the differential operator,

$$\begin{aligned} \mathscr{L} : H^1(1,1) &\to H^{-1}(-1,1) \\ w &\mapsto -\varepsilon\partial_{xx}w + p(x)\partial_x w + q(x)w, \end{aligned}$$

then, given a collection of grid-values x_j, v_j on Δ, there exists a unique continuous, piecewise smooth, $\mathscr{L}$-spline $\mathscr{S}_\Delta(x)$ such that: (see [46, 58] and [60, Chap. 10])

- the restriction of $\mathscr{S}_\Delta$ inside each interval (x_{j-1}, x_j) is of class C^2 and belongs to the kernel of the local operator $\mathscr{L}_{j-\frac{1}{2}}$ defined as,

$$\begin{cases} \mathscr{L}_{j-\frac{1}{2}} : w \mapsto -\varepsilon\partial_{xx}w + p_{j-\frac{1}{2}}(x)\partial_x w + q_{j-\frac{1}{2}}w, \\ p_{j-\frac{1}{2}}(x) = p'_{j-\frac{1}{2}}(x - x_{j-\frac{1}{2}}) + p_{j-\frac{1}{2}}, \; q_{j-\frac{1}{2}} \in \mathbb{R}^+, \end{cases} \tag{5}$$

  being piecewise-affine and -constant approximations of p and q, respectively,

$$p_{j-\frac{1}{2}} = \frac{1}{2}\big(p(x_j) + p(x_{j-1})\big), \qquad p'_{j-\frac{1}{2}} = \frac{p(x_j) - p(x_{j-1})}{\Delta x}$$

- for all $0 \leq j \leq J+1$, C^0 continuity holds: $\mathscr{S}_\Delta(x_j) = v_j$.

This meets with the object considered in e.g. [62, §2] in order to build a random-choice method for convection-diffusion equations. In case of a reaction-diffusion problem, $p(x) \equiv 0$ and $0 \leq q(x) \in C^0$, $\mathscr{L}$ factorizes into $\mathscr{L} = L^*L$ with $L = -\sqrt{\varepsilon}\partial_x + \sqrt{q}$ and its (formal) adjoint L^*, so a definition given in [40] (see also [19, 59]) can be recalled: *A $\mathscr{L}$-spline with knots x_j is a C^0 function X for which $L^*L(X) = 0$ on each open interval. If we were to require $X \in C^1$, then $L^*L(X) = 0$ would hold everywhere: thus, an $\mathscr{L}$-spline can be thought of being the "most differentiable" function satisfying $L^*L(X) = 0$ on the appropriate open intervals without satisfying it everywhere.* According to [19, 40, 59, 60], $\mathscr{L}$-splines exist for linear differential operators L of mth order with smooth coefficients: in such a context, each is endowed with $C^{2(m-1)}$ smoothness and may be delicate to compute. Here, $m = 1$, so the maximal regularity is C^0 and our $\mathscr{L}$-spline is easy to derive.

2.2 $\mathscr{L}$-Membrane Fitting of 2D Discrete Values

For later use, denote $\mathscr{L}_\alpha[\phi] := -\Delta\phi + \alpha\phi$, for any constant $\alpha \in \mathbb{R}_*^+$. A uniform Cartesian computational griding of Ω is characterized by a constant parameter $\Delta x = \Delta y > 0$: for all convenient indexes $i, j \in \mathbb{Z}$, piecewise constant approximations read $u_{i,j}^n \simeq u(t^n, x_i, y_j)$ with $t^n = n\Delta t$ (Δt a time-step, $n \in \mathbb{N}$) and $x_i = i\Delta x$, $y_j = j\Delta x$. Following Gartland's ideas [25], we first study radial solutions of $\mathscr{L}_\alpha[\phi] = \bar{f}, \bar{f} \in \mathbb{R}^+$, which results from "freezing" parameters in (3) at $x_{i-\frac{1}{2}}, y_{j-\frac{1}{2}}$:

$$\begin{cases} -\Delta\phi(x,y) + \alpha(x_{i-\frac{1}{2}}, y_{j-\frac{1}{2}})\phi = f(x_{i-\frac{1}{2}}, y_{j-\frac{1}{2}}), \\ (x,y) \in D := \left\{ |x - x_{i-\frac{1}{2}}|^2 + |y - y_{j-\frac{1}{2}}|^2 \leq \Delta x^2/2 \right\}, \\ \phi = \bar{\phi} \text{ on } \partial D. \end{cases} \tag{6}$$

Lemma 1 *Let $R = \Delta x/\sqrt{2}$: the unique bounded radial solution of (6) is of separated variables and rewrites as a Fourier series for $r \leq R, \theta \in [-\pi, \pi]$,*

$$\phi(r,\theta) = \frac{\bar{f}}{\bar{\alpha}} + A_0 \mathscr{I}_0(r\sqrt{\bar{\alpha}}) + \sum_{n \in \mathbb{N}_*} (A_n \cos(n\theta) + B_n \sin(n\theta)) \mathscr{I}_n(r\sqrt{\bar{\alpha}}). \tag{7}$$

where $\bar{\alpha} = \alpha(x_{i-\frac{1}{2}}, y_{j-\frac{1}{2}})$, $\bar{f} = f(x_{i-\frac{1}{2}}, y_{j-\frac{1}{2}})$ and $\mathscr{I}_n$ stands for the (monotonic, nonnegative) modified Bessel function of the first kind. In (7), Fourier coefficients A_n, B_n depend solely on Dirichlet boundary data $\bar{\phi}$ on ∂D.

Proof Given a general solution of $\mathscr{L}_\alpha[\phi] = 0$, one obtains a solution of (6) by subtracting $\frac{\bar{f}}{\bar{\alpha}}$ because both $\bar{\alpha}, \bar{f}$ are constants. Thus it suffices to solve $\mathscr{L}_\alpha[\phi] = 0$ with inhomogeneous Dirichlet data in a disc centered in $x_{i-\frac{1}{2}}, y_{j-\frac{1}{2}}$ of radius $\frac{\Delta x}{\sqrt{2}}$: expressing the Laplace operator in polar variables r, θ, (see [63, Chap. 10])

$$r^2 \frac{\partial^2 \phi}{\partial r^2} + r\frac{\partial \phi}{\partial r} + r^2 \bar{\alpha}\phi = -\frac{\partial^2 \phi}{\partial \theta^2}, \qquad \phi(r,\cdot) \quad 2\pi\text{-periodic.}$$

The operator on the right-hand side admits a spectral decomposition, so that $\phi(r,\theta) = F(r)G(\theta)$ and F should satisfy a (modified) Bessel differential equation,

$$r^2 F''(r) + rF'(r) - (\bar{\alpha}r^2 + n^2)F(r) = 0, \quad G(\theta) = \cos(n\theta) \text{ or } \sin(n\theta), \quad n \in \mathbb{N}.$$

As we only have piecewise constant grid samples, it is necessary to truncate (7):

Definition 1 For any $i, j \in \mathbb{Z}^2$ such that $x_i, y_j \in (-1,1)^2$, a 2D $\mathscr{L}$-spline (or $\mathscr{L}$-membrane) is the radial function corresponding to the first Fourier modes,

$$\tilde{\phi}(r,\theta) = \frac{\bar{f}}{\bar{\alpha}} + A_0 \mathscr{I}_0(r\sqrt{\bar{\alpha}}) + (A_1 \cos\theta + B_1 \sin\theta)\mathscr{I}_1(r\sqrt{\bar{\alpha}}),$$

such that the matching conditions hold,

$$\tilde{\phi}(R,0) = u^n_{i,j}, \qquad \tilde{\phi}(R,\frac{\pi}{2}) = u^n_{i-1,j}, \qquad \tilde{\phi}(R,-\frac{\pi}{2}) = u^n_{i,j-1}.$$

Fourier coefficients A_0, A_1, B_1 satisfy an invertible (for $\bar{\alpha} > 0$) linear system:

$$\begin{pmatrix} \mathscr{I}_0(r\sqrt{\bar{\alpha}}) & \mathscr{I}_1(r\sqrt{\bar{\alpha}}) & 0 \\ \mathscr{I}_0(r\sqrt{\bar{\alpha}}) & 0 & \mathscr{I}_1(r\sqrt{\bar{\alpha}}) \\ \mathscr{I}_0(r\sqrt{\bar{\alpha}}) & 0 & -\mathscr{I}_1(r\sqrt{\bar{\alpha}}) \end{pmatrix} \begin{pmatrix} A_0 \\ A_1 \\ B_1 \end{pmatrix} = \begin{pmatrix} u^n_{i,j} - \frac{\bar{f}}{\bar{\alpha}} \\ u^n_{i-1,j} - \frac{\bar{f}}{\bar{\alpha}} \\ u^n_{i,j-1} - \frac{\bar{f}}{\bar{\alpha}} \end{pmatrix}. \tag{8}$$

If $f = 0$ and $\alpha \to 0$, the functions $\mathscr{I}_n(r)$ in (7) converge to r^n, so that the usual centered, formally second-order accurate, scheme is (unexpectedly?) recovered.

2.3 C^1-Smoothness Characterization of Steady-States

Accordingly, let a knot $x_j \in \Delta$, with its two neighbors $x_{j\pm1}$, their corresponding values $v_{j\pm1}$ and both restrictions to (x_{j-1}, x_j) and (x_j, x_{j+1}) of the $\mathscr{L}$-spline, $\mathscr{S}_\Delta$:

Corollary 1 *Let $\{U_{j\pm\frac{1}{2}}(x), V_{j\pm\frac{1}{2}}(x)\}$ be fundamental systems for each of $\mathscr{L}_{j\pm\frac{1}{2}}$ in (x_{j-1}, x_j) and (x_j, x_{j+1}), respectively. Then v_j belongs to $v(x)$, the unique stationary curve defined on (x_{j-1}, x_{j+1}) such that $v(x_{j\pm1}) = v_{j\pm1}$ if and only if $\mathscr{S}_\Delta$ is C^1 at x_j:*

$$\boxed{\left\langle \begin{pmatrix} v_{j-1} \\ v_j \end{pmatrix}, M^{-T}_{j-\frac{1}{2}} \begin{pmatrix} U'_{j-\frac{1}{2}}(x_j) \\ V'_{j-\frac{1}{2}}(x_j) \end{pmatrix} \right\rangle = \left\langle \begin{pmatrix} v_j \\ v_{j+1} \end{pmatrix}, M^{-T}_{j+\frac{1}{2}} \begin{pmatrix} U'_{j+\frac{1}{2}}(x_j) \\ V'_{j+\frac{1}{2}}(x_j) \end{pmatrix} \right\rangle,} \tag{9}$$

where $\langle \cdot, \cdot \rangle$ denotes the $\mathbb{R}^2$ scalar product and the 2×2 matrices $M_{j\pm\frac{1}{2}}$ stand for:

$$M_{j-\frac{1}{2}} = \begin{pmatrix} U_{j-\frac{1}{2}}(x_{j-1}) & V_{j-\frac{1}{2}}(x_{j-1}) \\ U_{j-\frac{1}{2}}(x_j) & V_{j-\frac{1}{2}}(x_j) \end{pmatrix}, \quad M_{j+\frac{1}{2}} = \begin{pmatrix} U_{j+\frac{1}{2}}(x_j) & V_{j+\frac{1}{2}}(x_j) \\ U_{j+\frac{1}{2}}(x_{j+1}) & V_{j+\frac{1}{2}}(x_{j+1}) \end{pmatrix}.$$

Condition (9) is a simple manifestation of Steklov-Poincaré (or Dirichlet-to-Neumann) operator: starting from a set of discrete pointwise values (x_j, v_j), each side of (9) produces a (left or right) "normal derivative" at x_j, see [21].

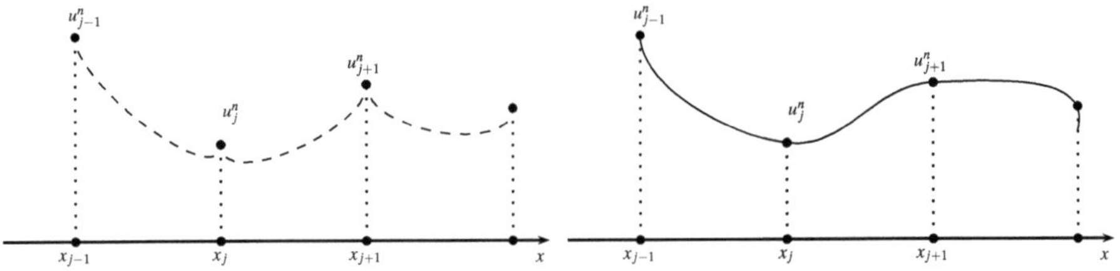

Fig. 1 "Lipschitz corners" (*left*) vs. smooth transmission (*right*) for transient vs. steady solutions

Proof There are two assertions to justify, an equivalence and the formula (9):

- Observe that the "lumped operator" $\mathscr{L}_j$, defined in (x_{j-1}, x_{j+1}) and obtained from $\mathscr{L}_{j\pm\frac{1}{2}}$, displays coefficients still satisfying the regularity requirements of Theorem 1: given $v_{j\pm1}$, there exists a unique H^1 integral curve of $\mathscr{L}_j$ with these boundary data in $x_{j\pm1}$. So, if $\mathscr{S}_\Delta$ is C^1 at x_j, then v_j belongs to this integral curve. Reciprocally, it's obvious that if v_j belongs to this unique integral curve at x_j, then it is in particular $W^{2,\infty}$ at this knot, so C^1, too.
- Since each $\mathscr{L}_{j\pm\frac{1}{2}}$ is linear, it is endowed with a two-dimensional fundamental system of solutions, $\{U_{j\pm\frac{1}{2}}, V_{j\pm\frac{1}{2}}\}$. By expressing $\mathscr{S}_\Delta$ in the basis,

$$\mathscr{S}_\Delta(x) = A_{j\pm\frac{1}{2}} U_{j\pm\frac{1}{2}}(x) + B_{j\pm\frac{1}{2}} V_{j\pm\frac{1}{2}}(x), \qquad (A_{j\pm\frac{1}{2}}, B_{j\pm\frac{1}{2}}) \in \mathbb{R}^4,$$

  related to each interval around x_j, along with its x-derivative, one sees that the C^1 matching conditions at the knot x_j just reduce to (9).

Formula (9) lies behind the derivation of the "El-Mistikawy-Werle scheme" (see e.g. [12] and [54, p. 63]): it exploits the fact that checking C^1 regularity is easy to reformulate as a finite-difference scheme and it's enough to secure uniqueness.

Remark 1 By imposing $p = q = f \equiv 0$ in (1), each fundamental system in (9) boils down to the simple one: $\{1, x\}$. Accordingly, usual finite-differences for the heat equation are recovered as a special case involving a piecewise-linear $\mathscr{L}$-spline.

Toward a time-marching algorithm, the most direct strategy is to *define the discrete time-derivative as the defect of C^1 smoothness (9) at each knot x_j of the grid Δ.*

Definition 2 Under the hypotheses of Corollary 1, the explicit well-balanced scheme for (1) with $f(x) \equiv 0$ reads: (see Fig. 1)

$$\boxed{\forall (x_j, n) \in \Delta \times \mathbb{N}, \qquad u_j^{n+1} = u_j^n - \frac{\varepsilon \Delta t}{\Delta x}\left[L^n_{j+\frac{1}{2}} - R^n_{j-\frac{1}{2}}\right].} \qquad (10)$$

with the numerical fluxes given by,

$$\left.\begin{aligned} R^n_{j-\frac{1}{2}} &= \left\langle \begin{pmatrix} u^n_{j-1} \\ u^n_j \end{pmatrix}, M^{-T}_{j-\frac{1}{2}} \begin{pmatrix} U'_{j-\frac{1}{2}}(x_j) \\ V'_{j-\frac{1}{2}}(x_j) \end{pmatrix} \right\rangle, \\ L^n_{j-\frac{1}{2}} &= \left\langle \begin{pmatrix} u^n_{j-1} \\ u^n_j \end{pmatrix}, M^{-T}_{j-\frac{1}{2}} \begin{pmatrix} U'_{j-\frac{1}{2}}(x_{j-1}) \\ V'_{j-\frac{1}{2}}(x_{j-1}) \end{pmatrix} \right\rangle. \end{aligned}\right\} \tag{11}$$

Since both p, q are independent of t, the matrices $M_{j-\frac{1}{2}}$ don't depend on n.

Lemma 2 *Numerical fluxes (11) don't depend on a specific choice* $U_{j\pm\frac{1}{2}}, V_{j\pm\frac{1}{2}}$ *of the fundamental system for* $\mathscr{L}_{j\pm\frac{1}{2}}$.

Proof Consider $\tilde{U}, \tilde{V}$ (indexes $j \pm \frac{1}{2}$ are dropped for readability) such that,

$$\begin{pmatrix} \tilde{U}(x_\ell) \\ \tilde{V}(x_\ell) \end{pmatrix} = A \begin{pmatrix} U(x_\ell) \\ V(x_\ell) \end{pmatrix}, \qquad A = \begin{pmatrix} a_{11} & a_{12} \\ a_{21} & a_{22} \end{pmatrix} \text{ being invertible },$$

for both $\ell = j - 1$ and $\ell = j$. The corresponding 2×2 matrix $\tilde{M} = M A^T$, so that $\tilde{M}^{-T} = M^{-T} A^{-1}$. Accordingly, the coefficient $\tilde{R}$ satisfies,

$$\begin{aligned} \tilde{R}^n_{j-\frac{1}{2}} &= \left\langle \begin{pmatrix} u^n_{j-1} \\ u^n_j \end{pmatrix}, \tilde{M}^{-T}_{j-\frac{1}{2}} \begin{pmatrix} \tilde{U}'_{j-\frac{1}{2}}(x_j) \\ \tilde{V}'_{j-\frac{1}{2}}(x_j) \end{pmatrix} \right\rangle \\ &= \left\langle \begin{pmatrix} u^n_{j-1} \\ u^n_j \end{pmatrix}, M^{-T}_{j-\frac{1}{2}} A^{-1} A \begin{pmatrix} U'_{j-\frac{1}{2}}(x_j) \\ V'_{j-\frac{1}{2}}(x_j) \end{pmatrix} \right\rangle = R^n_{j-\frac{1}{2}}, \end{aligned}$$

along with a similar computation for $\tilde{L}$: this completes the proof.

2.4 Transmission Conditions and Steklov-Poincaré Operator

Accordingly, **a convenient way to build a time-marching scheme for (3) around x_i, y_j consists in, after all parameters got "frozen" at each node of the computational grid, computing the residual of C^0 continuity of normal (that is, radial) derivatives of each $\mathscr{L}$-spline touching this point**: see Fig. 2. Hence, in general, a Steklov-Poincaré (or Dirichlet-to-Neumann) mapping $\mathscr{S}$, see [8, 21, 51] allows to derive the necessary radial derivatives for each disc D, as specified in (6),

$$\mathscr{S} : H^{\frac{1}{2}}(\partial D) \to H^{-\frac{1}{2}}(\partial D) \qquad \bar{\phi} = \phi(R, \cdot) \mapsto \frac{\partial \phi}{\partial r}(R, \cdot).$$

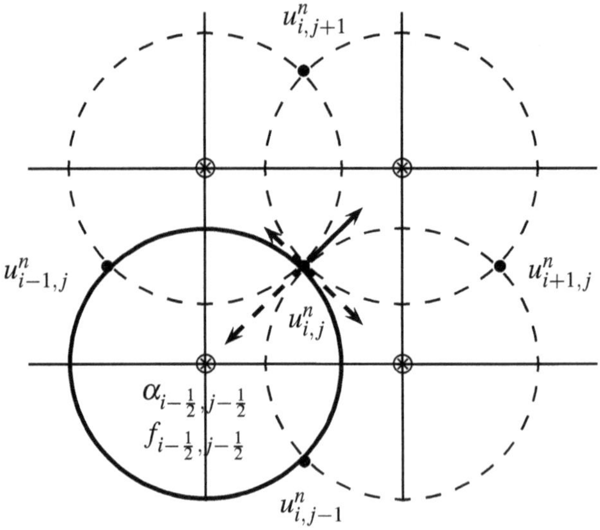

Fig. 2 Transmission conditions at $u^n_{i,j}$ given by Steklov-Poincaré operators

function $\phi(r,\theta)$ is referred to as to the "$H^1(D)$-extension of $\bar{\phi}(\theta)$"; applying differentiation rules of modified Bessel functions, $\mathscr{I}_0' = \mathscr{I}_1$, $\mathscr{I}_1' = \frac{\mathscr{I}_0+\mathscr{I}_2}{2}$,

$$\frac{1}{\sqrt{\bar{\alpha}}}\frac{\partial\tilde{\phi}}{\partial r}(R,\theta) = A_0\mathscr{I}_1(R\sqrt{\bar{\alpha}}) + (A_1\cos\theta + B_1\sin\theta)\frac{\mathscr{I}_0+\mathscr{I}_2}{2}(R\sqrt{\bar{\alpha}}), \tag{12}$$

and then inserting the Fourier coefficients satisfying (8), a 3-points approximation of the radial derivative is obtained, which is expected to be related to a discretization of the Laplace-Beltrami operator on the circle [1], that is, $\partial^2/\partial\theta^2$,

$$\frac{1}{\sqrt{\bar{\alpha}}}\frac{\partial\tilde{\phi}}{\partial r}(R,0) = \left\{ u^n_{i,j}\underbrace{\frac{\mathscr{I}_0+\mathscr{I}_2}{2\mathscr{I}_1}}_{\geq 0} + \frac{u^n_{i-1,j}+u^n_{i,j-1}}{2}\underbrace{\left[\frac{\mathscr{I}_1}{\mathscr{I}_0} - \frac{\mathscr{I}_0+\mathscr{I}_2}{2\mathscr{I}_1}\right]}_{\leq 0} - \frac{\bar{f}}{\bar{\alpha}^{\frac{3}{2}}}\underbrace{\frac{\mathscr{I}_1}{\mathscr{I}_0}}_{\geq 0}\right\} \times (R\sqrt{\bar{\alpha}}).$$

A crucial feature is that the aforementioned sign properties hold uniformly in $\bar{\alpha} > 0$.

Remark 2 The case where $\alpha(x,y) \equiv 0$ in the domain Ω, but $f(x,y) \neq 0$, can be handled in two slightly different manners, which yield essentially the same scheme:

- (split) when forcing both constants $\alpha_{i-\frac{1}{2},j-\frac{1}{2}} = 0$ and $f_{i-\frac{1}{2},j-\frac{1}{2}} \neq 0$ into (6), one can split the solution like $\phi = \psi + \xi$ such that,

$$-\Delta\psi = 0, \quad \psi\Big|_{\partial D} = \bar{\phi}; \qquad -\Delta\xi = f_{i-\frac{1}{2},j-\frac{1}{2}}, \quad \xi\Big|_{\partial D} = 0.$$

Then, on the one hand, $\psi(r,\theta)$ rewrites as a Fourier series (7) with r^n in place of $\mathscr{I}_n$. On the other hand, being $f_{i-\frac{1}{2},j-\frac{1}{2}}$ a constant in the disc D, the function ξ is purely radial, so that $r \mapsto \xi(r) \in C^2(D)$ solves,

$$\xi''(r) + \frac{1}{r}\xi'(r) = f_{i-\frac{1}{2},j-\frac{1}{2}}, \qquad \xi\left(R = \frac{\Delta x}{\sqrt{2}}\right) = 0.$$

Accordingly, since ξ' is bounded in $r = 0$,

$$(r \cdot \xi'(r))' = r \cdot f_{i-\frac{1}{2},j-\frac{1}{2}}, \quad \text{gives} \quad \xi'(R) = \frac{R}{2} \cdot f_{i-\frac{1}{2},j-\frac{1}{2}}.$$

- (potential) by introducing the steady "curl-free" potential $\Phi(x,y)$ such that,

$$-\Delta\Phi = f, \ \text{in } \Omega, \qquad \Phi|_{\partial\Omega} = 0,$$

Eq. (3) with $\varepsilon = 1$ rewrites,

$$\partial_t v - \Delta v = 0, \qquad v(t,x,y) = u(t,x,y) - \Phi(x,y),$$

so that numerical fluxes follow by applying the Fourier series (7) directly to v.

3 Time-Marching One-Dimensional Numerical Scheme

3.1 *Exponential $\mathscr{L}$-Splines for Constant Velocity Field p*

In that case, inside (x_{j-1}, x_j), the stationary equation yields, at any $t^n = n\Delta t$:

$$\begin{aligned} \mathscr{L}_{j-\frac{1}{2}}(w) &= -\varepsilon\partial_{xx}w + p\partial_x w + q_{j-\frac{1}{2}}w, \\ q_{j-\frac{1}{2}} &= \tfrac{1}{2}\big[q(x_{j-1}) + q(x_j)\big], \end{aligned} \qquad w(x_{j-1}) = u^n_{j-1},\ w(x_j) = u^n_j,$$

so that the general solution of $\mathscr{L}_{j-\frac{1}{2}}(w) = 0$ reads, for $x \in (x_{j-1}, x_j)$,

$$w(x) = A\exp(r_+ x) + B\exp(r_- x), \qquad r_\pm = \frac{1}{2\varepsilon}\left(p \pm \sqrt{p^2 + 4\varepsilon q_{j-\frac{1}{2}}}\right). \tag{13}$$

Lemma 3 *Let $p(x) \equiv p$ and $f \equiv 0$, the numerical fluxes (11) read:*

$$R^n_{j-\frac{1}{2}} = u^n_{j-1}\frac{D\exp(r_-\Delta x)}{1-\exp(-D\Delta x)} - u^n_j\left(\frac{r_+}{1-\exp(-D\Delta x)} - \frac{r_-}{\exp(D\Delta x)-1}\right)$$

$$L^n_{j-\frac{1}{2}} = u^n_{j-1}\left(\frac{r_+}{\exp(D\Delta x)-1} - \frac{r_-}{1-\exp(-D\Delta x)}\right) - u^n_j\frac{D\exp(-r_-\Delta x)}{\exp(D\Delta x)-1}$$

where $D = r_+ - r_-$, $r_\pm$ being the roots of the characteristic polynomial in (x_{j-1}, x_j).

Proof From the expression (13), it follows that $M_{j-\frac{1}{2}}$ reads,

$$M_{j-\frac{1}{2}} = \begin{pmatrix} \exp(r_+ x_{j-1}) & \exp(r_- x_{j-1}) \\ \exp(r_+ x_j) & \exp(r_- x_j) \end{pmatrix},$$

which determinant is nonnegative and vanish only if $D = 0$:

$$\det(M_{j-\frac{1}{2}}) = 2\sinh(D\Delta x/2)\exp\left((r_+ + r_-)x_{j-\frac{1}{2}}\right), \qquad x_{j-\frac{1}{2}} = \frac{1}{2}(x_{j-1} + x_j).$$

Computing cofactors of $M_{j-\frac{1}{2}}$ and scalar products in (11) yields the result.

The time-marching scheme from Lemma 3 is still usable [12, 54] when $|p'(x)| \ll |p(x)|$, the slowly-varying regime, despite asking for C^1 regularity at each interface in presence of discontinuous transport coefficients $p_{j-\frac{1}{2}}$ isn't fully rigorous.

3.2 Consistency, Monotonicity and Uniform Stability

Two limiting regimes are examined for (10):

- a diffusion-dominant one, $\varepsilon \simeq 1$ and $\Delta x \to 0$ under a CFL restriction;
- a complementary, and more singular, limit, i.e. sending $\varepsilon \to 0$ while $\Delta x \simeq 1$.

The second limit is referred to as to the "Asymptotic-Preserving" (AP) property.

Theorem 2 *In the diffusion-dominant regime, $\Delta x/\varepsilon \to 0$, and for $p(x) \equiv p$,*

1. *the fluxes in Lemma 3 are consistent with (1); the resulting explicit scheme (10) is nondecreasing with respect to its entries under the CFL restriction,*

$$\boxed{\Delta t \max_{x_j \in \Delta}\left(|p| + \frac{\varepsilon D_{j-\frac{1}{2}}}{\tanh(D_{j-\frac{1}{2}}\Delta x/2)}\right) \leq \Delta x, \quad D = r_+ - r_- \geq 0,}$$

2. *as* $\varepsilon \to 0$*, these fluxes corresponding to* $p(x) \equiv p > 0$ *yield a well-balanced scheme in the sense of [30, 31] for a reduced equation* $\partial_t u + p\partial_x u + q(x)u = 0$*,*

$$u_j^{n+1} = u_j^n - p\frac{\Delta t}{\Delta x}\left(u_j^n - \exp\left[-\frac{q_{j-\frac{1}{2}}\,\Delta x}{p}\right]u_{j-1}^n\right).$$

Proof Consistency with (1) as $\Delta x/\varepsilon \to 0$ is shown by doing Taylor expansions:

1. First, $D = r_+ - r_- \geq 0$, so every denominator is nonnegative. Then,

$$\begin{aligned} R_{j-\frac{1}{2}}^n &= \frac{D\exp(r_-\Delta x)}{1-\exp(-D\Delta x)}u_{j-1}^n - u_j^n\left(\frac{r_+}{1-\exp(-D\Delta x)} - \frac{r_-}{\exp(D\Delta x)-1}\right)\\ &= \frac{u_{j-1}^n\exp(r_+\Delta x)D - u_j^n(r_+\exp(D\Delta x)-r_-)}{\exp(D\Delta x)-1}\\ &= \frac{u_{j-1}^n\exp((r_++r_-)\Delta x)D - u_j^n(r_+\exp(r_+\Delta x)-r_-\exp(r_-\Delta x))}{\exp(r_+\Delta x)-\exp(r_-\Delta x)}\\ &= \frac{D\;\exp((r_++r_-)\Delta x)}{\exp(r_+\Delta x)-\exp(r_-\Delta x)}u_{j-1}^n\\ &\qquad -(r_+ + r_-)\frac{\exp(r_+\Delta x)-\exp(r_-\Delta x)}{\exp(r_+\Delta x)-\exp(r_-\Delta x)}u_j^n\\ &\qquad\qquad + \frac{r_-\exp(r_+\Delta x)-r_+\exp(r_-\Delta x)}{\exp(r_+\Delta x)-\exp(r_-\Delta x)}u_j^n. \end{aligned}$$

Each term is treated individually: as $\Delta x/\varepsilon \to 0$,

$$\begin{aligned} \frac{D\;\exp((r_++r_-)\Delta x)}{\exp(r_+\Delta x)-\exp(r_-\Delta x)} &\to \frac{1}{\Delta x} + (r_+ + r_-),\\ \frac{r_-\exp(r_+\Delta x)-r_+\exp(r_-\Delta x)}{\exp(r_+\Delta x)-\exp(r_-\Delta x)} &\to \frac{-1}{\Delta x} + \frac{r_+r_-\Delta x}{2}. \end{aligned}$$

Similarly,

$$\begin{aligned} L_{j-\frac{1}{2}}^n &= -u_{j-1}^n\left(\frac{r_+}{\exp(D\Delta x)-1} - \frac{r_-}{1-\exp(-D\Delta x)}\right) + u_j^n\frac{D\exp(-r_-\Delta x)}{\exp(D\Delta x)-1}\\ &= \frac{-u_{j-1}^n(r_+\exp(r_-\Delta x)-r_-\exp(r_+\Delta x))}{\exp(r_+\Delta x)-\exp(r_-\Delta x)} + \frac{u_j^n\;D}{\exp(r_+\Delta x)-\exp(r_-\Delta x)} \end{aligned}$$

Since $r_+ + r_- = p/\varepsilon$ and $r_+r_- = -q/\varepsilon$, by assembling (10) with $p > 0$,

$$\begin{aligned} u_j^{n+1} &= u_j^n - \frac{\varepsilon\Delta t}{\Delta x}\left[L_{j+\frac{1}{2}}^n - R_{j-\frac{1}{2}}^n\right]\\ &\simeq u_j^n + \frac{\varepsilon\Delta t}{\Delta x^2}\left[u_{j+1}^n - 2u_j^n + u_{j-1}^n\right]\\ &\qquad -\frac{\Delta t}{\Delta x}p\left[u_j^n - u_{j-1}^n\right] - \Delta t\left(\frac{q_{j-\frac{1}{2}} + q_{j+\frac{1}{2}}}{2}\right)u_j^n. \end{aligned}$$

The CFL restriction is obtained by computing coefficients acting on u_j^n,

$$\frac{r_+ \exp(D\Delta x/2) - r_- \exp(-D\Delta x/2)}{\exp(D\Delta x/2) - \exp(-D\Delta x/2)} = \frac{r_+ + r_-}{2} + \frac{D}{2}\coth(D\frac{\Delta x}{2}).$$

2. First, let's recall that characteristic roots $r_\pm$ behave as follows for $\varepsilon \ll 1$,

$$r_+ = \frac{p + \sqrt{p^2 + 4\varepsilon q}}{2\varepsilon} \simeq \frac{p}{\varepsilon} + \frac{q}{p}, \qquad r_- = \frac{p - \sqrt{p^2 + 4\varepsilon q}}{2\varepsilon} \simeq -\frac{q}{p},$$

so that $D = r_+ - r_- = \frac{p}{\varepsilon} + \frac{2q}{p} \to +\infty$ and all the terms divided by $\exp(D\Delta x)$ in the numerical fluxes vanish. What remains is only,

$$\begin{aligned} R^n_{j-\frac{1}{2}} &= u^n_{j-1} \tfrac{D\exp(r_-\Delta x)}{1-\exp(-D\Delta x)} - u^n_j \left(\tfrac{r_+}{1-\exp(-D\Delta x)} - \tfrac{r_-}{\exp(D\Delta x)-1} \right) \\ &\simeq -u^n_j \left(\tfrac{p}{\varepsilon} + \tfrac{q}{p} \right) + \left(\tfrac{p}{\varepsilon} + \tfrac{2q}{p} \right) \exp(-\tfrac{q}{p}\Delta x)\, u^n_{j-1} \\ L^n_{j-\frac{1}{2}} &= u^n_{j-1} \left(\tfrac{r_+}{\exp(D\Delta x)-1} - \tfrac{r_-}{1-\exp(-D\Delta x)} \right) - u^n_j \tfrac{D\exp(-r_-\Delta x)}{\exp(D\Delta x)-1} \\ &\simeq \tfrac{q}{p} u^n_{j-1}, \end{aligned}$$

which now leads to,

$$\begin{aligned} u_j^{n+1} &= u_j^n - \frac{\varepsilon \Delta t}{\Delta x}\left[L^n_{j+\frac{1}{2}} - R^n_{j-\frac{1}{2}} \right] \\ &\simeq u_j^n - p\frac{\Delta t}{\Delta x}\Bigg[u_j^n \left(1 + \frac{\varepsilon}{p^2}(q_{j-\frac{1}{2}} + q_{j+\frac{1}{2}}) \right) \\ &\qquad - \left(1 + \frac{2\varepsilon q_{j-\frac{1}{2}}}{p^2} \right) \exp(-\frac{q_{j-\frac{1}{2}}\Delta x}{p})\, u^n_{j-1} \Bigg] \\ &\to u_j^n - p\frac{\Delta t}{\Delta x}\left(u_j^n - \exp(-\frac{q_{j-\frac{1}{2}}}{p}\Delta x)\, u^n_{j-1} \right) \quad \text{as } \varepsilon \to 0. \end{aligned}$$

It remains to look at sequences $\mathscr{S}^\varepsilon_\Delta(x)$ of $\mathscr{L}$-splines, as $\varepsilon \to 0$ with Δ fixed:

Corollary 2 *Under the assumptions of Theorem 2, the sequence of C^0-continuous $\mathscr{L}$-splines $\mathscr{S}^\varepsilon_\Delta$ converges, as $\varepsilon \to 0$, to a discontinuous, piecewise C^2, BV interpolation. Between two knots, say (x_{j-1}, x_j), it satisfies, for any $n \in \mathbb{N}$,*

$$\boxed{p\,\partial_x w(x) + q_{j-\frac{1}{2}} w(x) = 0, \qquad w(x_{j-1}) = u^n_{j-1}.}$$

Proof Arguments follow classical results of singular perturbation in second-order ODE's without turning points, see for instance [48, Chap. 3]. Pick a knot $x_j \in \Delta$ and consider $w^\varepsilon \in C^2(x_{j-1}, x_j)$ solution of the boundary-value problem $\mathscr{L}^\varepsilon$ defined in (5) with $p'_{j-\frac{1}{2}} = 0$, so that $p_{j-\frac{1}{2}}(x) \equiv p > 0$. It rewrites in the form,

$$w^{\varepsilon}(x) = W(x, \varepsilon) + \eta(\frac{x_j - x}{\varepsilon}, \varepsilon), \qquad \eta \to 0 \text{ as } \tau = \frac{x_j - x}{\varepsilon} \to +\infty,$$

where both the "outer solution" and the "terminal layer correction" expand into,

$$W(x, \varepsilon) = \sum_{k=0}^{\infty} W_k(x)\varepsilon^k, \qquad \eta(\tau, \varepsilon) = \sum_{k=0}^{\infty} \eta_k(\tau)\varepsilon^k,$$

and $\tau = \frac{x_j - x}{\varepsilon}$ is the "stretched variable". Being W a smooth solution of (5) with "initial condition" $W(x_{j-1}, \varepsilon) = u^n_{j-1}$ for some $n \in \mathbb{N}$, it's enough to limit the ε-expansion to the first term which satisfies,

$$\begin{cases} pW'_0 + q_{j-\frac{1}{2}} W_0 = 0, \; W_0(x_{j-1}) = u^n_{j-1}, \\ \frac{d^2\eta_0}{d\tau^2} + p\frac{d\eta_0}{d\tau} = 0, \; \eta_0(0) = u^n_j - W(x_j, \varepsilon), \end{cases}$$

so that, with η supported inside the "terminal layer" of width $O(\varepsilon)$,

$$\eta_0(\tau) = \exp(-p\,\tau)[u^n_j - W_0(x_j)], \qquad w^{\varepsilon}(x) = W_0(x) + \eta(\tau) + O(\varepsilon).$$

In the limit $\varepsilon \to 0$, the $\mathscr{L}$-spline is a piecewise smooth curve displaying discontinuities at each knot $j > 0$. Let a sequence $0 \le u^n_j, j \in J$ be of bounded total-variation; for $\varepsilon > 0$, the H^1 regularity ensures that $\mathscr{S}^{\varepsilon}_{\Delta}$ is bounded in $W^{1,1}(-1,1)$ because $L^2(-1,1) \subset L^1(-1,1)$. In the limit $\varepsilon \to 0$, each local operator $\mathscr{L}_{j-\frac{1}{2}}$ (5) reduces to $p\partial_x w + q_{j-\frac{1}{2}} w = 0$, $w(x_{j-1}) = u^n_{j-1}$. The total variation of the $\mathscr{L}$-spline is,

$$TV(\mathscr{S}_{\Delta}) \le \sum_{j \in J} \left| u^n_j - u^n_{j-1} \exp(-\frac{q_{j-\frac{1}{2}} \Delta x}{p}) \right| + |u^n_{j-1}| \left(1 - \exp(-\frac{q_{j-\frac{1}{2}} \Delta x}{p}) \right).$$

3.3 An External Force $f(x)$ in a Reaction-Diffusion Equation

Hereafter, the aim is to solve the initial-boundary problem,

$$\partial_t u - \varepsilon \partial_{xx} u + q(x)u = f(x), \quad u(t, \pm 1) = u^{\pm}, \quad x \in (-1, 1), \tag{14}$$

so that the resulting scheme recovers exactly the collection of points $v(x_j)$, for $x_j \in \Delta$ the nodes of the grid. So, at each time-step t^n, the "$\mathscr{L}$-spline" interpolation of the data u^n_j proceeds by solving the stationary equation between u^n_j and u^n_{j+1}. Since, in (x_j, x_{j+1}), both $\bar{q}$ and $\bar{f}$ are constant, each "local profile" is obtained by a standard "variation of constants" technique involving exponential functions:

- let $r_{j-\frac{1}{2}} = \sqrt{\bar{q}/\varepsilon} > 0$ in (x_{j-1}, x_j) and $M_{j-\frac{1}{2}}$ be the matrix in,

$$\begin{pmatrix} u_{j-1}^n - \frac{\bar{f}}{\bar{q}} \\ u_j^n - \frac{\bar{f}}{\bar{q}} \end{pmatrix} = \begin{pmatrix} \exp(r_{j-\frac{1}{2}} x_{j-1}) & \exp(-r_{j-\frac{1}{2}} x_{j-1}) \\ \exp(r_{j-\frac{1}{2}} x_j) & \exp(-r_{j-\frac{1}{2}} x_j) \end{pmatrix} \begin{pmatrix} A \\ B \end{pmatrix}.$$

  Yet the determinant $|M_{j-\frac{1}{2}}| = -2\sinh(r_{j-\frac{1}{2}} \Delta x) \neq 0$.
- By mimicking Corollary 1, inside (x_{j-1}, x_j),

$$v'(x) = r_{j-\frac{1}{2}} \left(A \exp(r_{j-\frac{1}{2}} x) - B \exp(-r_{j-\frac{1}{2}} x) \right),$$

  and since $M_{j\pm\frac{1}{2}}$ are invertible, C^1 smoothness is just:

$$\begin{aligned} & r_{j-\frac{1}{2}} \left\langle \begin{pmatrix} u_{j-1}^n - \frac{\bar{f}}{\bar{q}} \\ u_j^n - \frac{\bar{f}}{\bar{q}} \end{pmatrix}, M_{j-\frac{1}{2}}^{-T} \begin{pmatrix} \exp(r_{j-\frac{1}{2}} x_j) \\ -\exp(-r_{j-\frac{1}{2}} x_j) \end{pmatrix} \right\rangle \\ & = r_{j+\frac{1}{2}} \left\langle \begin{pmatrix} u_j^n - \frac{\bar{f}}{\bar{q}} \\ u_{j+1}^n - \frac{\bar{f}}{\bar{q}} \end{pmatrix}, M_{j+\frac{1}{2}}^{-T} \begin{pmatrix} \exp(r_{j+\frac{1}{2}} x_j) \\ -\exp(-r_{j+\frac{1}{2}} x_j) \end{pmatrix} \right\rangle, \end{aligned} \tag{15}$$

  where we have used the adjoint of the inverse matrices $M_{j\pm\frac{1}{2}}^{-1}$ so that we don't need the values of A, B. Of course, the $\left(\frac{\bar{f}}{\bar{q}}\right)_{j\pm\frac{1}{2}}$ are different on both sides of the equality. By developing (15),

$$\begin{aligned} \frac{u_j^{n+1} - u_j^n}{\Delta t} - \frac{\varepsilon}{\Delta x} \Bigg[& \frac{r_{j+\frac{1}{2}}}{\sinh(r_{j+\frac{1}{2}} \Delta x)} \left(u_{j+1}^n - \cosh(r_{j+\frac{1}{2}} \Delta x) u_j^n \right) \\ & - \frac{r_{j-\frac{1}{2}}}{\sinh(r_{j-\frac{1}{2}} \Delta x)} \left(\cosh(r_{j-\frac{1}{2}} \Delta x) u_j^n - u_{j-1}^n \right) \\ & + r_{j+\frac{1}{2}} \frac{\cosh(r_{j+\frac{1}{2}} \Delta x) - 1}{\sinh(r_{j+\frac{1}{2}} \Delta x)} \cdot \left(\frac{\bar{f}}{\bar{q}} \right)_{j+\frac{1}{2}} \\ & + r_{j-\frac{1}{2}} \frac{\cosh(r_{j-\frac{1}{2}} \Delta x) - 1}{\sinh(r_{j-\frac{1}{2}} \Delta x)} \cdot \left(\frac{\bar{f}}{\bar{q}} \right)_{j-\frac{1}{2}} \Bigg] = 0. \end{aligned}$$

  Consistency can be established proceeding like in Theorem 2.

Remark 3 Considering (1) with $p(x) = q(x) \equiv 0$, or even (3) with $\alpha \equiv 0$, but $f \neq 0$ leads to the example presented in [43, §3.1–2] (where nonlinearity essentially reduces to sign penalization). Such a situation is simple enough so that (11) reduces to the simple finite-differences, as considered by Lions and Mercier.

$$u_j^{n+1} = u_j^n - \frac{\varepsilon \Delta t}{\Delta x}\left[\frac{(u_{j+1}^n - \frac{x_{j+1}^2}{2\varepsilon} f_{j+\frac{1}{2}}) - (u_j^n - \frac{x_j^2}{2\varepsilon} f_{j+\frac{1}{2}})}{\Delta x}\right.$$

$$\left. - \frac{(u_j^n - \frac{x_j^2}{2\varepsilon} f_{j-\frac{1}{2}}) - (u_{j-1}^n - \frac{x_{j-1}^2}{2\varepsilon} f_{j-\frac{1}{2}})}{\Delta x} + \frac{x_j}{\varepsilon}(f_{j+\frac{1}{2}} - f_{j-\frac{1}{2}})\right]$$

$$= u_j^n - \frac{\varepsilon \Delta t}{\Delta x}\left[\frac{u_{j+1}^n - u_j^n}{\Delta x} - \frac{u_j^n - u_{j-1}^n}{\Delta x} - \frac{\Delta x}{2\varepsilon}(f_{j+\frac{1}{2}} + f_{j-\frac{1}{2}})\right].$$

Their hyperbolic obstacle problem [43, §3.3] is much more related to [6, §2.2]: as one would expect nowadays, Lions and Mercier (p. 978) observe that numerical upwind schemes involving a centered treatment of rough x-dependent source terms in an advection equation produce excessive numerical viscosity at steady-state.

4 Two-Dimensional Algorithm Involving Bessel Functions

4.1 Definition of the Resulting 2D Scheme

A two-dimensional analogue of Definition 2 goes as follows:

Definition 3 Let $x_i, y_j \in (-1, 1)^2$ be a grid-point, $R = \Delta x/\sqrt{2}$, and consider its four adjacent $\mathscr{L}$-membranes, along with their respective boundary data,

$$\begin{cases} \tilde{\phi}_{SW} \to \tilde{\phi}_{SW}(R, \frac{\pi}{2}) = u_{i-1,j}^n, \tilde{\phi}_{SW}(R, 0) = u_{i,j}^n, \tilde{\phi}_{SW}(R, -\frac{\pi}{2}) = u_{i,j-1}^n, \\ \tilde{\phi}_{NW} \to \tilde{\phi}_{NW}(R, \pi) = u_{i-1,j}^n, \tilde{\phi}_{NW}(R, -\frac{\pi}{2}) = u_{i,j}^n, \tilde{\phi}_{NW}(R, 0) = u_{i,j+1}^n, \\ \tilde{\phi}_{NE} \to \tilde{\phi}_{NE}(R, \frac{\pi}{2}) = u_{i,j+1}^n, \tilde{\phi}_{NE}(R, \pi) = u_{i,j}^n, \tilde{\phi}_{NE}(R, -\frac{\pi}{2}) = u_{i+1,j}^n, \\ \tilde{\phi}_{SE} \to \tilde{\phi}_{SE}(R, 0) = u_{i+1,j}^n, \tilde{\phi}_{SE}(R, \frac{\pi}{2}) = u_{i,j}^n, \tilde{\phi}_{SE}(R, \pi) = u_{i,j-1}^n, \end{cases}$$

and their normal (radial) derivatives (12). For a uniform 2D Cartesian grid, let $\bar{R} = \sqrt{\Delta x^2 + \Delta y^2} = \Delta x\sqrt{2} = 2R$, so that the time-marching scheme reads:

$$u_{i,j}^{n+1} = u_{i,j}^n - \frac{\Delta t}{\bar{R}}\left[\left(\frac{\tilde{\phi}_{NE}}{\partial r}(R, \pi) + \frac{\tilde{\phi}_{SW}}{\partial r}(R, 0)\right) + \left(\frac{\tilde{\phi}_{NW}}{\partial r}(R, -\frac{\pi}{2}) + \frac{\tilde{\phi}_{SE}}{\partial r}(R, \frac{\pi}{2})\right)\right].$$

Each sum between parentheses corresponds to a finite-difference along a diagonal. This scheme is actually a generalization of the former (10) because it proceeds by computing the residual of all available transmission conditions at each grid point.

4.2 Consistency with the Continuous Problem

It will prove convenient to define the following functions, $\mathscr{F} \geq 0, \mathscr{H} \geq 0, \mathscr{G} \leq 0$,

$$\frac{\mathscr{F}(\alpha)}{\sqrt{\alpha}} = \frac{\mathscr{I}_0 + \mathscr{I}_2}{2\mathscr{I}_1}\left(\sqrt{\frac{\alpha}{2}}\Delta x\right), \quad \frac{\mathscr{H}(\alpha)}{\sqrt{\alpha}} = \frac{\mathscr{I}_1}{\mathscr{I}_0}\left(\sqrt{\frac{\alpha}{2}}\Delta x\right), \quad \mathscr{G}(\alpha) = \frac{\mathscr{H}(\alpha) - \mathscr{F}(\alpha)}{2},$$

so that, according to Definition 3, our 2D explicit scheme reads, for any $n \in \mathbb{N}$,

$$\begin{aligned}
u_{i,j}^{n+1} = u_{i,j}^n - \frac{\Delta t}{\Delta x\sqrt{2}} \Big\{ & u_{i,j}^n \left(\mathscr{F}(\bar{\alpha}_{i-\frac12,j-\frac12}) + \mathscr{F}(\bar{\alpha}_{i+\frac12,j-\frac12}) + \mathscr{F}(\bar{\alpha}_{i+\frac12,j+\frac12}) + \mathscr{F}(\bar{\alpha}_{i-\frac12,j+\frac12})\right) \\
& + u_{i-1,j}^n \left(\mathscr{G}(\bar{\alpha}_{i-\frac12,j-\frac12}) + \mathscr{G}(\bar{\alpha}_{i-\frac12,j+\frac12})\right) + u_{i,j-1}^n \left(\mathscr{G}(\bar{\alpha}_{i-\frac12,j-\frac12}) + \mathscr{G}(\bar{\alpha}_{i+\frac12,j-\frac12})\right) \\
& + u_{i+1,j}^n \left(\mathscr{G}(\bar{\alpha}_{i+\frac12,j-\frac12}) + \mathscr{G}(\bar{\alpha}_{i+\frac12,j+\frac12})\right) + u_{i,j+1}^n \left(\mathscr{G}(\bar{\alpha}_{i+\frac12,j+\frac12}) + \mathscr{G}(\bar{\alpha}_{i-\frac12,j+\frac12})\right) \\
& - \frac{\bar{f}_{i-\frac12,j-\frac12}}{\bar{\alpha}_{i-\frac12,j-\frac12}} \mathscr{H}(\bar{\alpha}_{i-\frac12,j-\frac12}) - \frac{\bar{f}_{i+\frac12,j-\frac12}}{\bar{\alpha}_{i+\frac12,j-\frac12}} \mathscr{H}(\bar{\alpha}_{i+\frac12,j-\frac12}) \\
& - \frac{\bar{f}_{i+\frac12,j+\frac12}}{\bar{\alpha}_{i+\frac12,j+\frac12}} \mathscr{H}(\bar{\alpha}_{i+\frac12,j+\frac12}) - \frac{\bar{f}_{i-\frac12,j+\frac12}}{\bar{\alpha}_{i-\frac12,j+\frac12}} \mathscr{H}(\bar{\alpha}_{i-\frac12,j+\frac12}) \Big\}. \quad (16)
\end{aligned}$$

The well-balanced character of (16) holds by construction: by uniqueness of the smooth stationary solution of (3) with inhomogeneous Dirichlet conditions, and assuming that $|\nabla\alpha|$ remains small, each $\mathscr{L}$-spline obtained by (6) is very close to the restriction of u^∞ to each disc D. Accordingly, numerical fluxes in (16) vanish.

Remark 4 Discretization (16) may be set up *mutatis mutandis* even in the accretive case $\alpha(x, y) < 0$ (in which the Fourier decomposition (7) holds with $\mathscr{J}_n$), so that $\sqrt{\alpha}$ is purely imaginary. According to the relations between Bessel functions $\mathscr{I}_n$ and $\mathscr{J}_n$ for $n = 0, 1, 2$, the scheme (16) produces real numbers, along with a consistent approximation of (3). Monotonicity properties may be lost, though.

Lemma 4 *The numerical scheme (16) is consistent with (3) as $\Delta x \to 0$.*

Proof As $z \to 0$, modified Bessel functions of the first kind behave like $\mathscr{I}_n(z) \simeq \frac{1}{n!}\left(\frac{z}{2}\right)^n$. Accordingly, as $\alpha > 0$, $\Delta x \to 0$, the functions $\mathscr{F}.\mathscr{G}, \mathscr{H}$ are such that,

$$\begin{cases} \frac{\mathscr{F}(\alpha)}{\Delta x\sqrt{2}} \simeq \frac{\sqrt{\alpha}}{\Delta x\sqrt{2}}\left(\frac{\sqrt{2}}{\Delta x\sqrt{\alpha}} + \sqrt{\frac{\alpha}{2}}\frac{\Delta x}{8}\right) = \frac{1}{\Delta x^2} + \frac{\alpha}{16}, \\ \frac{\mathscr{H}(\alpha)}{\Delta x\sqrt{2}} \simeq \frac{\alpha}{4}, \quad \mathscr{G}(\alpha) \simeq \frac{1}{2}\left(\frac{\alpha}{4} - \frac{1}{\Delta x} - \frac{\alpha}{16}\right) = -\frac{1}{2\Delta x} + \frac{3\alpha}{32}. \end{cases}$$

The terms in $\frac{1}{\Delta x^2}$ easily yield the discrete 5-points Laplace operator; the ones in $\frac{\alpha}{4}$ give consistency with the external forcing term. Concerning the damping term,

$$4 \times \frac{\alpha}{16} + 4 \times 2 \times \frac{3\alpha}{32} = \frac{\alpha}{4} + \frac{12\alpha}{16} = \alpha.$$

Remark 5 Lemma 4 gives that both the forcing and damping terms are upwinded on each of the 5 points belonging to the stencil according to weights with a definite sign: such a construction still applies with 9 points [14], but resulting radial derivatives display coefficients which aren't unconditionally of the correct signs.

4.3 Monotonicity, Positivity-Preserving and Convergence

Lemma 5 *The scheme (16) realizes a monotone function of each of its 5 entries under the (parabolic) CFL condition (which admits an approximation for small Δx),*

$$4 \max_{i,j} \mathscr{F}(\alpha_{i-\frac{1}{2},j-\frac{1}{2}})\Delta t \leq \sqrt{2}\Delta x, \qquad \Delta t \left(\frac{4}{\Delta x^2} + \max_{i,j} \frac{\alpha_{i-\frac{1}{2},j-\frac{1}{2}}}{4} \right) \leq 1. \qquad (17)$$

Proof Assuming (17), the scheme (16) is a non-negative combination of each of its arguments, hence a monotone, positivity-preserving approximation of (3).

Lax's equivalence theorem states that a stable and consistent scheme converges:

Theorem 3 *Under the CFL (17), the sequence of piecewise constant approximations $u^{\Delta x}$, defined by $u^{\Delta x}(t^n, x_i, y_j) = u^n_{i,j}$, is such that, for any $p \geq 1$,*

$$\forall t^n = n\Delta t > 0, \qquad \|u^{\Delta x}(t^n, \cdot, \cdot)\|_{L^p} \leq \|u^{\Delta x}(t = 0, \cdot, \cdot)\|_{L^p} + t^n \, \|f\|_{L^p}.$$

It converges weakly as $\Delta x \to 0$ towards the unique weak solution of (3).

Proof Thanks to the scheme's monotonicity (it isn't shift-invariant because of the space-dependence of both parameters α, f), the dissipation estimate follows from Jensen's inequalities for L^p norms (which are convex functions) and the fact that,

$$4\mathscr{F}(\alpha) + 8\mathscr{G}(\alpha) = 4\mathscr{H}(\alpha) \geq 0, \qquad \frac{4\Delta t}{\Delta x}\sqrt{\frac{\alpha}{2}} \frac{\mathscr{I}_1}{\mathscr{I}_0}\left(\sqrt{\frac{\alpha}{2}}\Delta x\right) \simeq \alpha \Delta t,$$

coming from standard properties of Bessel functions. Weak compactness follows.

5 Relations with Scharfetter-Gummel's Discretizations

5.1 Scharfetter-Gummel as a 1D $\mathscr{L}$-Spline Scheme

Consider a one-dimensional drift-diffusion equation, (see e.g. [29, pp. 4–5])

$$\partial_t \rho + \partial_x J = 0, \qquad J = p\rho - \varepsilon \partial_x \rho, \quad p, \varepsilon > 0.$$

In order to preserve local mass conservation at the discrete level,

$$\rho_j^{n+1} - \rho_j^n + \frac{\Delta t}{\Delta x}\left(J^n_{j+\frac{1}{2}} - J^n_{j-\frac{1}{2}}\right) = 0,$$

and all the difficulty lies in "choosing wisely" the numerical flux at each interface, say $J^n_{j-\frac{1}{2}}$. The insight given in [57, Appendix D] consists in taking advantage of the constancy of the mass flux at steady state, $\partial_x J = 0$, in order to define each interfacial flux by means of an (exponential) integrating factor,

$$J^n_{j-\frac{1}{2}} = -\varepsilon \partial_x \bar{\rho} + p\bar{\rho} = -\varepsilon \partial_x \left(\bar{\rho}(x) \exp(-px/\varepsilon)\right) \exp(px/\varepsilon).$$

As $\bar{\rho}(x)$ is intended to be a stationary curve joining x_{j-1} to x_j, one immediately recognizes an (exponential-type) $\mathscr{L}$-spline interpolation. Yet,

$$J^n_{j-\frac{1}{2}} \exp(-px/\varepsilon) = -\varepsilon \partial_x \left(\bar{\rho}(x) \exp(-px/\varepsilon)\right) \exp(px/\varepsilon),$$

so that, by elementary integration,

$$\forall j, n, \qquad J^n_{j-\frac{1}{2}} = p \frac{\rho_j^n \exp(-px_j/\varepsilon) - \rho_{j-1}^n \exp(-px_{j-1}/\varepsilon)}{\exp(-px_j/\varepsilon) - \exp(-px_{j-1}/\varepsilon)}.$$

Such a flux formula, involving so-called "Bernoulli functions", is endowed with many remarkable properties; in particular its accuracy doesn't degrade as $\varepsilon \ll \Delta x$. However, being its construction rather similar to the one presented in Sect. 3.1, one may wonder whether this well-known scheme realizes a special occurrence of the fluxes written in Lemma 3 (when $q = f \equiv 0$, (1) becomes a drift-diffusion equation). Accordingly, let $p > 0$: $r_+ = \frac{p}{\varepsilon}$, $r_- = 0$, so that $D = r_+$. Expressions (11) reduce to:

$$R^n_{j-\frac{1}{2}} = \frac{-r_+}{1 - \exp(-r_+ \Delta x)}(u_j^n - u_{j-1}^n) = \exp(r_+ \Delta x) L^n_{j-\frac{1}{2}}.$$

By applying the formalism (10)–(11) and $r_\pm$ being constants, it comes,

$$\begin{aligned}
u_j^{n+1} &= u_j^n - \frac{\varepsilon \Delta t}{\Delta x}\left[L^n_{j+\frac{1}{2}} - R^n_{j-\frac{1}{2}}\right].\\
&= u_j^n - \frac{\varepsilon \Delta t}{\Delta x}\left[\exp(r_+\Delta x)R^n_{j+\frac{1}{2}} - R^n_{j-\frac{1}{2}}\right].\\
&= u_j^n - \frac{p\,\Delta t}{\Delta x}\Bigg[\left(\frac{u_j^n}{1-\exp(-p\Delta x/\varepsilon)} - \frac{u_{j+1}^n}{\exp(p\Delta x/\varepsilon)-1}\right)\\
&\qquad - \left(\frac{u_{j-1}^n}{1-\exp(-p\Delta x/\varepsilon)} - \frac{u_j^n}{\exp(p\Delta x/\varepsilon)-1}\right)\Bigg],
\end{aligned}$$

and this is one manner of expressing Scharfetter-Gummel's numerical fluxes.

5.2 *An Extension Toward Two-Dimensional Equations*

A main obstacle lying on the way to 2D problems is that the stationary equation $\partial_x J = 0$ just becomes $\operatorname{div}(\mathbf{J}(x,y)) = 0$ and this is clearly insufficient to carry on with the aforementioned derivation. However, we can invoke the following result:

Theorem 4 *Let $V(x,y) \in C^2(\mathbb{R}^2)$ be a smooth potential and consider $\rho(t,x,y)$ the nonnegative solution of the conservative drift-diffusion equation,*

$$\partial_t \rho - \varepsilon\, div\left[\exp(\frac{V}{\varepsilon})\nabla\left(\exp(-\frac{V}{\varepsilon})\rho\right)\right] = 0. \tag{18}$$

Then, the auxiliary unknown $u(t,x,y) = \exp(-\frac{V(x,y)}{2\varepsilon})\rho(t,x,y)$ satisfies,

$$\partial_t u - \varepsilon\Big(\Delta u + \alpha_\varepsilon(x,y)u\Big) = 0, \qquad \alpha_\varepsilon(x,y) = \left|\frac{\nabla V}{2\varepsilon}\right|^2 + \frac{\Delta V}{2\varepsilon}. \tag{19}$$

Fortunately, the auxiliary unknown u can be handled thanks to the 2D construction based on Steklov-Poincaré operator, which leads to the scheme (16). This furnishes a roadmap to derive a 2D extension of Scharfetter-Gummel's idea because in 1D, it matches our $\mathscr{L}$-spline construction, and in 2D, very similar procedures yield (16). An interesting question is therefore to determine whether such an approach, when applied to the simplest 1D drift-diffusion equation, allows again to recover Scharfetter-Gummel's discretization: accordingly, consider, for $\varepsilon = 1$,

$$\partial_t \rho - \partial_x\Big(\partial_x \rho - p\rho\Big) = 0 = \partial_t\rho - \partial_x\Big(\exp(px)\partial_x\big(\exp(-px)\rho\big)\Big).$$

Applying Theorem 4 to an auxiliary unknown, $u(t,x) = \exp(-px/2)\rho(t,x)$, comes a damped heat equation, $\partial_t u - \partial_{xx} u + \frac{p^2}{4} u = 0$. Moreover, from (10)–(11),

$$\begin{aligned} u_j^{n+1} &= u_j^n - \tfrac{\Delta t}{\Delta x}\left[L^n_{j+\frac12} - R^n_{j-\frac12}\right], \quad r^{\pm} = \pm\tfrac{p}{2}, \quad D = p,\\ R^n_{j-\frac12} &= u^n_{j-1}\tfrac{p\exp(-p\Delta x/2)}{1-\exp(-p\Delta x)} - u_j^n\left(\tfrac{p/2}{1-\exp(-p\Delta x)} + \tfrac{p/2}{\exp(p\Delta x)-1}\right)\\ L^n_{j+\frac12} &= u_j^n\left(\tfrac{p/2}{\exp(p\Delta x)-1} + \tfrac{p/2}{1-\exp(-p\Delta x)}\right) - u^n_{j+1}\tfrac{p\exp(p\Delta x/2)}{\exp(p\Delta x)-1} \end{aligned}$$

In order to recover a $\mathscr{L}$-spline scheme acting on the conserved unknown ρ, it suffices to define $\rho_j^n = \exp(px_j/2)u_j^n$. Since there are quantities in $x_{j\pm1}$, one finds that certain exponential terms act on $\rho^n_{j\pm1}$: by rearranging the expression,

$$\begin{aligned} \rho_j^{n+1} = \rho_j^n - p\cdot\tfrac{\Delta t}{\Delta x}\Bigg[&\left(\tfrac{\rho_j^n}{1-\exp(-p\Delta x)} - \tfrac{\rho^n_{j+1}}{\exp(p\Delta x)-1}\right)\\ &-\left(\tfrac{\rho^n_{j-1}}{1-\exp(-p\Delta x)} - \tfrac{\rho_j^n}{\exp(p\Delta x)-1}\right)\Bigg], \end{aligned}$$

being precisely the 1D Scharfetter-Gummel scheme in its simplest expression.

Accordingly, in order to solve numerically a Cauchy problem for (18) in, say, $(-1,1)^2$, given some smooth potential $V(x,y)$ and Dirichlet boundary conditions, the following procedure applies:

- at time $t=0$, form auxiliary initial data, $u^0(x_i,y_j) = \exp(-\frac{V(x_i,y_j)}{2\varepsilon})\rho^0(x_i,y_j)$;
- set up the marching scheme (16) for the resulting 2D damped heat equation;
- at final time, $t = n\Delta t > 0$, return to the conserved unknown with

$$\forall i,j, \qquad \rho(n\Delta t, x_i, y_j) = \exp\left(\frac{V(x_i,y_j)}{2\varepsilon}\right)u(n\Delta t, x_i, y_j).$$

Clearly, in presence of a weakly nonlinear coupling involving a Poisson equation, like for instance $-\Delta V = \rho$ with Dirichlet boundary conditions, the exponential factor must be applied at every time-step; however, the computation of the Laplacian, ΔV appearing when forming the coefficient α, becomes straightforward.

6 Two-Dimensional Numerical Validations

Equations like (18) admit several types of steady-states, like for instance $\bar\rho = C\exp(\frac{V}{\varepsilon})$; hence, an easy way to scrutinize the accuracy of any numerical scheme on a bounded domain, say $(-1,1)^2$ consists in imposing boundary data compatible with $\bar\rho$, then, starting from any non-negative initial data, compute the deviation of the stationary numerical solution with the restriction to $(-1,1)^2$ of the exact one $\bar\rho$.

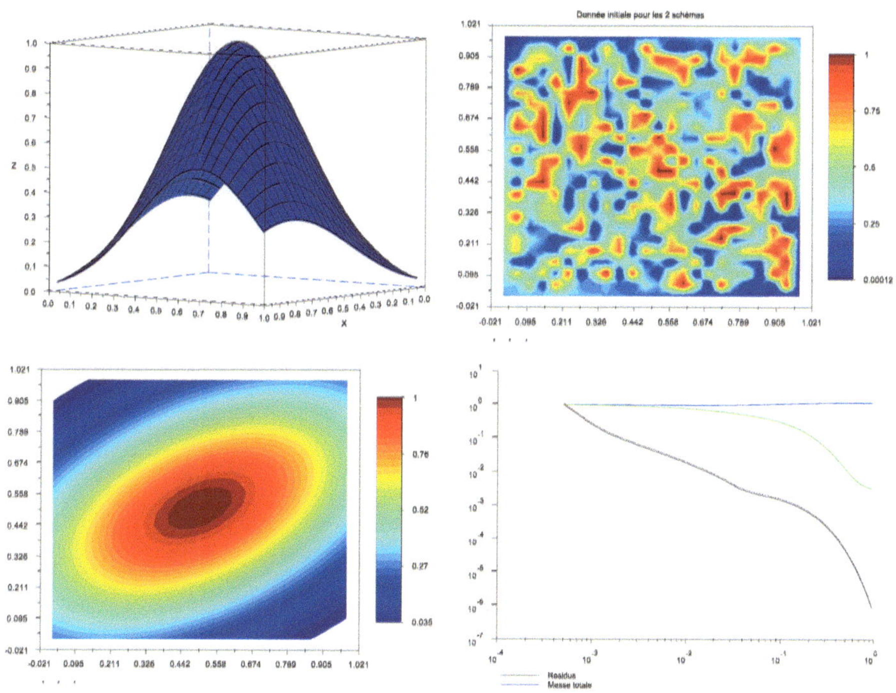

Fig. 3 Exact profile $\bar{\rho}$, initial data, stationary numerical solution and residues decay

Table 1 Evolution of pointwise errors at steady-state for 2D vs. FD schemes

Grid points	Maximum 2D error	Maximum FD error
$N = 2^4$	0.0030357	0.0942922
$N = 3 \cdot 2^3$	0.0015190	0.0706629
$N = 2^5$	0.0009252	0.0562096

6.1 *Quadratic Potential* $V(x, y)$ *and Gaussian Steady Solution*

A first benchmark consists in setting up (18) with the potential,

$$V(x, y) = -(\frac{3}{2}|x'|^2 + 5|y'|^2), \qquad (x', y') = R_\theta(x, y),$$

being R_θ the usual rotation matrix of angle $\theta = \frac{\pi}{6}$ and $\varepsilon = 0.65$. Discrete boundary conditions are prescribed by discretizing the steady-state solution $\bar{\rho}(x, y)$ with $C = 1$, and perturbing it by random noise only in the interior of the computational domain: see top row in Fig. 3. The numerical stationary regime emerges at time $t \simeq 1$: it is displayed on the bottom row of Fig. 3 along with the corresponding residues' decay. A comparison of (pointwise) global errors generated by our 2D algorithm versus more conventional centered finite-differences is given on Table 1: the superiority of our 2D numerical algorithm for coarse grids appears neatly.

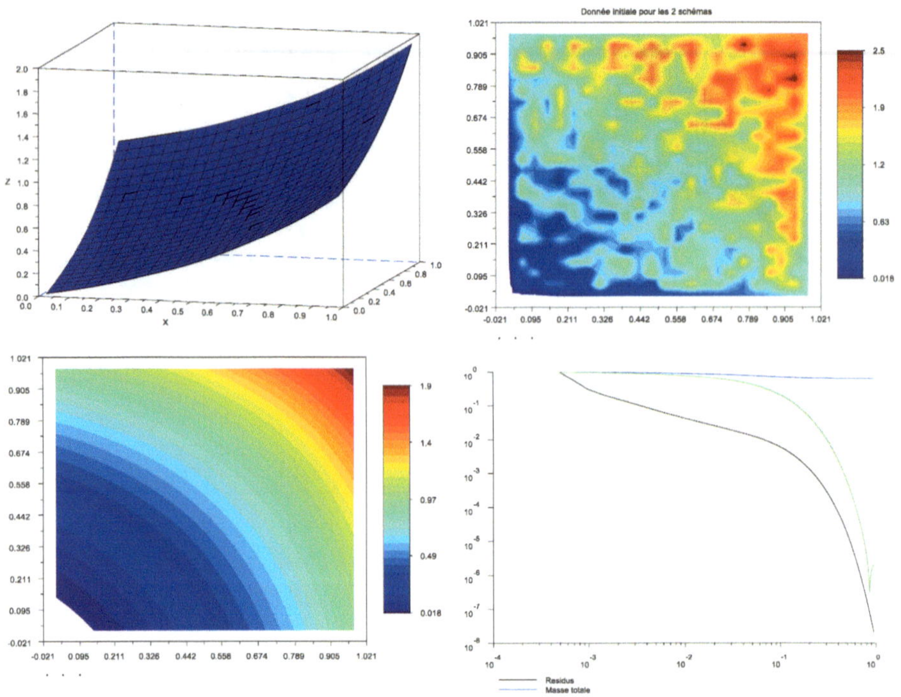

Fig. 4 Similar to Fig. 3, but with a linear potential V and an exponential layer

Table 2 Evolution of pointwise errors at steady-state for 2D vs. FD schemes

Grid points	Maximum 2D error	Maximum FD error
$N = 2^4$	0.0000048	0.0084850
$N = 3 \cdot 2^3$	0.0000018	0.0060502
$N = 2^5$	0.0000006	0.0046897

6.2 *Linear Potential $V(x, y)$, Advection-Diffusion and Layers*

A simpler, yet potentially delicate, potential is the linear one,

$$V(x, y) = x + y, \qquad \bar{\rho}(x, y) = \frac{\exp(x/\varepsilon) + \exp(y/\varepsilon) - 2}{\exp(1/\varepsilon) - 1}.$$

The steady-state is now endowed with a diagonal (i.e. not aligned with the computational grid's axes) exponential layer: numerical results are displayed in Fig. 4 with 24×24 discrete points. Again, it's interesting to scrutinize the behavior of pointwise errors for our 2D scheme based on Bessel functions (and not on exponentials, oppositely to 1D) with the ones obtained out of standard finite-differences. On Table 2, such a comparison is made for several coarse grids: the presence of an oblique exponential layer reveals that finite-differences, corresponding to local linear approximations of the solution, produce errors several orders of magnitude bigger than the 2D scheme. This behavior is reminiscent of the derivation of numerical schemes based on "coincident modes", see [42].

6.3 Linear Solutions of an Advection-Diffusion Equation

Linear potentials may be dangerous because the resulting drift (or transport) equation admits linear solutions, too. Hence a good numerical scheme should be able to treat both these classes of solutions equally well ... Such an example reads:

$$V(x,y) = \frac{x}{2} + y, \qquad \bar{\rho}(x,y) = 1 - (x - \frac{y}{2}).$$

More generally, given any linear potential

$$V(x,y) = x(\partial_x V) + y(\partial_y V), \qquad \bar{\rho}(x,y) = C + (\partial_y V)x - (\partial_x V)y, \quad C \in \mathbb{R},$$

is solution. In that case, the numerical scheme involving the "coincident modes" is actually the usual finite-differences one. Indeed, being the stationary solution a linear function of x, y, a local approximation of derivatives based on first-degree polynomials is very well-suited (Fig. 5 and Table 3).

Accordingly, the behavior of pointwise errors at steady-state reveals that finite-differences perform better than the 2D scheme involving Bessel functions. However, even on such a favorable case, centered discretizations outperform only of a factor between 10 and 20, decreasing with the grid fineness. When compared to discrepancies measured on Tables 1 and 2, involving factors of more than 100, such a drawback doesn't appear to be very important.

7 Conclusion and Outlook

Along with all similarities it shares with the original Lax-Wendroff scheme for wave equations, involving "spherical means", and proposed in [23], there are many directions in which such a genuinely multi-dimensional framework relying on piecewise-steady (locally exact) solutions can be extended. Let us name a few of them:

- the most obvious one is the resolution of weakly non-linear 2D drift-diffusion equations, that is, (18) in which V corresponds to a Poisson potential deduced from ρ via a mean-field approximation. Such interactions appear for instance in

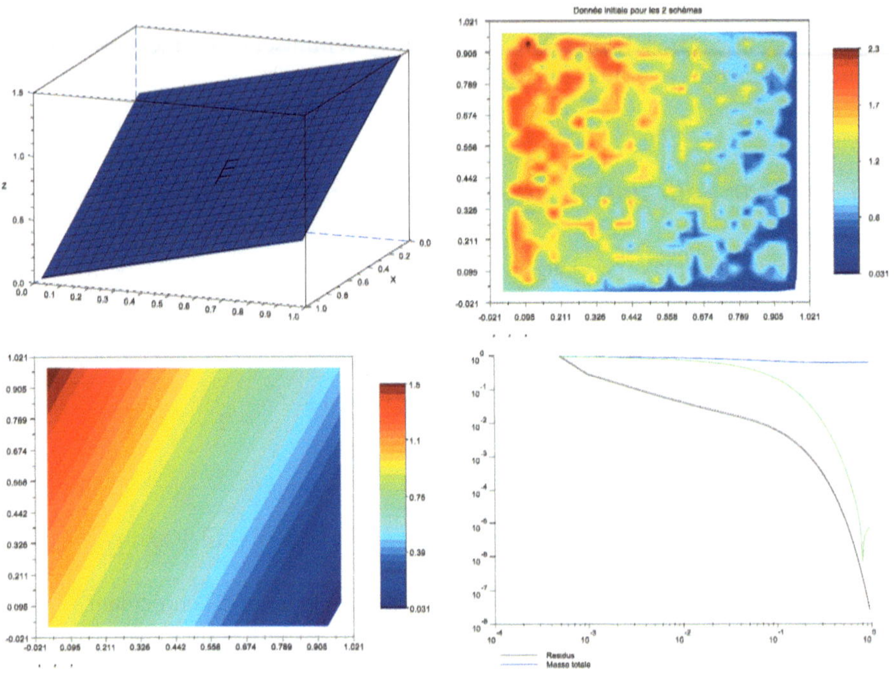

Fig. 5 Similar to Fig. 4, but now a linear potential inducing a linear solution

Table 3 Evolution of pointwise errors at steady-state for 2D vs. FD schemes

Grid points	Maximum 2D error	Maximum FD error
$N = 2^4$	0.0000143	0.0000009
$N = 3 \cdot 2^3$	0.0000060	0.0000006
$N = 2^5$	0.0000027	0.0000003

electrostatic (repulsive, see [15]) and chemotaxis (attractive) models. An issue for chemotaxis simulations lies in an adequate implementation of Neumann-type boundary conditions in (16).

- another interesting challenge would be to set up (16) in order to compute transients of 2D incompressible Navier-Stokes equations in the stream/vorticity functions formulation. This meets with some ideas originally proposed in Gartland [25], later rephrased in [32] (see also [35]). Similarly as Theorem 4, an exponential change of variable allows to switch from an advection-diffusion equation to a damped heat equation, so that the same machinery may apply successfully in a context of incompressible turbulence, too.

References

1. Achdou, Y., Nataf, F.: Local approximations of the Steklov-Poincaré operator for a class of elliptic operators, and domain decomposition methods. CMAP Internal Report 94/303, Ecole Polytechnique, Palaiseau
2. Adam, D., Felgenhauer, A., Roos, H.-G., Stynes, M.: A nonconforming finite element method for a singularly perturbed boundary value problem. Computing **54**, 1–25 (1995)
3. Ahlberg, J.H., Nilson, E.N., Walsh, J.L.: Fundamental properties of generalized splines. Proc. Natl. Acad. Sci. U. S. A. **52**(6), 1412–1419 (1964)
4. Ainsworth, M., Dorfler, W.: Fundamental systems of numerical schemes for linear convection-diffusion equations and their relationship to accuracy. Computing **66**, 199–229 (2001)
5. Allen, D.N. De G.: A suggested approach to finite difference representation of differential equations. Q. J. Mech. Appl. Math. **15**, 11–33 (1962)
6. Amadori, D., Gosse, L.: Error Estimates for Well-Balanced Schemes on Simple Balance Laws: One-Dimensional Position-Dependent Models. BCAM Springer Briefs in Mathematics. Springer, Berlin (2015)
7. Andrade, M.G., Do Val, J.B.R.: A numerical scheme based on mean value solutions for the Helmholtz equation on triangular grids. Math. Comput. **66**, 477–493 (1997)
8. Arendt, W., ter Elst, A.F.M., Kennedy, J.B., Sauter, M.: The Dirichlet-to-Neumann operator via hidden compactness. J. Funct. Anal. **266**, 1757–1786 (2014)
9. Babuska, I., Osborn, J.E.: Generalized finite element methods: their performance and their relation to mixed methods. SIAM J. Numer. Anal. **20**, 510–536 (1983)
10. Bagrinovskii, K.A., Godunov, S.K.: Difference schemes for multidimensional problems. Dokl. Akad. Nauk USSR **115**, 431–433 (1957)
11. Beirao da Veiga, L., Brezzi, F., Cangiani, A., Manzini, G., Marini, L.D., Russo, A.: Basic principles of virtual element methods. Math. Models Methods Appl. Sci. **23**, 199–214 (2013)
12. Berger, A.E., Solomon, J.M., Ciment, M.: An analysis of a uniformly accurate difference method for a singular perturbation problem. Math. Comput. **37**, 79–94 (1981)
13. Berger, A.E., Han, H., Kellogg, R.B.: A priori estimates and analysis of a numerical method for a turning point problem. Math. Comput. **42**, 465–492 (1984)
14. Birkhoff, G., Gulati, S.: Optimal few points discretizations of linear source problems. SIAM J. Numer. Anal. **11**, 700–728 (1974)
15. Brezzi, F., Marini, D., Pietra, P.: Two-dimensional exponential fitting and applications to drift-diffusion models. SIAM J. Numer. Anal. **26**, 1342–1355 (1989)
16. Buffa, A., Monk, P.: Error estimates for the Ultra Weak Variational Formulation of the Helmholtz equation. ESAIM: Math. Model. Numer. **42**, 925–940 (2008)
17. Canosa, J., Gomes De Oliveira, R.: A new method for the solution of the Schrodinger equation. J. Comput. Phys. **5**, 188–207 (1970)
18. Cessenat, O., Despres, B.: Application of an ultra-weak variational formulation of elliptic PDE's to the two-dimensional Helmholtz problem. SIAM J. Numer. Anal. **35**, 255–299 (1998)
19. Champion, R., Lenard, C.T., Mills, T.M.: A variational approach to splines. ANZIAM J. **42**, 119–135 (2000)
20. Cheung, Y.K., Jin, W.G., Zienkiewicz, O.C.: Direct solution procedure for solution of harmonic problems using complete, non-singular, Trefftz functions. Commun. Appl. Numer. Methods **5**, 159–169 (1989)
21. Daners, D.: Non-positivity of the semigroup generated by the Dirichlet-to-Neumann operator. Positivity **18**, 235–256 (2014)
22. Despres, B., Buet, Ch.: The structure of well-balanced schemes for Friedrichs systems with linear relaxation. Appl. Math. Comput. **272**, 440–459 (2016)
23. Fan, D., Roe, P.L.: Investigations of a new scheme for wave propagation. In: Proceedings of 22nd AIAA Computational Fluid Dynamics Conference, Dallas, TX, 22–26 June 2015. doi:10.2514/6.2015-2449

24. Farrell, P.A., Gartland, E.C. Jr.: A uniform convergence result for a turning point problem. In: BAIL V (Shanghai, 1988). Boole Press Conference Series, vol. 12, pp. 127–132 (1988)
25. Gartland, E.C. Jr.: Discrete weighted mean approximation of model convection-diffusion equation. SIAM J. Sci. Stat. Comput. **3**, 460–472 (1982)
26. Gartland, E.C. Jr.: Strong stability of compact discrete boundary value problems via exact discretizations. SIAM J. Numer. Anal. **25**, 111–123 (1988)
27. Gartland, E.C. Jr.: On the uniform convergence of the Scharfetter-Gummel discretization in one dimension. SIAM J. Numer. Anal. **30**, 749–758 (1993)
28. Glimm, J., Sharp, D.H.: An S-matrix theory for classical nonlinear physics. Found. Phys. **16**, 125–141 (1986)
29. Gosse, L.: Computing qualitatively correct approximations of balance laws. In: Exponential-Fit, Well-Balanced and Asymptotic-Preserving. SIMAI Springer Series. Springer, Berlin (2013). ISBN: 978-88-470-2891-3
30. Gosse, L., LeRoux, A.Y.: Un schéma-équilibre adapté aux lois de conservation scalaires non-homogènes. C.R. Acad. Sci. Paris Série I **323**, 543–546 (1996)
31. Greenberg, J., LeRoux, A.Y.: A well balanced scheme for the numerical processing of source terms in hyperbolic equations. SIAM J. Numer. Anal. **33**, 1–16 (1996)
32. Han, H., Huang, Z., Kellogg, R.B.: A tailored finite point method for a singular perturbation problem on an unbounded domain. J. Sci. Comput. **36**, 243–261 (2008)
33. Hennart, J.-P.: On the numerical analysis of analytical nodal methods. Numer. Methods Partial Differ. Equ. **4**, 233–254 (1988)
34. Herrera, I.: The algebraic theory approach for ordinary differential equations: highly accurate finite differences. Numer. Methods Partial Differ. Equ. **3**, 199–218 (1987)
35. Hofreither, C., Langer, U., Pechstein, C.: A non-standard finite element method for convection-diffusion-reaction problems on polyhedral meshes. In: Application of Mathematics in Technical and Natural Sciences. AIP Conference Proceedings, vol. 1404, pp. 397–404 (2011). doi:10.1063/1.3659943
36. Huang, L., Liu, T.P.: A conservative, piecewise-steady difference scheme for transonic nozzle flow. Comput. Math. Appl. **12A**, 377–388 (1986)
37. Isaacson, E., Temple, B.: Convergence of the 2×2 Godunov method for a general resonant nonlinear balance law. SIAM J. Appl. Math. **55**, 625–640 (1995)
38. Jerome, J.W.: On uniform approximation by certain generalized spline functions. J. Approx. Theory **7**, 143–154 (1973)
39. Jerome, J.W.: Analysis of Charge Transport: Mathematical Theory of Semiconductor Devices. Springer, Berlin (1996)
40. Kimmeldorf, G.S., Wahba, G.: The correspondence between Bayesian estimation on stochastic processes and smoothing by splines. Ann. Math. Stat. **41**, 495–502 (1970)
41. Kita, E., Kamiya, N.: Trefftz method: an overview. Adv. Eng. Softw. **24**, 3–12 (1995)
42. Kuo, C.C.J., Levy, B.: Mode-dependent finite-difference discretization of linear homogeneous differential equations. SIAM J. Sci. Comput. **9**, 992–1015 (1988)
43. Lions, P.L., Mercier, B.: Splitting algorithms for the sum of two nonlinear operators. SIAM J. Numer. Anal. **16**, 964–979 (1979)
44. Liu, T.P.: Quasilinear hyperbolic systems. Commun. Math. Phys. **68**, 141–172 (1979)
45. Makarov, V.L., Samarskii, A.A.: Application of exact difference schemes to estimation of the convergence rate of the method of lines. Sem. Inst. Prikl. Mat. Dokl. **14**, 18–23 (1980)
46. McCartin, B.J.: Computation of exponential splines. SIAM J. Sci. Comput. **11**, 242–262 (1990)
47. Niijima, K.: A uniformly convergent difference scheme for a semilinear singular perturbation problem. Numer. Math. **43**, 175–198 (1984)
48. O'Malley, R.: Singular Perturbation Methods for Ordinary Differential Equations. Springer, Berlin (1991)
49. Osher, S.: Nonlinear singular perturbation problems and one-sided difference schemes. SIAM J. Numer. Anal. **18**, 129–144 (1981)
50. Pruess, S.: Estimating the eigenvalues of Sturm-Liouville problems by approximating the differential equation. SIAM J. Numer. Anal. **10**, 55–68 (1973)

51. Quarteroni, A., Valli, A.: Theory and application of Steklov-Poincaré operators for boundary-value problems. In: Spigler, R. (ed.) Applied and Industrial Mathematics. Venice-1, 1989, vol. 56, pp. 179–203. Kluwer, Dordrecht (1991)
52. Roos, H.-G.: Global uniformly convergent schemes for a singularly perturbed boundary-value problem using patched base spline-functions. J. Comput. Appl. Math. **29**, 69–77 (1990)
53. Roos, H.-G.: An analytically oriented discretization technique for boundary value problems. Abh. Math. Sem. Univ. Hamburg **61**, 139–152 (1991)
54. Roos, H.-G., Stynes, M., Tobiska, L.: Robust numerical methods for singularly perturbed differential equations. In: Convection-Diffusion-Reaction and Flow Problems. Springer Series in Computational Mathematics, vol. 24, 2nd edn. Springer, Berlin (2008). ISBN: 978-3-540-34466-7
55. Salama, A.A., Zidan, H.Z.: Fourth-order schemes of exponential type for singularly perturbed parabolic partial differential equations. Rocky Mt. J. Math. **36**, 1049–1068 (2006)
56. Samarskii, A.A.: The Theory of Difference Schemes. Monographs and Textbooks in Pure and Applied Mathematics, vol. 240. Dekker, New York (2001). ISBN: 0-8247-0468-1
57. Scharfetter, H.L., Gummel, H.K.: Large signal analysis of a silicon Read diode oscillator. IEEE Trans. Electron Devices **16**, 64–77 (1969)
58. Schmidt, E., Lancaster, P., Watkins, D.: Bases of splines associated with constant coefficient differential operators. SIAM J. Numer. Anal. **12**, 630–645 (1975)
59. Schultz, M.H., Varga, R.S.: *L*-splines. Numer. Math. **10**, 345–369 (1967)
60. Schumaker, L.L.: Spline Functions: Basic Theory, 3rd edn. Cambridge University Press, Cambridge (2007). ISBN: 978-0-521-70512-7
61. Shevaldin, V.-T.: Approximation by local *L*-splines corresponding to a linear differential operator of the second order. Proc. Steklov Inst. Math. **6**, S178–S196 (2006)
62. Sod, G.A.: A random choice method with application to reaction-diffusion systems in combustion. Comput. Math. Appl. **11**, 129–144 (1985)
63. Strauss, W.: Partial Differential Equations; An introduction, 2nd edn. Wiley, New York (2008). ISBN-13 978-0470-05456-7
64. Tikhonov, A.N., Samarskii, A.A.: Homogeneous difference schemes. USSR Comput. Math. Math. Phys. **1**, 5–67 (1962)
65. Tourin, A.: Splitting methods for Hamilton-Jacobi equations. Numer. Methods Partial Differ. Equ. **22**, 381–396 (2006)
66. Trefftz, E.: Ein gegenstuck zum Ritzschen verfahren. In: Proceedings of the 2nd International Congress for Applied Mechanics, Zurich, pp. 131–137 (1926)
67. van Leer, B.: On the relation between the upwind-differencing schemes of Godunov, Engquist-Osher and Roe. SIAM J. Sci. Stat. Comput. **5**, 1–20 (1984)
68. Wang, H.: Convergence of a numerical method for solving discontinuous Fokker–Planck equations. SIAM J. Numer. Anal. **45**, 1425–1452 (2007)
69. Weinan, E.: Homogenization of scalar conservation laws with oscillatory forcing terms. SIAM J. Appl. Math. **52**, 959–972 (1992)

Filtered Gradient Algorithms for Inverse Design Problems of One-Dimensional Burgers Equation

Laurent Gosse and Enrique Zuazua

Abstract Inverse design for hyperbolic conservation laws is exemplified through the 1D Burgers equation which is motivated by aircraft's sonic-boom minimization issues. In particular, we prove that, as soon as the target function (usually a N-wave) isn't continuous, there is a whole convex set of possible initial data, the backward entropy solution being possibly its centroid. Further, an iterative strategy based on a gradient algorithm involving "reversible solutions" solving the linear adjoint problem is set up. In order to be able to recover initial profiles different from the backward entropy solution, a filtering step of the backward adjoint solution is inserted, mostly relying on scale-limited (wavelet) subspaces. Numerical illustrations, along with profiles similar to F-functions, are presented.

1 An Elementary Modeling of Sonic Boom

The study of sound waves propagation in the atmosphere appears to trace back to Whitham's famous paper [46], where the so-called "F-function", representing an approximate, explicit pressure profile around a supersonic bullet, was derived. With some adjustments, that F-function was taken as the standard macroscopic pressure shape located in a close neighborhood of a plane jet: see for instance, [3, 19, 38, 39].

However, in order to propagate down to the ground the imprint a given, even approximate, pressure profile, one was still needing to solve inviscid Euler equations of gas dynamics. In Cleveland's Ph.D. thesis [13, Eq. (2.69), p. 36] (see also [40]),

L. Gosse (✉)
IAC, CNR, via dei Taurini, 19, 00185 Roma, Italy
e-mail: l.gosse@ba.iac.cnr.it

E. Zuazua
DeustoTech, Fundación Deusto, Avda Universidades, 24, 48007 Bilbao, Basque Country, Spain

Departamento de Matemáticas, Universidad Autónoma de Madrid, 28049 Madrid, Spain

Facultad Ingeniería, Universidad de Deusto, Avda. Universidades, 24, 48007 Basque Country, Spain
e-mail: enrique.zuazua@uam.es

L. Gosse, R. Natalini (eds.), *Innovative Algorithms and Analysis*,
Springer INdAM Series 16, DOI 10.1007/978-3-319-49262-9_7

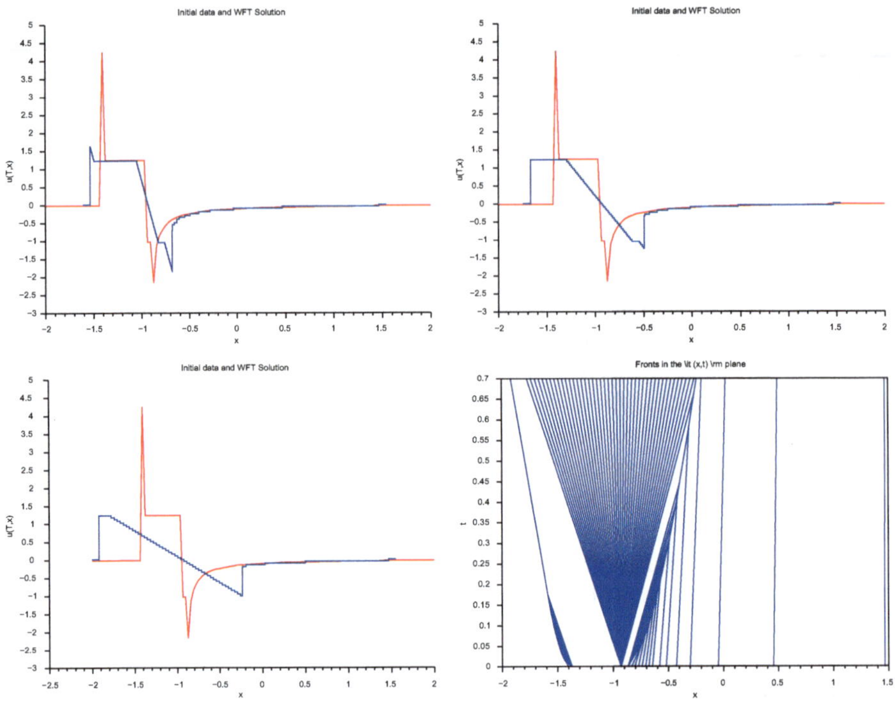

Fig. 1 Decay of a F-function onto a N-wave by solving a Burgers equation like (1)

a so-called "augmented Burgers equation" was derived, reading,

$$\partial_x P - P \partial_t P = \frac{1}{\Gamma} \frac{\partial^2 P}{\partial t^2} - \frac{1}{2A} \frac{\partial A}{\partial x} \cdot P, \tag{1}$$

the terms on the right-hand side accounting for absorption and ray-tube spreading, respectively. Hence, in the simplest case, it reduces to a Burgers equation with opposite velocity and the roles of space and time being exchanged. With empirical data of a standard F-function at hand, one can visualize the evolution induced by such an homogeneous, inviscid, Burgers equation, see Fig. 1; the simulation is performed by means of the wavefront-tracking algorithm [28], which doesn't contain any numerical viscosity, being not based on finite-differences. The dominant process is the spreading of the big rarefaction wave (induced by the entropy-violating downward jump initially located around $x \simeq -1$) separated by two shocks. Both the upward (in $x \simeq -1.5$) and downward ($x \simeq -0.8$) spikes, initially in the F-function, are quickly dissipated in order to let the N-wave develop. Such a N-wave stands for the self-similar large-time behavior (also called "intermediate asymptotics") of one-dimensional, genuinely nonlinear, scalar conservation laws.

This situation raises very delicate issues when it comes to trying to optimize the pressure profile around a jet plane (by slightly correcting its aerodynamic design) based on the constraint of minimizing its sound imprint on the ground. Such an optimal control problem is usually referred to as to both "sonic boom minimization", and the "inverse design problem". Despite the simplicity of the Burgers model (1), it appeared that the hopes of aiming at a rigorous inverse design procedure are discouragingly thin, as a consequence of irreversibility expressed by entropy dissipation beyond shock onset. This is illustrated, for instance, on Fig. 2, where very different initial data end up yielding the same shock at time $t \simeq 10$, even after quite complicated intermediate dynamics inside which a lot of entropy dissipates. Aware of that drawback, engineers decided to reduce some inverse design problems in aerodynamics to the more tractable optimization of a finite number of parameters inside the F-function, as advocated in [19] and followers, like e.g. [39]. Hereafter, we aim at taking a step back, by still planning to **perform a gradient-based optimization of a Burgers equation's initial data with respect to a given target at time $t = T$, but modulating its iterations by means of a specific filtering, driving the iterates toward minimizers belonging to one or another linear subspace of L^2**. Accordingly different shapes of initial profiles can be reconstructed: see for instance Fig. 7, where two different wavelet filters are applied during the totality of the optimization process in order to drive iterates onto rather different initial data, each one belonging to a different "scale-limited subspace" of L^2. Adequate filters can be drawn out of signal processing books, like [27, 34]: in particular, Prolate Spheroids [42, 47] and Father Wavelet functions [16, 36] furnish good candidates toward an efficient filtering of a gradient optimization algorithm. Strategies can be combined with one another in order to produce "hybridized algorithms": see for instance Fig. 9, where different wavelet filters are applied to the iterates. Moreover, PSWF and wavelets can even be merged into each other as in [45] (see also [25, §3.4]). Of course, applying filters in order to restore stability in the context of a control problem isn't new: see e.g. [23, 37]. Theoretical approaches to optimal design problems for hyperbolic equations in the presence of shocks are presented in [9, 10]. Application-driven references are e.g. [18, 20, 30, 31], whereas more specific results are provided in [5, 21, 22, 24].

2 Some Analytical Aspects of 1D Inverse Design

2.1 Irreversibility Versus Reverse Poincaré Inequality

As a preliminary step, consider the 1D linear wave equation,

$$\partial_{tt} y - \partial_{xx} y = 0, \qquad t > 0, \, x \in \mathbb{R},$$

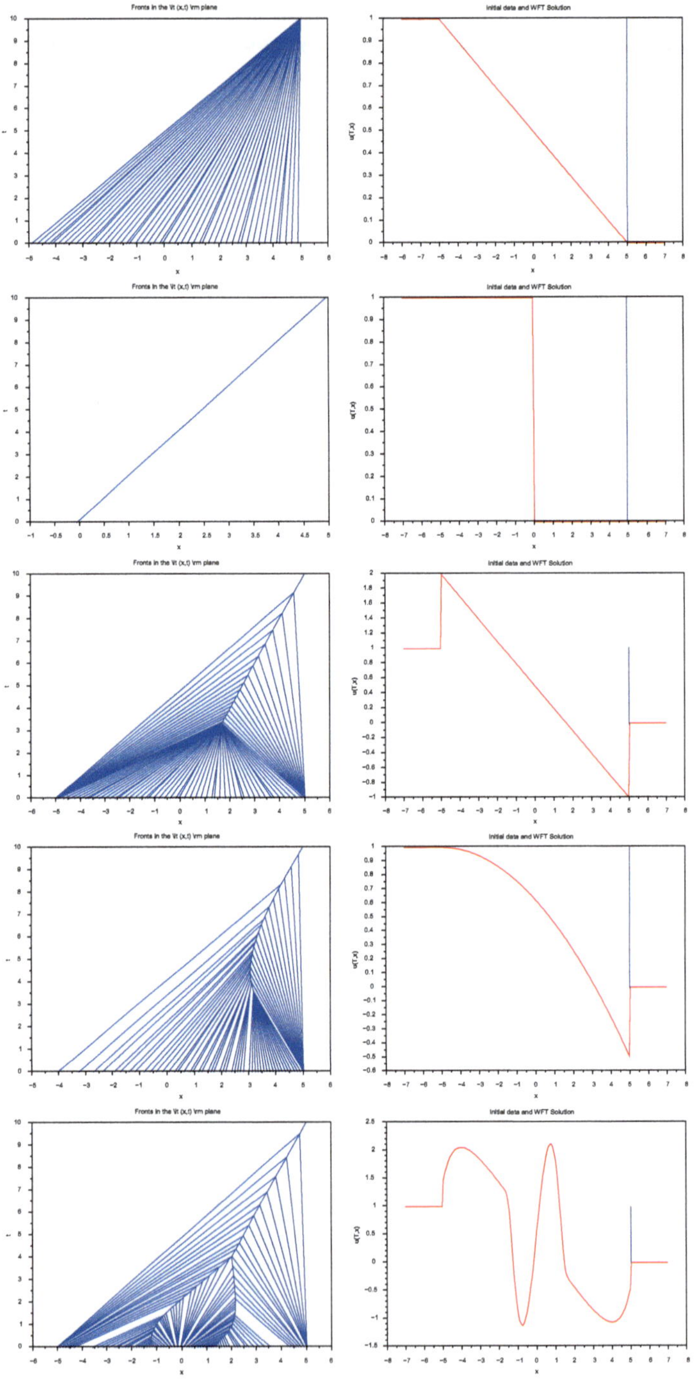

Fig. 2 Initial profiles (in *red*) yielding a single shock (*blue*) in $x = 5$ at $t = 10$

with enough decay at infinity so as to ensure conservation of energy:

$$E(t=T) = \int_{\mathbb{R}} |\partial_t y(T,x)|^2 + |\partial_x y(T,x)|^2 \, dx = E(t=0),$$

The equation, being well-posed in the backward sense, the inverse design problem has a unique solution, belonging to the same space as the prescribed target. Instead,

$$\partial_t y - \partial_{xx} y = 0, \qquad t > 0,\ x \in \mathbb{R}.$$

brings out a different type of energy identity since, in that case,

$$\frac{d}{dt}\int_{\mathbb{R}} \frac{y^2}{2} dx = -\int_{\mathbb{R}} \mid \partial_x y \mid^2 dx = -\Lambda(t) \parallel y(t) \parallel^2_{L^2(\mathbb{R})} . \tag{2}$$

The so-called "reverse Poincaré inequality" for the heat equation reads,

$$\forall t \geq 0, \qquad \Lambda(t) = \frac{\|\partial_x y(t)\|^2_{L^2}}{\|y(t)\|^2_{L^2}} \leq \Lambda(0) = \Lambda_0 ,$$

also called "Cacciopoli's inequality", [35, 44], and so that,

$$\frac{d}{dt}\int_{\mathbb{R}} \frac{y^2}{2} dx + \Lambda_0 \parallel y(t) \parallel^2_{L^2(\mathbb{R})} \geq 0,$$

hence the ratio of initial data to final (target) data grows exponentially with time:

$$\boxed{\forall t > 0, \qquad ||y(0)||^2 \leq \exp(2\Lambda_0\, t)||y(t)||^2,} \tag{3}$$

expressing strong ill-conditioning of the inverse design problem. This estimate is sharp, being the energy version of the Fourier series representation for an equation posed in a bounded domain. Most linear systems enjoy the property of backward uniqueness (except transport equations in bounded domains). For Burgers equation (more generally for genuinely nonlinear scalar laws) irreversibility manifests itself through gradient steepening and shock formation: as a (decreasing, $u^+ \leq u^-$) jump splits $\mathbb{R}$ into open sets $Q^{\pm}$, the pair (u, φ)= (flow solution, shock location) solves:

$$\begin{cases} \partial_t u + u\partial_x u = 0, & \text{in } Q^- \cup Q^+, \\ \dot{\varphi}(t) = \dfrac{u^+(\varphi(t),t) + u^-(\varphi(t),t)}{2}, & t \in (0,T), \\ \varphi(0) = \varphi^0 \in \mathbb{R},\ u(x,0) = u^0(x), & \text{in } \{x < \varphi^0\} \cup \{x > \varphi^0\}. \end{cases}$$

A less singular problem is the viscous Burgers equation, see (1), for which the Hopf-Cole transform [11, 14, 29] applies; consider $u^\epsilon(t,x)$ a classical solution of,

$$\partial_t u^\epsilon - \epsilon \partial_{xx} u^\epsilon + \partial_x |u^\epsilon|^2 = 0, \qquad | u^\epsilon(x,\, t) | + | \partial_x u^\epsilon(x,\, t) | \to 0 \text{ as } | x | \to \infty. \tag{4}$$

Since an antiderivative of u^ϵ,

$$v(x,\, t) = \int_{-\infty}^{x} u^\epsilon(s,\, t) ds, \qquad \text{solves} \qquad \partial_t v_t - \epsilon \partial_{xx} v + | \partial_x v |^2 = 0,$$

then, by rescaling the time variable, one gets that

$$z(x,t) = v(x,\, t/\epsilon)/\epsilon \qquad \text{solves} \qquad \partial_t z - \partial_{xx} z + | \partial_x z |^2 = 0.$$

At last, by defining

$$\eta(x,\, t) = \exp(-z(t,x)) = \exp(-v(x,\, t/\epsilon)/\epsilon),$$

one (maybe surprisingly!) recovers the linear heat equation,

$$\boxed{\partial_t \eta - \partial_{xx} \eta = 0. \quad \text{from which} \quad u^\epsilon(x,\, t) = -\epsilon \frac{\partial_x \eta(x,\, \epsilon t)}{\eta(x,\, \epsilon t)}.} \tag{5}$$

This classical argument suggests that, even in the absence of shocks as $u^\epsilon \in C^\infty$, the inverse design problem for (4) is exponentially ill-conditioned, too, as a consequence of (3). In the singular limit $\epsilon \to 0$, the situation deteriorates even more, because as soon as a shock appears, all the information about any initial profile which gave rise to it is irremediably lost: this loss of information is quantified through dissipation of entropy. In a numerical perspective, two issues emerge:

- Build efficient numerical solvers to find one (generally smooth) inverse design;
- Further, try to recover as many other possible inverse designs to study their set.

As an illustration, Fig. 2 displays several, quite different, initial profiles u_0 decaying, after having produced more or less intricate interaction patterns, onto a single entropy shock connecting 1 to 0 at time $t = 10$. Theoretical results are available by Adimurthi et al. [1], using the explicit Lax-Oleinik representation formula: see also related works by Ancona and Cannarsa [4] for Hamilton-Jacobi equations.

2.2 Convexity of the Set of Inverse Designs

Consider the simplest occurrence of (1), where, for ease of reading, space and time variables are recast into their usual setup: in smooth areas, $u(t,x)$ solves,

$$\partial_t u + u\partial_x u = 0, \qquad u(t=0,x) = u_0(x).$$

The method of characteristics stems on curves passing through each point (t,x),

$$\dot{X}(t) = u(t, X(t)), \quad \dot{u}(t, X(t)) = 0, \Rightarrow X(t) = X(t=0) + t \cdot u_0(X(t=0)).$$

However, as $u_0(X(t=0)) = u(t, X(t))$, the solution of Burgers equation reads,

$$u(t,x) = \frac{x-y}{t}, \qquad y \text{ the foot of the characteristic passing at } (t,x). \tag{6}$$

Actually, beyond shock onset, (6) still holds almost everywhere, [17, p. 145],

$$u(t,x) = \frac{x-\xi_u}{t}, \quad \inf_y \left(\frac{|x-y|^2}{2t} + \int^y u_0(s)ds \right) = \frac{|x-\xi_u|^2}{2t} + \int^{\xi_u} u_0(s)ds. \tag{7}$$

The functional which is minimized is the Lax function: according to [15, Chap. 11], given a point (t,x), the minimum on $\mathbb{R}$ is unique (and the backward characteristic is a shock-free straight line) when $u(t,x)$ is continuous. Oppositely, if $u(t,x)$ displays an entropy shock, there are two distinct minima $\xi_u^\pm$, with $\xi_u^- < \xi_u^+$, another manifestation of Oleinik's condition: as $u(t,x^\pm) = u_0(\xi^\pm)$, by differentiating,

$$\frac{u_0(\xi^+) - u_0(\xi^-)}{\xi^+ - \xi^-} = \frac{u(t,x^+) - u(t,x^-)}{\xi^+ - \xi^-} = \frac{1}{t}.$$

An illustration is provided by Fig. 3, for the special case of Burgers equation with initial data $u_0(x) = 0.1 - \tanh(x)$, so that the Lax function reads,

$$G(y;t,x) = \frac{|x-y|^2}{2t} - \log\cosh y + 0.1\, y.$$

The figure displays the different minima for $x \equiv 0.2$ with $t = 0.5, 1, 2, 4$. At time $t \simeq 2$, the shock appears and G is endowed with two distinct (global) minimizers.

The existence of inverse designs, given a specific target at time $T > 0$ supported inside an interval of $\mathbb{R}$, was studied by the authors of Adimurthi [1, Theorem 1.2]:

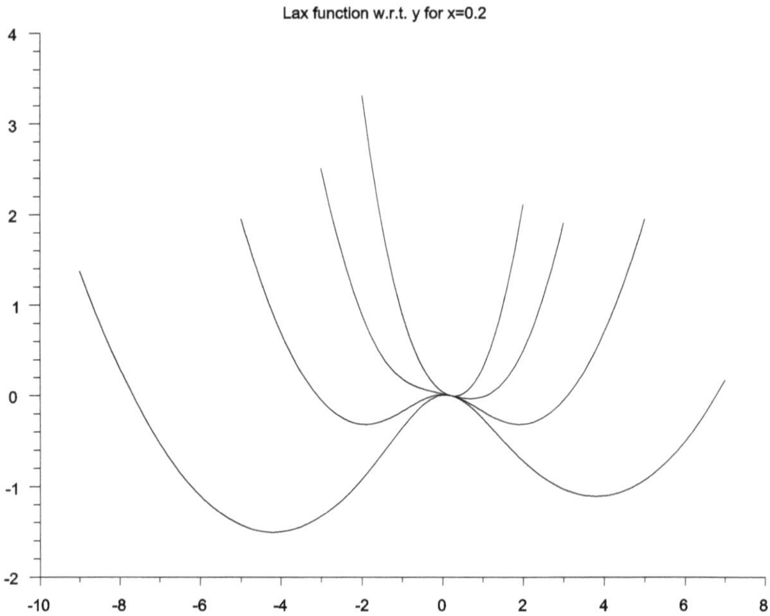

Fig. 3 Lax function with $u_0(x) = 0.1 - \tanh(x)$ and $x = 0.2$, $t = 0.5, 1, 2, 4$

Theorem 1 *Let ρ be a locally bounded, non-decreasing, possibly discontinuous function, and, given a time $T > 0$, consider the BV_{loc}, entropy admissible, target,*

$$\forall x \in (X_1, X_2), \qquad \bar{u}(x) = \frac{x - \rho(x)}{T}, \qquad \rho' \geq 0.$$

Then under the following condition,

$$\rho(X_1) \geq A_1 > Y_1, \qquad \rho(X_2) \leq A_2 < Y_2,$$

there exists an inverse design $u_0 \in L^\infty(Y_1, Y_2)$ yielding at time T an entropy solution of Burgers equation, whose restriction to the interval (X_1, X_2) is exactly $\bar{u}$.

Being a difference of two non-decreasing, locally bounded functions, $\bar{u} \in BV_{loc}(\mathbb{R})$; by construction it is entropy admissible because it cannot contain increasing shocks. Theorem 1 does not provide a characterisation of all the possible inverse designs, neither additional information about the structure of the set of inverse designs.

Theorem 2 *Let u_0, v_0 be two $L^\infty(\mathbb{R})$ inverse designs; assume that, at a given time $T > 0$, their respective, piecewise-continuous, targets $\bar{u}(\cdot)$ and $\bar{v}(\cdot)$ coincide, i.e. $u(T,x) = v(T,x)$. Then, for any $\alpha \in (0,1)$, any convex combination $w_0^\alpha = \alpha u_0 +$*

$(1-\alpha)v_0$ yields the same target at the same time, too,

$$\forall x \in \mathbb{R}, \qquad w^\alpha(T,x) = u(T,x) = v(T,x).$$

Given any reachable target, the union of its inverse designs is a convex set of $L^\infty(\mathbb{R})$.

If the target is a non-negative N-wave, the results in [33] ensure that a Dirac atom can be considered an inverse design (in a weaker sense of bounded measures), too.

Proof It splits into three main steps:

- First, there is no restriction in considering regularized, one-sided Lipschitz, initial data, defined like $u_0^\varepsilon = S_\varepsilon u^0$, $v_0^\varepsilon = S_\varepsilon u_0$ in order to have pointwise values at hand. For $u^\varepsilon(t,\cdot) = S_\varepsilon u_0$ and similar for $v^\varepsilon(t,\cdot)$, Kružkov contraction yields

$$\forall t > 0, \qquad \|u(t,\cdot) - u^\varepsilon(t,\cdot)\|_{L^1} \le \|u_0 - u_0^\varepsilon\|_{L^1} \to 0, \quad \varepsilon \to 0.$$

- Yet, consider the representation (7) for both $u(t,\cdot)$ and $v(t,\cdot)$: accordingly, ξ_u and ξ_v are mappings $(t,x) \mapsto \xi_u(t,x)$ or $\xi_v(t,x)$ (yielding the foot of the corresponding characteristic where the target is smooth). At any point $t = T$, $x \in \mathbb{R}$,

$$u(t,x) = v(t,x) \Rightarrow \xi_u(t,x) = \xi_v(t,x) = \text{ some value } \xi \in \mathbb{R}.$$

If T, x is a shock location, each minimum is reached at two distinct locations,

$$\xi_u^\pm(t=T,x) = \xi_v^\pm(t=T,x) = \xi^\pm \in \mathbb{R}^2.$$

- Now, applying again (7) to $w_0^\alpha = \alpha u_0 + (1-\alpha)v_0$ should bring $\xi_w = \xi$ with,

$$\begin{aligned}
\inf_y \left(\frac{|x-y|^2}{2t} + \int^y (\alpha u_0(s) + (1-\alpha)v_0(s))ds \right) \\
= \frac{|x-\xi_w|^2}{2t} + \int^{\xi_w} (\alpha u_0(s) + (1-\alpha)v_0(s))ds \quad . \\
= \alpha \left(\frac{|x-\xi_w|^2}{2t} + \int^{\xi_w} u_0(s) \right) + (1-\alpha) \left(\frac{|x-\xi_w|^2}{2t} + \int^{\xi_w} v_0(s))ds \right).
\end{aligned}$$

Consequently, since $\inf(A+B) \ge \inf(A) + \inf(B)$, where

$$A = \alpha \left(\frac{|x-y|^2}{2t} + \int^y u_0(s)ds \right), \qquad B = (1-\alpha) \left(\frac{|x-y|^2}{2t} + \int^y v_0(s)ds \right).$$

and both A, B are minimized at the same point $\xi_u = \xi_v = \xi$, then equality holds:

$$\inf_y \left(\frac{|x-y|^2}{2t} + \int^y (\alpha u_0(s) + (1-\alpha) v_0(s)) ds \right) \geq$$

$$\alpha \left(\frac{|x-\xi|^2}{2t} + \int^\xi u_0(s) \right) + (1-\alpha) \left(\frac{|x-\xi|^2}{2t} + \int^\xi v_0(s)) ds \right) =$$

$$\frac{|x-\xi|^2}{2t} + \int^\xi (\alpha u_0(s) + (1-\alpha) v_0(s)) ds.$$

That inequality shows that the value of $G_{w^\alpha}(t,x,\xi)$ is smaller than its infimum on y; hence $y = \xi$ is automatically a critical point for $w^\alpha(t,x)$. The same holds for $\xi^\pm$.

Besides Fig. 2, which already illustrates the variety of inverse designs associated to the simple target $\chi_{x<5}$, at time $T = 10$, Fig. 4 shows several convex combinations

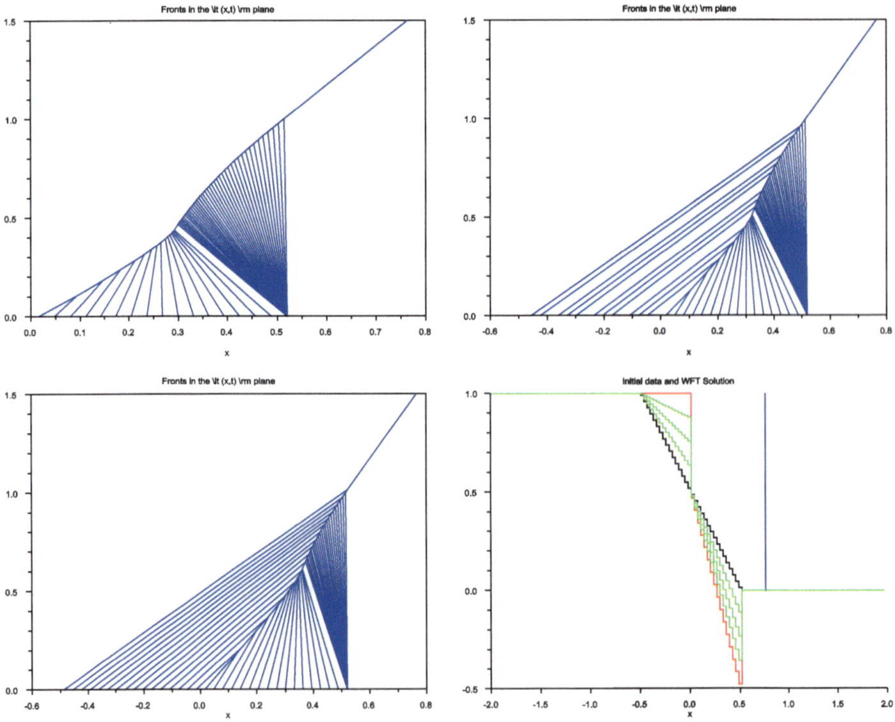

Fig. 4 Time evolution of convex combinations of continuous and discontinuous "ramps"

of two elementary inverse designs, (χ_A being the indicator of the set A)

$$u_0(x) = \chi_{x<0} + (\frac{1}{2} - 2x)\chi_{0<x<\frac{1}{2}}, \qquad v_0(x) = \min(1, \frac{1}{2} - x)\chi_{x<\frac{1}{2}},$$

which all reach the same target at time, namely a single shock at time $t \simeq 1.2$.

2.3 Backward Entropy Solution as the Centroid

An elementary observation is that, given a smooth target $\bar{u} \in C^\infty(\mathbb{R})$ and a time $T > 0$ small enough, the inverse design problem is solved by "tracing characteristics backwards". More precisely, such a solution is $u(t, X(t)) = \bar{u}(X(T))$ with,

$$\dot{X}(t) = u(t, X(t)), \text{ so that } X(t) = X(T) + (t - T)u(T, X(T)), \quad 0 \leq t < T.$$

Such a process is actually equivalent to solving Burgers equation backward in time,

$$\partial_t u - \partial_x (u^2/2) = 0, \qquad u(T, x) = \bar{u}(x). \tag{8}$$

This situation corresponds to the convex set of Theorem 2 being reduced to a single point. Yet, when the target $\bar{u}$ is an admissible entropy (hence one-sided Lipschitz) BV function, more elaborate techniques (generalized characteristics, Lax formula, ...) allow to derive an inverse design by means of the entropy solution of (8): see first row in Fig. 2. Clearly, as the flux function changes its sign, any entropy admissible (decreasing) jump in $\bar{u}$ becomes entropy-violating for (8) and is instantaneously converted in a (smooth) backward expansion wave. In contrast, the other solutions shown on Fig. 2 all display increasing (entropy-violating) jumps at time $t = 0$. This suggests that this particular inverse design from (8) may be the centroid of the convex set $K \in L^\infty(\mathbb{R})$ containing all the possible inverse designs of a discontinuous target: by definition, the centroid is the barycenter,

$$C = \frac{\int_K y\,dy}{\int_K dy}, \qquad C(x) \text{ an inverse design of } \bar{u}(x).$$

Let η stand for any convex entropy (for instance $\eta(u) = u^2$): by Jensen's inequality,

$$\int_{\mathbb{R}} \eta(C(x))dx = \int_{\mathbb{R}} \eta\left(\frac{\int_K y\,dy}{\int_K dy}\right) dx \leq \int_K \frac{\int_{\mathbb{R}} \eta(y(x))\,dx}{\int_K dy}\,dy,$$

so the centroid $C(x)$ is the inverse design of the discontinuous, entropy admissible, target $\bar{u}$ minimizing a convex entropy η with respect to all the others in K.

Remark 1 The relevance of the backward entropy solution of (8) is limited to **"small targets" or "small times"**, an assumption tacitly made in Theorem 1:

$$\frac{d\rho}{dx} \geq 0 \quad \Rightarrow \quad \frac{d\bar{u}}{dx} \leq \frac{1}{T} \text{ where the target } \bar{u}(x) = \frac{x - \rho(x)}{T}. \tag{9}$$

Indeed, if $\bar{u}$ contains entropy jumps, the backward evolution cannot induce wave cancellations (see first and second rows of Fig. 5), so the backward entropy solution

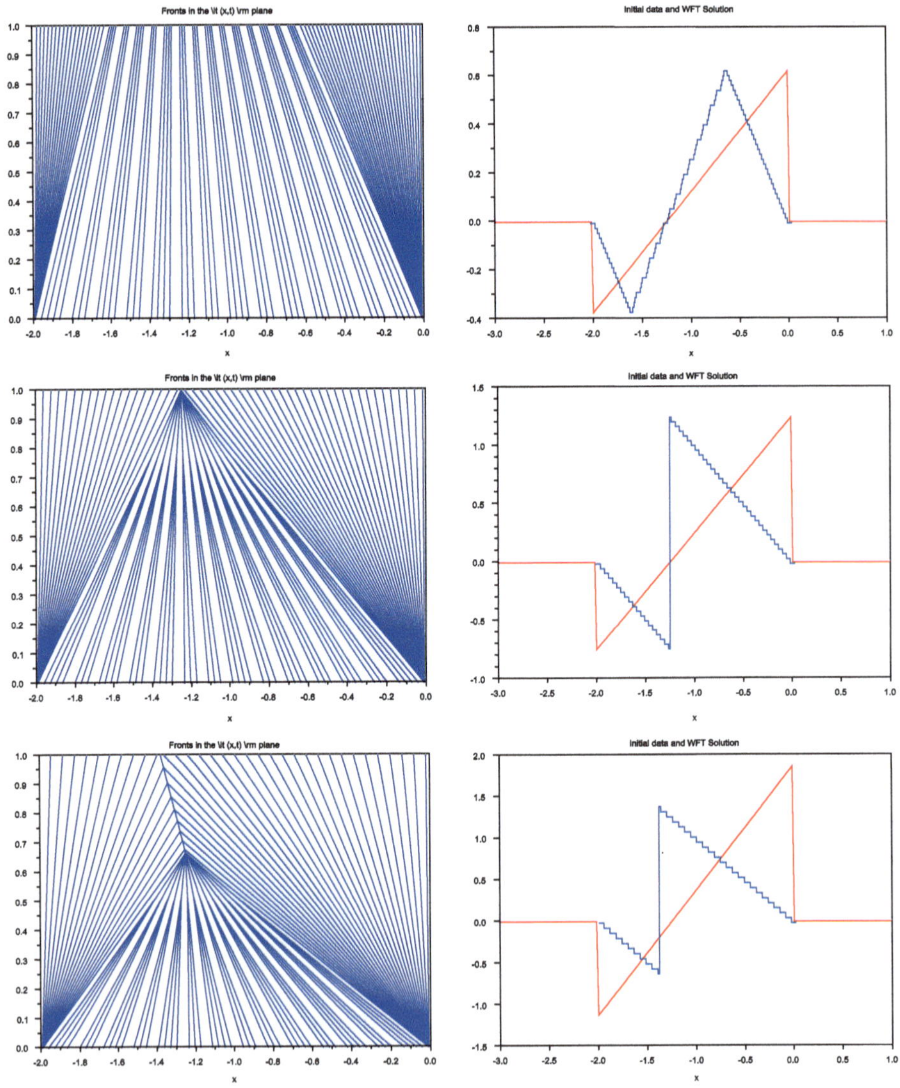

Fig. 5 Dissipation in backward entropy solutions of (8) with $\bar{u}(x)$ three bigger N-waves

realizes an inverse design. Oppositely, the third row of Fig. 5 shows that, when such a smallness assumption is broken, that solution isn't an inverse design anymore because too much entropy, like the L^2 norm, was dissipated in the discontinuity.

Corollary 1 *Under the assumptions of Theorem 1, let $u(t,x)$ be the "backward entropy solution" solving (8) with final data $u(T,\cdot) = \bar{u}$, then*

- *for all $t \in (0,T)$, $u(t,\cdot)$ is Lipschitz continuous;*
- *entropy is preserved during the backward evolution;*
- *$u(0,\cdot)$ is an inverse design, so that $u_0 = u(0,\cdot)$.*

However, $\bar{u}$ can display decreasing shocks, and u_0, increasing ones.

Proof First, as $\bar{u}$ is entropy-admissible, it may contain only decreasing discontinuities which are all instantaneously regularized as rarefaction waves in the backward evolution. Then, the smallness assumption (9) implies that the entropy solution of (8) cannot blow up in the open interval $(0,T)$ because its blowup time satisfies,

$$\bar{T} = \frac{1}{\sup_x \bar{u}(x)} \geq T.$$

A numerical discussion of inverse designs for "big targets" was given in [2, §6.4].

3 A PDE-Constraint Optimization Process

3.1 Reduced Objective Functional and Control-to-State Map

Inverse design consists in reconstructing an initial profile u_0 at time $t = 0$ able to match (or to get as close as possible to) some given target function $\bar{u}$ at time $t = T \in \mathbb{R}^+$. It can be formalized as a PDE-constrained optimization process for an "objective functional" $\mathcal{J}_T$ and a control $u_0 \in U$, namely

$$\inf_U \mathcal{J}_T(u) = \frac{1}{2}\int |u(T,x) - \bar{u}(x)|^2 dx, \qquad T > 0,$$

subject to the (presumably infinite set of) constraints,

$$\partial_t u + \partial_x(u^2/2) = 0, \qquad u(t=0,\cdot) = u_0, \qquad \partial_t \eta(u) + \partial_x q(u) \leq 0.$$

Here, the entropy η stands for any convex function and q, the entropy flux is such that $q'(u) = u \cdot \eta'(u)$. Since Burgers' flux is strictly convex, only one convex entropy is actually needed because just one entropy inequality yields all the other ones. So the optimization process contains only one equality and one inequality, say for $\eta(u) = u^2$. The PDE rewrites $e(u, u_0) = 0$ and is referred to as an "equation of state". That equation of state $e(\cdot,\cdot)$ rewrites as a mapping such that,

$$e : L^1 \cap BV(\mathbb{R}) \times L^1 \cap L^\infty(\mathbb{R}) \to \mathcal{M}(\mathbb{R}),$$

where, for any $\phi(t,x) \in C^1(\mathbb{R} \times \mathbb{R}^+)$ with compact support,

$$\int\int_{\mathbb{R}} u\partial_t\phi(t,x) + \frac{u^2}{2}\partial_x\phi(t,x)dx.dt + \int_{\mathbb{R}} u_0(x)\phi(t=0,x)dx = 0.$$

For any $t \geq 0$, let $u(t,\cdot) = S_t(u_0)$ stand for the entropy solution (in the sense of Kružkov) of the inviscid 1D scalar Burgers equation,

$$\partial_t u + \partial_x(u^2/2) = 0, \qquad \partial_t(u^2/2) + \partial_x(u^3/3) \leq 0, \qquad x \in \mathbb{R},\ t > 0.$$

Being a semi-group of L^1 contractions, it implies a BV estimate,

$$\forall h > 0, \qquad \|u(t,\cdot+h) - u(t,\cdot)\|_{L^1} \leq \|u_0(\cdot+h) - u_0(\cdot)\|_{L^1} \leq h \cdot TV(u_0),$$

where $TV(u_0)$ stands for the total variation of u_0. So, $BV \subset L^\infty$ is a stable space for this convex scalar law: $u_0 \in L^1 \cap BV(\mathbb{R}) \subset L^2$ (by classical interpolation inequalities) produces an entropy solution lying in $L^2(\mathbb{R})$, too. Accordingly, the "control-to-state map" S_T is defined along with a "reduced objective functional",

$$\tilde{\mathcal{J}}_T(u_0) = \frac{1}{2}\int |S_T(u_0)(x) - \bar{u}(x)|^2 dx = \mathcal{J}_T(\text{“}u(u_0)\text{''}), \tag{10}$$

inside which the PDE dynamics are already encoded (in the form of $S_T(u_0)$). In particular, for a given $T > 0$, one may consider the minimization problem,

$$\text{given } \bar{u}, \qquad \inf_{u_0 \in U} \tilde{\mathcal{J}}_T(u_0), \qquad U = L^1 \cap BV(\mathbb{R}) \subset L^2(\mathbb{R}),$$

which can be expected to admit many solutions (see Theorem 2) when the target function $\bar{u}$ contains entropy shocks.

3.2 *Regularized Objective Functionals and Duality Solutions*

Duality solutions is a peculiar notion of weak solutions to one-dimensional continuity equations enjoying both uniqueness and weak stability properties; they were proposed in [8], then applied to differentiability issues for convex scalar conservation laws in [9]. Finally their numerical analysis was tackled in [26]. Following both [12] and [9], one can consider the "sensitivity problem" for Burgers equation: let $u_0 \in L^1 \cap L^\infty(\mathbb{R})$ be initial data, and $u(T,\cdot) = S_T(u_0)$ its entropy solution at time $t = T > 0$. A small deviation δu_0 and a parameter $0 < \eta \ll 1$, yield new data,

$$u_0^\eta = u_0 + \eta \cdot \delta u_0, \qquad u^\eta(T,\cdot) = S_T(u_0^\eta) \dot{=} u(T,\cdot) + \eta \cdot \delta u,$$

Clearly, by defining $\delta u(t,\cdot)$ as the perturbation at time $0 \le t \le T$,

$$\delta u(t,\cdot) = \frac{S_t(u_0^\eta) - S_t(u_0)}{\eta}, \quad \partial_t(\delta u) + \partial_x(a(t,x)\cdot \delta u) = 0, \text{ with } a = \frac{|u^\eta|^2 - |u|^2}{2(u^\eta - u)},$$

one finds that it satisfies a linear continuity equation, with a possibly discontinuous coefficient, though, so that uniqueness is a delicate issue.

Definition 1

- Let $m(t = 0,\cdot) = m_0$ a bounded measure on $\mathbb{R}$ and a bounded velocity field $a(t,x) \in L^\infty$, a weak solution $m(t,\cdot)$ is a duality solution to

$$\partial_t m + \partial_x(a \cdot m) = 0, \qquad \partial_x a(t,\cdot) \le \alpha(t), \text{ with } \alpha \in L^1(]0,T[),$$

  if it satisfies, for any $0 < \tau \le T$ and any $p(t,x)$ reversible solution in $(0,\tau)\times\mathbb{R}$ with compact support in x,

$$\frac{d}{dt}\int_{\mathbb{R}} p(t,x)m(t,x)dx = 0, \qquad t \in [0,\tau]. \tag{11}$$

- under the same assumptions, given a Lipschitz final datum p^T, a reversible solution to the dual transport equation is a Lipschitz solution of,

$$\partial_t p + a \cdot \partial_x p = 0, \qquad p(T,\cdot) = p(\cdot),$$

  which is equivalently characterized by $TV(p(t,\cdot)) = TV(p^T)$ or $p = p_1 - p_2$, a difference of two non-decreasing Lipschitz solutions.

Thanks to these notions, Bouchut and James stated in [9] the following:

Theorem 3 *Let u_0 be entropy admissible, i.e. $\partial_x u_0(\cdot) \le C \in \mathbb{R}$, then as $\eta \to 0$, the perturbation $\delta u(t,\cdot) \to m(t,\cdot)$, which is the duality solution to,*

$$\partial_t m + \partial_x(S_t(u_0)\cdot m) = 0, \qquad m(t = 0,\cdot) = \delta u_0.$$

Accordingly, $m(t,\cdot)$ is the directional derivative of $S_t(u_0)$ in the direction δu_0.

Especially, that duality solution satisfies, for any (Lipschitz) reversible solution p,

$$\int_{\mathbb{R}} p(T,x)m(T,x)dx = \int_{\mathbb{R}} p(t=0,x)m(t=0,x)dx = \int_{\mathbb{R}} p(0,x)\delta u_0(x)dx. \tag{12}$$

Yet, following [12], let an objective functional and its Gâteaux derivative along δu_0,

$$\mathscr{J}_T(u^\eta) = \frac{1}{2}\int_{\mathbb{R}} |S_T(u_0^\eta) - \bar{u}(x)|^2 dx, \qquad \delta\mathscr{J}_T = \int_{\mathbb{R}} \delta u(T,\cdot)\cdot(S_T(u_0^\eta) - \bar{u})dx.$$

An essential property for inverse design feasibility is (12) holding in the case,

$$p(T,\cdot) = S_T(u_0^\eta) - \bar{u}, \text{ so that } \delta\mathscr{J}_T = \int_{\mathbb{R}} p(0,x)\delta u_0(x)dx = \langle p(0,\cdot), \delta u_0\rangle .$$

Theorem 3 states that, being δu in general a (measure) duality solution, such a property can be expected to hold mostly for p being a (Lipschitz) reversible solution, see (11). As $S_T(u_0^\eta) - \bar{u}$ is likely to be discontinuous, setting up "regularized objective functionals", involving some convolution kernel, secures the Lipschitz regularity of each $p(T,\cdot)$ and the crucial property (12), too: for instance, let

$$0 \le \mu^\varepsilon(x) = \frac{1}{\varepsilon\sqrt{\pi}}\exp(-|x/\varepsilon|^2) \in C^\infty(\mathbb{R}),$$

in order to define "ε-regularized objective functions",

$$\mathscr{J}_T^\varepsilon(u) = \frac{1}{2}\int \left|[\mu^\varepsilon * u(T,\cdot)] - \bar{u}(x)\right|^2 dx = \frac{1}{2}\int \left|[\mu^\varepsilon * S_T(u_0)] - \bar{u}(x)\right|^2 dx. \tag{13}$$

Clearly, for $\varepsilon \ll 1$, this type of perturbation of the L^2 norm remains negligible, especially when it comes to numerics. More precisely, for most applications, a grid size $\Delta x > 0$ is fixed, so that the correction (13) becomes transparent if $\varepsilon \ll 1$.

3.3 *Critical Points of Reduced Objective Functionals*

Even in the favorable case where $\mathscr{J}_T$ is strictly convex on $L^2(\mathbb{R})$, a numerical algorithm remains too costly as long as it needs to compute gradients in all the directions corresponding to any Hilbertian basis functions $\varphi_i(x)$. So, an operator $\nabla\tilde{\mathscr{J}}_T$ (a linear form in the dual of $L^2(\mathbb{R})$) expressing the first variation of $\mathscr{J}_T$,

$$\forall\varphi_i, \qquad \delta\mathscr{J}_T(u)[\varphi_i] = \left\langle \nabla\tilde{\mathscr{J}}_T(u_0), \varphi_i\right\rangle = \int_{\mathbb{R}} \nabla\tilde{\mathscr{J}}_T(u_0)\,\varphi_i(x)dx,$$

would be very useful. Formally, using (10), it comes,

$$\delta \mathscr{J}_T(u = S_T(u_0))[\varphi_i] = \left\langle \nabla_u \mathscr{J}_T(u), \nabla S_T(\varphi_i) \right\rangle,$$

by means of the chain rule. By taking the adjoint in the duality bracket,

$$\delta \mathscr{J}_T(S_T(u_0))[\varphi_i] = \left\langle (\nabla S_T)^*[\nabla_u \mathscr{J}_T(u)], \varphi_i \right\rangle,$$

so that our desired operator reads $\nabla \tilde{\mathscr{J}}_T(u_0) = (\nabla S_T)^*[\nabla_u \mathscr{J}_T(u)]$. In this expression, ∇S_T stands for the linearized evolution operator associated to the Burgers equation, that is, the solution of the transport equation

$$\partial_t v + u(t,x)\partial_x v = 0, \qquad t \in (0,T).$$

Yet, the adjoint $(\nabla S_T)^*[\nabla_u \mathscr{J}_T(u)]$ corresponds to the backward problem,

$$-\partial_t p - u(t,x)\partial_x p = 0, \quad p(T,x) = u(T,x) - \bar{u}(x) = S_T(u_0)(x) - \bar{u}(x).$$

In particular, in the dual space of $L^2(\mathbb{R})$, $(\nabla S_T)^*[\nabla_u \mathscr{J}_T(u)] = p(t=0,\cdot)$,

$$\boxed{\forall \varphi_i, \qquad \left\langle \nabla \tilde{\mathscr{J}}_T(u_0), \varphi_i \right\rangle = \langle p(t=0,\cdot), \varphi_i \rangle .} \tag{14}$$

Accordingly, $\nabla \tilde{\mathscr{J}}_T^\varepsilon$ is a linear form on $L^2(\mathbb{R})$ such that,

$$\forall \varphi_i, \qquad \delta \mathscr{J}_T^\varepsilon(u)[\varphi_i] = \left\langle \nabla \tilde{\mathscr{J}}_T^\varepsilon(u_0), \varphi_i \right\rangle.$$

As μ^ε is an **even function**, similar computations yield for any φ_i,

$$\begin{aligned} \delta \mathscr{J}_T^\varepsilon(u^\varepsilon = \mu^\varepsilon * S_T(u_0))[\varphi_i] &= \left\langle \nabla_u \mathscr{J}_T(u^\varepsilon), \mu^\varepsilon * [\nabla S_T(\varphi_i)] \right\rangle \\ &= \left\langle \mu^\varepsilon * [\nabla_u \mathscr{J}_T(u^\varepsilon)], \nabla S_T(\varphi_i) \right\rangle \\ &= \left\langle (\nabla S_T)^*[\mu^\varepsilon * \nabla_u \mathscr{J}_T(u^\varepsilon)], \varphi_i \right\rangle, \end{aligned}$$

and finally, the expression of the modified gradient operator follows,

$$\nabla \tilde{\mathscr{J}}_T^\varepsilon(u_0) = (\nabla S_T)^*[\mu^\varepsilon * \nabla_u \mathscr{J}_T(u^\varepsilon)], \qquad u^\varepsilon(T,x) = [\mu^\varepsilon * S_T(u_0)](x).$$

The backward equation remains unchanged, except for its final data, so

$$\boxed{\left\langle \nabla \tilde{\mathscr{J}}_T^\varepsilon(u_0), \varphi_i \right\rangle = \langle p^\varepsilon(0,\cdot), \varphi_i \rangle \text{ with } p^\varepsilon(T,\cdot) = \mu^\varepsilon * [\mu^\varepsilon * S_T(u_0) - \bar{u}]} \tag{15}$$

The **velocity field** $u(t,\cdot)$ **in** $\partial_t p^\varepsilon + u(t,x)\partial_x p^\varepsilon = 0$ **is not regularized**: by Oleinik's form of the entropy condition, it may display downward jumps, so that, to ensure stability, $p^\varepsilon(t,x)$ should be interpreted as a "reversible solution", see Definition 1.

4 Minimizing Sequences in Filtered Gradient Algorithms

4.1 Large-Time Behavior of Numerical Schemes

Let us consider now numerical approximation schemes for the inviscid problem:

$$\begin{cases} u_j^{n+1} = u_j^n - \dfrac{\Delta t}{\Delta x}\big(g(u_j^n, u_{j+1}^n) - g(u_{j-1}^n, u_j^n)\big), & j \in \mathbb{Z}, \quad n > 0. \\ u_j^0 = \frac{1}{\Delta x}\int_{(j-1/2)\Delta x}^{(j+1/2)\Delta x} u_0(x)dx, & j \in \mathbb{Z}. \end{cases}$$

The approximate, piecewise-constant, numerical solution $u_\Delta(t,x)$ is given by

$$u_\Delta(t,x) = u_j^n, \qquad (j-1/2)\Delta x < x < (j+1/2)\Delta x, \qquad n\Delta t \le t < (n+1)\Delta t,$$

Are corresponding large-tine dynamics discrete analogues of continuous ones?

1. Lax-Friedrichs

$$g^{LF}(u,v) = \frac{u^2+v^2}{4} - \frac{\Delta x}{\Delta t}\left(\frac{v-u}{2}\right),$$

2. Engquist-Osher

$$g^{EO}(u,v) = \frac{u(u+|u|)}{4} + \frac{v(v-|v|)}{4}, \tag{16}$$

3. Godunov

$$g^{G}(u,v) = \begin{cases} \min\limits_{w\in[u,v]} \frac{w^2}{2}, & \text{if } u \le v, \\ \max\limits_{w\in[v,u]} \frac{w^2}{2}, & \text{if } v \le u. \end{cases}$$

These methods converge in $L^1_{loc}(\mathbb{R}\times\mathbb{R}^+)$**, i.e. on all compact subsets: do they behave correctly as** $t \to \infty$**? Such a question was already raised in [41, §6.4]**. It has consequences for numerical inverse design problems because a gradient descent method needs an iterative resolution of both forward and adjoint dynamics repeatedly, possibly over long time intervals. Three-point monotone schemes rewrite like

$$\frac{u_j^{n+1}-u_j^n}{\Delta t}+\frac{(u_{j+1}^n)^2-(u_{j-1}^n)^2}{4\Delta x}=R(u_j^n,u_{j+1}^n)-R(u_{j-1}^n,u_j^n)$$

where the numerical viscosity R is defined in a unique manner as

$$R(u,v)=\frac{Q(u,v)(v-u)}{2}=\frac{1}{2\Delta x}\Big(\frac{u^2}{2}+\frac{v^2}{2}-2g(u,v)\Big).$$

Accordingly, the following results were proved in [30]

Theorem 4 *Let $u_0 \in L^1\cap L^\infty(\mathbb{R})$ and $\Delta t\|u^n\|_{\infty,\Delta}\le \Delta x$: for any $p\in[1,\infty)$,*

- *the numerical solution u_Δ given by the Lax-Friedrichs scheme satisfies*

$$\lim_{t\to\infty} t^{\frac{1}{2}(1-\frac{1}{p})}\|u_\Delta(t)-w(t)\|_{L^p(\mathbb{R})}=0, \tag{17}$$

where the diffusion wave $w=w_{M_\Delta}$, $M_\Delta=\int_{\mathbb{R}}u_\Delta^0$, is the unique solution of

$$\begin{cases}\partial_t w+\partial_x\Big(\frac{w^2}{2}\Big)=\frac{(\Delta x)^2}{2\Delta t}\partial_{xx}w, & x\in\mathbb{R},\quad t>0,\\ w(t=0,x)=M_\Delta\delta(x), & \text{the Dirac mass in } x=0.\end{cases} \tag{18}$$

- *the ones given by both Engquist-Osher and Godunov schemes satisfy a similar asymptotic behavior with an $N-$ wave, $w=w_{p_\Delta,q_\Delta}$, being unique solution of*

$$\begin{cases}\partial_t w+\partial_x\Big(\frac{w^2}{2}\Big)=0, \quad x\in\mathbb{R},\quad t>0,\\ w(t=0,x)=M_\Delta\delta(x),\quad \lim\limits_{t\to0}\displaystyle\int_0^x w(t,z)dz=\begin{cases}0, & x<0,\\ -p_\Delta, & x=0,\\ q_\Delta-p_\Delta, & x>0,\end{cases}\end{cases} \tag{19}$$

with $M_\Delta=\int_{\mathbb{R}}u_\Delta^0$, $p_\Delta=-\min_{x\in\mathbb{R}}\int_{-\infty}^x u_\Delta^0(z)dz$, and $q_\Delta=\max_{x\in\mathbb{R}}\int_x^\infty u_\Delta^0(z)dz$.

These quantitative results confirm, in an accurate manner, the observation formerly made in [41, §6.4]: “*Neither of these two results [Error Estimates in Theorems 6.4.1 and 6.4.2] is uniform with respect to the time and indeed they could not be*”.

4.2 Reversible Solutions of the Adjoint Problem

A classical tool to build a minimizing sequence $(u_0^n)_{n\in\mathbb{N}}$ is the fixed-step gradient algorithm: let $0 < \eta \ll 1$ be small enough as an iteration step, (see e.g. [6, 7])

$$u_0^{n+1} = u_0^n - \eta \nabla \tilde{\mathscr{J}}_T(u_0^n) \quad \Rightarrow \quad \tilde{\mathscr{J}}_T(u_0^{n+1}) \leq \tilde{\mathscr{J}}_T(u_0^n).$$

What saves the day is that, again on $(0, T)$, and under the condition that the function $p_0 := p(t = 0, \cdot)$ can be well defined, the minimizing sequence is just,

$$u_0^{n+1} = u_0^n - \eta \cdot p_0^n, \qquad \text{in } L^2(\mathbb{R})^*,$$

where $p_0^n = p^n(t = 0, \cdot) = \nabla \tilde{\mathscr{J}}(u_0^n)$ as seen in (14). It stops when $p_0^n \simeq 0$. For definiteness, let $g(u, v)$ stand for the Engquist-Osher numerical flux (16) for the (forward) Burgers equation, along with n, k, j being the minimizing sequence, time and space indexes, respectively: an inverse design comes from iterating in $n \in \mathbb{N}$,

$$\left.\begin{array}{l}
\forall j, \qquad u_j^{n=0,k=0} = 0; \\[2ex]
\begin{cases} u_j^{n,k+1} = u_j^{n,k} - \frac{\Delta t}{\Delta x}\left[g(u_j^{n,k}, u_{j+1}^{n,k}) - g(u_{j-1}^{n,k}, u_j^{n,k})\right], \\ \text{for } k = 0, 1, \dots, \frac{T}{\Delta t} - 1; \end{cases} \\[3ex]
\begin{cases} \forall j, \qquad p_j^{n,k=T/\Delta t} = u_j^{n,k=T/\Delta t} - \bar{u}_j, \\ \text{or its regularized value given in (15);} \end{cases} \\[3ex]
\begin{cases} p_j^{n,k} = p_j^{n,k+1} + \frac{\Delta t}{\Delta x}\Big[\frac{\partial g}{\partial u}(u_j^{n,k}, u_{j+1}^{n,k})(p_{j+1}^{n,k} - p_j^{n,k}) \\ \qquad\qquad + \frac{\partial g}{\partial v}(u_{j-1}^{n,k}, u_j^{n,k})(p_j^{n,k} - p_{j-1}^{n,k})\Big], \\ \text{for } k = \frac{T}{\Delta t} - 1, \dots, 1, 0; \end{cases} \\[3ex]
\begin{cases} \forall j, \qquad u_j^{n+1,0} = u_j^{n,0} - \eta\, p_j^{n,0}, \\ \text{or a filtered version, see (22).} \end{cases}
\end{array}\right\} \quad (20)$$

An issue in (20) is that the backward solution p may not be unique when u admits discontinuities. However, as long as the final data p^T remains Lipschitz continuous, both uniqueness and stability hold for "reversible solutions" because $u(t, \cdot)$ satisfies a one-sided Lipschitz condition (OSLC). Moreover, usual finite-difference (dual) schemes converge toward reversible solutions as the grid size $\Delta x \to 0$.

Theorem 5 *(See [26, Theorem 3.7]) Let p^T be a Lipschitz continuous function with Lipschitz constant $Lip(p^T)$. Let the backward scheme be consistent, satisfy a uniform OSLC and that monotonicity-preservation is ensured by a CFL restriction.*

Then the sequence $p^{\Delta x}$ of piecewise-constant functions converges as $\Delta x \to 0$ strongly in $L^1_{loc}(\mathbb{R} \times \mathbb{R}^+)$ and almost everywhere toward the reversible solution.

Obviously, as $\mu^\varepsilon \in C^\infty(\mathbb{R})$, $\mu^\varepsilon * [u^\varepsilon(T,\cdot) - \bar{u}] \in C^\infty$ is also a Lipschitz function. So $p^\varepsilon(t,\cdot)$ is rigorously a reversible solution of the adjoint equation because the velocity field $u(t,\cdot)$ satisfies Oleinik's OSLC. Accordingly, most of the monotonicity-preserving backward finite-difference schemes approximating $p^\varepsilon(t,\cdot)$ converge toward this unique and stable reversible solution under the CFL restriction,

$$\Delta t \max_{j,k,n} |u_j^{n,k}| \leq \frac{\Delta x}{2}.$$

Remark 2 One may wonder why (20) wasn't initialized with the entropy solution of the "backward Burgers equation" (8) with $\bar{u}$ as its final data. Indeed, consider again the situation displayed on Fig. 2: that backward solution is the "ramp" displayed on the first row. Starting with these data, (20) produces the entropy shock, so that the velocity field acting on $p(T,\cdot) \equiv 0$ (for ε small enough if the regularization (13) is applied) is discontinuous. By definition of reversible solutions, $p(t,\cdot)$ will remain null ($TV(p(t,\cdot)) = TV(p(T,\cdot))$) in the cone of dependence of the shock, so $p(t = 0,\cdot) \equiv 0$, too, and the algorithm will always keep the "ramp" as a steady inverse design. Hence, initializing with (8) may not be a good strategy when seeking different, more oscillating inverse designs (belonging to K).

4.3 Projection and Component-Wise Gradient Algorithm

The previously introduced method allows to iteratively recover an initial profile which image by the evolution semi-group S_T matches the target $\bar{u}$, just by solving a simple transport equation for p^ε (for which supposedly many schemes may be available). Indeed, it's very easy to numerically check that trying to devise a "non-entropic backward evolution" for (8) starting from $\bar{u}$ is strongly ill-posed since entropy dissipation in shocks cancels many details (oscillations), so the backward evolution algorithm would have to guess all those missing data. Accordingly, a majority of (stable) finite-difference schemes converge to the (unique) reversible solution: hence only one initial profile (probably the centroid of the convex set of inverse designs) is generally recovered. Unfortunately different, generally more sharply-varying, profiles are sought in the realm of concrete applications: it may be necessary to circumvent the stability of the reversible solution p^ε by projecting it onto a closed linear subspace V of the Hilbert space $L^2(\mathbb{R})$. Let P_V stand for the orthogonal projection onto V and, still, $0 < \eta \ll 1$ so that,

$$\tilde{\mathscr{J}}^\varepsilon_T(\tilde{u}_0^{n+1}) = \tilde{\mathscr{J}}^\varepsilon_T(\tilde{u}_0^n) - \eta \left\langle \nabla \tilde{\mathscr{J}}^\varepsilon_T(\tilde{u}_0^n), P_V \nabla \tilde{\mathscr{J}}^\varepsilon_T(\tilde{u}_0^n) \right\rangle + o(\eta) \leq \tilde{\mathscr{J}}^\varepsilon_T(\tilde{u}_0^n),$$

because

$$\left\langle \nabla \tilde{\mathscr{J}}_T^{\varepsilon}(\tilde{u}_0^n), P_V \nabla \tilde{\mathscr{J}}_T^{\varepsilon}(\tilde{u}_0^n) \right\rangle = \|P_V \nabla \tilde{\mathscr{J}}_T^{\varepsilon}(\tilde{u}_0^n)\|^2 \geq 0. \tag{21}$$

Clearly, the definition (14) is quite convenient, because it suffices to apply it to φ_i, $i \in \mathbb{N}$, the Hilbertian base of the linear subspace V. Imposing that the control belongs to the subspace V, i.e., $u_0 \in L^1 \cap L^\infty \subset V \subset L^2(\mathbb{R})$,

$$u_0(x) = \sum_i \alpha_i \varphi_i(x), \qquad (\alpha_i) \in \ell^2(\mathbb{Z}),$$

at each iteration n, by taking all the duality brackets with $p_0^n = \sum_i \beta_i \varphi_i$,

$$\boxed{\left\langle u_0^{n+1}, \varphi_i \right\rangle = \langle u_0^n, \varphi_i \rangle - \eta \left\langle \nabla \tilde{\mathscr{J}}_T^{\varepsilon}(u_0^n), \varphi_i \right\rangle \quad \Leftrightarrow \quad \alpha_i^{n+1} = \alpha_i^n - \eta \cdot \beta_i^n.} \tag{22}$$

Instead of (22), one may apply filtering to p^T when V contains Lipschitz functions.

Remark 3 Let K be the convex (see Theorem 2) associated to a discontinuous target $\bar{u}$; the inverse design in V, obtained through the filtering (22), is not (in general) the projection of the centroid $C(x)$ of K. Since P_V is an orthogonal projection,

$$\int_{\mathbb{R}} |P_V(C)(x)|^2\, dx = \|P_V(C)\|^2 = \|C\|^2 - \|(Id - P_V)(C)\|^2 \leq \int_{\mathbb{R}} |C(x)|^2\, dx.$$

So, being $\eta(u) = |u|^2$ a convex entropy for Burgers equation, $P_V(C)$ would dissipate more entropy than C: in general, the minimizer generated by (20)–(22) is the projection onto V of an inverse design in K, different from its centroid, see Fig. 7.

4.4 F-Functions or Scale-Limited Subspaces of L^2

A natural choice for V is one of the nested scaled-limited subspaces of a multi-resolution analysis of $L^2(\mathbb{R})$, spanned by a so-called "scaling function".

Definition 2 Nested subspaces V_j form a *Multi-Resolution Analysis* of $L^2(\mathbb{R})$ if: $\{0\} \subset \cdots \subset V_{-1} \subset V_0 \subset V_1 \subset \cdots \subset L^2(\mathbb{R})$ and moreover,

- for all $f \in L^2(\mathbb{R})$, $\|P_{V_j} f - f\|_{L^2} \to 0$ as $j \to +\infty$ also, $P_{V_j} f \to 0$ as $j \to -\infty$.
- if $f(x) \in V_j$, then $f(x/2) \in V_{j-1}$ and for all $k \in \mathbb{Z}, f(x - 2^j k) \in V_j$.
- there exists a shift-invariant orthonormal base of V_0 given by the scaling function $\varphi_i(x) = \varphi(x - i)$ for $i \in \mathbb{Z}$.

In this definition, P_{V_j} stands for the orthogonal projector onto the subspace V_j. Intuitively, it asks for the V_j's to be linear subspaces of functions with increasing spatial resolution: when j decreases, functions in V_j tend to become constants. Oppositely, when j increases, they are allowed to oscillate with high instantaneous frequency. The wavelet spaces W_j are defined as the orthogonal complement of V_j inside V_{j+1}: for all $j \in \mathbb{Z}$, $V_{j+1} = V_j \oplus W_j$. From φ_i, the base of V_0, a base of V_j is deduced by simple dilatation,

$$\varphi_{j,n}(x) = \sqrt{2^j}\varphi_n(2^j x) = \sqrt{2^j}\varphi(2^j x - n). \tag{23}$$

Thus, the orthogonal projection of f onto V_j reads:

$$P_{V_j} f = \sum_{n\in\mathbb{Z}} < f, \varphi_{j,n} > \varphi_{j,n}, \qquad < f, \varphi_{j,n} >= \int_{\mathbb{R}} f(x)\varphi_{j,n}(x).dx, \tag{24}$$

which is the best approximation of f in V_j in the least-squares sense. At this level, one has all the necessary elements to devise a "scale-limited, relaxed gradient algorithm", in component-wise form, by plugging each $\varphi_{j,n}$ into (22) for a given scale $j \in \mathbb{Z}$. Clearly, different choices for φ exist: the Haar function, the multi-fractal Daubechies-4, the smoother Symmlet, etc ...

For specific applications to aerodynamics problems related to sonic-boom imprints, so-called "F-functions" are preferred initial profiles (hence controls). A F-function can be represented by a finite set of parameters,

$$F(x) = \sum_{i=1}^{5} \alpha_i F_i(x), \qquad < F_i, F_j >= \delta_{i,j}.$$

Hence the former algorithm (22) applies as soon as the basis functions F_i are orthogonalized, for instance by means of a Gram-Schmidt procedure: Fig. 1 shows forward.in-time evolution of F with the control-to-state map S_T.

5 Preliminary Numerical Illustrations

Hereafter, the time at which the target is meant to be reached by the (forward) entropy solution of Burgers equation is normalized at unity: $T = 1$. The smallness assumption raised in Remark 1 corresponds to $\|\bar{u}\|_\infty \leq 1$. Our examples are treated with $2^8 = 256$ points, and an adaptive step-size η^n is used in the gradient algorithm:

$$\forall n, k, \qquad \eta^n = 2\left(\frac{\sum_j |p_j^{n,k=0} p_j^{n,k=T/\Delta t}|^2}{\sum_j |p_j^{n,k=0}|^2 \sum_j |p_j^{n,k=T/\Delta t|^2}}\right)^{\frac{1}{2}}.$$

5.1 Filtered Gradient and Hybrid Methods

The simplest filtering is obtained through the Haar scaling function, which is just the indicator of $x \in (0, 1)$: see Fig. 6 for the inverse design of a nonnegative N-wave with increasing resolutions. Differently, Fig. 7 displays the inverse design of the same target, but with scaling functions corresponding to Daubechies-4 (left) and Symmlet-6 (right). For completeness, forward evolution by wavefront tracking, highlighting forward wave interactions (and cancellations), is displayed below.

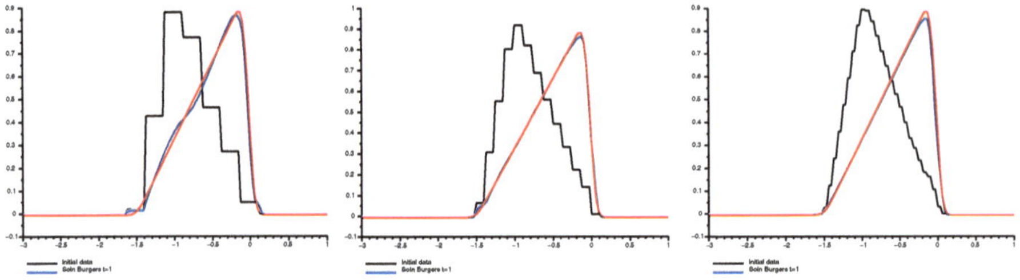

Fig. 6 Haar filters of increasing resolution; no filtering corresponds to maximal resolution

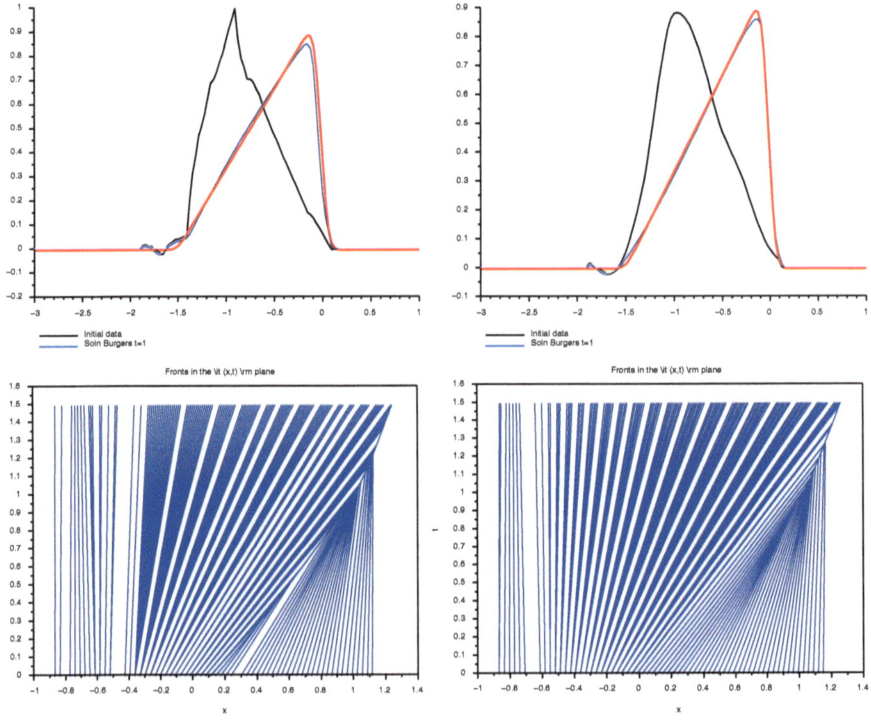

Fig. 7 Inversions with Daubechies-4 (*left*) and Symmlet-6 (*right*); forward evolution by WFT

5.2 A Target Given by a (Small) Sine Wave

When there exists only one inverse design to a smooth target $\bar{u} \in C^\infty$, wavelet filtering may be used in order to ease the iterations of the gradient algorithm: selecting scaling functions $\varphi_{j,n}$ with many vanishing moments, the inverse design at $T = 1$ of

$$\bar{u}(x) = \pm 0.35 \sin(2\pi \frac{x+1}{3}) \chi_{|x+1|<\frac{3}{2}}, \qquad T \leq \frac{30}{7\pi} \simeq 1.36,$$

can be recovered, see Fig. 8. Despite a limited amount of grid points, the decay of the cost functional $\mathscr{J}_T$ is close to a factor 10^3 with 100 iterations of the adaptive filtered gradient algorithm. Of course, as the inverse design is unique, it matches the smooth solution of the backward Burgers equation (8) with final data $\bar{u}$.

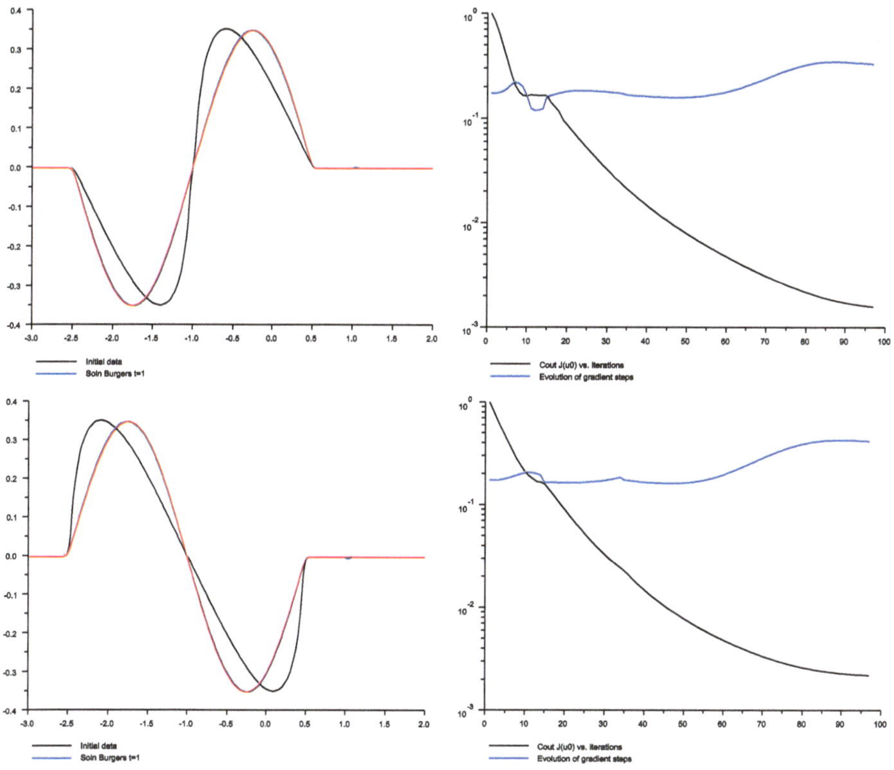

Fig. 8 Inverse design of sine waves with an hybrid method involving Coiflet-3 and Symmlet-6

5.3 Targets Given by Several N-Waves

N-waves realize the large-time behavior (so-called "intermediate asymptotics") of compactly supported initial data: one shouldn't expect uniqueness of inverse designs when prescribing such type of (discontinuous) targets. Hereafter, we consider the final data used in Fig. 5, that is,

$$\bar{u}(x) = \beta(\frac{5}{4} + x)\chi_{|x+1|<1}, \qquad T \leq \frac{1}{\beta}, \qquad \beta > 0.$$

On Fig. 9, inverse designs corresponding to $\beta = 0.45, 0.75$ and 0.95 are displayed (top to bottom) for $T = 1$, along with the decay of the objective functional $\mathscr{J}_T$ and the values taken by the adaptive step-size. Filtering was made so that the first iterations are done with a smooth Coiflet-3 scaling function, but the subsequent ones (details at small scale) are with a multi-fractal Daubechies-4. Switching from one another is done between 10 and 20 iterations of the gradient algorithm, as indicated by means of the brutal variation of the adaptive step-size. General convergence of the algorithm deteriorates with the increasing values of β, so that our inversion time $T = 1$ gets closer to the critical time prescribed by the existence Theorem 1. For bigger inversion times, the target ceases to be reachable, and numerical inverse designs display spikes, see also [2, §6.4]. Of course, by using different scaling functions, it is possible to get alternative inverse designs, see Fig. 10, along with all convex combinations which can be built from each two of them.

5.4 F-Functions as Combinations of "Haar and Spikes"

By looking at Fig. 1, one may hope to recover interesting inverse designs starting from truncated N-waves at time $T = 1$ like,

$$\bar{u}(x) = \frac{3}{4}\min\left(\frac{4}{5}, \max\left(-\frac{1}{2}, \frac{5}{4} + x\right)\right)\chi_{|x+1|<1}, \qquad T \leq \frac{4}{3}.$$

Our (mostly empirical, inspired by Tropp [43]) computational approach consists in

- setting up a very coarse Haar scaling function during the 10 first iterations,
- then, switching to a slightly finer one to correct smoother regions,
- during the whole process, alternating between a standard wavelet filtering and a "spike filtering", in which cumulative integrals of corrections $\int_{x_0}^{x_1} p(0,x)dx$, are concentrated at a given location $x \in (x_0, x_1)$.

Preliminary results are displayed on Fig. 11, to be compared with Fig. 1.

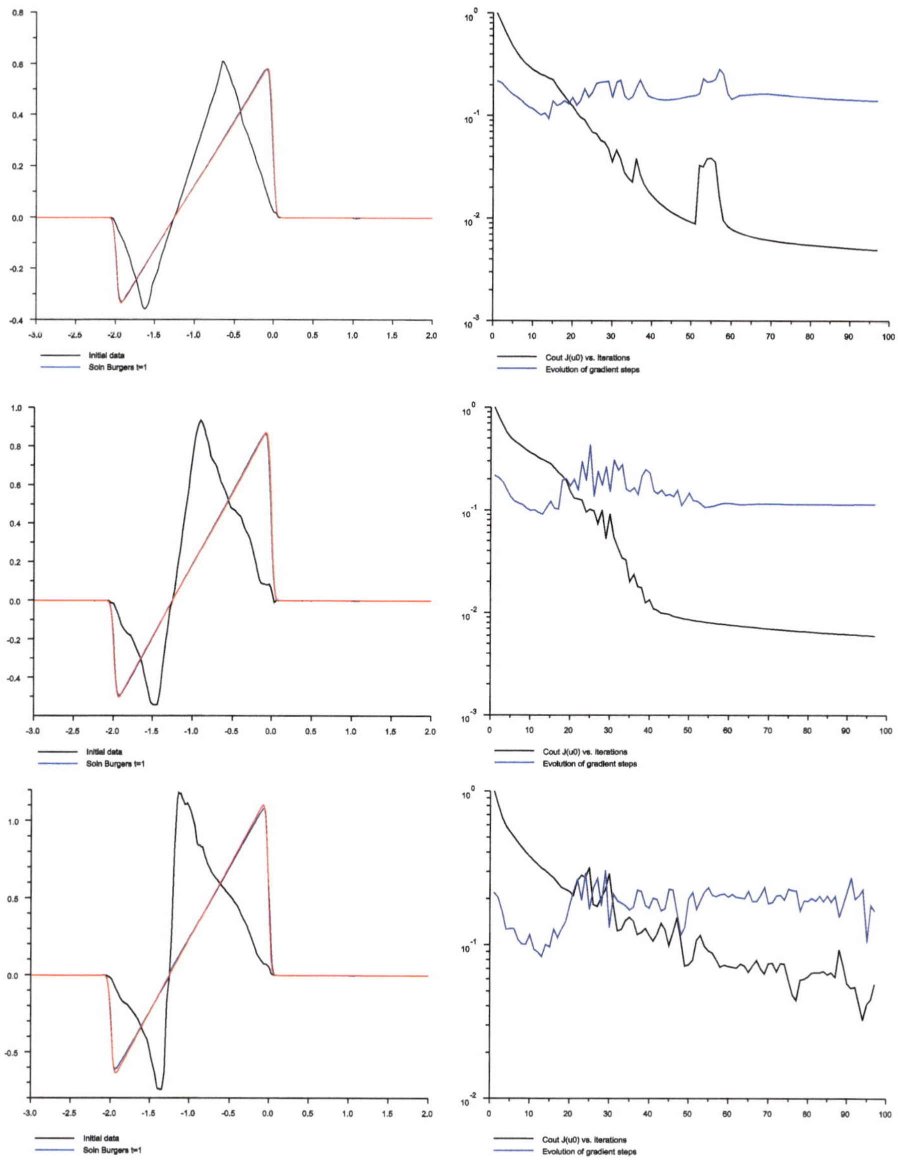

Fig. 9 Inverse design of N-waves with an hybrid method involving Coiflet-3 and Daubechies-4

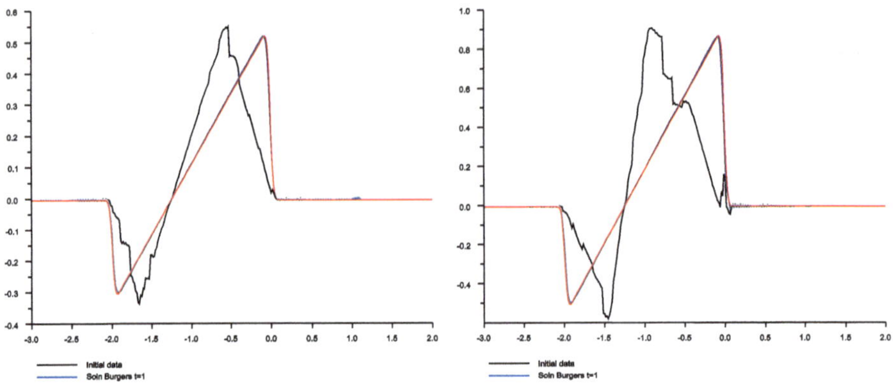

Fig. 10 Inverse design for $\beta = 0.45$ with hybrid methods involving Haar and Symmlet wavelets

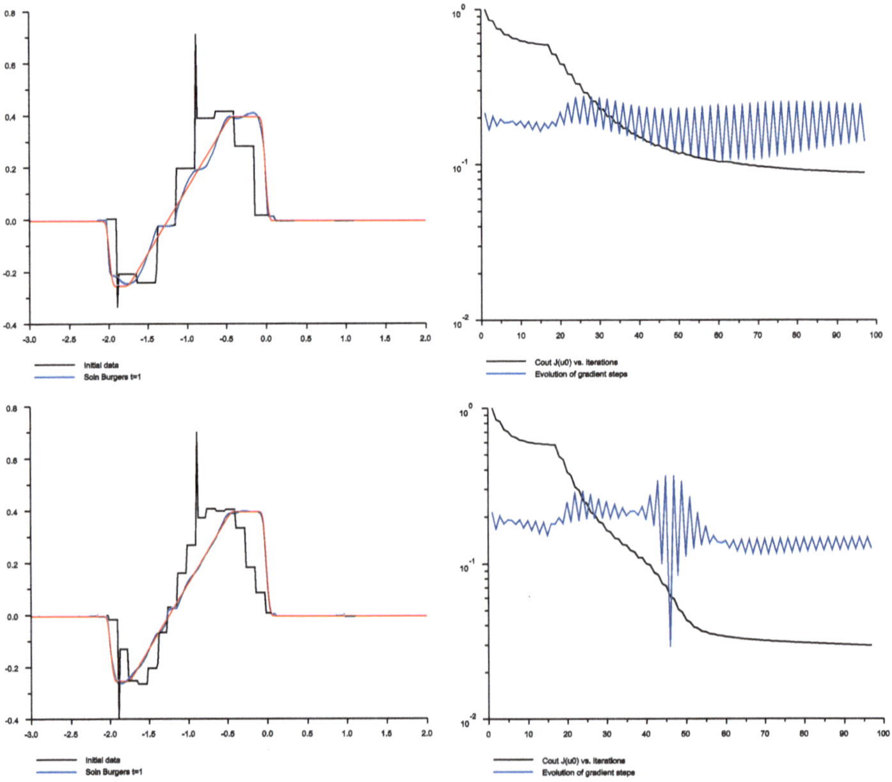

Fig. 11 Inverse designs similar to F-functions of a truncated N-wave with "Haar and spikes"

6 Conclusions

A few aspects of inverse design for convex scalar conservation laws, exemplified by 1D Burgers equation, were reviewed in the present note. In particular, the huge variety of initial data leading to an entropy-admissible, piecewise discontinuous target is revealed by Theorem 2. An iterative gradient algorithm relying on numerically usually produces the "backward entropy solution", mostly because numerically solving linearized backward transport equations brings out corresponding "reversible solutions". In order to circumvent that issue, and to be able to produce different, more steeply-varying numerical inverse designs, a filtering procedure was applied, by adequate wavelet projectors. The convergence of such a process progressively degrades as targets become close to a non-admissible profile.

Acknowledgements This work was partially supported by the Advanced Grant 694126-DYCON (Dynamic Control) of the European Research Council Executive Agency, ICON of the French ANR (2016-ACHN-0014-01), FA9550-15-1-0027 of AFOSR, A9550-14-1-0214 of the EOARD-AFOSR, and the MTM2014-52347 Grant of the MINECO (Spain). Both the authors thank Prof. Andreas Griewank who suggested to study convex combinations of inverse designs during a meeting in September 2015.

References

1. Adimurthi, A., Ghoshal, S.S., Gowda, G.D.V.: Exact controllability of scalar conservation laws with strict convex flux. Math. Control Relat. Fields **4**, 401–449 (2014)
2. Allahverdi, N., Pozo, A., Zuazua, E.: Numerical aspects of large-time optimal control of Burgers equation. Contemp. Math. **658**, 267–279 (2016)
3. Alonso, J.J., Colonno, M.R.: Multidisciplinary optimization with applications to sonic-boom minimization. Annu. Rev. Fluid Mech. **44**, 505–506 (2012)
4. Ancona, F., Cannarsa, P., Nguyen, K.T.: Quantitative compactness estimates for Hamilton-Jacobi equations. Arch. Ration. Mech. Anal. **219**(2), 793–828 (2016)
5. Bardos, C., Pironneau, O.: Data assimilation for conservation laws. Methods Appl. Anal. **12**(2), 103–134 (2005)
6. Barzilai, J., Borwein, J.M.: Two point step size gradient methods. IMA J. Numer. Anal. **8**, 141–148 (1988)
7. Berlinet, A.F., Roland, Ch.: Geometric interpretation of some Cauchy related methods. Numer. Math. **119**, 437–464 (2011)
8. Bouchut, F., James, F.: One-dimensional transport equations with discontinuous coefficients. Nonlinear Anal. Theory Methods Appl. **32**, 891–933 (1998)
9. Bouchut, F., James, F.: Differentiability with respect to initial data for a scalar conservation law. In: Fey, M., Jeltsch, R. (eds.) Proceedings of the 7th Conference on Hyperbolic Problems, Zürich, 1998. International Series of Numerical Mathematics, vol. 129, pp. 113–118. Birkhäuser, Basel (1999)
10. Bressan, A., Marson, A.: A variational calculus for discontinuous solutions of systems of conservation laws. Commun. Partial Differ. Equ. **20**, 1491–1552 (1995)
11. Burgers, J.M.: Application of a model system to illustrate some points of the statistical theory of free turbulence. Proc. Konink. Nederl. Akad. Wetensch. **43**, 2–12 (1940)

12. Castro, C., Palacios, F., Zuazua, E.: An alternative descent method for the optimal control of the inviscid Burgers equation in the presence of shocks. Math. Models Methods Appl. Sci. **18**(3), 369–416 (2008)
13. Cleveland, R.O.: Propagation of sonic-booms through a real stratified atmosphere. Ph.D. dissertation, The University of Texas at Austin (1995)
14. Cole, J.D.: On a quasi-linear parabolic equation occurring in aerodynamics. Q. Appl. Math. **9**, 225–236 (1951)
15. Dafermos, C.M.: Hyperbolic Conservation Laws in Continuum Physics. Grundlehren der Mathematischen Wissenschaften Series, vol. 325. Springer, Berlin (2010)
16. Daubechies, I.C., Gilbert, A.C.: Harmonic analysis, wavelets, and applications. In: Cafarelli, L., Weinan, E. (eds.) Hyperbolic Equations and Frequency Interactions. IAS/Park City Mathematics Series, vol. 5, pp. 159–226 (1998)
17. Evans, L.C.: Partial Differential Equations. Graduate Studies in Mathematics, vol. 19, 2nd edn. American Mathematical Society, Providence, RI (2010)
18. Fikl, A., Le Chenadec, V., Sayadi, T., Schmid, P.J.: A comprehensive study of adjoint-based optimization of non-linear systems with application to Burgers' equation. In: 46th AIAA Fluid Dynamics Conference (2016). doi:10.2514/6.2016-3805
19. George, A.R., Seebass, R.: Sonic boom minimization including both front and rear shocks. AIAA J. **9**(10), 2091–2093 (1971)
20. Giles, M.B., Pierce, N.A.: An introduction to the adjoint approach to design, turbulence and combustion. Flow Turbul. Combust. **65**, 393–415 (2001)
21. Giles, M., Ulbricht, S.: Convergence of linearized and adjoint approximations for discontinuous solutions of conservation laws. Part 1: linearized approximations and linearized output functionals. SIAM J. Numer. Anal. **48**(3), 882–904 (2010)
22. Giles, M., Ulbricht, S.: Convergence of linearized and adjoint approximations for discontinuous solutions of conservation laws. Part 2: adjoint approximations and extensions. SIAM J. Numer. Anal. **48**(3), 905–921 (2010)
23. Gilmore, P., Kelley, C.T.: An implicit filtering algorithm for optimization of functions with many local minima. SIAM J. Optim. **5**(2), 269–285 (1995)
24. Godlewski, E., Raviart, P.A.: The linearized stability of solutions of nonlinear hyperbolic systems of conservation laws. Math. Comput. Simul. **50**, 77–95 (1999)
25. Gosse, L.: A Donoho-Stark criterion for stable signal recovery in discrete wavelet subspaces. J. Comput. Appl. Math. **235**, 5024–5039 (2011)
26. Gosse, L., James, F.: Numerical approximation of linear one-dimensional conservation equations with discontinuous coefficients. Math. Comput. **69**, 987–1015 (2000)
27. Hogan, J.A., Lakey, J.D.: Time-Frequency and Time-Scale Methods. Birkhauser, Boston (2005)
28. Holden, H., Risebro, N.H.: Front Tracking for Hyperbolic Conservation Laws. Applied Mathematical Sciences, vol. 152. Springer, New York (2002)
29. Hopf, E.: The partial differential equation $u_t + uu_x = u_{xx}$. Commun. Pure Appl. Math. **3**, 20–23 (1950)
30. Ignat, L.I., Pozo, A., Zuazua, E.: Large-time asymptotics, vanishing viscosity and numerics for 1-D scalar conservation laws. Math. Comput. **84**, 1633–1662 (2015)
31. Lecaros, R., Zuazua, E.: Tracking control of the 1D scalar conservation laws in the presence of shocks. In: Trends in Contemporary Mathematics. Springer INdAM Series, vol. 8, pp. 195–219. Springer, Berlin (2014)
32. Li, Y., Osher, S., Tsai, R.: Heat source identification based on ℓ^1 constrained minimization. Inverse Prob. Imaging **8**, 199–221 (2014)
33. Liu, T.-P., Pierre, M.: Source-solutions and asymptotic behavior in conservation laws. J. Differ. Equ. **51**, 419–441 (1984)
34. Mallat, S.: A Wavelet Tour of Signal Processing. Academic, New York (1998)
35. Martinez, P., Vancostenoble, J.: Carleman estimates for one-dimensional degenerate heat equations. J. Evol. Equ. **6**(2), 325–362 (2006)
36. Meyer, Y.: Ondelettes sur l'intervalle. Riv. Matem. Iberoam. **7**, 115–133 (1991)

37. Negreanu, M., Matache, A.M., Schwab, C.: Wavelet filtering for exact controllability of the wave equation. SIAM J. Sci. Comput. **28**(5), 1851–1885 (2006)
38. Plotkin, K.J.: State of the art of sonic boom modeling. J. Acoust. Soc. Am. **111**(1), 530–536 (2002)
39. Plotkin, K.J., Rallabhandi, S.K., Li, W.: Generalized formulation and extension of sonic boom minimization theory for front and aft shaping. In: 47th AIAA Aerospace Sciences Meeting (Including The New Horizons Forum and Aerospace Exposition), Orlando, FL, 5–8 Jan 2009
40. Rallabhandi, S.K.: Advanced sonic boom prediction using augmented Burgers equation. J. Aircr. **48**, 1245–1253 (2011)
41. Serre, D.: Systems of Conservation Laws I: Hyperbolicity, Entropies, Shock Waves. Cambridge University Press, Cambridge (1999)
42. Slepian, D.: Some comments on Fourier analysis, uncertainty and modeling. SIAM Rev. **25**, 379–393 (1983)
43. Tropp, J.: On the linear independence of spikes and sines. J. Fourier Anal. Appl. **14**, 838–858 (2008)
44. Vancostenoble, J., Zuazua, E.: Null controllability for the heat equation with singular inverse-square potential. J. Funct. Anal. **254**, 1864–1902 (2008)
45. Walter, G.G., Shen, X.: Wavelet based on prolate spheroidal wave functions. J. Fourier Anal. Appl. **10**, 1–26 (2004)
46. Whitham, G.B.: The flow pattern of a supersonic projectile. Commun. Pure Appl. Math. **V**, 301–348 (1952)
47. Xiao, H., Rokhlin, V., Yarvin, N.: Prolate spheroidal wave functions, quadrature and interpolation. Inverse Prob. **17**, 805–838 (2001)

A Well-Balanced Scheme for the Euler Equations with Gravitation

Roger Käppeli

Abstract We present a novel well-balanced scheme for the Euler equations with gravitation. The scheme is designed to preserve exactly an isothermal hydrostatic equilibrium. It uses a novel equilibrium preserving reconstruction based on the Gibbs free energy. Moreover, the developed scheme is applicable beyond the ideal gas law which is crucial for many astrophysical applications. The scheme is second-order accurate in space and can easily be implemented into any existing code solving time explicitly or implicitly the Euler equations. We assess the performance of the scheme on two test problems involving a realistic equation of state for stellar matter.

1 Introduction

Many interesting astrophysical phenomena can be modeled by the Euler equations with gravitational source terms

$$\frac{\partial \rho}{\partial t} + \nabla \cdot (\rho \mathbf{v}) = 0 \tag{1}$$

$$\frac{\partial \rho \mathbf{v}}{\partial t} + \nabla \cdot (\mathbf{v} \rho \mathbf{v}) + \nabla p = -\rho \nabla \phi \tag{2}$$

$$\frac{\partial E}{\partial t} + \nabla \cdot [(E + p)\, \mathbf{v}] = -\rho \mathbf{v} \cdot \nabla \phi \tag{3}$$

expressing the conservation of mass, momentum and energy, respectively. Here ρ is the mass density, $\mathbf{v}$ the velocity and $E = \rho e + \rho v^2/2$ the total (fluid) energy density being the sum of internal and kinetic energy density. The system is closed by an equation of state (EoS) relating the pressure $p = p(\rho, e)$ to the density and specific internal energy.

The source terms on the right hand side of the momentum and energy equations model the effect of gravity forces onto the fluid. They are given in terms of the

R. Käppeli (✉)
ETH Zürich, Seminar for Applied Mathematics, Rämistrasse 101, CH-8092 Zürich, Switzerland
e-mail: roger.kaeppeli@sam.math.ethz.ch

L. Gosse, R. Natalini (eds.), *Innovative Algorithms and Analysis*,
Springer INdAM Series 16, DOI 10.1007/978-3-319-49262-9_8

gradient of the gravitational potential ϕ which may either be a given function or, in case of self-gravity, is the solution of the Poisson equation

$$\nabla^2 \phi = 4\pi G \rho, \tag{4}$$

where G is the gravitational constant.

Numerical methods for the Euler equations are in a major stage of development for a vast range of flow regimes. Among the most popular schemes are the so-called Godunov-type finite volume methods. We refer to e.g. [14, 16, 24] for a comprehensive introduction. In these schemes, cell averages of conserved variables are evolved in terms of numerical fluxes. The latter are determined by approximate or exact solutions of Riemann problems at each cell interface. High order spatial accuracy is provided by suitable non-oscillatory polynomial reconstruction procedures such as TVD, ENO and WENO (see e.g. [21]). High order temporal accuracy is achieved by appropriate ODE integrators such as the strong stability preserving (SSP) Runge-Kutta schemes (see e.g. [9]).

In astrophysics it often happens that the flow of interest is close to hydrostatic equilibrium

$$\nabla p = -\rho \nabla \phi. \tag{5}$$

Such conditions arise for example in the simulation of waves in the atmosphere of our sun, convective nuclear burning during stellar evolution and in the dramatic death of a star in a supernova. The numerical approximation of near equilibrium flows turns out to be challenging for standard numerical methods. The reason behind these difficulties is the fact that they do not necessarily satisfy a discrete version of the balance between the pressure gradient and gravity forces. Consequently, equilibrium states are not preserved exactly but are approximated with an error proportional to the truncation error. So if one is interested in simulating small perturbations of the equilibrium, the numerical resolution has to be increased to the point that truncation errors do not obscure the phenomena of interest. Especially in multiple dimensions, the required resolution may become prohibitively large.

In order to overcome this challenge, Greenberg and Leroux [10] introduced the concept of well-balanced schemes. In these schemes, a discrete balance is satisfied at steady states. For instance, many well-balanced schemes have been designed for the shallow water equations with bottom topography, see e.g. [1, 15, 20]. A comprehensive review on well-balanced schemes in many applications is also given in the book by Gosse [8]. Well-balanced schemes for hydrostatic equilibrium have received considerable attention in the recent literature [3, 5–7, 11, 12, 17–19, 25, 26]. However, it turns out that Eq. (4) only defines a mechanical equilibrium. A further assumption on a thermodynamic equilibrium is needed in order to uniquely specify the hydrostatic equilibrium. Two physically relevant cases are the isentropic and isothermal equilibria assuming constant entropy or temperature, respectively. In this contribution we focus on the latter one. Such conditions arise for example in the interior of white dwarf stars due to the high thermal conductivity of degenerate electrons (see e.g. [13]).

Well-balanced schemes for isothermal hydrostatic equilibrium have presented in [5, 7, 18, 19, 26] assuming an ideal gas EoS. This limits their applicability to more complex astrophysical scenario quite strongly. Therefore we present here an approach that works with any EoS. Our approach closely follows the one by Käppeli and Mishra [11]. Actually, only the thermodynamic potential for the local hydrostatic reconstruction has to be changed. To keep the present contribution compact and because the developments here follow the latter reference very closely, we present below only the strictly necessary modifications of [11].

The rest of the article is organized as follows: the well-balanced scheme is presented in Sect. 2, numerical results are reported in Sect. 3 and we conclude in Sect. 4.

2 Numerical Methods

For the sake of simplicity, we present the well-balanced scheme in a one-dimensional setting. Pointers to the literature for multidimensional extension are given at the end of the section.

In one spatial dimension, the Euler equations are given by

$$\frac{\partial \mathbf{u}}{\partial t} + \frac{\partial \mathbf{F}}{\partial x} = \mathbf{S} \tag{6}$$

where

$$\mathbf{u} = \begin{bmatrix} \rho \\ \rho v_x \\ E \end{bmatrix}, \quad \mathbf{F} = \begin{bmatrix} \rho v_x \\ \rho v_x^2 + p \\ (E+p)v_x \end{bmatrix} \quad \text{and} \quad \mathbf{S} = -\begin{bmatrix} 0 \\ \rho \\ \rho v_x \end{bmatrix} \frac{\partial \phi}{\partial x}, \tag{7}$$

are the vectors of conserved variables, fluxes and source terms, respectively. Furthermore, we denote the primitive variables by $\mathbf{w} = [\rho, v_x, p]^T$.

2.1 One-Dimensional Finite Volume Method

Numerical approximations to Eq. (6), proceed by discretizing the spatial domain into finite volumes (or cells) $I_i = [x_{i-1/2}, x_{i+1/2}]$. For ease of presentation, we take the cells I_i to be uniform, $\Delta x = x_{i+1/2} - x_{i-1/2}$, with centers $x_i = (x_{i-1/2} + x_{i+1/2})/2$. The integration of Eq. (6) over a cell I_i then results in a semi-discrete finite volume scheme for the evolution of the cell averaged conserved variables

$$\frac{d\mathbf{u}_i}{dt} = \mathbf{L}(\mathbf{u}) = -\frac{1}{\Delta x}\left(\mathbf{F}_{i+1/2} - \mathbf{F}_{i-1/2}\right) + \mathbf{S}_i, \tag{8}$$

where the $\mathbf{F}_{i\pm1/2}$ are the numerical fluxes at cell interfaces and $\mathbf{S}_i$ is the cell averaged gravity source term.

Aforementioned numerical fluxes are obtained by solving (approximately) Riemann problems at cell interfaces

$$\mathbf{F}_{i+1/2} = \mathscr{F}(\mathbf{w}_{i+1/2-}, \mathbf{w}_{i+1/2+}), \tag{9}$$

where both $\mathbf{w}_{i+1/2-}$ and $\mathbf{w}_{i+1/2+}$ denote extrapolated primitive variables at the cell interface. In our hydrostatic well-balanced scheme (elaborated below), we require standard properties such as consistency of the numerical flux, $\mathscr{F}(\mathbf{w}, \mathbf{w}) = \mathbf{F}(\mathbf{w})$, along with Lipschitz continuity. In forthcoming numerical examples, the HLLC Riemann solver with wave speed given by Batten et al. [2] is used.

In traditional schemes (at least the most common in astrophysics), the gravitational term is approximated by standard second-order central differences,

$$\mathbf{S}_i = -\begin{bmatrix} 0 \\ \rho_i \\ (\rho v_x)_i \end{bmatrix} \frac{\phi_{i+1} - \phi_{i-1}}{2\Delta x}, \tag{10}$$

where the ϕ_i are the cell centered values of the gravitational potential.

Cell interface extrapolated variables $\mathbf{w}_{i\pm1/2\mp}$ are generally obtained by some reconstruction procedure from the cell averages $\mathbf{u}_i$. Usually, non-oscillatory ones, such as TVD, ENO or WENO, are used (see e.g. [24] and references therein). Unfortunately, these reconstructions are not well suited for the case considered here. **The reason for that is simply the fact that the steady states of interest are generally not polynomial functions**. In the next section, an equilibrium-fit reconstruction is built, so that, along with a special momentum source discretization automatically fulfills a discrete version of certain steady states.

A fully discrete finite volume scheme is obtained by discretizing the temporal domain into suitably sized time steps $\Delta t^n = t^{n+1} - t^n$ where the superscript n labels the different time levels. For the schemes developed below, any adequate explicit or implicit time integrator for the system of ordinary differential Eq. (8) could be used. However, in the numerical results below, we use the explicit second-order strong stability preserving (SSP) Runge-Kutta method (see e.g. [9, Proposition 4.1]):

$$\begin{aligned} \mathbf{u}_i^{(1)} &= \mathbf{u}_i^n + \Delta t^n \mathbf{L}(\mathbf{u}^n) \\ \mathbf{u}_i^{(2)} &= \mathbf{u}_i^{(1)} + \Delta t^n \mathbf{L}(\mathbf{u}^{(1)}) \\ \mathbf{u}_i^{n+1} &= \tfrac{1}{2}\left(\mathbf{u}_i^n + \mathbf{u}_i^{(2)}\right), \end{aligned} \tag{11}$$

where $\mathbf{L}$ is defined in Eq. (8). The time step Δt^n is fixed by some CFL condition.

2.2 Local Isothermal Hydrostatic Equilibrium Reconstruction

In the following we first derive an explicit expression for isothermal hydrostatic equilibrium.

We start with the well-known thermodynamic relation

$$dg = \frac{dp}{\rho} - s dT, \tag{12}$$

where g is the specific Gibbs free energy

$$g = e + \frac{p}{\rho} - Ts, \tag{13}$$

T the temperature and s the specific entropy. Assuming isothermal conditions and the thermodynamic relation Eq. (12), we can express hydrostatic equilibrium Eq. (5) as

$$\frac{\nabla p}{\rho} = \nabla g = -\nabla \phi \tag{14}$$

which can be trivially integrated to yield

$$g + \phi = const. \tag{15}$$

Equation (15) defines a local equilibrium reconstruction within each cell I_i,

$$g_{0,i}(x) = g_i + \phi_i - \phi(x) \tag{16}$$

where g_i is simply the Gibbs free energy evaluated from the cell averaged conserved variables $\mathbf{u}_i$ and ϕ_i is the point value of the gravitational potential at cell center. The gravitational potential $\phi(x)$ may either be known as a given function or has to be obtained by some suitable interpolation.

From the local equilibrium reconstruction of the Gibbs free energy $g_{0,i}(x)$, an equilibrium density $\rho_{0,i}(x)$ and pressure $p_{0,i}(x)$ reconstruction can be obtained through the EoS. This is achieved by noting that $g = g(T, p)$ and solving the following equation

$$g_{0,i}(x) = g(T_i, p_{0,i}(x)). \tag{17}$$

This defines implicitly the equilibrium for any $x \in I_i$.

In case of an ideal gas EoS, $p = \rho RT$, explicit expressions can be given for the equilibrium pressure

$$p_{0,i}(x) = p_i e^{-\frac{1}{RT_i}(\phi(x)-\phi_i)} \tag{18}$$

and density

$$\rho_{0,i}(x) = \frac{p_i}{RT_i} e^{-\frac{1}{RT_i}(\phi(x)-\phi_i)}. \tag{19}$$

There, R is the gas constant and T_i (p_i) is the temperature (pressure) evaluated from the conserved variables $\mathbf{u}_i$. Here we recover the same expressions as [5–7, 26].

However, the strength of our approach is that it can be used with any EoS, even a tabulated one: this is highly relevant for computational astrophysics.

In this case, the equilibrium pressure and density distribution have to be obtained by some iterative procedure. In order to get the equilibrium pressure distribution $p_{0,i}(x)$, the equation to solve is the following

$$f(p_{0,i}(x)) = g(p_{0,i}(x), T_i) - g_{0,i}(x) = 0. \tag{20}$$

The latter equation can be solved iteratively with a Newton-Raphson method

$$\begin{aligned} p_{0,i}^{(k+1)}(x) &= p_{0,i}^{(k)}(x) - f(p_{0,i}(x))/f'(p_{0,i}(x)) \\ &= p_{0,i}^{(k)}(x) - \frac{g(p_{0,i}^{(k)}(x), T_i) - g_{0,i}(x)}{\rho(T_i, p_{0,i}^{(k)}(x))}. \end{aligned} \tag{21}$$

In the last equation, the density $\rho = \rho(T, p)$ has to be computed given a temperature and a pressure through the EoS. That equilibrium density $\rho_{0,i}(x)$ can also be recovered from the equilibrium pressure $p_{0,i}(x)$. In the numerical example below, a tabulated EoS, that is widely used in astrophysics, is used.

It remains to specify an adequate source term discretization. For the momentum source term, we follow the approach suggested by Audusse et al. [1], Botta et al. [3], and Fuchs et al. [7] and define

$$S_{\rho v,i} = \frac{p_{0,i}(x_{i+1/2}) - p_{0,i}(x_{i-1/2})}{\Delta x} \tag{22}$$

where $p_{0,i}(x_{i\pm1/2})$ are the equilibrium pressures at cell interfaces. We leave the energy source term discretization as in Eq. (10). It can be shown as in [11] that the above expression is a consistent and spatially second-order accurate approximation of the momentum source term. It is then straightforward to show that the semi-discrete finite volume scheme, Eq. (8), with the above local equilibrium reconstruction and source term discretization is well-balanced with respect to any isothermal hydrostatic equilibrium (see [11]). Such a scheme may be extended to (formal) second-order spatial accuracy, multiple dimensions (by dimensional-splitting) and curvilinear meshes; see [11, 12].

3 Numerical Results

In this section a series of numerical experiments demonstrates the performance of the proposed well-balanced scheme. For comparison, we also show results obtained with a standard (unbalanced) one. Moreover, we only present results obtained at second-order accuracy (first-order schemes have little practical use in computational astrophysics).

For all the computations we use the Helmholtz EoS of Timmes and Swesty [23]. It is an astrophysically meaningful EoS including contributions of (photon) radiation, nuclei, electrons and positrons. The radiation is treated as a black body in local thermodynamic equilibrium and the nuclei are modeled by an ideal gas. The electrons and positrons contributions are obtained from a table with a thermodynamically consistent interpolation procedure. The electrons/positrons may have speeds arbitrarily close to relativistic limits and an arbitrary degree of degeneracy. We use the publicly available version of this EoS [22].

3.1 *Isothermal Hydrostatic Atmosphere*

As a first numerical experiment, we consider the very simple setting of a one-dimensional isothermal hydrostatic atmosphere in a constant gravitational field

$$\frac{\mathrm{d}p}{\mathrm{d}x} = -\rho g \tag{23}$$

where g is the constant gravitational acceleration.[1] The gravitational potential is a linear function $\phi(x) = gx$.

The computational domain is $D = [0, L]$, where $L = 3 \times 10^8$ cm, and is regularly discretized by N cells. Hence, the grid size is $\Delta x = L/N$, cell interfaces $x_{i+1/2} = i\Delta x$ and cell centers $x_i = (x_{i-1/2} + x_{i+1/2})/2$ for $i = 1, \ldots, N$. For the resolution, we choose $N = 32, 64, 128, 256, 512, 1024, 2048$.

In the following we consider **typical white dwarf like conditions**. Due to the high thermal conductivity of stellar matter at these conditions, the isothermal hydrostatic equilibrium is an appropriate model. Accordingly, gravitational acceleration is $g = 5 \times 10^8$ cm/s^2. Ground-level density and pressure at $x = 0$ are $\rho_0 = 10^7$ g/cm^3 and $p_0 = 8.53 \times 10^{23}$ erg/cm^3. Moreover, the chemical composition is set to half carbon ^{12}C and half oxygen ^{16}O.

The isothermal hydrostatic atmosphere is initialized over the whole domain D with Eq. (15). Its density and pressure profile are shown in the left panel of Fig. 1. The resulting atmosphere has then a sound crossing time $\tau_{\text{sound}} \approx 2$ s, i.e. a sound wave goes through from base to top and back in a time τ_{sound}. This reflects a time

[1] The gravitational acceleration shouldn't be confused with Gibbs free energy from Sect. 2.2.

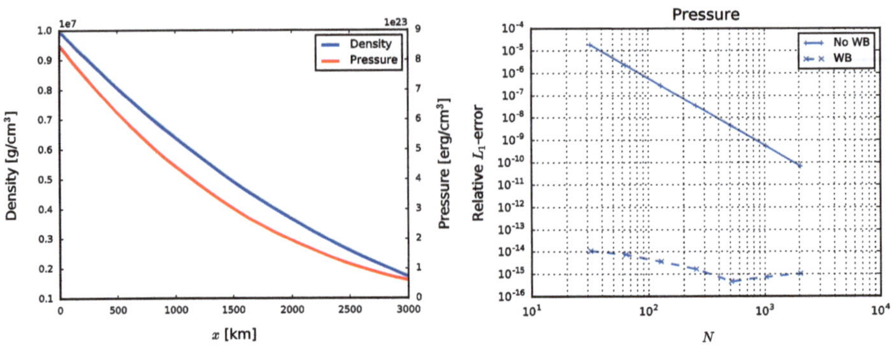

Fig. 1 *Left panel*: Density and pressure profile of the isothermal hydrostatic atmosphere. *Right panel*: Relative L_1-norm of the difference between the initial and final state of the simulation. The *solid and dashed lines* represent the results from the standard and the well-balanced scheme, respectively

scale on which the equilibrium reacts on perturbations. On boundaries hydrostatic equilibrium extrapolations for the density and pressure are used; see [11]. For the velocity we use a simple piecewise linear extrapolation, in order to minimize spurious reflection effects as much as possible.

We begin by numerically verifying the well-balancing properties of the scheme. For this purpose, we evolve the isothermal atmosphere with the well-balanced and standard (unbalanced) second-order schemes for two sound crossing times, i.e. the final time of the simulation is $t_f = 2\tau_{\text{sound}}$. At the end of the simulation, we compute the L_1-error, i.e. the L_1-norm of the difference between the initial and final state of the simulation. This reflects how well the discrete equilibrium is preserved by the numerical scheme.

In the right panel of Fig. 1 is shown the relative L_1-error for the pressure. The results for the density are similar and are not displayed. From the figure it is clear that the well-balanced scheme is able to maintain the hydrostatic equilibrium down to machine precision. In contrast, the standard scheme is clearly not able to preserve the stationary state. Indeed the error is proportional to its truncation error and scales with the second power of the mesh width, i.e. in accord with its design accuracy.

Next we test the ability of the proposed scheme to propagate waves up the atmosphere. For this we impose a periodic velocity perturbation at its base,

$$v_{x,m}^n = A \sin\left(6\frac{2\pi t^n}{t_f}\right), \tag{24}$$

where $m = -1, 0$ is the boundary cell index and A is the wave amplitude. For the amplitude we choose a small one with $A = 10^3$ cm/s and a large one with

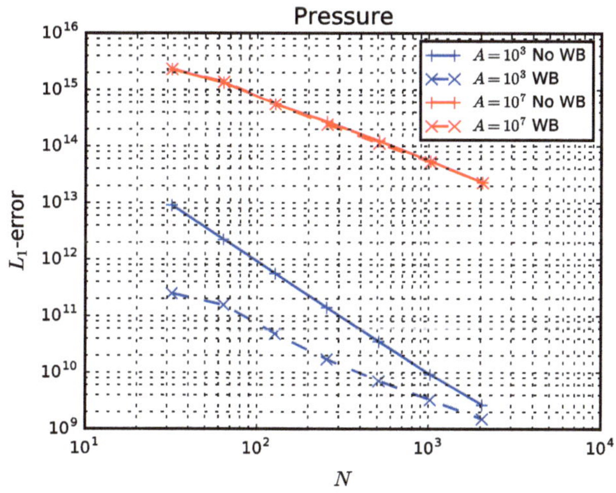

Fig. 2 L_1-norm error in pressure for small (*blue lines*) and large (*red lines*) amplitude A perturbations. The reference solutions were computed with the well-balanced scheme and a resolution $N = 8192$. The *solid and dashed lines* have been obtained with the standard and well-balanced scheme, respectively. Note that here the absolute and not the relative error is shown

$A = 10^7$ cm/s. We then evolve this setting for a time $t_f = 0.45\tau_{\text{sound}}$, i.e. shortly before the waves arrive at the top of the atmosphere.

In Fig. 2 are shown the errors in pressure which have been computed from a reference solution obtained with the well-balanced scheme at resolution $N = 8192$. The errors in density and velocity are similar and not shown. We observe that for the low amplitude case the errors obtained with the well-balanced scheme (dashed blue) are much smaller than the ones obtained with the standard scheme (solid blue). For instance, the well-balanced scheme needs roughly a factor 3–4 less resolution than the standard scheme to reach a given accuracy (for $N \leq 512$). Note that this saving becomes particularly meaningful in a multi-dimensional setting.

The velocity profiles for the small amplitude test are shown in the left panel of Fig. 3. The blue solid line is the reference solution and the solid/dashed red lines are the solutions obtained with the standard and well-balanced schemes at $N = 256$, respectively. From the figure it is clear that the standard scheme suffers some spurious deviations.

For the large amplitude case, the errors in pressure are displayed in Fig. 2 with the red lines. The solid and dashed lines have been computed with the standard and well-balanced schemes, respectively. Here both schemes do equally well and show a first-order convergence. This is expected as large amplitude waves steepen into shock waves as they propagate up the atmosphere. This is displayed for the velocity in the right panel of Fig. 3: the well-balanced scheme does not suffer any deterioration in robustness.

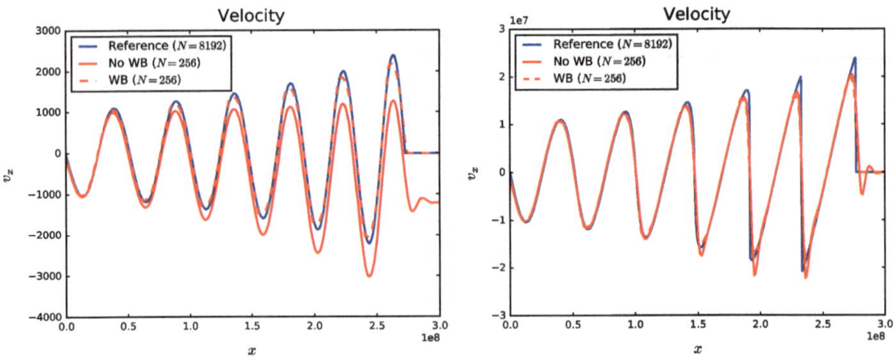

Fig. 3 Velocity profile for the small (*left*) and large (*right*) amplitude waves propagating up the isothermal atmosphere. The *solid/dashed red lines* have been obtained with the standard/well-balanced scheme at resolution $N = 256$. The *blue solid line* is a reference solution obtained with the well-balanced scheme at resolution $N = 8192$

3.2 Rayleigh-Taylor Instability

In this example, we simulate the development of a Rayleigh-Taylor instability [4]. We consider the two-dimensional Euler equations (1)–(3) in polar coordinates. Hence, this test demonstrates the ability of the presented scheme to generalize to multiple dimensions and curvilinear coordinates.

The computational domain is $D = [r_{in}, r_{out}] \times [0, 2\pi]$, where $r_{in} = 10^8$ cm and $r_{out} = 4 \times 10^8$ cm. The gravitational potential is purely radial and linear $\phi = \phi(r) = gr$ with $g = 1 \times 10^8$ cm/s^2. At the base, we set the density $\rho_0 = 10^5$ g/cm^3 and the temperature $T_0 = 10^8$ K. At $r_i = 2.5 \times 10^8$ cm we introduce a temperature jump by setting the temperature to $T_1 = 5 \times 10^7$ K for $r \geq r_i$. Again, we set the chemical composition of the gas to half carbon and half oxygen. The complete equilibrium profile is then obtained by integrating the equation of hydrostatic equilibrium. The resulting profile then has a density jump at r_i that is unstable to the Rayleigh-Taylor instability.

To trigger the instability, we set a velocity perturbation around the interface at r_i

$$v_r(r, \varphi) = 10^{-3} c_s \cos(40\varphi) \exp\left[-100\left(\frac{r - r_i}{r_{out} - r_{in}}\right)^2\right], \tag{25}$$

where c_s is the speed of sound at the interface. The azimuthal velocity v_φ is set to zero everywhere.

We discretize the computational domain D uniformly by $N_r = 100$ radial and $N_\varphi = 300$ azimuthal cells and evolve the initial conditions until final time $t_f = 8$ s. Moreover, for this test we use reflecting boundaries in radial direction and periodic boundaries in azimuthal direction.

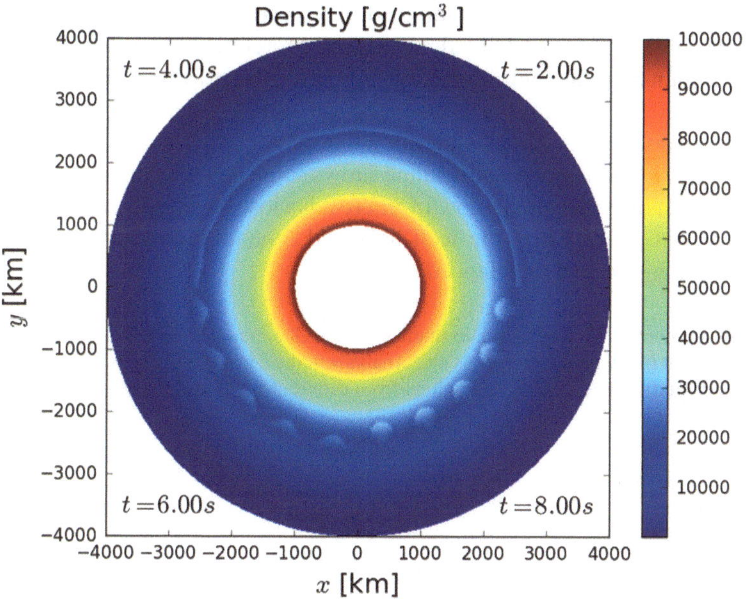

Fig. 4 Density for the Rayleigh-Taylor test problem at different times

The results computed with the well-balanced scheme are shown in Fig. 4. The figure shows the density contours where each quadrant depicts a different time. As seen from the figure, the instability develops slowly and is barely visible at times $t \leq 4$ s. At final time t_f, the Rayleigh-Taylor mushrooms are clearly developed. Although not shown, the standard scheme has some difficulties especially at following the early development of the instability as it suffers from spurious velocity perturbations due to its inability to resolve hydrostatic equilibrium, i.e. the truncation errors overwhelm the very small velocity perturbation, Eq. (25), imposed to trigger the physical instability. Moreover, at later times the standard scheme also presents some parasitic deviations near the upper and lower radial boundaries again due to its inability to maintain the close to hydrostatic state. This again confirms that well-balanced schemes are superior when the dynamics of interest happen on top of an equilibrium.

4 Conclusion

We have presented a well-balanced scheme for the Euler equations with gravity. The scheme is able to maintain up to machine precision any isothermal hydrostatic equilibrium. The well-balanced property is achieved by a novel equilibrium preserving reconstruction based on the Gibbs free energy together with a matching momentum sour term discretization. The advantage of the present approach is that it can easily handle complex EoS beyond the ideal gas law. Moreover, it can also be extended to multiple dimensions and curvilinear meshes in a straightforward manner.

The scheme's performance is assessed on astrophysically relevant test cases involving a complex EoS applicable for solar matter in a wide range of conditions. The tests show that the proposed scheme shows some clear advantages over standard unbalanced schemes when the flow of interest is close to an isothermal equilibrium. Away from an equilibrium, the well-balanced scheme recovers the accuracy and robustness of a standard unbalanced scheme.

Acknowledgements We would like to thank Laurent Gosse and Roberto Natalini for organizing the workshop on Innovative Algorithms and Analysis in the enchanting city of Rome. This resulted in many fruitful discussions with researchers from different fields and triggered many new ideas.

References

1. Audusse, E., Bouchut, F., Bristeau, M.O., Klein, R., Perthame, B.: A fast and stable well-balanced scheme with hydrostatic reconstruction for shallow water flows. SIAM J. Sci. Comput. **25**(6), 2050–2065 (2004)
2. Batten, P., Clarke, N., Lambert, C., Causon, D.M.: On the choice of wavespeeds for the HLLC Riemann solver. SIAM J. Sci. Comput. **18**(6), 1553–1570 (1997)
3. Botta, N., Klein, R., Langenberg, S., Lützenkirchen, S.: Well balanced finite volume methods for nearly hydrostatic flows. J. Comput. Phys. **196**(2), 539–565 (2004)
4. Chandrasekhar, S.: Hydrodynamic and Hydromagnetic Stability. International Series of Monographs on Physics (Oxford, England). Dover Publications, New York (1981)
5. Chandrashekar, P., Klingenberg, C.: A second order well-balanced finite volume scheme for euler equations with gravity. SIAM J. Sci. Comput. **37**(3), B382–B402 (2015)
6. Desveaux, V., Zenk, M., Berthon, C., Klingenberg, C.: A well-balanced scheme for the Euler equation with a gravitational potential. Springer Proc. Math. Stat. **77**, 217–226 (2014)
7. Fuchs, F., McMurry, A., Mishra, S., Risebro, N., Waagan, K.: High order well-balanced finite volume schemes for simulating wave propagation in stratified magnetic atmospheres. J. Comput. Phys. **229**(11), 4033–4058 (2010)
8. Gosse, L.: Computing Qualitatively Correct Approximations of Balance Laws. Springer Milan, Dordrecht (2013)
9. Gottlieb, S., Shu, C.W., Tadmor, E.: Strong stability-preserving high-order time discretization methods. SIAM Rev. **43**(1), 89–112 (2001)
10. Greenberg, J., Leroux, A.: A well-balanced scheme for the numerical processing of source terms in hyperbolic equations. SIAM J. Numer. Anal. **33**(1), 1–16 (1996)
11. Käppeli, R., Mishra, S.: Well-balanced schemes for the euler equations with gravitation. J. Comput. Phys. **259**(0), 199–219 (2014)
12. Käppeli, R., Mishra, S.: A well-balanced finite volume scheme for the Euler equations with gravitation. The exact preservation of hydrostatic equilibrium with arbitrary entropy stratification. Astron. Astrophys. **587**, A94 (2016)
13. Kippenhahn, R., Weigert, A., Weiss, A.: Stellar Structure and Evolution. Astronomy and Astrophysics Library. Springer, Berlin (2012)
14. Laney, C.B.: Computational Gasdynamics. Cambridge University Press, Cambridge (1998)
15. LeVeque, R.J.: Balancing source terms and flux gradients in high-resolution Godunov methods: the quasi-steady wave-propagation algorithm. J. Comput. Phys. **146**(1), 346–365 (1998)

16. LeVeque, R.J.: Finite Volume Methods for Hyperbolic Problems. Cambridge Texts in Applied Mathematics, 1st edn. Cambridge University Press, Cambridge (2002)
17. LeVeque, R.J., Bale, D.S.: Wave propagation methods for conservation laws with source terms. In: Hyperbolic Problems: Theory, Numerics, Applications, Vol. II (Zürich, 1998). International Series of Numerical Mathematics, vol. 130, pp. 609–618. Birkhäuser, Basel (1999)
18. Li, G., Xing, Y.: Well-balanced discontinuous Galerkin methods for the euler equations under gravitational fields. J. Sci. Comput. **67**, 1–21 (2015)
19. Li, G., Xing, Y.: High order finite volume {WENO} schemes for the euler equations under gravitational fields. J. Comput. Phys. **316**, 145–163 (2016)
20. Noelle, S., Pankratz, N., Puppo, G., Natvig, J.R.: Well-balanced finite volume schemes of arbitrary order of accuracy for shallow water flows. J. Comput. Phys. **213**(2), 474–499 (2006)
21. Shu, C.W.: High Order ENO and WENO Schemes for Computational Fluid Dynamics, pp. 439–582. Springer, Berlin/Heidelberg (1999)
22. Timmes, F.X.: http://cococubed.asu.edu/code_pages/eos.shtml (2013)
23. Timmes, F.X., Swesty, F.D.: The accuracy, consistency, and speed of an electron-positron equation of state based on table interpolation of the Helmholtz free energy. Astrophys. J. Suppl. **126**, 501–516 (2000)
24. Toro, E.F.: Riemann Solvers and Numerical Methods for Fluid Dynamics: A Practical Introduction. Springer, Berlin (1997)
25. Vides, J., Braconnier, B., Audit, E., Berthon, C., Nkonga, B.: A Godunov-type solver for the numerical approximation of gravitational flows. Commun. Comput. Phys. **15**(1), 46–75 (2014)
26. Xing, Y., Shu, C.W.: High order well-balanced WENO scheme for the gas dynamics equations under gravitational fields. J. Sci. Comput. **54**(2–3), 645–662 (2013)

Practical Convergence Rates for Degenerate Parabolic Equations

Kenneth H. Karlsen, Nils H. Risebro, and E.B. Storrøsten

Abstract We investigate the convergence rates of numerical schemes for degenerate convection diffusion equations. Recent results bound these rates as $1/3$ in one space dimension and $2/(19+d)$ in several space dimension. In our numerical experiments, we obtain much better rates, indicating that the theoretical bounds are not optimal.

1 Introduction

We are interested in convection-diffusion problems of the form

$$\begin{cases} u_t + \operatorname{div}(f(u)) = \Delta A(u), & (t,x) \in (0,T) \times \mathbb{R}^d, \\ u(0,x) = u_0(x), & x \in \mathbb{R}^d, \end{cases} \tag{1}$$

where T is some fixed final time and u is the unknown. The initial value u_0, the convective flux function $f : \mathbb{R} \to \mathbb{R}^d$ and the diffusion function $A : \mathbb{R} \to \mathbb{R}$ are given functions satisfying

$$u_0 \in L^1(\mathbb{R}) \cap BV(\mathbb{R}), \quad f, A \text{Lipschitz continuous; } A \text{ nondecreasing.} \tag{2}$$

Whenever $A'(u) = 0$ for all u in some interval $[\alpha, \beta] \subset \mathbb{R}$, the equation is referred to as *strongly degenerate*. Strongly degenerate equations will, in general, possess discontinuous (shock wave) solutions. As such (1) will have to be interpreted in the weak sense. However, as is well known for conservation laws ($A \equiv 0$), weak solutions are no longer unique and a second condition is needed, the so called *entropy condition*. More precisely, for $u_0 \in L^1 \cap L^\infty$, there exists a unique solution

K.H. Karlsen • N.H. Risebro (✉) • E.B. Storrøsten
Department of Mathematics, University of Oslo, Oslo, Norway
e-mail: kennethk@math.uio.no; nilshr@math.uio.no; erlendbs@math.uio.no

L. Gosse, R. Natalini (eds.), *Innovative Algorithms and Analysis*,
Springer INdAM Series 16, DOI 10.1007/978-3-319-49262-9_9

$u \in C([0,T];L^1) \cap L^\infty$ of (1) such that $\partial_x A(u) \in L^2$ and for all convex functions $S : \mathbb{R} \to \mathbb{R}$

$$\partial_t S(u) + \mathrm{div} Q_S(u) - \Delta R_S(u) \leq 0 \text{ weakly on } [0,T) \times \mathbb{R}^d,$$

where $Q_S' = f'S'$ and $R_S' = A'S'$. The satisfaction of this inequality for all convex S is the entropy condition, and a weak solution satisfying the entropy condition is an *entropy solution.* Entropy solutions are unique, and the initial value problem is well-posed in the class of entropy solutions, see [1]. If $u \mapsto A(u)$ is strictly monotone increasing, then the above entropy condition is not needed to show uniqueness, see the book of Wu and Jin [15]. In several dimensions and for equations with partial diffusivity, one can employ a different entropy solution concept, see [4].

To obtain a finite difference approximation of (1) the flux is discretized by a two-point monotone numerical flux and the diffusion term is approximated by a second order central difference. As an example, consider the implicit scheme in one spatial dimension and suppose for simplicity that $f' \geq 0$. Let $\Delta x > 0$, $\Delta t = T/N$ for $N \geq 1$ and suppose $U = \left\{U_j^n\right\}_{j \in \mathbb{Z}, 0 \leq n \leq N}$ satisfies

$$\frac{U_j^{n+1} - U_j^n}{\Delta t} + \frac{f(U_j^{n+1}) - f(U_{j-1}^{n+1})}{\Delta x} = \frac{A(U_{j+1}^{n+1}) - 2A(U_j^{n+1}) + A(U_{j-1}^{n+1})}{\Delta x^2}, \tag{3}$$

for $0 \leq n < N, j \in \mathbb{Z}$. Set $t_n = n\Delta t, x_j = j\Delta x$. For simplicity, in this paper, we assume that $f'(u) \geq 0$. therefore we use the upwind discretization for the derivative $f(u)_x$. In the general case, $f(U_j)$ can be replaced by any monotone two point numerical flux $F(U_j, U_{j+1})$. The piecewise constant interpolation u_Δ is defined by

$$u_\Delta(t,x) = U_j^n, \quad \text{whenever } (t,x) \in I_j^n = (t_{n-1}, t_n] \times [x_{j-1/2}, x_{j+1/2}). \tag{4}$$

In general we let Δt be a suitable function of Δx, i.e., $\Delta t \sim \Delta x^\gamma$ for some $\gamma > 0$. We are interested in the largest possible $r > 0$ for which there exists a constant C, independent of Δx, such that for all $0 \leq t \leq T$

$$\|u_\Delta(t,\cdot) - u(t,\cdot)\|_{L^1} \leq C\Delta x^r. \tag{5}$$

The number r is then referred to as the *rate of convergence*. Regarding convergence rates we restrict our attention to the BV setting, $u_0 \in BV$.

In the case of conservation laws ($A \equiv 0$) it is a classical result due to Kuznetsov that monotone schemes converges with rate $1/2$ for general $d \geq 1$ [8], and this is optimal. In general, adding diffusion to the equation yields solutions of higher regularity. One way this extra regularity manifests itself is that if $d = 1$, $x \mapsto A(u(t,x))_x$ is of bounded variation for $t > 0$. This means that $A(u)$ is continuous, and if A is strictly increasing, so is $u(\cdot, t)$. As the approximation of a derivative by a finite difference is fundamentally dependent on the regularity of the solution, one would expect a higher rate of convergence than in the completely degenerate case

$A \equiv 0$. This is indeed the case for parabolic equations, i.e., when $A' > \eta$ for some $\eta > 0$. Under this assumption, solutions to (1) are smooth, and convergence takes place according to the formal order of the scheme. Thus, naively one would expect a rate for degenerate equations to be no lower than for conservation laws. However, the presence of a term with two derivatives may be "incompatible" with solutions containing discontinuities (belong to BV) and thereby lower the convergence rate.

Regarding error estimates for weak solutions of degenerate parabolic equations without nonlinear convection, see e.g. [11, 12]. In [3], Ebmeyer considered finite element approximation in the case $f \equiv 0$ and $A(u) = u\,|u|^{m-1}$, and obtained the convergence rate $\|u - u_{\Delta x}\|_{L^2((0,T);L^2(\Omega))} \leq \mathcal{O}\left(\Delta x^{1/6}\right)$ independently of m. This result holds where Ω is a bounded domain in $\mathbb{R}^d$ with Dirichlet boundary conditions.

Concerning degenerate problems with a nonlinear convective term, it was recently shown in [6] that (5) holds true with $r = 1/3$ in the case $d = 1$. Subsequently, in [7] a semidiscrete scheme (time-continuous) was considered for general $d \geq 1$ using a cartesian grid. Therein, (5) was shown with $r = 2/(19+d)$, for $d > 1$ and $r = 1/3$ for $d = 1$. Let us point out that this result holds true for the implicit scheme as well. The discrepancy between naive intuition and the theoretical results, and between $d = 1$ and $d > 1$ raises the question of whether or not the rates obtained in [6] and [7] are optimal. We remark that the results of Ebmeyer [3] are not in conflict with those in [6, 7] since the norms are different, and Ebmeyer considers weak solutions with Dirichlet conditions on a bounded domain whereas [6, 7] treat entropy solutions of the Cauchy problem.

In the purely hyperbolic case Teng and Zhang [14] demonstrated that for monotone schemes, the convergence rate was $r = 1$ if the entropy solution consisted of a finite number of shocks. For the general case, Sabac [13] proved that the rate $r = 1/2$ is optimal for monotone schemes, even for strictly nonlinear flux functions. Also, for strictly convex flux functions, Nessyahu and Tadmor [9] and Nessyahu et al. [10], showed that in a weaker norm, better convergence rates may be obtained. More precisely,

$$\|u - u_{\Delta x}\|_{W^{s,p}(\mathbb{R})} \leq \mathcal{O}\left(\Delta x^{(1-sp)/2p}\right), \qquad -1 \leq s \leq 1/p.$$

For $(s, p) = (0, 1)$ this is the "usual L^1 rate" of $r = 1/2$. This result uses that for entropy solutions to scalar conservation laws with a strictly convex flux, $\|u(t, \cdot)\|_{\mathrm{Lip}^+} \leq \mathcal{O}(1/t)$. In the presence of nonlinear diffusion, this does not hold, as simple examples will show. Hence, trying to generalize the results of [9, 10] to degenerate parabolic problems does not seem straightforward.

In this paper we seek to shed some light on this question by way of numerical experiments. In most (all but one) of these experiments we have a formula for the exact solution, with isolated discontinuities in the function or its derivative. In all of these cases we find that the actual (numerical) convergence rate is much better than the theoretical bound. In order to investigate this we use the numerical truncation error. For more information about monotone schemes for strongly degenerate diffusion equations, we refer to [5], and to [2] for an overview of error estimation theory for conservation laws.

1.1 Formal Rates of Convergence

Under the assumption of a smooth solution, it is relatively easy to obtain error estimates by considering the truncation error. Due to the low regularity of the entropy solution, it is clear that this does not give a rigorous estimate. Nonetheless, in the sequel, several numerical examples are presented where the exact solution is known. In these cases the truncation error may shed some light on the observed rate of convergence. Let us therefore derive a formal error estimate for the scheme (3).

Let D_+, D_- denote the forward and backward difference with respect to Δx acting in the x-variable respectively. That is, for any function $u = u(x)$

$$D_+u(x) = \frac{u(x+\Delta x) - u(x)}{\Delta x}, \qquad D_-u(x) = \frac{u(x) - u(x-\Delta x)}{\Delta x}.$$

The difference quotients D^t_+ and D^t_- are defined analogously, but with respect to Δt and acting with respect to the t-variable. The piecewise constant approximation u_Δ, defined by (4), satisfies

$$D^t_- u_\Delta + D_- f(u_\Delta) = D_- D_+ A(u_\Delta), \qquad \text{for } 0 \le t \le T. \tag{6}$$

Let u be the entropy solution to (1). The truncation error $\mathscr{T}_\Delta$ is defined by

$$D^t_- u + D_- f(u) = D_- D_+ A(u) + \mathscr{T}_\Delta, \qquad (\Delta t \le t \le T). \tag{7}$$

Subtract (6) from (7), multiply by sign $(u - u_\Delta)$ and integrate to arrive at

$$\begin{aligned}\int_{\mathbb{R}} \operatorname{sign}(u-u_\Delta)\, D^t_-(u-u_\Delta)\,dx = -\int_{\mathbb{R}} &\operatorname{sign}(u-u_\Delta)\,(\mathscr{A}(u) - \mathscr{A}(u_\Delta))\,dx \\ &+ \int_{\mathbb{R}} \operatorname{sign}(u-u_\Delta)\,\mathscr{T}_\Delta\,dx,\end{aligned}$$

where $\mathscr{A}(u) = D_-(f(u) - D_+A(u))$. A fundamental observation regarding the scheme (3) is the fact that $\mathscr{A}$ is an accretive operator, see e.g., [7, Lemma A.1]. This yields

$$\int_{\mathbb{R}} \operatorname{sign}(u-u_\Delta)\,(\mathscr{A}(u) - \mathscr{A}(u_\Delta))\,dx \ge 0.$$

Let $w = u - u_\Delta$ and observe that

$$D^t_-\,|w(t)| = \operatorname{sign}(w(t))\,D^t_- w(t) - \underbrace{\frac{2}{\Delta t}\int_{w(t-\Delta t)}^{w(t)} \delta(z)(z - w(t-\Delta t))\,dz}_{\ge 0},$$

where $\delta(z)$ denotes the Dirac measure centered at $z = 0$. Hence, by the above observations

$$
\begin{aligned}
\int_{\mathbb{R}} |w(t_n)| - |w(t_0)| \, dx &= \Delta t \sum_{k=1}^{n} \int_{\mathbb{R}} D_-^t \, |w(t_k)| \, dx \\
&\leq \Delta t \sum_{k=1}^{n} \int_{\mathbb{R}} \operatorname{sign}(w(t_k)) \, D_-^t w(t_k) \, dx \\
&\leq \Delta t \sum_{k=1}^{n} \int_{\mathbb{R}} \operatorname{sign}(w(t_k)) \, \mathscr{T}_\Delta(t_k) \, dx.
\end{aligned}
$$

Consequently,

$$
\begin{aligned}
\|u(t_n) - u_\Delta(t_n)\|_{L^1(\mathbb{R})} \leq \left\| u_0 - u_\Delta^0 \right\|_{L^1(\mathbb{R})} \\
+ \Delta t \sum_{k=1}^{n} \underbrace{\int_{\mathbb{R}} \operatorname{sign}(u(t_k) - u_\Delta(t_k)) \, \mathscr{T}_\Delta(t_k) \, dx}_{R_\Delta(t_k)},
\end{aligned}
\tag{8}
$$

where u_Δ^0 denotes the initial value of the scheme. Now, let us under the assumption that u is smooth, write out an explicit expression for $\mathscr{T}_\Delta$. To this end, subtract (1) from (7) to arrive at

$$
(D_-^t u - \partial_t u) + (D_- f(u) - \partial_x f(u)) = (D_- D_+ A(u) - \partial_x^2 A(u)) + \mathscr{T}_\Delta.
$$

Using Taylor series with remainders we arrive at the expression

$$
\begin{aligned}
\mathscr{T}_\Delta(t,x) = & \underbrace{-\frac{1}{\Delta t} \int_{-\Delta t}^{0} \partial_t^2 u(t+\sigma, x)(\sigma + \Delta t) \, d\sigma}_{\mathscr{T}_\Delta^t} \\
& \underbrace{-\frac{1}{\Delta x} \int_{-\Delta x}^{0} \partial_x^2 f(u(t, x+\sigma))(\sigma + \Delta x) \, d\sigma}_{\mathscr{T}_\Delta^f(t,x)} \\
& \underbrace{+\frac{1}{6\Delta x^2} \int_{0}^{\Delta x} \partial_x^4 \left(A(u(t,x+\sigma)) - A(u(t,x-\sigma))\right) (\sigma - \Delta x)^3 \, d\sigma}_{\mathscr{T}_\Delta^A(t,x)} .
\end{aligned}
$$

The assumption that u is smooth yields the formal convergence rate of the scheme. Note that the flux-term is of order Δx, the time-discretization-term is of order Δt, while the diffusion-term is of order Δx^2. Regarding numerical experiments, it may occur, even with discontinuous solutions, that convergence takes place at the formal order. Let us therefore continue the discussion in order to establish some intuition regarding what happens in the case of an isolated discontinuity in u or its derivatives.

We first restrict our attention to the error associated with the spatial discretization. Let w represent either $f(u(t,\cdot))$ or $A(u(t,\cdot))$, for some fixed $t \in [0,T]$. Assume that w satisfies

$$w(x) = \begin{cases} w_r(x), & x > \xi, \\ w_l(x), & x < \xi, \end{cases}$$

where ξ is a constant, and where both w_r and w_l belong to $C_b^\infty(\mathbb{R})$ (the space of infinitely differentiable functions with bounded derivatives). The k-th derivative of w satisfies

$$\partial^k w(x) = \sum_{n=0}^{k-1} \underbrace{(\partial^n w_r(\xi) - \partial^n w_l(\xi))}_{[\![\partial^n w]\!]_\xi} \partial^{k-n-1}\delta(x-\xi) \\ + \underbrace{\mathbf{1}_{x<\xi}\partial^k w_l(x) + \mathbf{1}_{\xi<x}\partial^k w_r(x)}_{\mathfrak{S}(\partial^k w)},$$

in the distributional sense. For the term $\mathcal{T}_\Delta^f$ this yields

$$\begin{aligned} \mathcal{T}_\Delta^f(\cdot,x) = {} & [\![f(u)]\!]_\xi \tfrac{1}{\Delta x} \int_{-\Delta x}^0 \delta(x+\sigma-\xi)\,d\sigma \\ & - [\![\partial_x f(u)]\!]_\xi \tfrac{1}{\Delta x} \int_{-\Delta x}^0 \delta(x+\sigma-\xi)(\sigma+\Delta x)\,d\sigma \\ & - \tfrac{1}{\Delta x} \int_{-\Delta x}^0 \mathfrak{S}(\partial_x^2 f(u(\cdot,x+\sigma)))(\sigma+\Delta x)\,d\sigma. \end{aligned}$$

Let us split R_Δ defined in (8), according to (9), so that

$$R_\Delta = R_\Delta^t + R_\Delta^f + R_\Delta^A.$$

By the above, we obtain

$$R_\Delta^f = [\![f(u)]\!]_\xi \frac{1}{\Delta x} \int_\xi^{\xi+\Delta x} \operatorname{sign}(u - u_\Delta)\,dx + \mathcal{O}(\Delta x).$$

Consider $\mathcal{T}_\Delta^A$. Inserting the expression for the distributional derivative yields

$$\begin{aligned} \mathcal{T}_\Delta^A(\cdot,x) = {} & [\![\partial_x A(u)]\!]_\xi \tfrac{1}{\Delta x^2} \int_0^{\Delta x} (\delta(x+\sigma-\xi) - \delta(x-\sigma-\xi))(\sigma-\Delta x)\,d\sigma \\ & - [\![\partial_x^2 A(u)]\!]_\xi \tfrac{1}{3\Delta x^2} \int_0^{\Delta x} (\delta(x+\sigma-\xi) - \delta(x-\sigma-\xi))(\sigma-\Delta x)^2\,d\sigma \\ & + [\![\partial_x^3 A(u)]\!]_\xi \tfrac{1}{6\Delta x^2} \int_0^{\Delta x} (\delta(x+\sigma-\xi) - \delta(x-\sigma-\xi))(\sigma-\Delta x)^3\,d\sigma \\ & + \tfrac{1}{6\Delta x^2} \int_0^{\Delta x} \mathfrak{S}\left(\partial_x^4 \left(A(u(\cdot,x+\sigma)) - A(u(\cdot,x-\sigma))\right)\right)(\sigma-\Delta x)^3\,d\sigma, \end{aligned}$$

where we applied the fact that $A(u(t,\cdot))$ is continuous. It follows that

$$R_\Delta^A = -[\![\partial_x A(u)]\!]_\xi \tfrac{1}{\Delta x^2} \int_{\mathbb{R}} \operatorname{sign}(u-u_\Delta)\, K_\Delta(x,\xi)\, dx \\ \underbrace{+ [\![\partial_x^2 A(u)]\!]_\xi \frac{1}{3\Delta x^2} \int_{\mathbb{R}} \operatorname{sign}(u-u_\Delta)\, K_\Delta^2(x,\xi)\, dx}_{\mathscr{O}(\Delta x)} + \mathscr{O}\left(\Delta x^2\right),$$

where

$$K_\Delta(x,\xi) = \begin{cases} x-(\xi-\Delta x) & \text{if } \xi-\Delta x \le x \le \xi, \\ x-(\xi+\Delta x) & \text{if } \xi \le x \le \xi+\Delta x, \\ 0 & \text{otherwise.} \end{cases}$$

Let us turn to the error associated with the discretization in time, i.e., $\mathscr{T}_\Delta^t$. Assume that $u = u(t,x)$ has a discontinuity at $t = \tau(x)$, of the same type as above. Then

$$\begin{aligned} \mathscr{T}_\Delta^t(t,x) = [\![u]\!]_\tau \tfrac{1}{\Delta t} &\int_{-\Delta t}^0 \delta(t+\sigma-\tau)\, d\sigma \\ -[\![\partial_t u]\!]_\tau \tfrac{1}{\Delta t} &\int_{-\Delta t}^0 \delta(t+\sigma-\tau)(\sigma+\Delta t)\, d\sigma \\ -\tfrac{1}{\Delta t} &\int_{-\Delta t}^0 \mathfrak{S}(\partial_t^2 u(t+\sigma,x))(\sigma+\Delta t)\, d\sigma, \end{aligned}$$

where $[\![u]\!]_\tau = u(\tau+) - u(\tau-)$, using the notation $u(\tau\pm) = \lim_{h\downarrow 0} u(\tau \pm h)$. It follows that

$$R_\Delta^t(t) = \frac{1}{\Delta t} \int_{\mathbb{R}} \operatorname{sign}(u-u_\Delta)\, [\![u]\!]_\tau \mathbf{1}_{t-\Delta t \le \tau \le t}\, dx + \mathscr{O}(\Delta t).$$

Suppose the discontinuity lies along an increasing curve $\xi(t)$, so that $\xi(\tau(x)) = x$, or equivalently $\tau(\xi(t)) = t$. Then

$$t - \Delta t \le \tau(x) \le t \Leftrightarrow \xi(t-\Delta t) \le x \le \xi(t),$$

and so

$$R_\Delta^t(t) = \frac{1}{\Delta t} \int_{\xi(t-\Delta t)}^{\xi(t)} \operatorname{sign}(u(t,x) - u_\Delta(t,x))\, [\![u]\!]_{\tau(x)}\, dx + \mathscr{O}(\Delta t).$$

What we now have is a description of how the presence of an isolated discontinuity in the solution will affect the truncation error, possibly causing a slower convergence than the formal order.

2 Numerical Experiments

Now we investigate the errors for a series of problems where we have an exact solution, or where we have a reference solution which we believe is sufficiently accurate. In the cases where we have a formula for the exact solution, we consider the expressions for the truncation error, and hopefully shed some light on the observed convergence rate.

In addition to the implicit scheme defined by (3), we test the Crank-Nicholson scheme defined by

$$D^t_- U_j^{n+1} + \frac{1}{2} D_- \left(f(U_j^{n+1}) + f(U_j^n) \right) = \frac{1}{2} D_- D_+ \left(A(U_j^{n+1}) + A(U_j^n) \right). \tag{9}$$

For both schemes we set $U_j^0 = \frac{1}{\Delta x} \int_{I_j} u_0(x)\, dx$. The Crank-Nicholson scheme is formally second order in time, and if $f \equiv 0$, second order in space as well.

We shall always measure the error as follows. Let u_r be some reference solution (either available as an expression, or calculated numerically). The relative error is defined as

$$E = 100 \times \frac{\|u_\Delta(t,\cdot) - u_r(t,\cdot)\|_{L^1}}{\|u_r(t,\cdot)\|_{L^1}}. \tag{10}$$

In the following examples we set $\Delta t = \Delta x/2$.

2.1 No Convection

Set

$$A(u) = \operatorname{sign}(u) \max(|u| - 1, 0), \tag{11}$$

then

$$u(t,x) = e^{-\pi^2 t} \cos(\pi x) + \operatorname{sign}(\cos(\pi x)) \tag{12}$$

is an exact solution of

$$u_t = A(u)_{xx}. \tag{13}$$

Note that A is piecewise linear and continuous, with $A'(u) = 0$ for $u \in (-1,1)$ and $A'(u) = 1$ outside this interval. Table 1 shows the errors of the approximate solutions generated by the implicit and Crank-Nicholson schemes for $t = 0.2$, $\Delta x = 2 \times 2^{-N}$ with $N = 3, \ldots, 12$. We note that for this example, both schemes gave

Table 1 Errors for the implicit and the Crank-Nicholson schemes for (11)–(13)

N	3	4	5	6	7	8	9	10	11	12	Rate
Implicit	7.54	4.11	2.17	1.13	0.70	0.35	0.18	0.09	0.05	0.02	0.93
CN	0.57	0.14	0.04	0.02	5e−3	1e−3	3e−4	9e−5	2e−5	5e−6	1.83

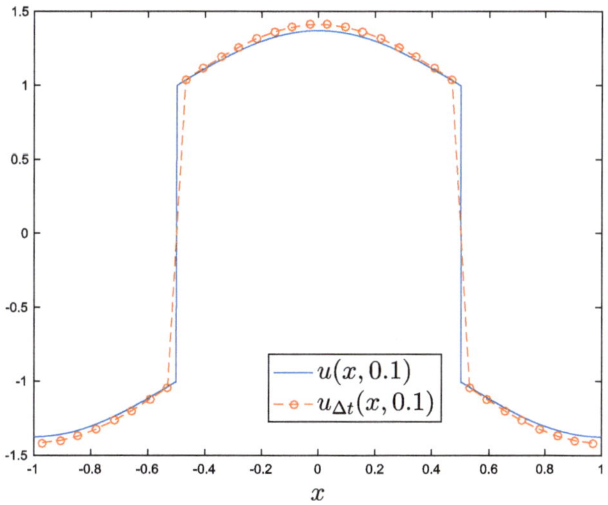

Fig. 1 The exact solution (12) and the approximation found with the implicit scheme with $\Delta x = 2^{-4}$ at $t = 0.1$ for (11)–(13)

errors which were consistent with their formal orders, and that the errors produced by the Crank-Nicholson scheme were significantly smaller than those of the implicit scheme. Figure 1 depicts the exact and the approximate solution using the implicit scheme with 2^5 grid point in the interval $[-1, 1]$.

The fact that convergence takes place according to the formal order follows from the truncation analysis. Consider the truncation error $\mathscr{T}_\Delta = \mathscr{T}_\Delta^t + \mathscr{T}_\Delta^A$ for the implicit scheme defined in (7). By (9)

$$\mathscr{T}_\Delta^A(t,x) = e^{-\pi^2 t}\frac{\pi^4}{6\Delta x^2}\int_0^{\Delta x} \underbrace{\big(\cos(\pi(x+\sigma)) - \cos(\pi(x-\sigma))\big)}_{-2\sin(\pi x)\sin(\pi\sigma)}(\sigma - \Delta x)^3\, d\sigma,$$

and so $\mathscr{T}_\Delta^A(t,x)$ is of order Δx^3. Similarly

$$\mathscr{T}_\Delta^t(t,x) = -\frac{\pi^4}{\Delta t}e^{-\pi^2 t}\cos(\pi x)\int_{-\Delta t}^0 e^{-\pi^2\sigma}(\sigma + \Delta t)\, d\sigma,$$

and so $\mathscr{T}_\Delta^t(t,x)$ is of the order Δt. The comparatively large error connected to the time discretization with versus the spatial discretization explains the improved results obtained by the Crank-Nicholson method. A plot of the local truncation error sign $(u - u_\Delta)\,\mathscr{T}_\Delta$, for $t = 0.1$, is given in Fig. 2. It is clear that $\mathscr{T}_\Delta^t$ is the dominant term.

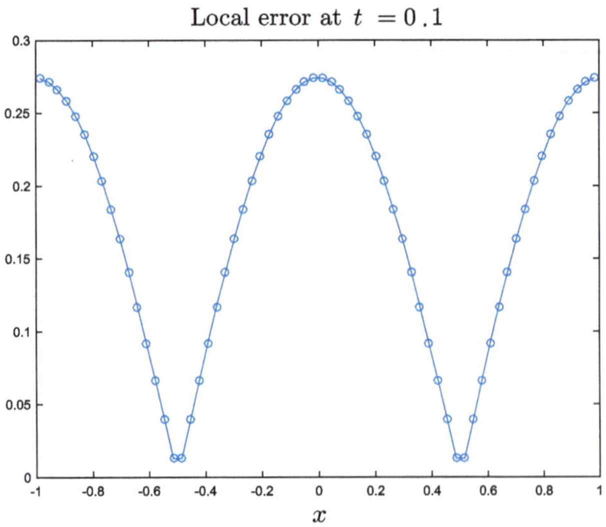

Fig. 2 The truncation error $\text{sign}\,(u - u_\Delta)\,\mathscr{T}_\Delta$ for the implicit scheme with $\Delta x = 2^{-5}$ at $t = 0.1$ for (11)–(13)

This convergence can also be explained by the fact that the schemes "do not see" the degenerate diffusion. If we set $\mu = \Delta t/\Delta x^2$, $\omega = 4\mu(\cos^2(\pi\Delta x/2) - 1)$ and

$$V_j^n = (1-\omega)^{-n}\cos(\pi j\Delta x),$$

then $U_j^n = V_j^n + \text{sign}\left(V_j^0\right)$ will solve the implicit scheme. Similarly, if

$$W_j^n = \left(\frac{2+\omega}{2-\omega}\right)^n \cos(\pi j\Delta x),$$

then $W_j^n + \text{sign}\left(W_j^0\right)$ is a solution of the Crank-Nicholson scheme. We observe that $\left\{V_j^n\right\}$ is a solution of the implicit scheme for the linear heat equation $v_t = v_{xx}$, hence

$$\begin{aligned}\|U^n - u(n\Delta t,\cdot)\|_{L_1} &= \left\|V^m + \text{sign}\left(V^0\right) - e^{-\pi n\Delta t}\cos(\pi\,\cdot) - \text{sign}\left(\cos(\pi\,\cdot)\right)\right\|_{L^1(\mathbb{R})}\\ &= \left\|V^m - e^{-\pi n\Delta t}\cos(\pi\,\cdot)\right\|_{L^1(\mathbb{R})} \le \mathscr{O}\left(\Delta x\right).\end{aligned}$$

Similarly $\left\{W_j^n\right\}$ is a solution of the Crank-Nicholson scheme for the linear heat equation, thus $\|w^n - v(n\Delta t,\cdot)\|_{L^1} = \mathscr{O}\left(\Delta x^2\right)$.

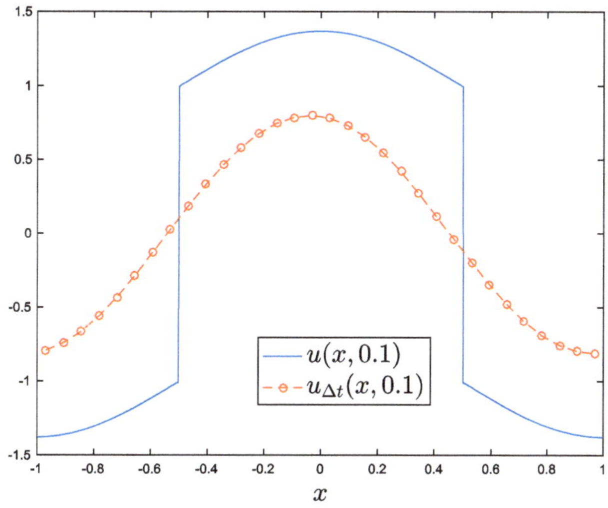

Fig. 3 The exact solution and the approximation found with the implicit scheme with $\Delta x = 2^{-4}$ at $t = 0.1$ for Example 2.2

Table 2 Errors for the implicit and the Crank-Nicholson schemes for (14)

N	3	4	5	6	7	8	9	10	11	12	Rate
Implicit	99.64	82.62	58.91	38.19	23.16	13.31	7.32	3.89	2.02	1.03	0.75
CN	90.77	68.91	46.24	28.77	16.88	9.44	5.09	2.66	1.37	0.69	0.80

2.2 Linear Convection

If u is given by (12), then $v(t,x) = u(t, x - st)$ solves the equation

$$v_t + sv_x = A(v)_{xx}, \tag{14}$$

where A is given by (11). We use this example with $s = 20$, and compute approximate solutions at $t = 0.1$, by which time the discontinuities in the initial data will have traveled one period. See Fig. 3 for a depiction of the approximate solution.

Table 2 shows the errors produced by the implicit and the Crank-Nicholson methods for this example. In this case the errors are comparable for the two methods, and the order has been reduced to about 0.8.

Since the discontinuities in the solution are moving, they will be detected by the local truncation error, which is shown in Fig. 4. Note that the error is much larger, and has two pronounced spikes located at the discontinuities. These spikes may be explained by the unbounded terms in the truncation error. That is, terms of the order $\frac{1}{\Delta x}$. Note that the solution is discontinuous along $\xi_k(t) = x_k + st$, $x_k = k + \frac{1}{2}$ for

Fig. 4 The truncation error $\text{sign}\,(u - u_\Delta)\,\mathscr{T}_\Delta$ for the implicit scheme with $\Delta x = 2^{-5}$ at $t = 0.1$ for Example 2.2

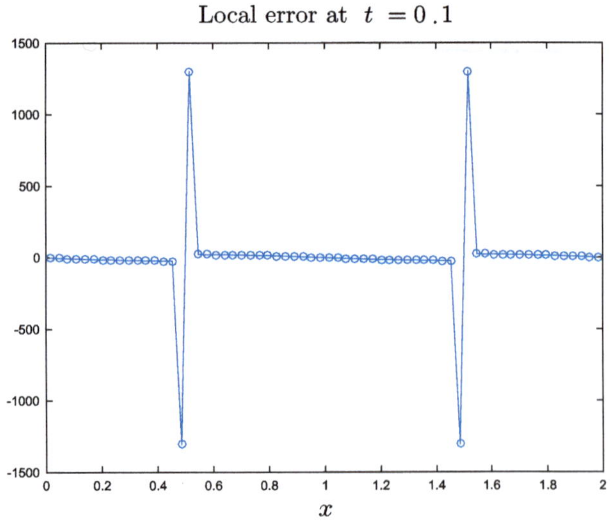

$k \in \mathbb{Z}$. Then

$$\mathscr{T}^f_\Delta(t,x) = \sum_k [\![su]\!]_{\xi_k} \frac{1}{\Delta x} \mathbf{1}_{\xi_k(t) \le x \le \xi_k(t) + \Delta x} + \mathscr{O}(1),$$

$$\mathscr{T}^t_\Delta(t,x) = \sum_k [\![u]\!]_{\tau_k} \frac{1}{\Delta t} \mathbf{1}_{\xi_k(t-\Delta t) \le x \le \xi_k(t)} + \mathscr{O}(1),$$

where $\tau_k = \frac{1}{s}(x - x_k)$. The size of the spikes fits very well with the $\mathscr{T}^f_\Delta$ term. It seems harder to explain the apparently "perfect" change of the sign, yielding a cancellation when integrated in space.

2.3 A Traveling Wave Solution

Define

$$w(y) = \begin{cases} 2(1 - e^{y/2}) & y \le 0, \\ 0 & y > 0. \end{cases} \tag{15}$$

Now $u = w(x - 2t)$ is an entropy solution of

$$u_t + \left(u^2\right)_x = \left(u^2\right)_{xx}. \tag{16}$$

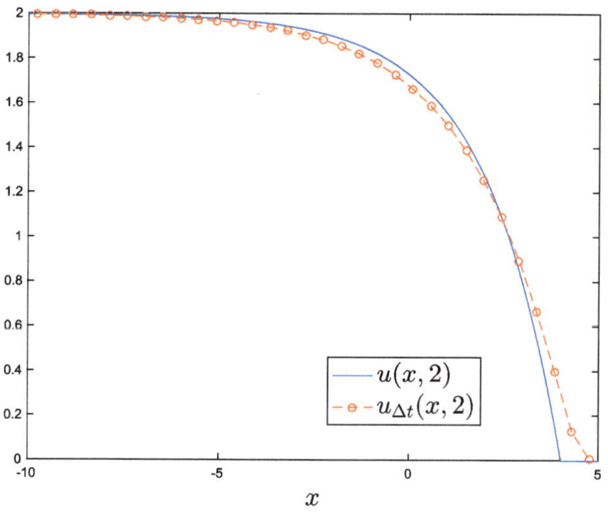

Fig. 5 The exact solution and the approximation found with the implicit scheme with $\Delta x = 15 \cdot 2^{-5}$ at $t = 2$ for (16)

Table 3 Errors for the implicit and the Crank-Nicholson schemes for (15)

N	3	4	5	6	7	8	9	10	11	12	Rate
Implicit	8.87	5.18	2.49	1.31	0.67	0.34	0.17	0.09	0.04	0.02	0.98
CN	7.23	4.20	2.00	1.04	0.53	0.27	0.13	0.07	0.04	0.02	0.98

Observe that in this case u is continuous, and in fact a classical solution except on the line $x = 2t$. We tested this on the interval $x \in [-10, 5]$ with 2^N equally spaced grid points, for $N = 3, \dots, 12$. Here we used Dirichlet boundary conditions, with the exact solution $u(t, -10)$ and $u(t, 5)$ as boundary values. Figure 5 shows the exact and the approximate solution found by the implicit scheme.

At $t = 2$ the two schemes gave the errors reported in Table 3. For this example, both schemes converge with order 1. In this case, there are no discontinuities, but a kink appears at $x = 2t$. By truncation error analysis, the only possible violation of the formal order convergence could arise due to a jump in $\partial_x A(u)$. However,

$$\partial_x A(u(t,x)) = \begin{cases} 2w(x-2t)w'(x-2t) & \text{for } x \le 2t, \\ 0, & \text{for } x > 2t, \end{cases}$$

and so, since $w(0) = 0$, there is no jump in $\partial_x A(u)$.

2.4 A Moving Discontinuity

The function

$$u(t,x) = \begin{cases} 2 - \exp(\frac{1}{4}(x - \frac{3}{4}t)), & x < \frac{3}{4}t, \\ 0, & x \ge \frac{3}{4}t, \end{cases} \tag{17}$$

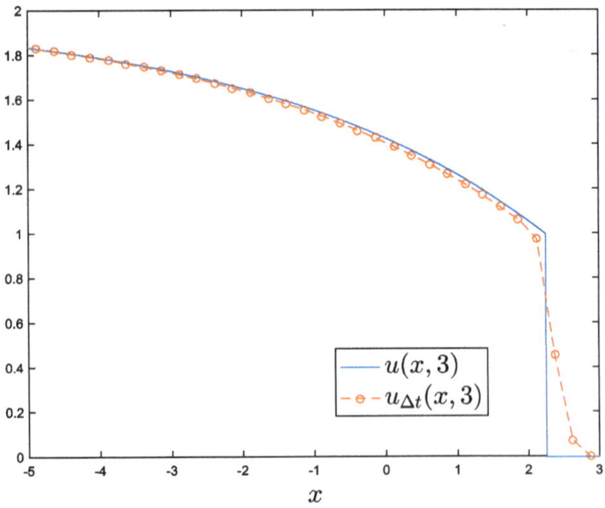

Fig. 6 The exact solution and the approximation found with the implicit scheme with $\Delta x = 8 \cdot 2^{-5}$ at $t = 3$ with the flux and diffusion given by (18)

Table 4 Errors for the implicit and Crank-Nicholson schemes with solution (17), and convective and diffuse terms specified by (18)

N	3	4	5	6	7	8	9	10	11	12	Rate
Implicit	8.36	5.93	2.12	1.08	0.54	0.27	0.14	0.07	0.03	0.02	1.0
CN	6.22	4.49	1.33	0.67	0.33	0.17	0.08	0.04	0.02	0.01	1.0

is an entropy solution to (1) if

$$f(u) = \begin{cases} \frac{1}{2}u^2, & u < 1, \\ u - \frac{1}{2}, & u \geq 1, \end{cases} \qquad A(u) = \begin{cases} 0, & u < 1, \\ u - 1, & u \geq 1. \end{cases} \tag{18}$$

We tested this with 2^N equally spaced grid points in the in interval $[-5, 3]$. See Fig. 6 for a depiction of the exact and approximate solutions at $t = 3$. The errors at $t = 3$ are reported in Table 4. Again, the convergence takes place according to the formal order of the scheme. In Fig. 7 the local error is shown. As expected the truncation error blows up at the discontinuity. As in (2.2) the sign of $(u - u_\Delta)$ seems to change exactly at the spike, making the total error small.

2.5 *Another Moving Discontinuity*

Let $v = v(y)$ satisfy the differential equation

$$\frac{d}{dy}(v-1)^2 = -u_l(v - u_l) + v^2 - u_l^2, \quad y < 0,$$

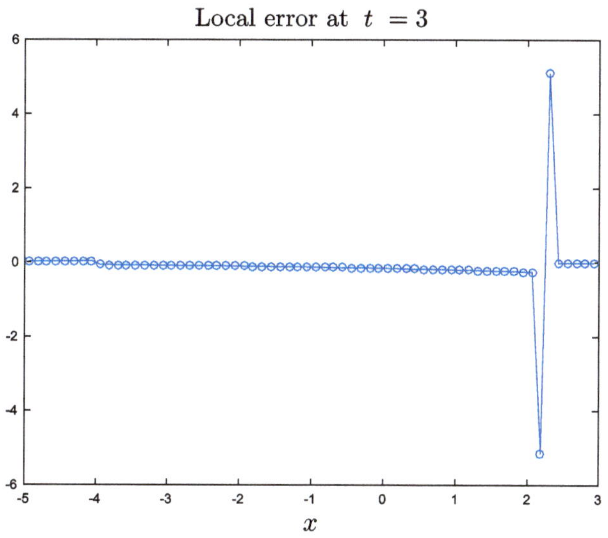

Fig. 7 The truncation error sign $(u - u_\Delta)\,\mathscr{T}_\Delta$ for the implicit scheme with $\Delta x = 2^{-5}$ at $t = 3$ for Example 2.4

and $v(0-) = 1$, $u_l > 1$. Remark that this equation can be integrated to yield

$$v(y)\,(v(y) - u_l)^{u_l - 1} = (1 - u_l)^{u_l - 1} e^{(u_l/2)y},$$

so if $u_l = 2$ we get the solution

$$v(y) = 1 + \sqrt{1 - e^y}, \quad y < 0.$$

Then the function

$$u(x,t) = \begin{cases} v(x - 2t), & x \leq 2t, \\ 0, & x > 2t, \end{cases} \tag{19}$$

is an entropy solution to the equation

$$u_t + \left(u^2\right)_x = \left(\chi_{u \geq 1}(u)(u - 1)^2\right)_{xx}. \tag{20}$$

We tested the schemes for this example with x in the interval $[-3, 3]$ and calculated the solution up to $t = 1$. The exact and the approximate solutions are depicted in Fig. 8. Table 5 shows the errors in the case. Again we observe that both schemes converge with order 1, and that there is little difference between the Crank-Nicholson and the implicit scheme. The numerical convergence rate here is close to one for both schemes. In Fig. 9 we have plotted the local truncation error at $t = 1$. We observe that it is much smaller than that of Example 2.2, while it has

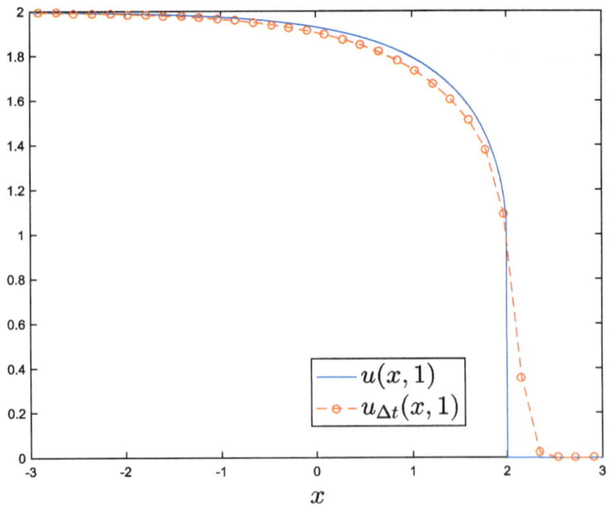

Fig. 8 The exact solution and the approximation found with the implicit scheme with $\Delta x = 6 \cdot 2^{-5}$ at $t = 1$ for (19)–(20)

Table 5 Errors for the implicit and the Crank-Nicholson schemes with solution (20), with solution (19) for $t = 1$

N	3	4	5	6	7	8	9	10	11	12	Rate
Implicit	7.62	6.52	2.11	1.69	0.53	0.43	0.13	0.11	0.03	0.03	1.0
CN	6.09	5.39	1.58	1.36	0.39	0.35	0.09	0.09	0.02	0.02	1.0

an oscillatory spike located at the discontinuity $\xi(t) = 2t$. Here, $A(u)$ has a kink, while u and $f(u)$ has a discontinuity. Hence, we expect the truncation error to blow up close to $\xi(t)$. Furthermore,

$$\partial_x f(u(t,x)) = \begin{cases} 2v(x-2t)v'(x-2t) & \text{for } x \le 2t, \\ 0 & \text{for } x > 2t, \end{cases} \qquad v'(y) = -\frac{e^y}{2\sqrt{1-e^y}},$$

yielding that $\partial_x f(u)$ is unbounded as $x \uparrow \xi$. Similarly $\partial_t u$ is unbounded as $t \downarrow x/2$. This could, in comparison with the other examples, explain the apparent asymmetric of the spike. Again, it seems that the change of sign of $u - u_\Delta$ is a key factor, preventing the total error from becoming large. In fact, we note that the singularity does not seem to obstruct the convergence much, while this appear hard to explain based on the truncation error.

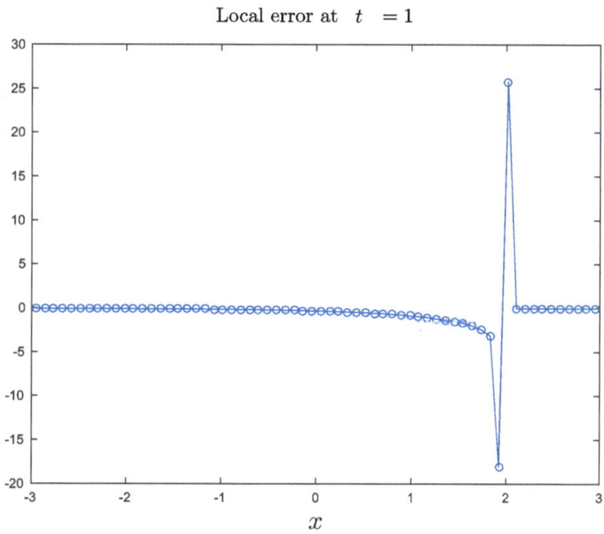

Fig. 9 The truncation error for Example 2.5 at $t = 1$ using 2^6 grid points in the interval $[-3, 3]$

2.6 *A Problem Motivated by Flow in Porous Media*

Inspired by models of two-phase flow in porous media, we investigated the convergence rate of our schemes in the case where

$$f(u) = \frac{u^2}{u^2 + (1-u)^2}, \qquad A(u) = \frac{1}{10}(f(u))^2. \tag{21}$$

with initial data

$$u_0(x) = \begin{cases} 1 & x \in [0.2, 1], \\ 0 & \text{otherwise}. \end{cases} \tag{22}$$

Figure 10 shows the Crank-Nicholson approximation $u_{\Delta x}(0.5, \cdot)$ calculated using 2^9 equally spaced grid points in the interval $[0, 2]$. For this problem, we have no solution on closed form, and to test a convergence rate, we compared our approximations to a reference solution calculated using the Crank-Nicholson scheme on a grid with 2^{12} grid points.

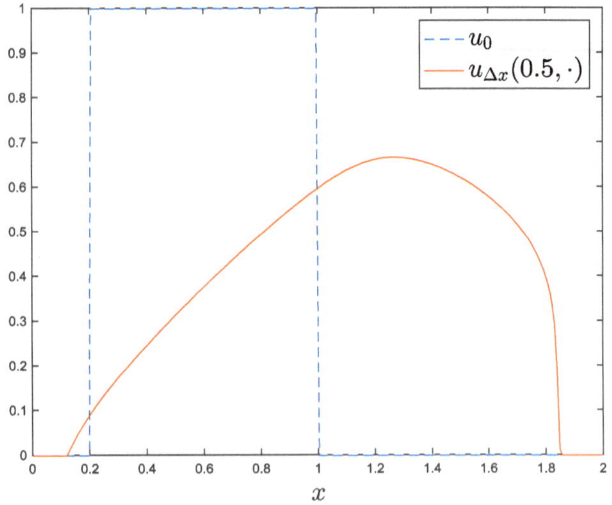

Fig. 10 The approximate solution to (1), with f and A given by (21), initial data given by (22) at $t = 0.5$, using 2^9 equally spaced grid points

Table 6 Errors for the implicit and the Crank-Nicholson schemes with convective and diffuse terms specified by (21)

N	3	4	5	6	7	8	9	10	Rate
Implicit	27.8	17.8	9.4	4.9	2.7	1.4	0.6	0.2	0.97
CN	27.1	17.5	9.2	4.8	2.7	1.4	0.6	0.2	0.97

Table 6 shows the estimated errors for the two methods applied to this example using 2^N equally spaced grid points in the interval $[0, 2]$.

2.7 *No Convection in Two Dimensions*

We now test the scheme on the two dimensional problem

$$u_t = \Delta A(u), \tag{23}$$

where A is given by (11). In this case an exact solution is $u(t, x, y) = e^{-\pi^2 t} v(x, y) + \operatorname{sign}(v(x, y))$, with

$$v(x, y) = \sin(\pi x) + \sin(\pi y).$$

Table 7 shows the error generated by the implicit and the Crank-Nicholson schemes at $t = 1/4$ on the domain $[-1, 1] \times [-1, 1]$ and $t = 0.2$ with $\Delta x = \Delta y = 2^{-N}$ for $N = 4, \ldots, 10$. Again, as in the one dimensional case (2.1), convergence takes place according to the formal order of the scheme. As in the one dimensional case, this is not a surprise as $(t, x, y) \mapsto A(u(t, x, y))$ is smooth.

Table 7 Errors for the implicit and the Crank-Nicholson schemes for (23)

N	4	5	6	7	8	9	10	Rate
Implicit	6.24	3.05	1.51	0.75	0.37	0.18	0.09	1.0
CN	0.3872	0.0935	0.0232	0.0058	0.0014	0.0004	0.0001	2.0

Table 8 Errors for the implicit and the Crank-Nicholson schemes for (24)

N	4	5	6	7	8	9	10	Rate
Implicit	39.0	28.5	18.7	12.9	8.9	6.0	4.1	0.55
CN	32.9	23.7	14.9	10.2	6.9	4.7	3.2	0.56

2.8 *Linearly Convected Discontinuities*

If u denotes the exact solution of the previous example, then for all constants a and b, $\tilde{u}(t,x,y) = u(t, x - at, y - bt)$ will solve

$$\tilde{u}_t + a\tilde{u}_x + b\tilde{u}_y = \Delta A(\tilde{u}), \tag{24}$$

where A is given by (11). Table 8 shows the errors obtained by the implicit and the Crank-Nicholson methods for this example on the domain $[-1,1] \times [-1,1]$ with $a = \pi$ and $b = 1$ and $t = 0.2$.

Here, the rate is close to $1/2$ which is considerably lower than the formal order. Due to the smoothness of the mapping $(t,x,y) \mapsto A(u(t,x,y))$, the lowered rate is certainly related to the transport rather than the degenerate diffusion.

3 Conclusions

We have presented a series of numerical experiments aiming at calculating the convergence rate for the implicit (and for the Crank-Nicholson) scheme for nonlinear degenerate convection diffusion equations. In one space dimension, the best known theoretical bound for the convergence rate is $1/3$. The technical reason for this rather low bound is the degenerate diffusion, not the non-linear flux term. Our examples do not exhibit this low convergence rate, indeed, in most of them, we record first order convergence or better. However, the examples where we see the "best" convergence rates, are examples where the scheme does not see the degeneracy in the diffusion. These examples have no convective term, and adding a linear convective term reduces the convergence rates to that of the linear convection equation (without diffusive terms). This seems to indicate that the nonlinear degenerate diffusion is not an obstacle to getting the same convergence rate as for scalar conservation laws. Nevertheless, most of our examples compare the numerical approximation with an exact solution on closed form, and we see that there is some cancellation in the local truncation error. The cause of this may be that the data are compatible with the equation, and we would not see this cancellation with other initial data.

Unfortunately, it is more difficult to measure errors (in particular truncation errors) when we have no exact solution to compare with. In Sect. 2.6 we try to do this, and observe first order convergence. Since we have no exact solution, this first order convergence is only self-convergence of the scheme. Also, the nonlinear diffusion in this example is strictly increasing, meaning that the solution is continuous which may not be the worst case scenario.

Unfortunately, we conclude that we lack a complete understanding of the mechanisms leading to the better than theoretical convergence rates observed here, and that whether $1/3$ is the optimal rate in one space dimension remains an open problem.

Acknowledgements This work was supported by the Research Council of Norway via grants no. 250674/F20 and 214495.

References

1. Carrillo, J.: Entropy solutions for nonlinear degenerate problems. Arch. Ration. Mech. Anal. **147**(4), 269–361 (1999). MR 1709116 (2000m:35132)
2. Cockburn, B.: Continuous dependence and error estimation for viscosity methods. Acta Numer. **12**, 127–180 (2003). MR 2249155
3. Ebmeyer, C.: Error estimates for a class of degenerate parabolic equations. SIAM J. Numer. Anal. **35**(3), 1095–1112 (1998) (electronic). MR 1619863
4. Escobedo, M., Vázquez, J.L., Zuazua, E.: Entropy solutions for diffusion-convection equations with partial diffusivity. Trans. Am. Math. Soc. **343**(2), 829–842 (1994). MR 1225573
5. Evje, S., Karlsen, K.H.: Monotone difference approximations of BV solutions to degenerate convection-diffusion equations. SIAM J. Numer. Anal. **37**(6), 1838–1860 (2000) (electronic). MR 1766850 (2001g:65110)
6. Karlsen, K.H., Risebro, N.H., Storrøsten, E.B.: L^1 error estimates for difference approximations of degenerate convection-diffusion equations. Math. Comput. **83**(290), 2717–2762 (2014). MR 3246807
7. Karlsen, K.H., Risebro, N.H., Storrøsten, E.B.: On the convergence rate of finite difference methods for degenerate convection-diffusion equations in several space dimensions. ESAIM Math. Model. Numer. Anal. **50**(2), 499–539 (2016). MR 3482553
8. Kuznetsov, N.N.: The accuracy of certain approximate methods for the computation of weak solutions of a first order quasilinear equation. Ž. Vyčisl. Mat. i Mat. Fiz. **16**(6), 1489–1502, 1627 (1976). MR 0483509 (58 #3510)
9. Nessyahu, H., Tadmor, E.: The convergence rate of approximate solutions for nonlinear scalar conservation laws. SIAM J. Numer. Anal. **29**(6), 1505–1519 (1992). MR 1191133
10. Nessyahu, H., Tadmor, E., Tassa, T.: The convergence rate of Godunov type schemes. SIAM J. Numer. Anal. **31**(1), 1–16 (1994). MR 1259963
11. Nochetto, R.H.: Finite element methods for parabolic free boundary problems. In: Advances in Numerical Analysis (Lancaster, 1990), vol. I, pp. 34–95. Oxford Science Publications/Oxford University Press, New York (1991). MR 1138471
12. Radu, F.A., Pop, I.S., Knabner, P.: Error estimates for a mixed finite element discretization of some degenerate parabolic equations. Numer. Math. **109**(2), 285–311 (2008). MR 2385655
13. Şabac, F. : The optimal convergence rate of monotone finite difference methods for hyperbolic conservation laws. SIAM J. Numer. Anal. **34**(6), 2306–2318 (1997). MR 1480382

14. Teng, Z.-H., Zhang, P.-W.: Optimal L^1-rate of convergence for viscosity method and monotone scheme to piecewise constant solutions with shocks. In: Advances in Numerical Mathematics. Proceedings of the Second Japan-China Seminar on Numerical Mathematics (Tokyo, 1994). Lecture Notes in Numerical and Applied Analysis, vol. 14, pp. 209–225. Kinokuniya, Tokyo (1995). MR 1469007
15. Wu, Z., Zhao, J., Yin, J., Li, H.: Nonlinear Diffusion Equations. World Scientific Publishing Co., Inc., River Edge (2001); Translated from the 1996 Chinese original and revised by the authors. MR 1881297

Analysis and Simulation of Nonlinear and Nonlocal Transport Equations

Frédéric Lagoutière and Nicolas Vauchelet

Abstract This article is devoted to the analysis of some nonlinear conservative transport equations, including the so-called aggregation equation with pointy potential, and numerical method devoted to its numerical simulation. Such a model describes the collective motion of individuals submitted to an attractive potential and can be written as a continuity transport equation with a velocity field computed through a self-consistent interaction potential. In the strongly attractive setting, L^p solutions may blow up in finite time, then a theory of existence of weak measure solutions has been defined. In this approach, we show the existence of Filippov characteristics allowing to define solutions of the aggregation initial value problem as a pushforward measure. Then numerical analysis of an upwind type scheme is proposed allowing to recover the dynamics of aggregates beyond the blowup time.

1 Introduction

This paper is devoted to existence and uniqueness, and numerical approximation of measure valued solutions to the following nonlinear and nonlocal transport equation in d space dimensions,

$$\partial_t \rho + \operatorname{div}\big((V * \rho)\rho\big) = 0, \qquad t > 0, \quad x \in \mathbb{R}^d, \tag{1}$$

complemented with the initial condition $\rho(0,x) = \rho^{ini}$. This equation governs the dynamics of a density of individuals, ρ at time $t > 0$, position $x \in \mathbb{R}^d$. The interaction between individuals is modelled by a given function $V : \mathbb{R}^+ \times \mathbb{R}^d \mapsto \mathbb{R}^d$. One motivation is the so-called aggregation equation where $V = \nabla_x W$ derives from

F. Lagoutière (✉)
Institut Camille Jordan UMR CNRS 5208, Université Lyon 1, 43 boulevard du 11 novembre 1918, 69622 Villeurbanne Cedex, France
e-mail: lagoutiere@math.univ-lyon1.fr

N. Vauchelet
LAGA - UMR 7539, Institut Galilée, Université Paris 13, 99, avenue Jean-Baptiste Clément, 93430 Villetaneuse, France
e-mail: vauchelet@math.univ-paris13.fr

L. Gosse, R. Natalini (eds.), *Innovative Algorithms and Analysis*,
Springer INdAM Series 16, DOI 10.1007/978-3-319-49262-9_10

an interaction potential W whose gradient $\nabla_x W(x-y)$ measures the relative force exerted by a unit mass localized at a point y onto a unit mass located at a point x.

The aggregation equation appears in many applications in physics and population dynamics. One may cite for instance applications in granular media [3], crowd motion [13], collective migration of cells by swarming [29, 30, 33], bacterial chemotaxis [18, 19, 25]. In many of these examples, the potential W is usually mildly singular, i.e. W has a weak singularity at the origin. Due to this weak regularity, finite time blowup of (weak L^p) solutions has been observed for such systems and has gained the attention of several authors (see e.g. [4, 7, 10, 28]). Finite time concentration is sometimes considered as a very simple mathematical way to mimick aggregation of individuals, as opposed to diffusion. Since finite time blowup of regular solutions occurs, a natural framework to study the existence of global in time solutions is to work in the space of probability measures. Two strategies have been proposed in the literature. In [10], the aggregation equation is seen as *a gradient flow minimizing the interaction energy in a Wasserstein space*. In [12, 25, 27], this system is considered as *a conservative transport equation with velocity field* $\nabla_x W * \rho$. Then a flow Z can be constructed allowing to define the solution as a pushforward measure by the flow: $\rho = Z_\# \rho^{ini}$. See also [5] for a similar definition. To be able to define such a flow, some assumptions on the potential are needed that allows for mild singularity of the potential. The usual assumption consists in considering *pointy* potentials with singularity at the origin, such as the Morse potential $W(x) = e^{-|x|}$, or $W(x) = -|x|$. In this paper, we extend this assumption to a more general class of potentials. Here we list the assumptions that will be used:

- **One-sided Lipschitz Condition**: there exists a function λ such that

$$\langle V(t,x) - V(t,y), x-y\rangle \leq \lambda(t)|x-y|^2, \quad \lambda \in L^1_{loc}(\mathbb{R}^+), \tag{2}$$

  where $\langle \cdot, \cdot \rangle$ denotes the Euclidean inner product. We notice for the case of the aggregation equation, i.e. when $V(t,x) = \nabla_x W(x)$, this assumption is satisfied provided the interaction potential $W : \mathbb{R}^d \to \mathbb{R}$ is λ-concave, i.e. $x \mapsto W(x) - \frac{\lambda}{2}|x|^2$ is concave for some constant $\lambda \geq 0$.
- For the sake of simplicity of the presentation, we only consider **bounded velocity fields**, then we assume that there exists a nonnegative constant v_∞ such that,

$$\text{for a.e. } t \in \mathbb{R}^+ \text{ and } x \in \mathbb{R}^d, \qquad |V(t,x)| \leq v_\infty. \tag{3}$$

- An interesting issue is raised when V **is discontinuous**, since, as already mentioned, it may imply blowup in finite time of weak L^p ($p > 1$) solutions. We may assume that there exists a finite set of discontinuity points. More precisely, there exists a finite set of points in $\mathbb{R}^d$, denoted $\xi_1, \ldots, \xi_L$, such that

$$V(t,x) = \sum_{\ell=1}^{L} \big(V_\ell(t, x-\xi_\ell) + V_\ell(t, x+\xi_\ell)\big) + V_r(t,x), \tag{4}$$

where for all $\ell = 1, \ldots, L$, we assume

$$V_\ell \in L^\infty_{loc}(\mathbb{R}^+, C(\mathbb{R}^d \setminus \{0\}), \quad V_\ell(t,x) = -V_\ell(t,-x) \\ \text{and } V_r \in L^\infty_{loc}(\mathbb{R}^+, W^{1,\infty}(\mathbb{R}^d)). \tag{5}$$

We notice also that since V_r is Lipschitz-continuous in space, then it satisfies (2) with a constant $\lambda(t) = \|DV_r(t)\|_\infty \in L^\infty_{loc}(\mathbb{R}^+)$. Thus, $V_s := V - V_r$ verifies also (2), and $V_s(t,x) = \sum_{\ell=1}^{L} \big(V_\ell(t, x-\xi_\ell) + V_\ell(t, x+\xi_\ell)\big)$ is odd.

Thus, to summarize, with respect to the preceding established results in [10, 12], we avoid three assumptions: we do not assume that V has only 1 singularity, we do not assume that V is odd, and we do not assume that V is the gradient of a potential. As a consequence, gradient flow solutions with respect to Wasserstein metric may not be defined. Then we must adapt some proofs in [12]. However, we still restrict to bounded velocity fields (see [5] for an approach in the radially symmetric case with non bounded velocity fields). Although extremely accurate numerical schemes were developed to study the blowup profile for smooth solutions, see e.g. [23, 24] for the aggregation equation, very few numerical schemes have been proposed to simulate the behaviour of solutions beyond blowup time. The so-called sticky particle method was shown to be convergent in [10] and used to obtain qualitative properties of the solutions such as the finite time total collapse. However, this method is not that practical to deal with the behavior of solutions after blowup in dimensions larger than one. In one dimension, this task has been performed in [25]. Recently, in higher dimensions, particle methods have been proposed and studied in [9, 14] but only the convergence for smooth solutions, before the blowup time, has been proved.

Finite volume schemes were developed, too, and the present paper stands in this frame. Note that the difficulty here is twofold: first, the velocity is not smooth (and only one-sided Lipschitz-continuous), and second, it is a nonlinear problem. In the linear case and when the given velocity field is only one-sided Lipschitz-continuous:

- in [21], the convergence of dissipative schemes is proven in dimension 1 (weak convergence in the sense of measures),
- in [16], the convergence of upwind-type or, more generally, some dissipative schemes, at order $\frac{1}{2}$ in Wasserstein distance, was obtained (in any dimension).

In the fully nonlinear context, in [26], a finite volume scheme is proposed allowing to simulate the behaviour of the solution to the one dimensional aggregation Eq. (1) after blowup, and the authors prove its convergence. A finite volume method for a large class of PDEs including in particular (1) has been also proposed in [11] but no convergence result has been given. Finally, a finite volume scheme of Rusanov (or Lax-Friedrichs) type for general measures as initial data has been proposed and studied in [12]. Numerical simulations of solutions in dimension greater than one have been obtained, allowing to observe the behaviour after blowup. Its weak convergence in the sense of measure is proven. We propose in this paper to extend this result to the upwind scheme for a more general class of equations, that is system (1) with an interaction function V satisfying assumptions (2)–(5) only. This scheme is based on an idea developed in [25] and used later in [12, 26] which consists in using a careful discretization of the macroscopic velocity such that its product with the measure solution ρ is well-defined.

The outline of this paper is the following. In the next section, we recall briefly the theory of existence of solutions to the conservative transport equation with discontinuous velocity field. In Sect. 3, we establish the existence and uniqueness result. Section 4 is devoted to the numerical discretization. We show in particular the convergence of the numerical scheme towards the measure valued solution. Finally, we conclude this paper with some numerical illustration in Sect. 5.

2 Transport Equation with Discontinuous Velocity Field

2.1 Notations

All along the paper, we will make use of the following notations. We denote $C_0(\mathbb{R}^d)$ the space of continuous functions in $\mathbb{R}^d$ that tend to 0 at infinity. We denote $\mathcal{M}_b(\mathbb{R}^d)$ the space of Borel measures whose total variation is finite. For $\rho \in \mathcal{M}_b(\mathbb{R}^d)$, we denote by $|\rho|(\mathbb{R}^d)$ its total variation. From now on, $\mathcal{M}_b(\mathbb{R}^d)$ is always endowed with the weak topology $\sigma(\mathcal{M}_b(\mathbb{R}^d), C_0(\mathbb{R}^d))$. For $T > 0$, we note $\mathcal{S}_{\mathcal{M}} := C([0,T]; \mathcal{M}_b(\mathbb{R}^d) - \sigma(\mathcal{M}_b(\mathbb{R}^d), C_0(\mathbb{R}^d)))$. For ρ a measure in $\mathcal{M}_b(\mathbb{R}^d)$ and Z a measurable map, we denote by $Z_\# \rho$ the pushforward measure of ρ by Z; it satisfies, for any continuous function ϕ,

$$\int_{\mathbb{R}^d} \phi(x)\, Z_\#\rho(dx) = \int_{\mathbb{R}^d} \phi(Z(x))\, \rho(dx).$$

We denote by $\mathcal{P}(\mathbb{R}^d)$ the subset of $\mathcal{M}_b(\mathbb{R}^d)$ of probability measures. We define the space of probability measures with finite second order moment by

$$\mathcal{P}_2(\mathbb{R}^d) = \left\{ \mu \in \mathcal{P}(\mathbb{R}^d),\ \int_{\mathbb{R}^d} |x|^2 \mu(dx) < \infty \right\}.$$

Here and in the following, $|\cdot|^2$ stands for the Euclidean norm that derives from the Euclidean inner product $\langle \cdot, \cdot \rangle$. This space is endowed with the Wasserstein distance d_W defined by (see e.g. [1, 34, 35])

$$d_W(\mu, \nu) = \inf_{\gamma \in \Gamma(\mu,\nu)} \left\{ \int_{\mathbb{R}^d \times \mathbb{R}^d} |y - x|^2\, \gamma(dx, dy) \right\}^{1/2} \tag{6}$$

where $\Gamma(\mu, \nu)$ is the set of measures on $\mathbb{R}^d \times \mathbb{R}^d$ with marginals μ and ν. From a minimization argument, we know that in the definition of d_W the infimum is actually a minimum (see [32, 35]). A map that realizes the minimum in the definition (6) of d_W is called an *optimal plan*, the set of which is denoted by $\Gamma_0(\mu, \nu)$. Then for all $\gamma_0 \in \Gamma_0(\mu, \nu)$, we have

$$d_W(\mu, \nu)^2 = \int_{\mathbb{R}^d \times \mathbb{R}^d} |y - x|^2 \, \gamma_0(dx, dy).$$

2.2 Weak Measure Solutions for Conservative Transport Equation

We recall in this section some useful results on weak measure solutions to the conservative linear transport equation with given velocity field b,

$$\partial_t u + \operatorname{div}(bu) = 0; \qquad u(t = 0) = u^0. \tag{7}$$

We start by the following definition of characteristics [20]:

Definition 1 Let us assume that $b = b(t, x) \in \mathbb{R}^d$ is a vector field defined on $[0, T] \times \mathbb{R}^d$ with $T > 0$. A Filippov characteristic $X(t; s, x)$ that stems from $x \in \mathbb{R}^d$ at time s is a continuous function $X(\cdot; s, x) \in C([0, T], \mathbb{R}^d)$ such that $\frac{\partial}{\partial t} X(t; s, x)$ exists for a.e. $t \in [0, T]$ and satisfies

$$\frac{\partial}{\partial t} X(t; s, x) \in \big\{Convess(b)(t, \cdot)\big\}(X(t; s, x)), \quad \text{a.e. } t \in [0, T]; \qquad X(s; s, x) = x.$$

From now on, we will use the notation $X(t, x) = X(t; 0, x)$.

In this definition $Convess(E)$ denotes the essential convex hull of a set E. We recall its definition for the sake of completeness, see [2, 20] for more details. We denote by $Conv(E)$ the classical convex hull of E, i.e., the smallest closed convex set containing E. Given the vector field $b(t, \cdot) : \mathbb{R}^d \longrightarrow \mathbb{R}^d$, the essential convex hull at point x is defined as

$$\{Convess(b)(t, \cdot)\}(x) = \bigcap_{r>0} \bigcap_{N \in \mathcal{N}_0} Conv\left[b\left(t, B(x, r) \setminus N\right)\right],$$

where $\mathcal{N}_0$ is the set of zero Lebesgue measure sets.

At this stage there is no smoothness assumption on b. Existence and uniqueness of a flow is classically ensured if b is smooth. A possible way to go beyond this, and use possibly discontinuous velocity fields, is to introduce the so-called one-sided Lipschitz continuity, see (8) below. The following existence and uniqueness result of Filippov characteristics ensures that the solution does not depend on the representative of b that is chosen.

Theorem 1 ([20]) *Let $T > 0$. Let us assume that the vector field $b \in L^1_{loc}(\mathbb{R}; L^\infty(\mathbb{R}^d))$ satisfies the OSL condition, that is for a.e. x and y in $\mathbb{R}^d$, a.e. $t \in [0,T]$,*

$$\langle b(t,x) - b(t,y), x - y\rangle \le \alpha(t)\|x-y\|^2, \qquad \text{for } \alpha \in L^1(0,T). \tag{8}$$

Then there exists a unique Filippov characteristic X associated to this vector field.

An important consequence of this result is the existence and uniqueness of weak measure solutions for the conservative linear transport equation. This result has been proved by Poupaud and Rascle [31, Theorem 3.3]:

Theorem 2 ([31]) *Let $T > 0$. Let $b \in L^1([0,T], L^\infty(\mathbb{R}^d))$ be a vector field satisfying the OSL condition (8). Then for any $u_0 \in \mathcal{M}_b(\mathbb{R}^d)$, there exists a unique measure solution u in $\mathcal{S}_\mathcal{M}$ to the conservative transport Eq. (7) such that $u(t) = X(t)_\# u_0$, where X is the unique Filippov characteristic, i.e. for any $\phi \in C_0(\mathbb{R}^d)$, we have*

$$\int_{\mathbb{R}^d} \phi(x) u(t,dx) = \int_{\mathbb{R}^d} \phi(X(t,x)) u_0(dx), \qquad \text{for } t \in [0,T].$$

In one dimension, such solutions are equivalent to duality solutions defined by Bouchut and James in [8]. A pioneering numerical investigation of this equation in one dimension is provided in [21]. A numerical investigation of measure valued solutions defined in Theorem 2 with a convergence order proof is proposed in [16]. Finally, we recall the following stability result for the Filippov characteristics which has been established by Bianchini and Gloyer [6, Theorem 1.2]

Theorem 3 ([6]) *Let $T > 0$. Assume that the sequence of vector fields b_n converges weakly to b in $L^1([0,T], L^1_{loc}(\mathbb{R}^d))$. Then the Filippov flow X_n generated by b_n converges locally in $C([0,T] \times \mathbb{R}^d)$ to the Filippov flow X generated by b.*

3 Filippov Characteristic Flow for the Aggregation Equation

This section is devoted to the existence of a Filippov flow for the aggregation equation (1) as it has been stated in [12] in a slightly less large context.

Under assumption (4), we define for $\rho \in C([0,T], \mathcal{P}_2(\mathbb{R}^d))$ the velocity field $\hat{a}_\rho$ by

$$\hat{a}_\rho(t,x) = \sum_{\ell=1}^{L} \int_{\mathbb{R}^d} \big(\widehat{V}_\ell(t, x-y-\xi_\ell) + \widehat{V}_\ell(t, x-y+\xi_\ell)\big)\rho(t,dy) + \int_{\mathbb{R}^d} V_r(t, x-y)\rho(t,dy), \tag{9}$$

where $\widehat{V}_\ell$ is defined for $\ell = 1, \ldots, L$ by

$$\widehat{V}_\ell(t,x) = \begin{cases} V_\ell(t,x), & \text{when } x \neq 0; \\ 0, & \text{otherwise.} \end{cases}$$

From now on, we will use the notation

$$\widehat{V}(t,x) := \sum_{\ell=1}^{L} \big(\widehat{V}_\ell(t,x-\xi_\ell) + \widehat{V}_\ell(t,x+\xi_\ell)\big) + V_r(t,x). \tag{10}$$

The following theorem states existence and uniqueness of a solution defined by the Filippov characteristics. Its proof in the case where $V(t,x) = \nabla_x W(x)$ with $W \in C^1(\mathbb{R}^d \setminus \{0\})$ and λ-convex has been obtained in [12, Theorems 2.5 and 2.9]. In the present article we extend it to the case at hand.

Theorem 4 *Let V satisfy assumptions* (2)–(5) *and let ρ^{ini} be given in $\mathscr{P}_2(\mathbb{R}^d)$. There exists a unique solution $\rho \in C([0,+\infty), \mathscr{P}_2(\mathbb{R}^d))$ that satisfies in the sense of distributions the aggregation equation*

$$\partial_t \rho + \operatorname{div}(\hat{a}_\rho \rho) = 0, \qquad \rho(0,\cdot) = \rho^{ini},$$

where $\hat{a}_\rho$ is defined by (9). *This solution is the pushforward measure $\rho := Z_\# \rho^{ini}$ where Z is the unique Filippov characteristic flow associated to the velocity field $\hat{a}_\rho$.*

3.1 Sketch of the Proof of Theorem 4

The proof of the existence and uniqueness result in Theorem 4 follows the ideas developed in [12]. For the sake of completeness, we recall the main steps of the proof and detail below the main changes to extend it to the case at hand.

3.1.1 Step 1: Definition of the Macroscopic Velocity

A difficulty when we want to deal with measure valued solutions to transport equation is that the velocity field should be defined carefully to be able to give a sense to the product in the divergence term in (1) when ρ is a measure. Here we use the definition (9) for the velocity field. This definition is motivated by Lemma 2 which is stated and proved below. Indeed, from assumptions (4)–(5), we have $V(t,x) = \sum_{\ell=1}^{L}(V_\ell(t,x-\xi_\ell) + V_\ell(t,x+\xi_\ell)) + V_r(t,x)$, with $V_\ell \in C^1(\mathbb{R}^d \setminus \{0\})$ for $\ell = 1, \ldots, L$. Then, Lemma 2 implies that if we regularize V by a sequence V_n, for instance by taking the convolution of each V_ℓ with mollifiers, and if for all $t \geq 0$, $\rho_n(t)$ is a sequence of probability measures in $\mathscr{P}_2(\mathbb{R}^d)$ such that $\rho_n \rightharpoonup \rho$ in the sense of measures, then $V_n \star \rho_n \rightharpoonup \hat{a}_\rho \rho$ in the sense of measures (see Lemma 2 in Sect. 3.2).

Moreover, we have the following one-sided Lipschitz estimate:

Lemma 1 *Let $\rho \in L^\infty(0, T, \mathcal{M}_b(\mathbb{R}^d))$ be nonnegative. Then under assumptions (2)–(5), the function $(t, x) \mapsto \hat{a}_\rho(t, x)$ defined in (9) satisfies the one-sided Lipschitz (OSL) estimate*

$$\langle \hat{a}_\rho(t, x) - \hat{a}_\rho(t, y), x - y\rangle \leq \lambda(t)|\rho(t)|(\mathbb{R}^d)\|x - y\|^2. \tag{11}$$

Proof This result is an easy consequence of assumption (2) on V. Indeed, by definition (9), we have

$$\hat{a}_\rho(t, x) - \hat{a}_\rho(t, y) = \int_{\mathbb{R}^d} \big(\widehat{V}(t, x - z) - \widehat{V}(t, y - z)\big)\rho(dz),$$

where $\widehat{V}(t, x) = \sum_{\ell=1}^{L}(\widehat{V}_\ell(t, x - \xi_\ell) + \widehat{V}_\ell(t, x + \xi_\ell)) + V_r(t, x)$ is defined in (10). The conclusion follows directly from assumption (2) and the nonnegativity of ρ.

3.1.2 Step 2: Approximation with Dirac Masses

We use the idea of atomization consisting in approximating the solution by a finite sum of Dirac masses: let us consider that for all integer $N > 0$,

$$\rho^{ini,N} = \sum_{i=1}^{N} m_i\delta(x - x_i^0), \qquad x_i^0 \neq x_j^0 \text{ for } i \neq j, \quad \sum_{i=1}^{N} m_i = 1, \quad \sum_{i=1}^{N} m_i|x_i^0|^2 < +\infty.$$

Then we look for a solution of the aggregation equation given by

$$\rho^N(t, x) = \sum_{i=1}^{N} m_i\delta(x - x_i(t)).$$

By definition (9), $\hat{a}_{\rho^N}(t, x) = \sum_{i=1}^{N} m_i\widehat{V}(t, x - x_i(t))$, with $\widehat{V}$ defined in (10). From Lemma 1, $\hat{a}_{\rho^N}$ satisfies the OSL condition. Applying Theorem 1, it allows to define uniquely a Filippov characteristic, denoted $\widehat{X}^N$, associated to the velocity field $\hat{a}_{\rho^N}$. By construction, from Theorem 2, the Poupaud-Rascle pushforward measure $\rho_{PR} := \widehat{X}^N{}_{\#}\rho^{ini,N}$ is the unique measure valued solution to the conservative linear transport equation

$$\partial_t \rho_{PR}^N + \operatorname{div}(\hat{a}_{\rho^N}\rho_{PR}^N) = 0, \qquad \rho_{PR}^N(t = 0) = \rho^{ini,N}.$$

Moreover, by definition of the pushforward measure,

$$\hat{a}_{\rho_{PR}^N} = \int_{\mathbb{R}^d} \widehat{V}(t, x-y)\rho_{PR}^N(t, dy) = \int_{\mathbb{R}^d} \widehat{V}(t, x - \widehat{X}^N(t, y))\rho^{ini,N}(dy)$$
$$= \sum_{i=1}^{N} m_i \widehat{V}(t, x - \widehat{X}^N(t, x_i^0)) = \widehat{a}_{\rho^N}(t, x).$$

Thus $\rho_{PR}^N = \rho^N$. It gives the existence result for initial data given by a finite sum of Dirac masses.

3.1.3 Step 3: Passing to the Limit $N \to +\infty$

Making use of stability results, we may pass to the limit $N \to +\infty$ in the above construction. This step is the same as in [12]; for the sake of completeness, we recall the ideas but omit details. We assume that $\rho^{ini,N} \rightharpoonup \rho^{ini}$ as $N \to +\infty$. Then, since the velocity field $\hat{a}_{\rho^N}$ is uniformly bounded, thanks to (3), we may extract a subsequence that converges in L^∞ weak-$\star$. Using the stability result of Theorem 3, we deduce that $\widehat{X}^N \to \widehat{X}$. As a consequence, we get the weak convergence $\rho^N \rightharpoonup \rho := \widehat{X}_{\#}\rho^0$. Finally, we apply the stability result of Lemma 2 to conclude the proof of existence.

3.1.4 Step 4: Uniqueness

Uniqueness is deduced from the contraction estimate in Wasserstein distance d_W in Proposition 1 below. Indeed, if we take $\rho^0 = \tilde{\rho}^0 = \rho^{ini}$ in the estimate of Proposition 1, then we deduce that $\rho = \tilde{\rho}$. This proves the uniqueness of the solution that is a pushforward of the initial datum by the flow. Uniqueness of the distributional solution is deduced from Theorem 8.2.1 in [1, Chap. 8.2]. Indeed, in the present case where the flow is unique, this result tells that the distributional solution is necessarily the pushforward of the initial datum by the flow.

3.2 The Macroscopic Velocity

In the first step above, we have defined a macroscopic velocity for which the product in the divergence term in (1) has a sense. This definition relies on the following stability result.

Lemma 2 *Let V be a velocity field satisfying $V \in L^\infty_{loc}(\mathbb{R}^+, C(\mathbb{R}^d \setminus \{0\}))$, (2), (3) and $V(-x) = -V(x)$. Let $(V_n)_{n\in\mathbb{N}^*}$ be a sequence of odd functions in $C^1(\mathbb{R}^+ \times \mathbb{R}^d)$, uniformly bounded by v_∞ and such that for all $t \in \mathbb{R}^+$,*

$$sup_{x\in\mathbb{R}^d\setminus B(0,\frac{1}{n})} \big|V_n(t,x) - V(t,x)\big| \leq \frac{1}{n}, \qquad \textit{for all } n \in \mathbb{N}^*. \tag{12}$$

Let $\rho(t)$ be a probability measure for all $t \geq 0$. Let $(\rho_n(t))_n$ be a sequence of probability measures such that $\rho_n \rightharpoonup \rho$ weakly as measures as $n \to +\infty$, then for any $T > 0$, for every $\phi \in C_0([0,T] \times \mathbb{R}^d)$ and any $\xi \in \mathbb{R}^d$, we have

$$\begin{aligned}\lim_{n\to+\infty} \int_0^T \iint_{\mathbb{R}^d\times\mathbb{R}^d} & \phi(t,x)V_n(t,x-y-\xi)\rho_n(t,dx)\rho_n(t,dy)dt \\ &= \int_0^T \iint_{\mathbb{R}^d\times\mathbb{R}^d} \phi(t,x)\widehat{V}(t,x-y-\xi)\rho(t,dx)\rho(t,dy)dt.\end{aligned}$$

Proof We first introduce some notations that simplify the computations:

$$\begin{aligned}\mu_n(t) &:= \rho_n(t) \otimes \rho_n(t,\cdot-\xi) - \rho_n(t,\cdot-\xi) \otimes \rho_n(t), \\ \mu(t) &:= \rho(t) \otimes \rho(t,\cdot-\xi) - \rho(t,\cdot-\xi) \otimes \rho(t), \\ D_n &:= \left\{(x,y) \in \mathbb{R}^d \times \mathbb{R}^d, x \neq y, |x-y| < \frac{1}{n}\right\}.\end{aligned} \tag{13}$$

We recall that since $\rho_n \rightharpoonup \rho$ weakly as measures, we have that $\rho_n(t) \otimes \rho_n(t,\cdot-\xi) \rightharpoonup \rho(t) \otimes \rho(t,\cdot-\xi)$ and $\mu_n \rightharpoonup \mu$ weakly in the sense of measures.

Let us fix $t \in \mathbb{R}^+$ and let $\varepsilon > 0$. By definition of μ and D_n in (13) there exists $N \in \mathbb{N}$ such that $\forall n \geq N$,

$$|\mu|(t, D_n) \leq \varepsilon. \tag{14}$$

For such N, we observe that, for all $n \geq N$, $D_n \subset D_N$, and

$$|\mu_n|(t,D_n) \leq |\mu_n|(t,D_N) \leq |\mu_n - \mu|(t,D_N) + |\mu|(t,D_N). \tag{15}$$

From the weak convergence $\mu_n \rightharpoonup \mu$, we deduce that for n large enough, we have $|\mu_n - \mu|(t,D_N) \leq \varepsilon$. Injecting in (15), and using also (14), we deduce that for N large enough and $\forall\, n \geq N$, we have

$$|\mu_n|(t,D_n) \leq \varepsilon, \qquad \text{and} \qquad |\mu_n - \mu|(t,D_n) \leq \varepsilon. \tag{16}$$

For $\phi \in C_0([0,T] \times \mathbb{R}^d)$, we note

$$A_n(t) := \iint_{\mathbb{R}^d \times \mathbb{R}^d} V_n(t, x-y-\xi)\phi(t,x)\rho_n(t) \otimes \rho_n(t)(dx, dy) - \iint_{\mathbb{R}^d \times \mathbb{R}^d} \widehat{V}(t, x-y-\xi)\phi(t,x)\rho(t) \otimes \rho(t)(dx, dy).$$

After a change of variable, we may write

$$\begin{aligned} A_n(t) &= \iint_{\mathbb{R}^d \times \mathbb{R}^d} \phi(t,x)\Big(V_n(t,x-y)\rho_n(t,x)\rho_n(t,y-\xi) \\ &\qquad - \widehat{V}(t,x-y)\rho(t,x)\rho(t,y-\xi)\Big)\,dxdy \\ &= \frac{1}{2}\iint_{\mathbb{R}^d \times \mathbb{R}^d} \Big[V_n(t,x-y)\Big(\phi(t,x)\rho_n(t,x)\rho_n(t,y-\xi) \\ &\qquad - \phi(t,y)\rho_n(t,y)\rho_n(t,x-\xi)\Big) \\ &\qquad - \widehat{V}(t,x-y)\Big(\phi(t,x)\rho(t,x)\rho(t,y-\xi) \\ &\qquad - \phi(t,y)\rho(t,y)\rho(t,x-\xi)\Big)\Big]dxdy, \end{aligned}$$

where we have used the symmetry assumption $V_n(-x) = -V_n(x)$ and $V(-x) = -V(x)$ for the last equality. We may rewrite

$$\begin{aligned} A_n(t) =& \frac{1}{2}\iint_{\mathbb{R}^d \times \mathbb{R}^d} (\phi(t,x) - \phi(t,y))\Big[\big(V_n(t,x-y) - \widehat{V}(t,x-y)\big)\rho_n(t,x)\rho_n(t,y-\xi) \\ &+ \widehat{V}(t,x-y)\big(\rho_n(t,x)\rho_n(t,y-\xi) - \rho(t,x)\rho(t,y-\xi)\big)\Big]\,dxdy + \\ &\frac{1}{2}\iint_{\mathbb{R}^d \times \mathbb{R}^d} \phi(t,y)\big((V_n(t,x-y) - \widehat{V}(t,x-y))\mu_n + \widehat{V}(t,x-y)(\mu_n - \mu)\big)\,dxdy \\ =& I_n + II_n + III_n + IV_n. \end{aligned} \tag{17}$$

We bound each term of the right and side separately. Let us consider the **first term**,

$$I_n := \iint_{\mathbb{R}^d \times \mathbb{R}^d} (\phi(t,x) - \phi(t,y))\big(V_n(t,x-y) - \widehat{V}(t,x-y)\big)\rho_n(t,x)\rho_n(t,y-\xi)dxdy.$$

Using assumption (12) and the bound $\|V_n\|_\infty \leq v_\infty$, the latter integral on $\mathbb{R}^d \times \mathbb{R}^d \setminus D_n$ is bounded by a term of order $\frac{1}{n}$. We are left with the integral over D_n. Since $\phi \in C_0([0,T] \times \mathbb{R}^d)$, there exists a compact $K \subset \mathbb{R}^d$ such that for all $x \in \mathbb{R}^d \setminus K$, $|\phi(x)| \leq \varepsilon$. On the compact K, ϕ is uniformly continuous, then there exists $N \in \mathbb{N}$ such that for all $n \geq N$ and all $x, y \in D_n \cap K$, $|\phi(x) - \phi(y)| \leq \varepsilon$. Then, we deduce

$$|I_n(t)| \leq \frac{2}{n}\|\phi\|_\infty + 2v_\infty\varepsilon. \tag{18}$$

For the **second term**,

$$II_n := \iint_{\mathbb{R}^d\times\mathbb{R}^d} (\phi(t,x) - \phi(t,y))\widehat{V}(t,x-y) \big(\rho_n(t,x)\rho_n(t,y-\xi) - \rho(t,x)\rho(t,y-\xi)\big)dxdy.$$

We use the fact that the function $(x,y) \mapsto (\phi(t,x) - \phi(t,y))\widehat{V}(t,x-y)$ is continuous and the weak convergence in the sense of measures of ρ_n to deduce that

$$\lim_{n\to+\infty} II_n(t) = 0. \tag{19}$$

Considering now the **third term**,

$$III_n := \iint_{\mathbb{R}^d\times\mathbb{R}^d} \phi(t,y)(V_n(t,x-y) - \widehat{V}(t,x-y))\mu_n(t,x,y)\, dxdy.$$

We split the integral between the one on $\mathbb{R}^d \times \mathbb{R}^d \setminus D_n$ and the one on D_n. We get

$$|III_n| \leq \frac{2}{n}\|\phi\|_\infty + 2v_\infty\|\phi\|_\infty|\mu_n|(D_n) \leq \Big(\frac{2}{n} + 2v_\infty\varepsilon\Big)\|\phi\|_\infty, \tag{20}$$

where we use (16) for the last inequality. The **fourth term** reads,

$$\begin{aligned}
IV_n &:= \iint_{\mathbb{R}^d\times\mathbb{R}^d} \phi(t,y)\widehat{V}(t,x-y)(\mu_n - \mu)(t,x,y)\, dxdy \\
&= \iint_{\mathbb{R}^d\times\mathbb{R}^d} \phi(t,y)\big(\widehat{V}(t,x-y) - V_N(t,x-y)\big)(\mu_n - \mu)(t,x,y)\, dxdy \\
&\quad + \iint_{\mathbb{R}^d\times\mathbb{R}^d} \phi(t,y)V_N(t,x-y)(\mu_n - \mu)(t,x,y)\, dxdy,
\end{aligned}$$

where $N \in \mathbb{N}$ will be chosen large enough. The second term of the right hand side converges to 0 as n goes to $+\infty$ since $\mu_n \rightharpoonup \mu$ and $(x,y) \mapsto \phi(t,y)V_N(t,x-y)$ is continuous. Then it is bounded by ε for n large enough. We bound the first term of the right hand side as for the term III_n, we obtain, for n large enough

$$|IV_n| \leq \left(\frac{4}{N} + 2v_\infty\varepsilon\right)\|\phi\|_\infty + \varepsilon. \tag{21}$$

Finally, by injecting (18)–(21) into (17), we deduce the a.e. convergence $A_n(t) \to 0$ as $n \to +\infty$. Since $|A_n(t)| \leq 2v_\infty\|\phi\|_\infty$ holds uniformly, Lebesgue's dominated convergence theorem yields,

$$\int_0^T A_n(t)dt \to 0, \qquad n \to +\infty,$$

which concludes the proof.

3.3 Contraction Estimate

Proposition 1 *Under assumptions (2)–(5) on V, let ρ^0 and $\tilde{\rho}^0$ be given in $\mathscr{P}_2(\mathbb{R}^d)$. Then, there exists a nonnegative constant C such that the corresponding solutions $\rho = Z_\#\rho^0$ and $\tilde{\rho} = \widetilde{Z}_\#\tilde{\rho}^0$ verify*

$$d_W(\rho(t), \tilde{\rho}(t)) \leq e^{C(t+\int_0^t \lambda(s)ds)} d_W(\rho^0, \tilde{\rho}^0).$$

Proof Let us consider γ an optimal map with marginals ρ^0 and $\tilde{\rho}^0$ such that,

$$d_W(\rho^0, \tilde{\rho}^0)^2 = \iint_{\mathbb{R}^d\times\mathbb{R}^d} |x_1 - x_2|^2 \gamma(dx_1, dx_2).$$

We compute, formally,

$$\begin{aligned}
&\frac{d}{dt}\iint_{\mathbb{R}^d\times\mathbb{R}^d} |Z(t,x_1) - \widetilde{Z}(t,x_2)|^2\gamma(dx_1,dx_2) \\
&= 2\iint_{\mathbb{R}^d\times\mathbb{R}^d} \langle \hat{a}_\rho(t,Z(t,x_1)) - \hat{a}_{\tilde{\rho}}(t,\widetilde{Z}(t,x_2)), Z(t,x_1) - \widetilde{Z}(t,x_2)\rangle\gamma(dx_1,dx_2) \\
&= 2\iiiint \langle \widehat{V}(t, Z(t,x_1) - Z(t,y_1)) - \widehat{V}(t, \widetilde{Z}(t,x_2) - \widetilde{Z}(t,y_2)), \\
&\qquad Z(t,x_1) - \widetilde{Z}(t,x_2)\rangle\, \gamma(dx_1,dx_2)\gamma(dy_1,dy_2),
\end{aligned}$$

where $\widehat{V}$ is defined in (10). From assumption (4), we decompose

$$\frac{d}{dt}\iint_{\mathbb{R}^d\times\mathbb{R}^d} |Z(t,x_1) - \widetilde{Z}(t,x_2)|^2\gamma(dx_1,dx_2) = I_s + I_r,$$

with, using the notation $V_s = \sum_{\ell=1}^{L} \big(\widehat{V}_\ell(\cdot - \xi_\ell) + \widehat{V}_\ell(\cdot + \xi_\ell)\big)$,

$$I_s = 2 \iiiint \langle V_s(t, Z(t,x_1) - Z(t,y_1)) - V_s(t, \widetilde{Z}(t,x_2) - \widetilde{Z}(t,y_2)), \\ Z(t,x_1) - \widetilde{Z}(t,x_2)\rangle \, \gamma(dx_1, dx_2)\gamma(dy_1, dy_2)\,;$$

$$I_r = 2 \iiiint \langle V_r(t, Z(t,x_1) - Z(t,y_1)) - V_r(t, \widetilde{Z}(t,x_2) - \widetilde{Z}(t,y_2)), \\ Z(t,x_1) - \widetilde{Z}(t,x_2)\rangle \, \gamma(dx_1, dx_2)\gamma(dy_1, dy_2).$$

We treat each term separately. Using the symmetry of V_s (exchanging the role of x and y) we obtain

$$\begin{aligned} I_s = & \iiiint \langle V_s(t, Z(t,x_1) - Z(t,y_1)) - V_s(t, \widetilde{Z}(t,x_2) - \widetilde{Z}(t,y_2)), \\ & \qquad Z(t,x_1) - Z(t,y_1) - \widetilde{Z}(t,x_2) + \widetilde{Z}(t,y_2)\rangle \, \gamma(dx_1, dx_2)\gamma(dy_1, dy_2) \\ \le & \, \lambda(t) \iiiint |Z(t,x_1) - Z(t,y_1) - \widetilde{Z}(t,x_2) + \widetilde{Z}(t,y_2)|^2 \, \gamma(dx_1, dx_2)\gamma(dy_1, dy_2), \end{aligned}$$

where we use assumption (2) satisfied by V_s. Expanding the right hand side, we deduce straightforwardly

$$I_s \le 4\lambda(t) \iint_{\mathbb{R}^d \times \mathbb{R}^d} |Z(t,x_1) - \widetilde{Z}(t,x_2)|^2 \gamma(dx_1, dx_2). \tag{22}$$

For the term I_r, we deduce from the Lipschitz continuity of V_r that there exists a nonnegative constant C such that

$$\begin{aligned} |I_r| \le & \, C \iiiint |Z(t,x_1) - Z(t,y_1) - \widetilde{Z}(t,x_2) + \widetilde{Z}(t,y_2)| \\ & \qquad |Z(t,x_1) - \widetilde{Z}(t,x_2)| \, \gamma(dx_1, dx_2)\gamma(dy_1, dy_2), \\ \le & \, C \iint |Z(t,x_1) - \widetilde{Z}(t,x_2)|^2 \, \gamma(dx_1, dx_2) + \\ & \quad C \iiiint |Z(t,y_1) - \widetilde{Z}(t,y_2)||Z(t,x_1) - \widetilde{Z}(t,x_2)| \, \gamma(dx_1, dx_2)\gamma(dy_1, dy_2), \\ \le & \, C \iint |Z(t,x_1) - \widetilde{Z}(t,x_2)|^2 \, \gamma(dx_1, dx_2) + C \left(\iint |Z(t,x_1) - \widetilde{Z}(t,x_2)| \, \gamma(dx_1, dx_2) \right)^2 . \end{aligned}$$

Using a Cauchy-Schwarz inequality, we deduce

$$|I_r| \le 2C \iint_{\mathbb{R}^d \times \mathbb{R}^d} |Z(t,x_1) - \widetilde{Z}(t,x_2)|^2 \, \gamma(dx_1, dx_2). \tag{23}$$

Combining (22) and (23), we deduce

$$\frac{d}{dt}\iint_{\mathbb{R}^d\times\mathbb{R}^d}|Z(t,x_1)-\widetilde{Z}(t,x_2)|^2\gamma(dx_1,dx_2)$$
$$\leq (4\lambda(t)+2C)\iint_{\mathbb{R}^d\times\mathbb{R}^d}|Z(t,x_1)-\widetilde{Z}(t,x_2)|^2\gamma(dx_1,dx_2).$$

We conclude by a Gronwall argument and using the fact that

$$d_W(\rho(t),\tilde{\rho}(t))^2 \leq \iint_{\mathbb{R}^d\times\mathbb{R}^d}|Z(t,x_1)-\widetilde{Z}(t,x_2)|^2\gamma(dx_1,dx_2).$$

The above formal computation can be made rigorous by using a regularization of the function V and passing to the limit; see [12, Proposition 3.4]

4 Numerical Analysis

4.1 Definition of the Scheme

Let us first introduce an upwind type numerical scheme for the discretization of the aggregation equation. We denote by Δt the time step and consider a Cartesian grid with step Δx_i in the ith direction, $i = 1,\dots,d$, and $\Delta x = \max_i \Delta x_i$. We use standard notations for vectors $e_i = (0,\dots,1,\dots,0)$, $a = (a_1,\dots,a_d)$ and multi-indices,

$$J = (J_1,\dots,J_d) \in \mathbb{Z}^d, \quad x_J = (J_1\Delta x_1,\dots,J_d\Delta x_d).$$

Let $C_J = [(J_1-\frac{1}{2})\Delta x_1,(J_1+\frac{1}{2})\Delta x_1)\times\dots[(J_d-\frac{1}{2})\Delta x_d,(J_d+\frac{1}{2})\Delta x_d)$ be an elementary cell. For a given nonnegative measure $\rho^{ini}\in\mathscr{P}_2(\mathbb{R}^d)$, we define,

$$\forall J\in\mathbb{Z}^d, \qquad \rho_J^0 = \int_{C_J}\rho^{ini}(dx)\geq 0. \tag{24}$$

Since ρ^{ini} is a probability measure, the total mass of the system is $\sum_{J\in\mathbb{Z}^d}\rho_J^0 = 1$. We denote by ρ_J^n an approximation of the value $\rho(t^n,x_J)$, for $J\in\mathbb{Z}^d$. Assuming that an approximating sequence $(\rho_J^n)_{J\in\mathbb{Z}^d}$ is known at time t^n, then we compute the approximation at time t^{n+1} by the following scheme,

$$\rho_J^{n+1} = \rho_J^n - \sum_{i=1}^d\frac{\Delta t}{\Delta x_i}\big((a_{iJ}^n)^+\rho_J^n - (a_{iJ+e_i}^n)^-\rho_{J+e_i}^n - (a_{iJ-e_i}^n)^+\rho_{J-e_i}^n + (a_{iJ}^n)^-\rho_J^n\big). \tag{25}$$

The notation $(a)^+ = \max\{0, a\}$ stands for the positive part of the real number a and respectively $(a)^- = \max\{0, -a\}$ for the negative part. The discrete macroscopic velocity is computed thanks to the following discretization of Eq. (9),

$$a_{iJ}^n = \sum_{K\in\mathbb{Z}^d} \rho_K^n V_{iJ,K}^n, \quad \text{where} \quad V_{iJ,K}^n := \int_{t^n}^{t^{n+1}} \widehat{V}_i(s, x_J - x_K)ds, \tag{26}$$

where $\widehat{V}_i$ is the ith components of the velocity field $\widehat{V}$ defined in (10). We notice that this scheme may be interpreted as the Engquist-Osher scheme, see [21, Sect. 4.2].

Example 1 In one dimension, the scheme (25) reads

$$\rho_i^{n+1} = \rho_i^n - \frac{\Delta t}{\Delta x}\Big((a_i^n)^+\rho_i^n - (a_{i+1}^n)^-\rho_{i+1}^n - (a_{i-1}^n)^+\rho_{i-1}^n + (a_i^n)^-\rho_i^n\Big).$$

This scheme has the following interpretation. Defining $\rho_{\Delta x}^n = \sum_{i\in\mathbb{Z}} \rho_i^n \delta_{x_i}$, we construct the approximation at time t^{n+1} with the following two steps:

- The delta mass ρ_i^n located at position x_i, moves with velocity a_i^n to the position $x_i + a_i^n\Delta t$. Assuming a CFL condition $v_\infty \Delta t \le \Delta x$, the point $x_i + a_i^n\Delta t$ belongs to the interval $[x_i, x_{i+1}]$ if $a_i^n \ge 0$, and to the interval $[x_{i-1}, x_i]$ if $a_i^n \le 0$.
- Then we make a linear interpolation of the mass ρ_i^n between x_i and x_{i+1} if $a_i^n \ge 0$ and between x_{i-1} and x_i if $a_i^n \le 0$.

Finally, we emphasize that this scheme is not the standard finite volume upwind scheme for which the numerical velocity is computed at the interface $a_{i+1/2}^n$. This is due to the particular structure of the equation for which the product $\hat{a}_\rho\rho$ should be defined properly. If in the discretization we choose the velocity in a different grid point that the density, it creates a shift in the definition of the product and the numerical solution does not converge to the solution of Theorem 4. This point has been already noted in [22, 26] where numerical simulations emphasized the wrong behaviour of numerical solutions computed with the classical upwind scheme.

4.2 Convergence Analysis

In the following theorem, we establish the convergence of scheme (25) towards the unique solution of Theorem 4. More precisely the statement reads:

Theorem 5 *Let $\rho^{ini} \in \mathscr{P}_2(\mathbb{R}^d)$. Let us assume that V satisfies assumptions* (2)–(5). *Let $T > 0$ and $\rho = Z_\# \rho^{ini}$ be the unique measure solution on* $[0, T]$ *to the interaction* *equation* (1) *with initial data* ρ^{ini} *given by Theorem 4. Let us assume that the CFL condition holds:*

$$v_\infty \sum_{i=1}^{d} \frac{\Delta t}{\Delta x_i} \leq 1. \tag{27}$$

Let us define $(\rho_J^0)_{J\in\mathbb{Z}^d}$ *by* (24) *and define the reconstruction*

$$\rho_\Delta(t,x) = \sum_{J\in\mathbb{Z}^d} \rho_J^n \delta_{x_J}(x) \mathbf{1}_{[t^n,t^{n+1})}(t), \tag{28}$$

where the approximation sequence (ρ_J^n) *is computed thanks to the scheme* (25)–(26). *Then we have the weak convergence in the sense of measures* $\rho_\Delta \rightharpoonup \rho$ *in* $\mathcal{M}_b([0,T]\times\mathbb{R}^d)$ *as* Δt *and* Δx_i *go to* 0 *under the condition* (27).

Before going into the proof of this Theorem, we mention that this result extends to the upwind scheme and to the more general Eq. (1) at hand the convergence result stated in [12]. We mention also that an estimate of the order of convergence in the same spirit as [15, 16] is under progress [17]. We first recall the following well-known properties for the upwind scheme, whose proof is left to the reader.

Proposition 2 *Let us assume that* V *satisfies assumptions* (2)–(5) *and consider* $\rho^{ini} \in \mathcal{P}_2(\mathbb{R}^2)$. *Let us assume that the CFL condition* (27) *holds. We define* $(\rho_J^0)_{J\in\mathbb{Z}^d}$ *by* (24) *and the reconstruction* ρ_Δ *by* (28), *where the approximation sequences* (ρ_J^n) *and* (a_{iJ}^n) *are computed thanks to the scheme* (25)–(26). *Then, we have*

(i) Positivity: for all $J \in \mathbb{Z}^d$, $n \in \mathbb{N}$, $i = 1,\dots,d$, $\rho_J^n \geq 0$, $|a_{iJ}^n| \leq v_\infty$.

(ii) Mass conservation: for all $n \in \mathbb{N}^*$, *we have* $\sum_{J\in\mathbb{Z}^d} \rho_J^n = \sum_{J\in\mathbb{Z}^d} \rho_{ij}^0 = 1$.

(iii) Bound on the second moment: there is a constant $C > 0$ *such that for all* $n \in \mathbb{N}^*$,

$$M_2^n := \sum_{J\in\mathbb{Z}^d} |x_J|^2 \rho_J^n \leq e^{Cn\Delta t}\big(M_2^0 + C\big).$$

Proof of Theorem 5

Step 1: Weak Convergence. Under the CFL condition (27), we deduce from Proposition (2) that the sequence $(\rho_\Delta)_\Delta$ is a sequence of nonnegative bounded measures which satisfies for all $t \in [0,T]$, $|\rho_\Delta(t)|(\mathbb{R}^2) = 1$. Therefore, we can extract a subsequence, still denoted $(\rho_\Delta)_\Delta$, converging for the weak topology towards ρ as Δt, and Δx_i go to 0, satisfying (27).

Step 2: Identification of the Limit. We choose $\Delta t > 0$ and $N_T \in \mathbb{N}^*$ such that $T = \Delta t N_T$ and condition (27) holds. Let $\phi \in \mathcal{D}([0,T) \times \mathbb{R}^d)$, we multiply (25) by ϕ and integrate on $[t^n, t^{n+1}) \times \mathbb{R}^d$, using a discrete integration by parts, we get

$$\sum_{J\in\mathbb{Z}^d} (\rho_J^{n+1} - \rho_J^n)\phi_J^n = \sum_{i=1}^{d} \sum_{J\in\mathbb{Z}^d} \frac{\Delta t}{\Delta x_i} \Big[(a_{iJ}^n)^+ \rho_J^n (\phi_{J+e_i}^n - \phi_J^n) + (a_{iJ}^n)^- \rho_J^n (\phi_J^n - \phi_{J-e_i}^n)\Big],$$

with the notation $\phi_J^n = \int_{t^n}^{t^{n+1}} \phi(t, x_J)dt$. From a Taylor formula, we have

$$\phi_{J+e_i}^n = \phi_J^n + \partial_{x_i}\phi_J^n \Delta x_i + O(\Delta t \Delta x_i^2), \qquad \phi_{J-e_i}^n = \phi_J^n - \partial_{x_i}\phi_J^n \Delta x_i + O(\Delta t \Delta x_i^2).$$

Summing over n and using a discrete integration by parts, we deduce

$$\sum_{n=1}^{N_T} \sum_{J\in\mathbb{Z}^d} \rho_J^n(\phi_J^{n-1} - \phi_J^n) - \sum_{J\in\mathbb{Z}^d} \rho_J^0 \phi_J^0 = \sum_{n=0}^{N_T} \sum_{i=1}^{d} \sum_{J\in\mathbb{Z}^d} \Delta t a_{iJ}^n \rho_J^n \partial_{x_i}\phi_J^n + O(\Delta t \Delta x).$$

Finally, using also a Taylor formula for the first term of the left hand side, we deduce

$$\sum_{n=1}^{N_T} \sum_{J\in\mathbb{Z}^d} \rho_J^n \int_{t^n}^{t^{n+1}} \partial_t\phi(t, x_J)\, dt + \sum_{J\in\mathbb{Z}^d} \rho_J^0 \frac{\phi_J^0}{\Delta t} + \sum_{n=0}^{N_T} \sum_{i=1}^{d} \sum_{J\in\mathbb{Z}^d} a_{iJ}^n \rho_J^n \partial_{x_i}\phi_J^n = O(\Delta x + \Delta t). \tag{29}$$

Let us define the reconstruction for $i = 1, \ldots, d$,

$$a_{i\Delta}(t, x) = \sum_{J\in\mathbb{Z}^d} a_{iJ}^n \mathbf{1}_{[t^n, t^{n+1})\times C_J}(t, x).$$

Using also the definition (28), we may rewrite (29) as

$$\int_0^T \int_{\mathbb{R}^d} \rho_\Delta(t, x)\partial_t\phi(t, x)\, dtdx + \sum_{i=1}^{d} \int_0^T \int_{\mathbb{R}^d} a_{i\Delta}(t, x)\rho_\Delta(t, x)\partial_{x_i}\phi(t, x)\, dtdx$$
$$+ \int_{\mathbb{R}^d} \rho_\Delta^0(x)\phi(0, x)\, dx = O(\Delta x + \Delta t).$$

From the weak convergence of the sequence (ρ_Δ) (as a consequence of the first step), we may pass to the limit in the first term. Using also Lemma 2, we can pass to the limit in the second term. Then we deduce that the limit ρ is a solution in the sense of distributions of Eq. (1).

Step 3: Conclusion. We can extract, from the sequence $(\rho_\Delta)_\Delta$, a subsequence that converges weakly in the sense of measures towards a solution in the sense of distributions of the conservative transport Eq. (1). Moreover, from Theorem 4, there exists a unique pushforward measure that solves Eq. (1). The superposition principle (see Theorem 8.2.1 in [1, Chap. 8.2]) allows to conclude that the limit ρ is the pushforward measure of Theorem 4 (see also the end of Sect. 3.1 in the present paper). By uniqueness of such a solution, we deduce also that the whole sequence $(\rho_\Delta)_\Delta$ converges to ρ.

5 Numerical Simulations

5.1 One Dimensional Examples

We consider an interval $[-2.5, 2.5]$ discretized with a Cartesian grid of size step $\Delta x = \frac{1}{80}$. As an initial data, we choose

$$\rho^{ini}(x) = e^{-10(x-1)^2} + e^{-10(x+1)^2}.$$

Then we implement the numerical scheme presented in Sect. 4 for the function $V(t,x) = \partial_x W(x)$ where $W(x) = \frac{1}{2}|x| + \frac{1}{4}|x-\xi| + \frac{1}{4}|x+\xi|$ for $\xi = 0.5$. The times dynamics is plotted in Fig. 1. For the matter of comparison, we display in Fig. 2 the result obtained for the function $W(x) = |x|$, which corresponds to the case $\xi = 0$ in the previous example. We observe that in both case blowup in finite time occurs and that the solution concentrates in a Dirac delta in finite time. The only visible difference between the two graphs is the time of concentration which is smaller in the second case than in the first case.

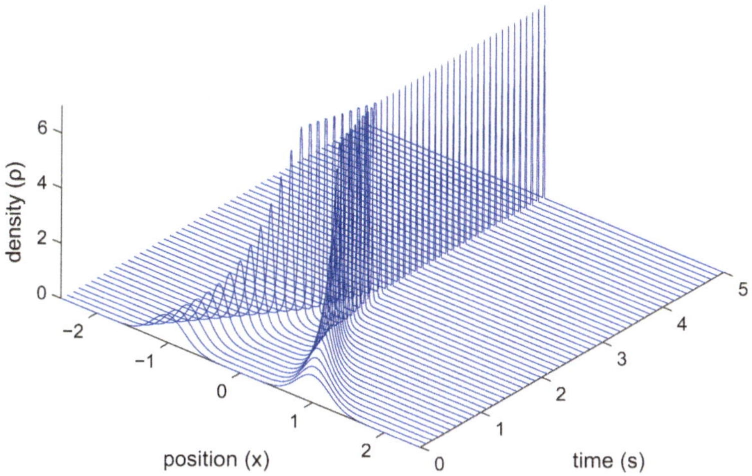

Fig. 1 Numerical simulation of Eq. (1) with $V(x) = \partial_x W(x)$ for $W(x) = \frac{1}{2}|x| + \frac{1}{4}|x - 0.5| + \frac{1}{4}|x + 0.5|$ and initial data compound of the sum of two bump functions

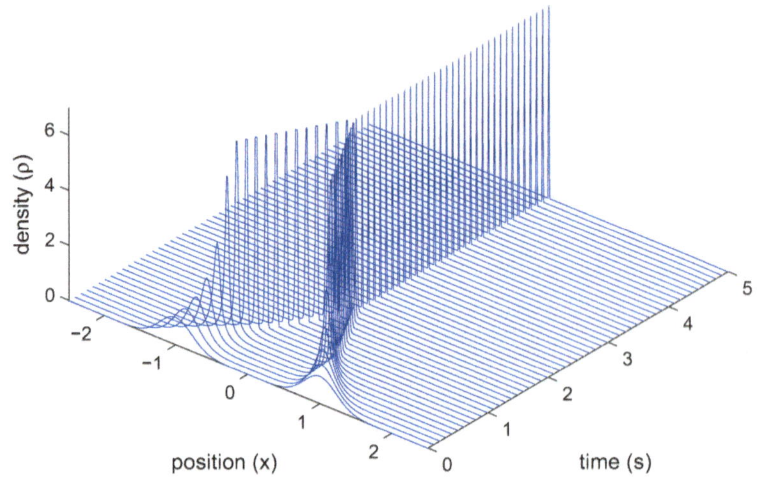

Fig. 2 Numerical simulation of Eq. (1) with $V(x) = \partial_x W(x)$ for $W(x) = |x|$ and initial data compound of the sum of two bump functions

5.2 Two Dimensional Examples

As an illustration, we propose now a numerical example in two dimensions. The spacial domain $[0, 1] \times [0, 1]$ is discretized with $N_x = 70$ nodes in the x-direction and $N_y = 70$ nodes in the y-direction and a time step $\Delta t = 10^{-3}$. We choose initial data,

$$\rho(t, x) = \mathbf{1}_{[0.2, 0.8]\times[0.2, 0.8]\setminus[0.3, 0.7]\times[0.3, 0.7]}\,.$$

We consider $V(x) = \nabla_x W(x)$ with the following interaction potentials: $W_1(x) = e^{-5|x|}$ and $W_2(x) = -5|x|$. For $|x|$ close to 0, we have that $\nabla_x W_1 \sim \nabla_x W_2$. Then the short range interaction is similar between both potential, but the long range interaction is different. The numerical results are displayed in Fig. 3 for the potential W_1 and in Fig. 4 for the potential W_2. As expected there is aggregation in finite time of ρ towards a Dirac delta in the center of the domain. It is also interesting to observe that the time dynamics during this step of concentration is different between both potentials. In both case the density ρ keeps a shape similar to the initial square shape which tighten as time increases. However, with the Morse potential (Fig. 3), a strong concentration occurs at the corners of the square, whereas in the case of the Newtonian potential (Fig. 4) the density concentration is homogeneous along the edges of the square with a slight concentration in the middle of the edges.

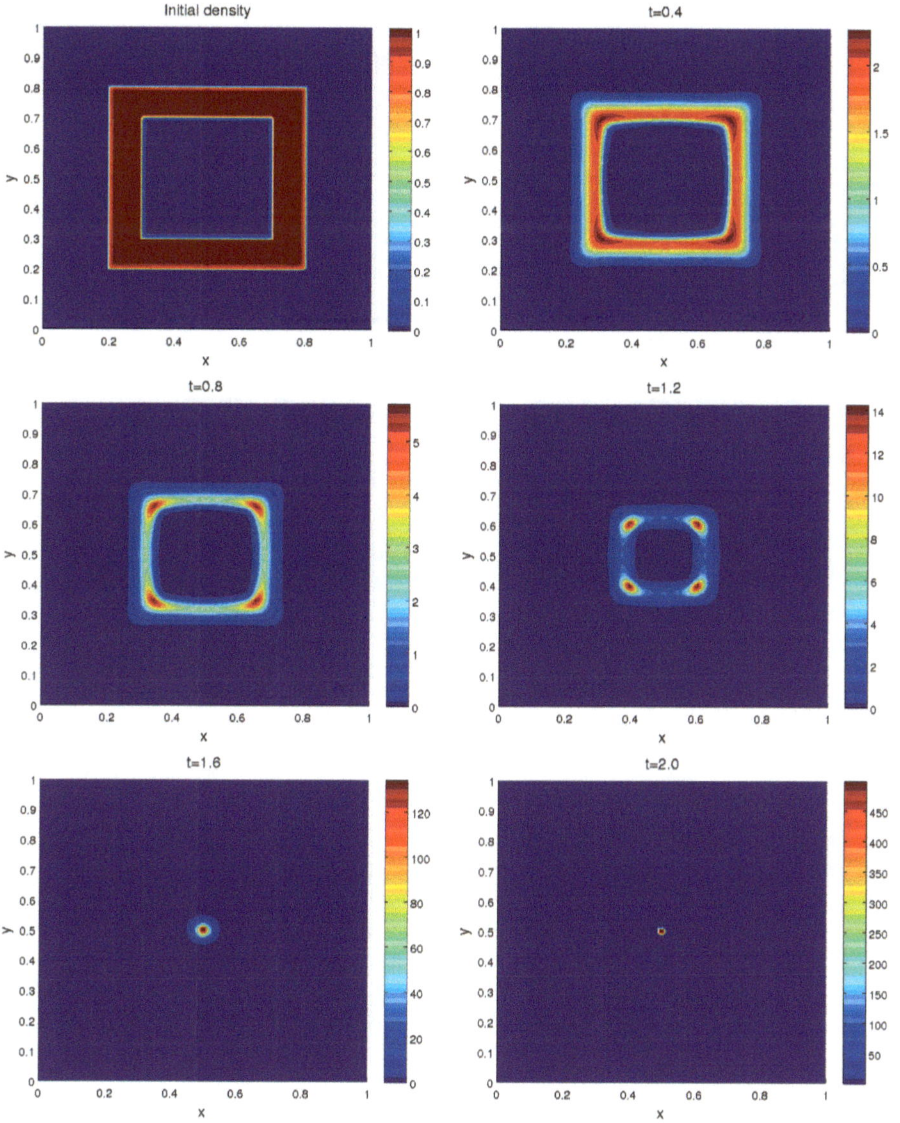

Fig. 3 Time dynamics of the numerical solution of the aggregation equation (1) with $V = \nabla_x W$, where $W(x) = e^{-5|x|}$ is a Morse potential. From *top left* to *bottom right*, the times considered are $t = 0$ (initial data), $t = 0.4$, $t = 0.8$, $t = 1.2$, $t = 1.6$ and $t = 2$

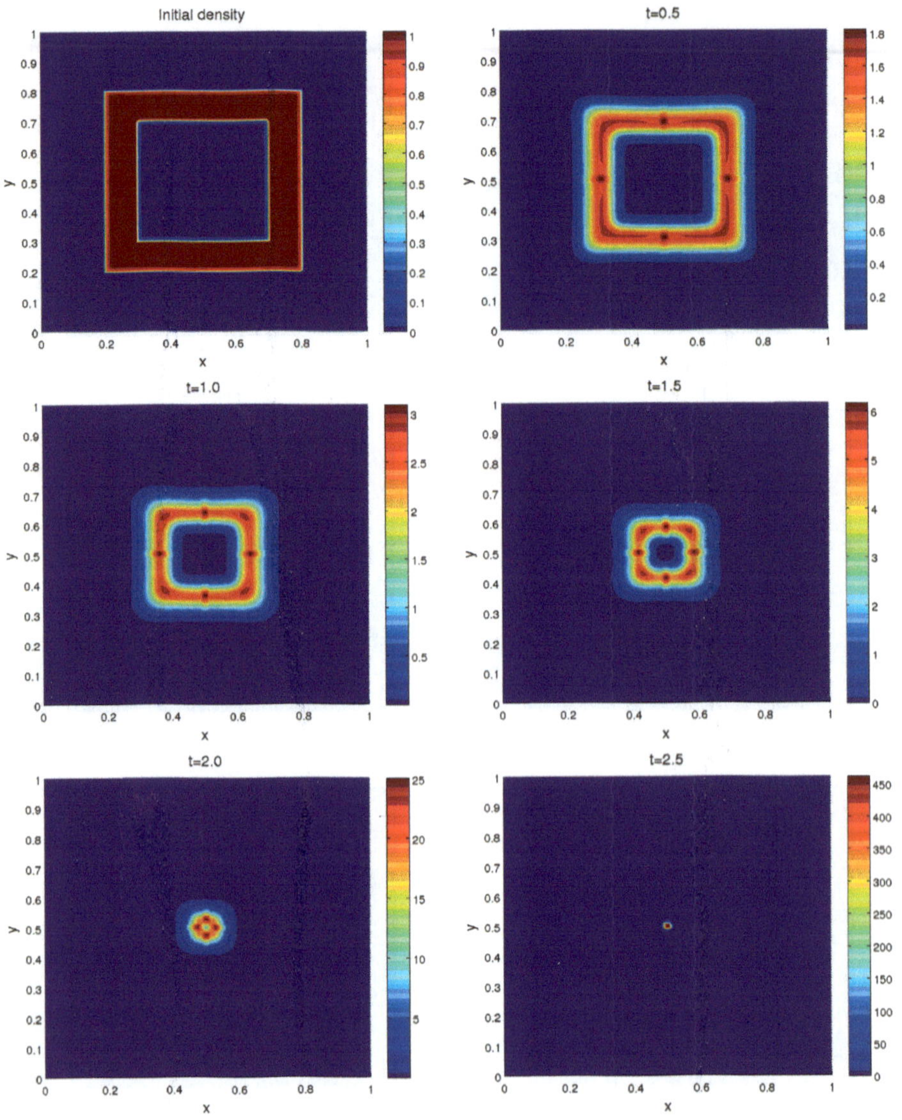

Fig. 4 Time dynamics of the numerical solution of the aggregation equation (1) with $V = \nabla_x W$, where $W(x) = -5|x|$ is a Newtonian potential. From *top left* to *bottom right*, the times considered are $t = 0$ (initial data), $t = 0.5$, $t = 1$, $t = 1.5$, $t = 2$ and $t = 2.5$

References

1. Ambrosio, L., Gigli, N., Savaré, G.: Gradient Flows in Metric Space of Probability Measures. Lectures in Mathematics. Birkäuser, Basel (2005)
2. Aubin, J.-P., Cellina, A.: Differential Inclusions: Set-Valued Maps and Viability Theory. Grundlehren der Mathematischen Wissenschaften [Fundamental Principles of Mathematical Sciences], vol. 264. Springer, Berlin (1984)
3. Benedetto, D., Caglioti, E., Pulvirenti, M.: A kinetic equation for granular media. RAIRO Model. Math. Anal. Numer. **31**, 615–641 (1997)
4. Bertozzi, A.L., Laurent, T., Rosado, J.: L^p theory for the multidimensional aggregation equation. Commun. Pure Appl. Math. **64**(1), 45–83 (2011)
5. Bertozzi, A.L., Garnett, J.B., Laurent, T.: Characterization of radially symmetric finite time blowup in multidimensional aggregation equations. SIAM J. Math. Anal. **44**(2), 651–681 (2012)
6. Bianchini, S., Gloyer, M.: An estimate on the flow generated by monotone operators. Commun. Partial Differ. Equ. **36**(5), 777–796 (2011)
7. Bodnar, M., Velázquez, J.J.L.: An integro-differential equation arising as a limit of individual cell-based models. J. Differ. Equ. **222**(2), 341–380 (2006)
8. Bouchut, F., James, F.: One-dimensional transport equations with discontinuous coefficients. Nonlinear Anal. TMA **32**(7), 891–933 (1998)
9. Campos Pinto, M., Carrillo, J.A., Charles, F., Choi, Y.-P.: Convergence of a linearly transformed particle method for aggregation equations. Preprint. https://arxiv.org/pdf/1507.07405
10. Carrillo, J.A., DiFrancesco, M., Figalli, A., Laurent, T., Slepčev, D.: Global-in-time weak measure solutions and finite-time aggregation for nonlocal interaction equations. Duke Math. J. **156**, 229–271 (2011)
11. Carrillo, J.A., Chertock, A., Huang, Y.: A finite-volume method for nonlinear nonlocal equations with a gradient flow structure. Commun. Comput. Phys. **17**(1), 233–258 (2015)
12. Carrillo, J.A., James, F., Lagoutière, F., Vauchelet, N.: The Filippov characteristic flow for the aggregation equation with mildly singular potentials. J. Differ. Equ. **260**(1), 304–338 (2016)
13. Colombo, R.M., Garavello, M., Lécureux-Mercier, M.: A class of nonlocal models for pedestrian traffic. Math. Models Methods Appl. Sci. **22**(4), 1150023, 34 (2012)
14. Craig, K., Bertozzi, A.L.: A blob method for the aggregation equation. Math. Comput. **85**(300), 1681–1717 (2016)
15. Delarue, F., Lagoutière, F.: Probabilistic analysis of the upwind scheme for transport equations. Arch. Ration. Mech. Anal. **199**, 229–268 (2011)
16. Delarue, F., Lagoutière, F., Vauchelet, N.: Convergence order of upwind type schemes for transport equations with discontinuous coefficients. J. Maths. Pures. Appl. (accepted)
17. Delarue, F., Lagoutière, F., Vauchelet, N.: Convergence order of upwind type schemes for nonlinear aggregation equation with pointy potential (preprint)
18. Dolak, Y., Schmeiser, C.: Kinetic models for chemotaxis: hydrodynamic limits and spatio-temporal mechanisms. J. Math. Biol. **51**, 595–615 (2005)
19. Filbet, F., Laurençot, Ph., Perthame, B.: Derivation of hyperbolic models for chemosensitive movement. J. Math. Biol. **50**, 189–207 (2005)
20. Filippov, A.F.: Differential equations with discontinuous right-hand side. Am. Math. Soc. Transl. (2) **42**, 199–231 (1964)
21. Gosse, L., James, F.: Numerical approximations of one-dimensional linear conservation equations with discontinuous coefficients. Math. Comput. **69**, 987–1015 (2000)
22. Gosse, L., Vauchelet, N.: Numerical high-field limits in two-stream kinetic models and 1D aggregation equations. SIAM J. Sci. Comput. **38**(1), A412–A434 (2016)
23. Huang, Y., Bertozzi, A.L.: Self-similar blowup solutions to an aggregation equation in $\mathbb{R}^n$. SIAM J. Appl. Math. **70**, 2582–2603 (2010)
24. Huang, Y., Bertozzi, A.L.: Asymptotics of blowup solutions for the aggregation equation. Discrete Contin. Dyn. Syst. Ser. B **17**, 1309–1331 (2012)

25. James, F., Vauchelet, N.: Chemotaxis: from kinetic equations to aggregation dynamics. Nonlinear Differ. Equ. Appl. **20**(1), 101–127 (2013)
26. James, F., Vauchelet, N.: Numerical method for one-dimensional aggregation equations. SIAM J. Numer. Anal. **53**(2), 895–916 (2015)
27. James, F., Vauchelet, N.: Equivalence between duality and gradient flow solutions for one-dimensional aggregation equations. Discrete Contin. Dyn. Syst. **36**(3), 1355–1382 (2016)
28. Li, H., Toscani, G.: Long time asymptotics of kinetic models of granular flows. Arch. Ration. Mech. Anal. **172**, 407–428 (2004)
29. Morale, D., Capasso, V., Oelschläger, K.: An interacting particle system modelling aggregation behavior: from individuals to populations. J. Math. Biol. **50**, 49–66 (2005)
30. Okubo, A., Levin, S.: Diffusion and Ecological Problems: Modern Perspectives. Springer, Berlin (2002)
31. Poupaud, F., Rascle, M.: Measure solutions to the linear multidimensional transport equation with discontinuous coefficients. Commun. Partial Differ. Equ. **22**, 337–358 (1997)
32. Santambrogio, F.: Optimal Transport for Applied Mathematicians. Optimal Transport for Applied Mathematicians. Birkhäuser/Springer, Cham (2015)
33. Topaz, C.M., Bertozzi, A.L.: Swarming patterns in a two-dimensional kinematic model for biological groups. SIAM J. Appl. Math. **65**, 152–174 (2004)
34. Villani, C.: Topics in Optimal Transportation. Graduate Studies in Mathematics, vol. 58. American Mathematical Society, Providence, RI (2003)
35. Villani, C.: Optimal Transport, Old and New. Grundlehren der Mathematischen Wissenschaften, vol. 338. Springer, Berlin (2009)

Semi-analytical Methods of Solution for the BGK-Boltzmann Equation Describing Sound Wave Propagation in Binary Gas Mixtures

Silvia Lorenzani

Abstract In the present paper, the propagation of high-frequency sound waves in binary gas mixtures is investigated by using the linearized Boltzmann equation based on a Bhatnagar-Gross-Krook (BGK)-type approach and diffuse reflection boundary conditions. It is shown that the semi-analytical methods of solution considered for the BGK-Boltzmann equations can be used to simplify the numerical procedures for solving our problem and gain insight into the qualitative behavior of the solution.

1 Introduction

In the low-frequency limit, the problem concerning the propagation of sound waves in gas mixtures is correctly described by the Navier-Stokes, Fourier, and Fick equations [16, 17]. If the excitation frequency ω of the sound wave becomes sufficiently high, the classical continuum approach fails even at ordinary densities, because ω^{-1} can become of the order of the molecular mean free time. Therefore, in the case of rarefied gas dynamics (or equivalently, in the case of a highly oscillatory phenomenon), there is no recourse but to turn to the kinetic theory and the Boltzmann equation [6, 10, 15, 19, 25]. In the sixties, Grad [12] conjectured that when a binary gas mixture is made up of species with very different molecular masses, under appropriate circumstances, the two components of such a mixture can support different temperatures, due to the slow exchange of kinetic energy between light and heavy species. Two-temperature continuum equations, necessary to replace those of hydrodynamics in part of the continuum regime, have been developed by a number of authors. From such "two-temperature hydrodynamics", Huck and Johnson [14] predicted the possibility of several simultaneous sound modes in a disparate-mass gas mixture (composed of very heavy plus very light molecules),

S. Lorenzani (✉)
Dipartimento di Matematica, Politecnico di Milano, Piazza Leonardo da Vinci 32, 20133 Milan, Italy
e-mail: silvia.lorenzani@polimi.it

L. Gosse, R. Natalini (eds.), *Innovative Algorithms and Analysis*,
Springer INdAM Series 16, DOI 10.1007/978-3-319-49262-9_11

for a narrow range of compositions at high enough frequencies. These predictions were later confirmed experimentally by Bowler [3]. However, the identity of these interfering modes has never been completely understood over the years, partly due to the inadequacy of the investigation tools used.

The kinetic description of a mixture of gases with different particle masses (and possibly with different internal energies) is not a trivial generalization of the classical Boltzmann theory for a single gas, since the collision operators have to take into account exchanges of momentum and energy among the different species (and also mass exchanges, in the case of reacting mixtures). Therefore, since the Boltzmann equations are quite awkward to deal with, consistent simpler models have been proposed in the literature. Among them, the McCormack model [20] has been over the years the most widely used to study a great variety of problems, such as Couette and Poiseuille flows, thermal creep, and heat transfer for mixtures, since all transport coefficients (i.e., viscosity, thermal conductivity, diffusion, and thermal diffusion ratio) can be correctly obtained from it applying the Chapman-Enskog procedure [21, 22]. Moreover, McCormack derived a model for the cross-collision operator of a general multicomponent monatomic mixture applicable to different intermolecular force laws.

Instead, in the present paper, we have chosen to perform our analysis by using the BGK relaxation model proposed by Andries et al. [1] and generalized by Groppi and Spiga [13] even to nonconservative collisions, that has been proven to be well posed from the mathematical point of view (correct Boltzmann collision invariants and Maxwellian equilibria are properly recovered, and the H theorem is fulfilled). Even if the BGK-Boltzmann equation for mixtures generally cannot match all the transport coefficients simultaneously, for the specific problem at hand, we will prove that the BGK model is able to produce results in close agreement with those obtained with more refined kinetic models. The advantage is that the equations which describe the BGK model are much simpler than those which describe, for instance, the McCormack model, allowing also a semi-analytical representation of the solution.

2 Linearized BGK Model

Let us consider a binary gaseous mixture confined between two flat, infinite, and parallel plates located at $z' = -d/2$ and $z' = d/2$. Both boundaries are held at the same constant temperature. The upper wall of the channel (located at $z' = d/2$) is fixed while the lower one (located at $z' = -d/2$) harmonically oscillates in the z'-direction (normal to the wall itself) with angular frequency ω' (the corresponding period being $T' = 2\pi/\omega'$). The velocity U_w' of the oscillating plate depends on time t' through the formula

$$U_w'(t') = U_0' \sin(\omega' t'), \tag{1}$$

where it is assumed that the amplitude U_0' is very small compared to the characteristic molecular velocity of the mixture given by

$$v_0 = \sqrt{2kT_0/m}, \tag{2}$$

with k being the Boltzmann constant, m being the mean molecular mass of the mixture and T_0 being the equilibrium temperature of the mixture. Under these conditions, the Boltzmann equation modeling the gaseous mixture motion inside the channel can be linearized by representing the distribution functions f^s ($s = 1, 2$) of both species as follows:

$$f^s = f_0^s(1 + h^s), \qquad |h^s| << 1\,, \tag{3}$$

with f_0^s being the Maxwellian configuration

$$f_0^s = n_0^s \left(\frac{m^s}{2\pi k T_0}\right)^{3/2} \exp\left[-\frac{m^s}{2kT_0}\boldsymbol{\xi}^2\right] \tag{4}$$

and h^s being the small perturbation with respect to the equilibrium state. In Eq. (4), n_0^s represents the equilibrium number density and m^s denotes the molecular mass. Using Eq. (3), the system of nonstationary BGK-Boltzmann equations reads as

$$\frac{\partial h^s}{\partial t'} + c_z \frac{\partial h^s}{\partial z'} = Lh^s_{BGK} \qquad s = 1, 2 \tag{5}$$

where Lh^s_{BGK} is the linearized BGK collision operator given by [2]

$$Lh^1{}_{BGK} = \nu_1 \Bigg\{ \rho^1 + 2(1 - \Gamma^1)\mathbf{c}\cdot\mathbf{v}^1 + 2\Gamma^1\mathbf{c}\cdot\mathbf{v}^2$$
$$+\left[1 - \frac{2\Gamma^1 M_{12}}{(1 + M_{12})}\right]\left(|\mathbf{c}|^2 - \frac{3}{2}\right)\tau^1 + \frac{2\Gamma^1 M_{12}}{(1 + M_{12})}\left(|\mathbf{c}|^2 - \frac{3}{2}\right)\tau^2 - h^1\Bigg\} \tag{6}$$

$$Lh^2{}_{BGK} = \nu_2 \Bigg\{ \rho^2 + 2(1 - \Gamma^2)\frac{\mathbf{c}}{M_{12}}\cdot\mathbf{v}^2 + 2\Gamma^2\frac{\mathbf{c}}{M_{12}}\cdot\mathbf{v}^1$$
$$+\left[1 - \frac{2\Gamma^2}{(1 + M_{12})}\right]\left(\frac{|\mathbf{c}|^2}{M_{12}} - \frac{3}{2}\right)\tau^2 + \frac{2\Gamma^2}{(1 + M_{12})}\left(\frac{|\mathbf{c}|^2}{M_{12}} - \frac{3}{2}\right)\tau^1 - h^2\Bigg\} \tag{7}$$

In Eqs. (6)–(7), ν_s are suitable collision frequencies (independent from the molecular velocity $\boldsymbol{\xi}$), $\mathbf{c} = \frac{\boldsymbol{\xi}}{\sqrt{2\frac{k}{m^1}T_0}}$, $M_{12} = m^1/m^2$, $\Gamma^1 = \frac{\nu_1^{12}\mu^{12}}{\nu_1 m^1}n_0^2$, $\Gamma^2 = \frac{\nu_1^{12}\mu^{12}}{\nu_2 m^2}n_0^1$, where $\mu^{12} = m^1 m^2/(m^1 + m^2)$ stands for the reduced mass, and ν_k^{sr} are suitably weighted integrals of the differential cross sections relevant to collisions between particles of species s and r [8]. The dimensionless macroscopic perturbed density ρ^s, velocity $\mathbf{v}^s$, temperature τ^s are obtained by taking the moments of h^s.

It is convenient now to rescale all variables appearing in Eq. (5) as follows:

$$t = t'/\theta_1, \quad z = z'/(v_0^1\theta_1), \tag{8}$$

with $\theta_1 = 1/\nu_1$, $v_0^1 = \sqrt{2kT_0/m^1}$. Furthermore, we define $\Theta_{12} = \theta_1/\theta_2 = \nu_2/\nu_1$ and $\delta = d/(v_0^1\theta_1)$, which is the dimensionless distance between the channel walls as well as the rarefaction parameter (inverse Knudsen number) of the species $s = 1$. Since the problem under consideration is one-dimensional in space, the unknown perturbed distribution functions h^s, as well as the overall quantities, depend only on the z coordinate. Likewise, we can reduce the dimensionality of the molecular-velocity space by introducing the projection procedure. First, we multiply Eq. (5) by $\frac{1}{\pi}\, e^{-({c_x}^2+{c_y}^2)}$, when $s = 1$, and by $\frac{1}{(\pi M_{12})}\, e^{-({c_x}^2+{c_y}^2)/M_{12}}$, when $s = 2$, and we integrate over all c_x and c_y. Then, we multiply Eq. (5) by $\frac{1}{\pi}\,(c_x^2 + c_y^2 - 1)\, e^{-({c_x}^2+{c_y}^2)}$, when $s = 1$, and by $\frac{1}{(\pi M_{12})}\,(\frac{c_x^2+c_y^2}{M_{12}} - 1)\, e^{-({c_x}^2+{c_y}^2)/M_{12}}$, when $s = 2$, and we integrate again over all c_x and c_y. The resulting equations after the projection are

$$\begin{aligned}\frac{\partial H^1}{\partial t} + c_z\,\frac{\partial H^1}{\partial z} + H^1 &= \rho^1 + 2(1-\Gamma^1)\,c_z\,v_z^1 + 2\Gamma^1 c_z\,v_z^2 \\ &+\left[1 - \frac{2\Gamma^1 M_{12}}{(1+M_{12})}\right](c_z^2 - \tfrac{1}{2})\,\tau^1 + \frac{2\Gamma^1 M_{12}}{(1+M_{12})}(c_z^2 - \tfrac{1}{2})\,\tau^2\end{aligned} \tag{9}$$

$$\begin{aligned}\frac{\partial H^2}{\partial t} + c_z\,\frac{\partial H^2}{\partial z} + \Theta_{12}H^2 &= \Theta_{12}\left\{\rho^2 + \frac{2(1-\Gamma^2)}{M_{12}}\,c_z\,v_z^2 + \frac{2\Gamma^2}{M_{12}}c_z\,v_z^1\right. \\ &\left.+\left[1 - \frac{2\Gamma^2}{(1+M_{12})}\right]\left(\frac{c_z^2}{M_{12}} - \frac{1}{2}\right)\tau^2 + \frac{2\Gamma^2}{(1+M_{12})}\left(\frac{c_z^2}{M_{12}} - \frac{1}{2}\right)\tau^1\right\}\end{aligned} \tag{10}$$

$$\frac{\partial \Psi^1}{\partial t} + c_z\,\frac{\partial \Psi^1}{\partial z} + \Psi^1 = \left[1 - \frac{2\Gamma^1 M_{12}}{(1+M_{12})}\right]\tau^1 + \frac{2\Gamma^1 M_{12}}{(1+M_{12})}\tau^2 \tag{11}$$

$$\frac{\partial \Psi^2}{\partial t} + c_z\,\frac{\partial \Psi^2}{\partial z} + \Theta_{12}\Psi^2 = \Theta_{12}\left\{\left[1 - \frac{2\Gamma^2}{(1+M_{12})}\right]\tau^2 + \frac{2\Gamma^2}{(1+M_{12})}\tau^1\right\} \tag{12}$$

where the reduced unknown distribution functions H^s and Ψ^s are defined as

$$H^1(z, c_z, t) = \frac{1}{\pi}\int_{-\infty}^{+\infty}\int_{-\infty}^{+\infty} h^1(z, \mathbf{c}, t)\, e^{-(c_x^2+c_y^2)}\, dc_x\, dc_y \tag{13}$$

$$H^2(z, c_z, t) = \frac{1}{\pi M_{12}}\int_{-\infty}^{+\infty}\int_{-\infty}^{+\infty} h^2(z, \mathbf{c}, t)\, e^{-(c_x^2+c_y^2)/M_{12}}\, dc_x\, dc_y \tag{14}$$

$$\Psi^1(z, c_z, t) = \frac{1}{\pi} \int_{-\infty}^{+\infty} \int_{-\infty}^{+\infty} (c_x^2 + c_y^2 - 1)\, h^1(z, \mathbf{c}, t)\, e^{-(c_x^2+c_y^2)}\, dc_x\, dc_y \tag{15}$$

$$\Psi^2(z, c_z, t) = \frac{1}{\pi M_{12}} \int_{-\infty}^{+\infty} \int_{-\infty}^{+\infty} \left(\frac{c_x^2 + c_y^2}{M_{12}} - 1 \right) h^2(z, \mathbf{c}, t)\, e^{-(c_x^2+c_y^2)/M_{12}}\, dc_x\, dc_y \tag{16}$$

The macroscopic fields appearing on the right-hand side of Eqs. (9)–(12) are given by

$$\rho^1(z, t) = \frac{1}{\sqrt{\pi}} \int_{-\infty}^{+\infty} H^1\, e^{-c_z^2}\, dc_z \tag{17}$$

$$\rho^2(z, t) = \frac{1}{\sqrt{\pi M_{12}}} \int_{-\infty}^{+\infty} H^2\, e^{-c_z^2/M_{12}}\, dc_z \tag{18}$$

$$v_z^1(z, t) = \frac{1}{\sqrt{\pi}} \int_{-\infty}^{+\infty} c_z\, H^1\, e^{-c_z^2}\, dc_z \tag{19}$$

$$v_z^2(z, t) = \frac{1}{\sqrt{\pi M_{12}}} \int_{-\infty}^{+\infty} c_z\, H^2\, e^{-c_z^2/M_{12}}\, dc_z \tag{20}$$

$$\tau^1(z, t) = \frac{1}{\sqrt{\pi}} \int_{-\infty}^{+\infty} \frac{2}{3} \left[(c_z^2 - 1/2)\, H^1 + \Psi^1 \right] e^{-c_z^2}\, dc_z \tag{21}$$

$$\tau^2(z, t) = \frac{1}{\sqrt{\pi M_{12}}} \int_{-\infty}^{+\infty} \frac{2}{3} \left[\left(\frac{c_z^2}{M_{12}} - \frac{1}{2} \right) H^2 + \Psi^2 \right] e^{-c_z^2/M_{12}}\, dc_z \tag{22}$$

Applying the same projection procedure to the linearized boundary conditions reported in [9, (Eqs. (17)–(18))] describing the diffuse scattering of gaseous particles on both walls of the microchannel, we are led to derive the following expressions

$$H^1(z = -\delta/2, c_z, t) = (\sqrt{\pi} + 2\, c_z)\, U_w - 2 \int_{\tilde{c}_z<0} d\tilde{c}_z\, \tilde{c}_z\, e^{-\tilde{c}_z^2}\, H^1(z = -\delta/2, \tilde{c}_z, t) \qquad c_z > 0 \tag{23}$$

$$H^2(z = -\delta/2, c_z, t) = \left(\sqrt{\frac{\pi}{M_{12}}} + \frac{2\, c_z}{M_{12}} \right) U_w - \frac{2}{M_{12}} \int_{\tilde{c}_z<0} d\tilde{c}_z\, \tilde{c}_z\, e^{-\tilde{c}_z^2/M_{12}}\, H^2(z = -\delta/2, \tilde{c}_z, t) \qquad c_z > 0 \tag{24}$$

$$\Psi^1(z=-\delta/2,c_z,t)=\Psi^2(z=-\delta/2,c_z,t)=0 \qquad c_z>0 \tag{25}$$

$$H^1(z=\delta/2,c_z,t)=2\int_{\tilde{c}_z>0} d\tilde{c}_z\,\tilde{c}_z\,e^{-\tilde{c}_z^2}\,H^1(z=\delta/2,\tilde{c}_z,t) \qquad c_z<0 \tag{26}$$

$$H^2(z=\delta/2,c_z,t)=\frac{2}{M_{12}}\int_{\tilde{c}_z>0} d\tilde{c}_z\,\tilde{c}_z\,e^{-\tilde{c}_z^2/M_{12}}\,H^2(z=\delta/2,\tilde{c}_z,t) \qquad c_z<0 \tag{27}$$

$$\Psi^1(z=\delta/2,c_z,t)=\Psi^2(z=\delta/2,c_z,t)=0 \qquad c_z<0 \tag{28}$$

In Eqs. (23) and (24), U_w is the dimensionless wall velocity given by

$$U_w(t)=U_0\,\sin(\omega\,t) \tag{29}$$

with $U_w = U'_w/v_0^1$, $U_0 = U'_0/v_0^1$, $\omega = \theta_1\,\omega'$, $T = 2\pi/\omega = T'/\theta_1$. The time-dependent problem described by Eqs. (9)–(12), with boundary conditions given by Eqs. (23)–(28), has been numerically solved by a deterministic finite-difference method. The region between the plates has been divided into a number of cells of the same size. For this particular application, high spatial resolution is required in order to capture properly the propagation of sound waves with short wavelength. It should be noted that, following previous numerical studies (see [18]), the emitter plate displacement is not taken into account. The gas feels the plate motion only through the boundary conditions (23)–(28). Similarly, a regular net of nodes has been arranged into a finite domain of the one-dimensional velocity space, large enough to contain the significant part of the distribution functions at any spatial location. In general, the extent of the velocity distribution function of species s ($s=1,2$) in the molecular velocity space is inversely proportional to $\sqrt{m^s}$. Thus, a separate grid for each species should be considered. Indeed, due to the normalization adopted in the present paper, we were able to use only one grid in the velocity space since the scaling factor: $\sqrt{m^1/m^2} = \sqrt{M_{12}}$ (which allows to switch between the grid associated with the first species and that associated with the second species) appears directly in the equations defining the reduced distribution functions H^s and Ψ^s (Eqs. (13)–(16)) and the macroscopic fields (Eqs. (17)–(22)). Then, the distribution functions are assumed to be constant within each cell of the phase space. Further details on the standard (upwind) implicit finite-difference scheme employed in the present computations are reported in [18]. The convergence of the numerical results presented has been extensively examined by modifying the number of grid points chosen in phase space. The variation of the components of the macroscopic fields due to the increase of the grid parameters was the criterion of the calculation accuracy. In addition, it has been checked that, allowing our data to be reduced to the case of a single monoatomic gas by taking $m^1 = m^2$, the results presented in [9] can be recovered.

In order to investigate more deeply the influence of the parameters peculiar to a binary gaseous mixture on the sound wave propagation, we present in Sects. 3 and 4 two semi-analytical methods of solution for the BGK-Boltzmann equations (9)–(12), which retain their usefulness since they are capable of providing an analytical representation of the solution and hence a picture of its qualitative behavior.

3 Integral Equations

Since the vibrations of the lower wall of the microchannel are generated by a time-harmonic forcing (of frequency ω) of the form $\sin(\omega\, t)$ (see Eq. (29)), we introduce the following expression $U_w = U_0\, e^{i\omega t}$ in Eqs. (23) and (24) and then we look for solutions of Eqs. (9)–(12) under the form

$$H^s(z, c_z, t) = \mathscr{H}^s(z, c_z)\, e^{i\,\omega\, t} \qquad s = 1, 2 \tag{30}$$

$$\Psi^s(z, c_z, t) = \Psi^s(z, c_z)\, e^{i\,\omega\, t} \qquad s = 1, 2. \tag{31}$$

The solutions of the original problem are then recovered by taking the imaginary part of H^s and Ψ^s. Putting Eqs. (30) and (31) into Eqs. (9)–(12) and then integrating the resulting equations along the trajectories of the molecules, we obtain an explicit expression for the distribution functions

$$\begin{aligned}\mathscr{H}^1(z, c_z) &= e^{-\frac{z}{c_z}(1+i\omega)}\, e^{-\frac{\delta\,\mathrm{sgn}c_z}{2\,c_z}(1+i\,\omega)}\, \mathscr{H}^1(-\delta/2\,\mathrm{sgn}c_z, c_z)\\ &+\int_{-\delta/2\,\mathrm{sgn}c_z}^{z} ds\, c_z^{-1}\Big\{\rho^1(s) + 2(1-\Gamma^1)\, c_z\, \mathrm{v}_z^1(s) + 2\Gamma^1 c_z\, \mathrm{v}_z^2(s)\\ &+\left[1-\frac{2\Gamma^1 M_{12}}{(1+M_{12})}\right](c_z^2-\tfrac{1}{2})\,\tau^1(s) + \frac{2\Gamma^1 M_{12}}{(1+M_{12})}(c_z^2-\tfrac{1}{2})\,\tau^2(s)\Big\}\\ &\times e^{-|z-s|\,\frac{(1+i\,\omega)}{|c_z|}}\end{aligned} \tag{32}$$

$$\begin{aligned}\mathscr{H}^2(z, c_z) &= e^{-\frac{z}{c_z}(\Theta_{12}+i\omega)}\, e^{-\frac{\delta\,\mathrm{sgn}c_z}{2\,c_z}(\Theta_{12}+i\,\omega)}\, \mathscr{H}^2(-\delta/2\,\mathrm{sgn}c_z, c_z)\\ &+\int_{-\delta/2\,\mathrm{sgn}c_z}^{z} ds\, \frac{\Theta_{12}}{c_z}\Big\{\rho^2(s) + \frac{2(1-\Gamma^2)}{M_{12}}\, c_z\, \mathrm{v}_z^2(s) + \frac{2\Gamma^2}{M_{12}} c_z\, \mathrm{v}_z^1(s)\end{aligned}$$

$$+\left[1-\frac{2\Gamma^2}{(1+M_{12})}\right]\left(\frac{c_z^2}{M_{12}}-\tfrac{1}{2}\right)\tau^2(s)+\frac{2\Gamma^2}{(1+M_{12})}\left(\frac{c_z^2}{M_{12}}-\frac{1}{2}\right)\tau^1(s)\Bigg\}$$

$$\times e^{-|z-s|\,\frac{(\Theta_{12}+i\omega)}{|c_z|}} \tag{33}$$

$$\Psi^1(z,c_z)=\int_{-\delta/2\,\mathrm{sgn}c_z}^{z} ds\, c_z{}^{-1}\left\{\left[1-\frac{2\Gamma^1 M_{12}}{(1+M_{12})}\right]\tau^1(s)\right.$$

$$\left.+\frac{2\Gamma^1 M_{12}}{(1+M_{12})}\tau^2(s)\right\}\times e^{-|z-s|\,\frac{(1+i\omega)}{|c_z|}} \tag{34}$$

$$\Psi^2(z,c_z)=\int_{-\delta/2\,\mathrm{sgn}c_z}^{z} ds\,\frac{\Theta_{12}}{c_z}\left\{\left[1-\frac{2\Gamma^2}{(1+M_{12})}\right]\tau^2(s)\right.$$

$$\left.+\frac{2\Gamma^2}{(1+M_{12})}\tau^1(s)\right\}\times e^{-|z-s|\,\frac{(\Theta_{12}+i\omega)}{|c_z|}} \tag{35}$$

Inserting in the definitions (17)–(22) the distribution functions (32)–(35), together with the boundary conditions, the integral equations for the macroscopic fields ρ^s, v_z^s, τ^s can be explicitly obtained. To solve the resulting system of six coupled integral equations, one can extend a finite difference technique first introduced in a paper by Cercignani and Daneri [5], by using the method reported in [7, 9]. In view of our subsequent considerations, we are interested in writing down the integral equations for the bulk velocities of the gas components:

$$\mathrm{v}_z^1(z)=\mathscr{K}_1(z,\gamma)+\frac{1}{\sqrt{\pi}}\int_{-\delta/2}^{\delta/2} ds\varrho^1(s)\,sgn(z-s)\,T_0(|z-s|\,\gamma)$$

$$-\frac{1}{\sqrt{\pi}}\int_{-\delta/2}^{\delta/2} ds\,\varrho^1(s)\,\mathscr{K}_2(z,s,\gamma)$$

$$+\frac{2}{\sqrt{\pi}}\int_{-\delta/2}^{\delta/2} ds\,[(1-\Gamma^1)\,\mathrm{v}_z^1(s)+\Gamma^1\,\mathrm{v}_z^2(s)]\,T_1(|z-s|\,\gamma)$$

$$-\frac{1}{\sqrt{\pi}}\int_{-\delta/2}^{\delta/2} ds\,[(1-\Gamma^1)\,\mathrm{v}_z^1(s)+\Gamma^1\,\mathrm{v}_z^2(s)]\,\mathscr{K}_3(z,s,\gamma)$$

$$+\frac{1}{\sqrt{\pi}}\int_{-\delta/2}^{\delta/2} ds\left[\left(1-\frac{2\,\Gamma^1 M_{12}}{(1+M_{12})}\right)\tau^1(s)+\frac{2\,\Gamma^1 M_{12}}{(1+M_{12})}\tau^2(s)\right]$$

$$\times\mathrm{sgn}(z-s)\,[T_2(|z-s|\,\gamma)-\tfrac{1}{2}T_0(|z-s|\,\gamma)]$$

$$-\frac{4}{\sqrt{\pi}}\int_{-\delta/2}^{\delta/2} ds\left[\left(1-\frac{2\,\Gamma^1 M_{12}}{(1+M_{12})}\right)\tau^1(s)+\frac{2\,\Gamma^1 M_{12}}{(1+M_{12})}\tau^2(s)\right]\mathscr{K}_4(z,s,\gamma)$$

(36)

$$\begin{aligned}
\mathrm{v}_z^2(z) &= \mathscr{K}_1(z,\tilde{\gamma})+\frac{\Theta_{12}}{\sqrt{\pi}}\int_{-\delta/2}^{\delta/2} ds\varrho^2(s)\,\mathrm{sgn}(z-s)\,T_0(|z-s|\,\tilde{\gamma})\\
&-\frac{\Theta_{12}}{\sqrt{\pi}}\int_{-\delta/2}^{\delta/2} ds\,\varrho^2(s)\,\mathscr{K}_2(z,s,\tilde{\gamma})\\
&+\frac{2\,\Theta_{12}}{\sqrt{\pi\,M_{12}}}\int_{-\delta/2}^{\delta/2} ds\,[(1-\Gamma^2)\,\mathrm{v}_z^2(s)+\Gamma^2\,\mathrm{v}_z^1(s)]\,T_1(|z-s|\,\tilde{\gamma})\\
&-\frac{\Theta_{12}}{\sqrt{\pi\,M_{12}}}\int_{-\delta/2}^{\delta/2} ds\,[(1-\Gamma^2)\,\mathrm{v}_z^2(s)+\Gamma^2\,\mathrm{v}_z^1(s)]\,\mathscr{K}_3(z,s,\tilde{\gamma})\\
&+\frac{\Theta_{12}}{\sqrt{\pi}}\int_{-\delta/2}^{\delta/2} ds\left[\left(1-\frac{2\,\Gamma^2}{(1+M_{12})}\right)\tau^2(s)+\frac{2\,\Gamma^2}{(1+M_{12})}\tau^1(s)\right]\\
&\quad\times\mathrm{sgn}(z-s)\,[T_2(|z-s|\,\tilde{\gamma})-\tfrac{1}{2}T_0(|z-s|\,\tilde{\gamma})]\\
&-\frac{4\,\Theta_{12}}{\sqrt{\pi}}\int_{-\delta/2}^{\delta/2} ds\left[\left(1-\frac{2\,\Gamma^2}{(1+M_{12})}\right)\tau^2(s)+\frac{2\,\Gamma^2}{(1+M_{12})}\tau^1(s)\right]\mathscr{K}_4(z,s,\tilde{\gamma})
\end{aligned}$$

(37)

where $\gamma=(1+i\,\omega)$, $\tilde{\gamma}=(\Theta_{12}+i\,\omega)/\sqrt{M_{12}}$, and the symbols $\mathscr{K}_i$ $(i=1,\ldots,4)$ stand for expressions involving products of the Abramowitz functions T_n defined by

$$T_n(x) := \int_0^{+\infty} s^n\,e^{-s^2-x/s}\,ds. \tag{38}$$

The explicit form of the functions $\mathscr{K}_i$ is given in [2]. In Sect. 5, we will show that some basic properties of sound wave propagation in different binary mixtures can be deduced simply by a qualitative analysis of the integral equations (36) and (37).

4 The Elementary Solutions

In order to simplify the numerical procedure needed to solve our problem, one can even improve the ansatz (30) and (31) through a separation of spatial and molecular velocity variables. This can be done by taking advantage of the explicit form of the distribution functions (H^s, Ψ^s) found in Sect. 3. The four coupled balance equations (9)–(12) can be rewritten in matrix notation as follows

$$\frac{\partial}{\partial t}\mathbf{Z}(z, c_z, t) + c_z \frac{\partial}{\partial z}\mathbf{Z}(z, c_z, t) + \Sigma\, \mathbf{Z}(z, c_z, t) = \Sigma \int_{-\infty}^{+\infty} \mathsf{K}(c_z, c_z')\, \mathbf{Z}(z, c_z', t)\, dc_z' \tag{39}$$

where $\mathbf{Z}(z, c_z, t) = (H^1, \Psi^1, H^2, \Psi^2)^T$ and $\Sigma = \text{diag}\Big(1, 1, \Theta_{12}, \Theta_{12}\Big)$. In addition, the components $k_{i,j}(c_z, c_z')$ of the kernel $\mathsf{K}(c_z, c_z')$ are listed explicitly in Appendix 1. The boundary conditions associated with Eq. (39) can be likewise expressed in the following form

$$\mathbf{Z}(z = -\delta/2, c_z, t) = \mathbf{R}\, U_w - 2\int_{\tilde{c}_z<0} d\tilde{c}_z\, \tilde{c}_z\, \mathsf{D}\, \mathbf{Z}(z = -\delta/2, \tilde{c}_z, t) \quad c_z > 0 \tag{40}$$

$$\mathbf{Z}(z = \delta/2, c_z, t) = 2\int_{\tilde{c}_z>0} d\tilde{c}_z\, \tilde{c}_z\, \mathsf{D}\, \mathbf{Z}(z = \delta/2, \tilde{c}_z, t) \quad c_z < 0 \tag{41}$$

where $\mathbf{R} = \Big((\sqrt{\pi} + 2c_z), 0, (\sqrt{\frac{\pi}{M_{12}}} + \frac{2c_z}{M_{12}}), 0\Big)^T$ and $\mathsf{D} = \text{diag}\Big(\mathrm{e}^{-\tilde{c}_z^2}, 0, \frac{1}{\mathrm{M}_{12}}\,\mathrm{e}^{-\tilde{c}_z^2/\mathrm{M}_{12}}, 0\Big)$. We seek solutions of Eq. (39) of the form

$$\mathbf{Z}(z, c_z, t) = e^{st}\, e^{-(s\mathsf{1}+\Sigma)\frac{z}{\nu}}\, \Phi(\nu, c_z) \qquad \text{with } s = \mathrm{i}\,\omega \tag{42}$$

where the separation constants ν and the elementary solutions $\Phi(\nu, c_z)$ (in general complex) are to be determined. Substituting Eq. (42) into Eq. (39), we find

$$\begin{aligned}(\nu - c_z)(s\mathsf{1} + \Sigma)\Phi(\nu, c_z) = \nu\, \Sigma \int_0^{+\infty} [&\mathsf{K}(c_z', c_z)\Phi(\nu, c_z') \\ &+ \mathsf{K}(-c_z', c_z)\Phi(\nu, -c_z')]\, dc_z'\end{aligned} \tag{43}$$

From Eq. (43), by replacing $c_z \to -c_z$, one has

$$\begin{aligned}(\nu + c_z)(s\mathsf{1} + \Sigma)\Phi(\nu, -c_z) = \nu\, \Sigma \int_0^{+\infty} [&\mathsf{K}(c_z', -c_z)\Phi(\nu, c_z') \\ &+ \mathsf{K}(-c_z', -c_z)\Phi(\nu, -c_z')]\, dc_z'\end{aligned} \tag{44}$$

Now, adding and subtracting Eqs. (43)–(44), one from the other, and taking into account the following identities

$$\mathsf{K}(c_z', -c_z) = \mathsf{K}(-c_z', c_z) \tag{45}$$

$$\mathsf{K}(-c_z', -c_z) = \mathsf{K}(c_z', c_z) \tag{46}$$

we conclude

$$\frac{1}{c_z^2}\left[\mathbf{V}(\nu,c_z) - (s\mathsf{1}+\Sigma)^{-1}\,\Sigma\int_0^{+\infty}\Gamma(c_z',c_z)\,\mathbf{V}(\nu,c_z')\,dc_z'\right] = \lambda\,\mathbf{V}(\nu,c_z) \tag{47}$$

and

$$\mathbf{U}(\nu,c_z) = \frac{\nu}{c_z}\left[\mathbf{V}(\nu,c_z) - (s\mathsf{1}+\Sigma)^{-1}\,\Sigma\int_0^{+\infty}\mathsf{K}_-(c_z',c_z)\,\mathbf{V}(\nu,c_z')\,dc_z'\right] \tag{48}$$

where

$$\mathbf{U}(\nu,c_z) = \Phi(\nu,c_z) + \Phi(\nu,-c_z) \tag{49}$$

$$\mathbf{V}(\nu,c_z) = \Phi(\nu,c_z) - \Phi(\nu,-c_z) \tag{50}$$

$$\lambda = \frac{1}{\nu^2} \tag{51}$$

$$\mathsf{K}_+(c_z',c_z) = \mathsf{K}(c_z',c_z) + \mathsf{K}(-c_z',c_z) \tag{52}$$

$$\mathsf{K}_-(c_z',c_z) = \mathsf{K}(c_z',c_z) - \mathsf{K}(-c_z',c_z) \tag{53}$$

$$\Gamma(c_z',c_z) = \mathsf{K}_-(c_z',c_z) + \frac{c_z}{c_z'}\,\mathsf{K}_+(c_z',c_z)$$
$$-\int_0^{+\infty}\frac{c_z}{c_z''}\,\mathsf{K}_+(c_z'',c_z)\,(s\mathsf{1}+\Sigma)^{-1}\,\Sigma\,\mathsf{K}_-(c_z',c_z'')\,dc_z'' \tag{54}$$

Equation (47) defines our eigenvalue problem. Once this eigenvalue problem is solved, we have the elementary solutions from

$$\Phi(\nu,c_z) = \frac{1}{2}[\mathbf{U}(\nu,c_z) + \mathbf{V}(\nu,c_z)] \tag{55}$$

$$\Phi(\nu,-c_z) = \frac{1}{2}[\mathbf{U}(\nu,c_z) - \mathbf{V}(\nu,c_z)] \tag{56}$$

Then, by taking into account the completeness theorems stated in [4], the general solution of Eq. (39) can be written as follows

$$\mathbf{Z}(z,c_z,t) = e^{st}\int_0^{+\infty} A(\nu)\left\{\Phi(\nu,c_z)\,e^{-(s\mathsf{1}+\Sigma)\left[\frac{z}{\nu}+\frac{\delta}{2\nu}\right]}\right.$$
$$\left.+\Phi(\nu,-c_z)\,e^{-(s\mathsf{1}+\Sigma)\left[-\frac{z}{\nu}+\frac{\delta}{2\nu}\right]}\right\}\,d\nu \tag{57}$$

where the symmetry condition

$$\Phi(\nu, c_z) = \Phi(-\nu, -c_z)$$

has been exploited. In Eq. (57) the arbitrary constant $A(\nu)$, to be determined from the boundary conditions, has been redefined by inserting a factor

$$e^{-(s1+\Sigma)\frac{\delta}{2|\nu|}},$$

according to the explicit form of the distribution functions found in Sect. 3. In order to solve numerically Eqs. (47), (48), (55)–(57), one can introduce a half-range quadrature scheme as done by Siewert et al. [10, 23, 24]. The algorithm requires only an eigenvalue/eigenvector routine and the solution of a system of linear algebraic equations. This allows to greatly reduce the computational effort if compared, for instance, with that required to solve a system of coupled integral equations as presented in Sect. 3. An alternative approach to handle numerically time-dependent problems in terms of elementary solutions has been recently proposed by Gosse in [11]. His scheme requires first the computation of the elementary solutions of the associated stationary boundary-value problem (which are always real functions) and then, in order to recover the time evolution, a well-balanced Godunov scheme is implemented.

5 Results and Discussion

The results presented in the following and obtained through a numerical integration of Eqs. (9)–(12) refer to the noble gaseous mixtures of He-Xe (that is, helium with molecular mass $m^1 = 4.0026$ au and xenon with molecular mass $m^2 = 131.29$ au) and Ne-Ar (that is, neon with molecular mass $m^1 = 20.179$ au and argon with molecular mass $m^2 = 39.948$ au) with the same ratio of the molar concentrations of the two species $N_{12} = n_0^1/n_0^2 = 1$. In order to inspect the physical processes going on in both gas mixtures, we are led to examine the macroscopic profiles of the mixture components in different regimes.

5.1 Near-Continuum Regime

We report in Figs. 1 and 2 the velocity and temperature profile of the mixture components as a function of the distance across the gap of the channel (at different stages during a period of oscillation) when $\delta = 10$. In order to prove the reliability of the results obtained from a numerical implementation of the BGK model presented in the present paper, we have also included in these pictures the outcomes derived

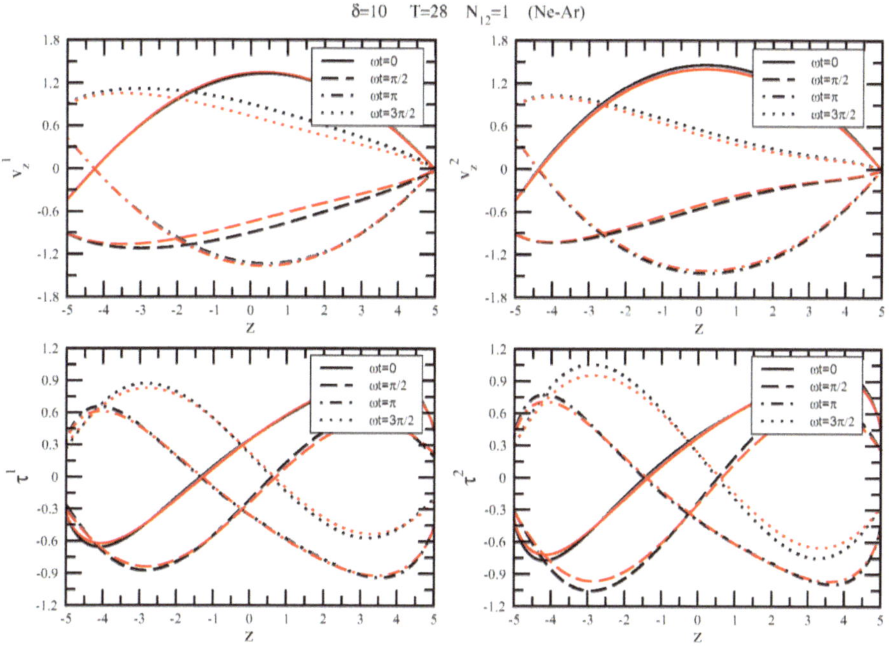

Fig. 1 Variation of the macroscopic velocities and temperatures of the Ne-Ar mixture components (with the same molar concentrations), in the z-direction across the gap of the channel for $\delta = 10$ and $T = 28$. Comparison between the BGK model (*black lines*) and the McCormack model (*red lines*). In each panel, the profiles of the macroscopic quantities are shown at different stages during a period of oscillation of the moving wall

from a numerical solution of the linearized McCormack model, widely considered in the literature to describe a gaseous mixture [23, 24]. For the sake of completeness, this model is briefly described in Appendix 2. In all the figures presented in the following, the lightest component of each mixture has been labelled with superscript 1, while the heaviest one with superscript 2.

In the case of the Ne-Ar mixture, both species have the same macroscopic fields (see Fig. 1), while for the He-Xe mixture, the pictures reveal that two different forced-sound modes are simultaneously present: a fast and a slow wave (see Fig. 2). The slow wave is a damped soundlike mode primarily carried by the Xe, while the fast wave should be associated to the He-component of the mixture. Looking at Figs. 3 and 4, where the velocity and temperature profiles of the components of the He-Xe mixture are shown for different values of the period T ($T = 2$ and $T = 70$), one can infer that the characteristics of the macroscopic fields, associated with the two sound modes, depend strongly on the excitation frequency of the modes themselves. At $T = 2$, both modes are damped, then, when T increases, they tend to equilibrate, even if, in a neighborhood of $T = 23$, the sound mode associated to the heaviest component of the mixture appears more damped than the other and with a

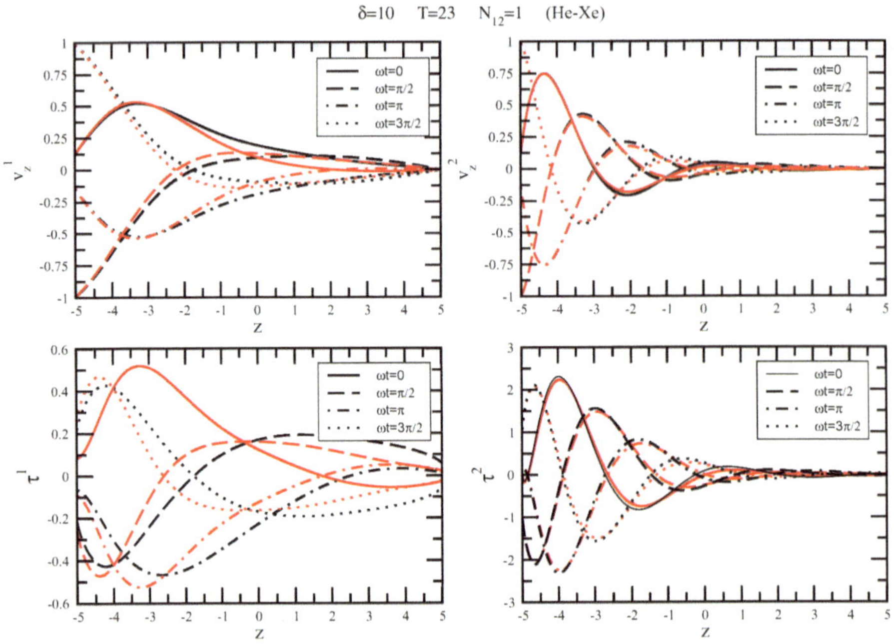

Fig. 2 Variation of the macroscopic velocities and temperatures of the He-Xe mixture components (with the same molar concentrations), in the z-direction across the gap of the channel for $\delta = 10$ and $T = 23$. Comparison between the BGK model (*black lines*) and the McCormack model (*red lines*). In each panel, the profiles of the macroscopic quantities are shown at different stages during a period of oscillation of the moving wall

very different speed of propagation. To explain quantitatively the different behavior of the two gaseous mixtures in terms of the parameters peculiar to a mixture itself, we can resort to an analysis of the integral equations given by Eqs. (36) and (37). In the near-continuum regime ($\delta \geq 10$), the integrals in Eqs. (36) and (37) give the main contribution to the velocities of gas components. For mixtures like Ne-Ar, whose constituents have comparable molecular mass, such that $\gamma \simeq \tilde{\gamma}$, $\Theta_{12} \simeq 1$, and $M_{12} \simeq 1$, the velocity profiles of the two species are symmetric by interchanging superscripts 1 and 2 (as the results reported in Fig. 1 show). In this case, Eq. (36) reduces to the equation for the velocity field of a single-component gas [9], since the terms $(-\Gamma^1 \mathrm{v}_z^1)$ and $(\Gamma^1 \mathrm{v}_z^2)$ cancel out as well as the terms: $-\left[\frac{2\,\Gamma^1 M_{12}}{(1+M_{12})}\right] \tau^1$ and $\left[\frac{2\,\Gamma^1 M_{12}}{(1+M_{12})}\right] \tau^2$. On the contrary, for mixtures like He-Xe, whose constituents have very different molecular mass, such that $\Theta_{12} << 1$, $M_{12} << 1$, and the real and imaginary parts of $\tilde{\gamma}$ are higher than those of γ, the field v_z^2, described only by the Abramowitz functions which depend on $\tilde{\gamma}$, approaches zero more quickly, as function of z, than v_z^1. This is true for each value of T. By comparing Eq. (36), when $M_{12} \simeq 0$, with the integral equation for the macroscopic velocity field of a single-

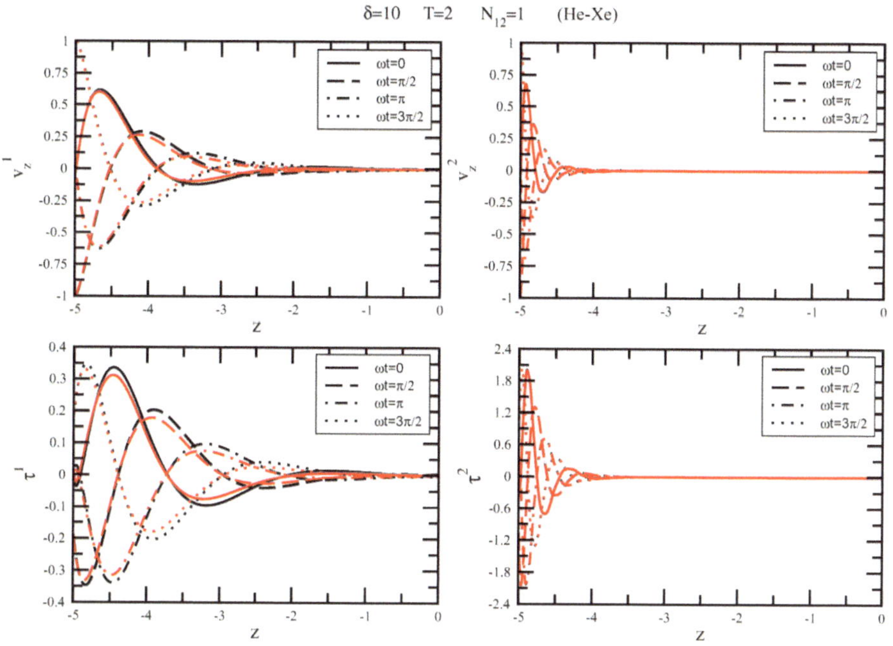

Fig. 3 Variation of the macroscopic velocities and temperatures of the He-Xe mixture components (with the same molar concentrations), in the z-direction across the gap of the channel for $\delta = 10$ and $T = 2$. Comparison between the BGK model (*black lines*) and the McCormack model (*red lines*). In each panel, the profiles of the macroscopic quantities are shown at different stages during a period of oscillation of the moving wall. In this figure we report only half of the spatial range since for $z \in [0, 5]$ both modes are completely damped and identically equal to zero

component gas, reported in [9, (Eq. (67))], we infer that the velocity field of the lightest species v_z^1 is the same as that reported in [9] except that, in Eq. (36), the influence of the integral term involving v_z^1 is weighted by the factor $(1 - \Gamma^1)$ instead of 1. This means that the wave associated to the He-component is not a single gas-based sound mode. Only when $\Gamma^1 \to 0$, that is, for $N_{12} >> 1$, Eq. (36) reduces to the equation for the velocity field of a single-component gas.

5.2 *Near-Free Molecular Flow Regime and Transitional Region*

In the near-free molecular flow regime ($\delta = 0.1$), the macroscopic field profiles of the lightest species of both mixtures coincide with those obtained for a single gas [9] (see Figs. 5 and 6). In the case of the Ne-Ar mixture, the velocity and temperature profiles of the two gas components are similar in shape, even if quantitatively different (see Fig. 5). In the case of the He-Xe mixture, Fig. 6 reveals that, for

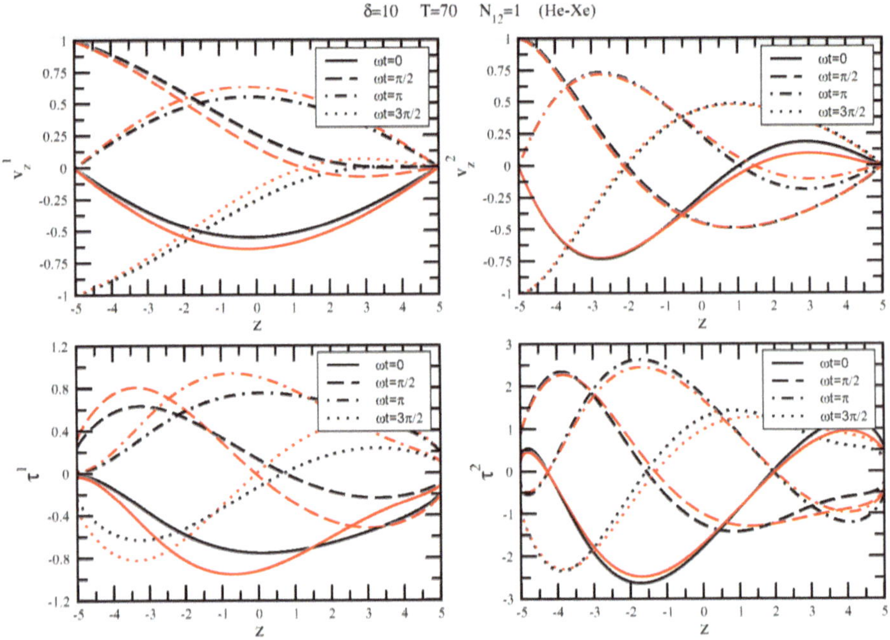

Fig. 4 Variation of the macroscopic velocities and temperatures of the He-Xe mixture components (with the same molar concentrations), in the z-direction across the gap of the channel for $\delta = 10$ and $T = 70$. Comparison between the BGK model (*black lines*) and the McCormack model (*red lines*). In each panel, the profiles of the macroscopic quantities are shown at different stages during a period of oscillation of the moving wall

$\delta = 0.1$, two different forced-sound modes are simultaneously present: a fast and a slow wave, with the same properties already analyzed in Sect. 5.1. The only difference with the near-continuum case is that, for $\delta = 0.1$, the wave associated to the He-component of the mixture appears to be a single component gas-based sound mode for each value of N_{12}, since the velocity field v_z^1 for helium is the same as that obtained for a simple monatomic gas. The results presented in Figs. 5 and 6 can be more deeply evaluated through an analysis of the integral equations given by Eqs. (36) and (37). In the near-free molecular flow regime, the integrals appearing in Eqs. (36) and (37) vanish and the velocity profiles of both components of the mixture are given by the same expression except that the Abramowitz functions depend on γ, in the case of v_z^1, and on $\tilde{\gamma}$, in the case of v_z^2. Therefore, for mixtures like Ne-Ar, whose constituents have comparable molecular mass, the parameters γ and $\tilde{\gamma}$ take close values and the macroscopic velocity fields v_z^1 and v_z^2 have a similar shape (as shown in Fig. 5). On the contrary, for disparate-mass gas mixtures like He-Xe, the Abramowitz functions, which depend on $\tilde{\gamma}$, approach zero more quickly as functions of z than those depending on γ, and the field v_z^2 represents a damped soundlike mode associated to the heaviest component of the mixture (as shown in Fig. 6). However, also in this case, the same remarks valid for $\delta = 10$ can be applied: when the period

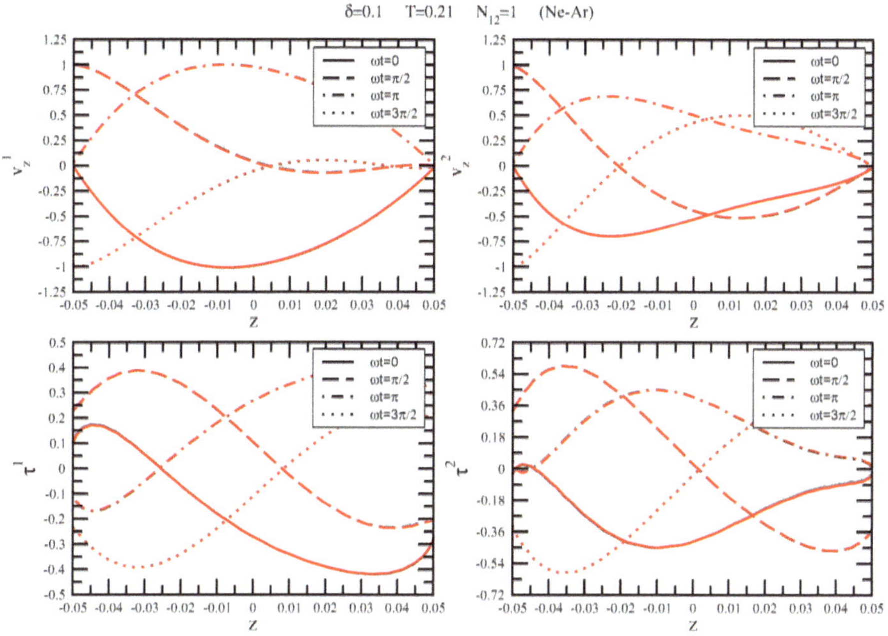

Fig. 5 Variation of the macroscopic velocities and temperatures of the Ne-Ar mixture components (with the same molar concentrations), in the z-direction across the gap of the channel for $\delta = 0.1$ and $T = 0.21$. Comparison between the BGK model (*black lines*) and the McCormack model (*red lines*). In each panel, the profiles of the macroscopic quantities are shown at different stages during a period of oscillation of the moving wall

T is very small (of the order of 10^{-2}) both modes associated to the two mixture components are damped, while when T increases, they tend to equilibrate. This result confirms that the equilibration of the modes associated with the two species is not related to the number of collisions between molecules, which, in the near free-molecular flow regime, is small. The same considerations valid for $\delta = 10$ and $\delta = 0.1$, concerning the characteristics of sound modes excited in the two different mixtures taken into account, hold also for each δ lying in the transitional region (that is, $0.1 < \delta < 10$).

6 Concluding Remarks

In the present paper, we have analyzed the high-frequency sound wave propagation in binary mixtures flowing through microchannels, within the framework of kinetic theory of rarefied gases. The results presented refer to mixtures whose constituents have comparable molecular mass (like Ne-Ar) as well as to disparate-mass gas mixtures (composed of very heavy plus very light molecules, like He-Xe). In order

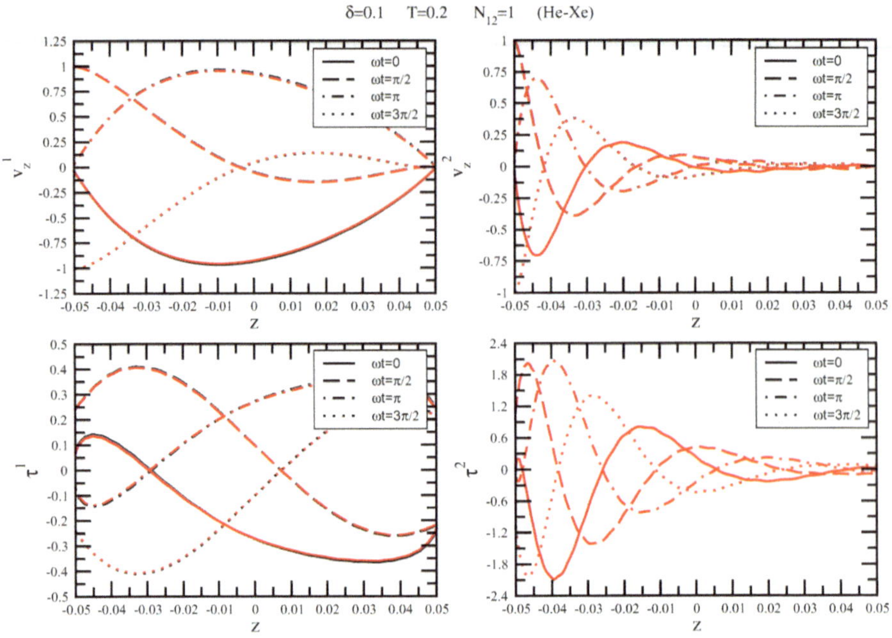

Fig. 6 Variation of the macroscopic velocities and temperatures of the He-Xe mixture components (with the same molar concentrations), in the z-direction across the gap of the channel for $\delta = 0.1$ and $T = 0.2$. Comparison between the BGK model (*black lines*) and the McCormack model (*red lines*). In each panel, the profiles of the macroscopic quantities are shown at different stages during a period of oscillation of the moving wall

to clarify the long standing open question about the precise nature of sound modes excited in binary gas mixtures, we have performed our analysis using the linearized Boltzmann equation based on a simplified kinetic model of BGK-type amenable to a relatively simple semi-analytical manipulation. In this way, through a direct comparison with the integral equations derived from the BGK model for a single-component gas, we were able to infer under what conditions the sound modes associated to the components of the mixture reduce to a single gas-based sound mode. In order to assess the reliability of our findings, since the BGK kinetic model considered in the present paper has been derived in closed-form by assuming Maxwell molecules, we have performed a comparison with the outputs of the McCormack model based on the more realistic assumption of intermolecular rigid-sphere interactions. The results presented in Figs. 1, 2, 3, 4, 5, and 6 show that there is a very good agreement between the outcomes of the BGK and McCormack models. A slight deviation can be observed for the He-Xe mixture in the near-continuum regime when the period of oscillation T increases, but the pictures reveal that the main conclusions drawn by using the BGK model are all confirmed.

Acknowledgements Silvia Lorenzani is supported by GNFM of INDAM, Italy.

Appendix 1

The components $k_{i,j}$ of the kernel $\mathsf{K}(c_z, c'_z)$ required in Eq. (39) read

$$k_{1,1} = \frac{2}{3}({c_z}^2 - \frac{1}{2})({c'_z}^2 - \frac{1}{2})\left[1 - \frac{(2\Gamma^1 M_{12})}{(1+M_{12})}\right]\frac{e^{-{c'_z}^2}}{\sqrt{\pi}} + \frac{e^{-{c'_z}^2}}{\sqrt{\pi}} + 2c_z c'_z(1-\Gamma^1)\frac{e^{-{c'_z}^2}}{\sqrt{\pi}}$$

$$k_{1,2} = \frac{2}{3}({c_z}^2 - \frac{1}{2})\left[1 - \frac{(2\Gamma^1 M_{12})}{(1+M_{12})}\right]\frac{e^{-{c'_z}^2}}{\sqrt{\pi}}$$

$$k_{1,3} = \frac{2}{3}({c_z}^2 - \frac{1}{2})\left(\frac{{c'_z}^2}{M_{12}} - \frac{1}{2}\right)\left[\frac{2\Gamma^1 M_{12}}{1+M_{12}}\right]\frac{e^{-{c'_z}^2/M_{12}}}{\sqrt{\pi M_{12}}} + 2c_z c'_z\,\Gamma^1\,\frac{e^{-{c'_z}^2/M_{12}}}{\sqrt{\pi M_{12}}}$$

$$k_{1,4} = \frac{2}{3}({c_z}^2 - \frac{1}{2})\left[\frac{2\Gamma^1 M_{12}}{1+M_{12}}\right]\frac{e^{-{c'_z}^2/M_{12}}}{\sqrt{\pi M_{12}}}$$

$$k_{2,1} = \frac{2}{3}({c'_z}^2 - \frac{1}{2})\left[1 - \frac{(2\Gamma^1 M_{12})}{(1+M_{12})}\right]\frac{e^{-{c'_z}^2}}{\sqrt{\pi}}$$

$$k_{2,2} = \frac{2}{3}\left[1 - \frac{(2\Gamma^1 M_{12})}{(1+M_{12})}\right]\frac{e^{-{c'_z}^2}}{\sqrt{\pi}}$$

$$k_{2,3} = \frac{2}{3}\left(\frac{{c'_z}^2}{M_{12}} - \frac{1}{2}\right)\left[\frac{2\Gamma^1 M_{12}}{1+M_{12}}\right]\frac{e^{-{c'_z}^2/M_{12}}}{\sqrt{\pi M_{12}}}$$

$$k_{2,4} = \frac{2}{3}\left[\frac{2\Gamma^1 M_{12}}{1+M_{12}}\right]\frac{e^{-{c'_z}^2/M_{12}}}{\sqrt{\pi M_{12}}}$$

$$k_{3,1} = \frac{2}{3}\left(\frac{{c_z}^2}{M_{12}} - \frac{1}{2}\right)({c'_z}^2 - \frac{1}{2})\left[\frac{2\Gamma^2}{1+M_{12}}\right]\frac{e^{-{c'_z}^2}}{\sqrt{\pi}} + 2\,c_z\,c'_z\,\frac{\Gamma^2}{M_{12}}\,\frac{e^{-{c'_z}^2}}{\sqrt{\pi}}$$

$$k_{3,2} = \frac{2}{3}\left(\frac{{c_z}^2}{M_{12}} - \frac{1}{2}\right)\left[\frac{2\Gamma^2}{1+M_{12}}\right]\frac{e^{-{c'_z}^2}}{\sqrt{\pi}}$$

$$k_{3,3} = \frac{2}{3}\left(\frac{{c_z}^2}{M_{12}} - \frac{1}{2}\right)\left(\frac{{c'_z}^2}{M_{12}} - \frac{1}{2}\right)\left[1 - \frac{(2\Gamma^2)}{(1+M_{12})}\right]\frac{e^{-{c'_z}^2/M_{12}}}{\sqrt{\pi M_{12}}} + \frac{e^{-{c'_z}^2/M_{12}}}{\sqrt{\pi M_{12}}}$$
$$+2\,c_z\,c'_z\,\frac{(1-\Gamma^2)}{M_{12}}\,\frac{e^{-{c'_z}^2/M_{12}}}{\sqrt{\pi M_{12}}}$$

$$k_{3,4} = \frac{2}{3}\left(\frac{{c_z}^2}{M_{12}} - \frac{1}{2}\right)\left[1 - \frac{(2\Gamma^2)}{(1+M_{12})}\right]\frac{e^{-{c'_z}^2/M_{12}}}{\sqrt{\pi M_{12}}}$$

$$k_{4,1} = \frac{2}{3}({c'_z}^2 - \frac{1}{2})\left[\frac{2\Gamma^2}{1+M_{12}}\right]\frac{e^{-{c'_z}^2}}{\sqrt{\pi}}$$

$$k_{4,2} = \frac{2}{3}\left[\frac{2\Gamma^2}{1+M_{12}}\right]\frac{e^{-c_z'^2}}{\sqrt{\pi}}$$

$$k_{4,3} = \frac{2}{3}\left(\frac{c_z'^2}{M_{12}} - \frac{1}{2}\right)\left[1 - \frac{(2\Gamma^2)}{(1+M_{12})}\right]\frac{e^{-c_z'^2/M_{12}}}{\sqrt{\pi M_{12}}}$$

$$k_{4,4} = \frac{2}{3}\left[1 - \frac{(2\Gamma^2)}{(1+M_{12})}\right]\frac{e^{-c_z'^2/M_{12}}}{\sqrt{\pi M_{12}}}$$

Appendix 2

In 1973, F. J. McCormack proposed in [20] a simple method of construction of linearized kinetic models of the Boltzmann equation for gas mixtures based upon the equivalence of moments of the Nth-order modeled collision operator and the moments of the full collision operator calculated with the Nth-order approximation to the distribution function. This method has been then used to construct a third-order model (which is the lowest order that gives a correct hydrodynamic description) of the linearized cross-collision operator for a general multicomponent monatomic gaseous mixture, known in the literature as 'McCormack model'. For the flow problem under investigation, the McCormack linearized collision operator reads

$$\begin{aligned}
Lh^1{}_{MC} = \gamma_1\Big\{&\rho^1 + 2(1-\eta^1_{1,2})\mathbf{c}\cdot\mathbf{v}^1 - 2\,\eta^2_{1,2}\,\mathbf{c}\cdot\mathbf{q}^1 + \left[1 - \frac{2\eta^1_{1,2}M_{12}}{(1+M_{12})}\right]\left(|\mathbf{c}|^2 - \frac{3}{2}\right)\tau^1 \\
&+ 2(1+\eta^4_{1,1}-\eta^3_{1,1}-\eta^3_{1,2})\,c_i c_j \tilde{P}^1_{ij} + \tfrac{8}{5}(1+\eta^6_{1,1}-\eta^5_{1,1}-\eta^5_{1,2})\left(|\mathbf{c}|^2 - \frac{5}{2}\right)\mathbf{c}\cdot\mathbf{q}^1 \\
&- \eta^2_{1,2}\left(|\mathbf{c}|^2 - \frac{5}{2}\right)\mathbf{c}\cdot\mathbf{v}^1 + 2\,\eta^1_{1,2}\,\mathbf{c}\cdot\mathbf{v}^2 + 2\,M_{12}\,\eta^2_{1,2}\,\mathbf{c}\cdot\mathbf{q}^2 + \eta^2_{1,2}\left(|\mathbf{c}|^2 - \frac{5}{2}\right)\mathbf{c}\cdot\mathbf{v}^2 \\
&+ \frac{2\,\eta^1_{1,2}M_{12}}{(1+M_{12})}\left(|\mathbf{c}|^2 - \frac{3}{2}\right)\tau^2 + \frac{2\,\eta^4_{1,2}}{M_{12}}\,c_i c_j \tilde{P}^2_{ij} + \frac{8\,\eta^6_{1,2}}{5\sqrt{M_{12}}}\left(|\mathbf{c}|^2 - \frac{5}{2}\right)\mathbf{c}\cdot\mathbf{q}^2 - h^1\Big\}
\end{aligned}$$

$$\begin{aligned}
Lh^2{}_{MC} = \gamma_2\Big\{&\rho^2 + \frac{2}{M_{12}}(1-\eta^1_{2,1})\mathbf{c}\cdot\mathbf{v}^2 - \frac{2\,\eta^2_{2,1}}{M_{12}}\,\mathbf{c}\cdot\mathbf{q}^2 + \left[1 - \frac{2\eta^1_{2,1}}{(1+M_{12})}\right]\left(\frac{|\mathbf{c}|^2}{M_{12}} - \frac{3}{2}\right)\tau^2 \\
&+ \frac{2}{M^2_{12}}(1+\eta^4_{2,2}-\eta^3_{2,2}-\eta^3_{2,1})\,c_i c_j \tilde{P}^2_{ij} + \frac{8}{5M_{12}}(1+\eta^6_{2,2}-\eta^5_{2,2}-\eta^5_{2,1})\left(\frac{|\mathbf{c}|^2}{M_{12}} - \frac{5}{2}\right)\mathbf{c}\cdot\mathbf{q}^2 \\
&- \frac{\eta^2_{2,1}}{M_{12}}\left(\frac{|\mathbf{c}|^2}{M_{12}} - \frac{5}{2}\right)\mathbf{c}\cdot\mathbf{v}^2 + \frac{2\,\eta^1_{2,1}}{M_{12}}\,\mathbf{c}\cdot\mathbf{v}^1 + \frac{2\,\eta^2_{2,1}}{M^2_{12}}\,\mathbf{c}\cdot\mathbf{q}^1 + \frac{\eta^2_{2,1}}{M_{12}}\left(\frac{|\mathbf{c}|^2}{M_{12}} - \frac{5}{2}\right)\mathbf{c}\cdot\mathbf{v}^1 \\
&+ \frac{2\,\eta^1_{2,1}}{(1+M_{12})}\left(\frac{|\mathbf{c}|^2}{M_{12}} - \frac{3}{2}\right)\tau^1 + \frac{2\,\eta^4_{2,1}}{M_{12}}\,c_i c_j \tilde{P}^1_{ij} + \frac{8\,\eta^6_{2,1}}{5\sqrt{M_{12}}}\left(\frac{|\mathbf{c}|^2}{M_{12}} - \frac{5}{2}\right)\mathbf{c}\cdot\mathbf{q}^1 - h^2\Big\}
\end{aligned}$$

where

$$\eta^k_{s,r} = \frac{\nu^k_{s,r}}{\gamma_s} \quad (s, r = 1, 2 \quad k = 1, \dots, 6)$$

and the collision frequencies $\gamma_s \quad (s = 1, 2)$ are expressed as

$$\gamma_1 = (\psi_1\, \psi_2 - \nu^4_{1,2}\, \nu^4_{2,1})\, (\psi_2 + \nu^4_{1,2})^{-1}$$
$$\gamma_2 = (\psi_1\, \psi_2 - \nu^4_{1,2}\, \nu^4_{2,1})\, (\psi_1 + \nu^4_{2,1})^{-1}$$

with

$$\psi_1 = \nu^3_{1,1} + \nu^3_{1,2} - \nu^4_{1,1}$$
$$\psi_2 = \nu^3_{2,2} + \nu^3_{2,1} - \nu^4_{2,2}$$

In addition

$$\nu^1_{s,r} = \frac{16}{3}\frac{\mu^{sr}}{m^s}\, n^r\, \Omega^{11}_{s,r},$$
$$\nu^2_{s,r} = \frac{64}{15}\left(\frac{\mu^{sr}}{m^s}\right)^2 n^r \left(\Omega^{12}_{s,r} - \frac{5}{2}\Omega^{11}_{s,r}\right),$$
$$\nu^3_{s,r} = \frac{16}{5}\left(\frac{\mu^{sr}}{m^s}\right)^2 \frac{m^s}{m^r}\, n^r \left(\frac{10}{3}\Omega^{11}_{s,r} + \frac{m^r}{m^s}\Omega^{22}_{s,r}\right),$$
$$\nu^4_{s,r} = \frac{16}{5}\left(\frac{\mu^{sr}}{m^s}\right)^2 \frac{m^s}{m^r}\, n^r \left(\frac{10}{3}\Omega^{11}_{s,r} - \Omega^{22}_{s,r}\right),$$
$$\nu^5_{s,r} = \frac{64}{15}\left(\frac{\mu^{sr}}{m^s}\right)^3 \frac{m^s}{m^r}\, n^r\, \Gamma^5_{s,r},$$
$$\nu^6_{s,r} = \frac{64}{15}\left(\frac{\mu^{sr}}{m^s}\right)^3 \left(\frac{m^s}{m^r}\right)^{3/2} n^r\, \Gamma^6_{s,r}$$

with

$$\Gamma^5_{s,r} = \Omega^{22}_{s,r} + \left(\frac{15\, m^s}{4\, m^r} + \frac{25\, m^r}{8\, m^s}\right)\Omega^{11}_{s,r} - \left(\frac{m^r}{2\, m^s}\right)(5\, \Omega^{12}_{s,r} - \Omega^{13}_{s,r}),$$
$$\Gamma^6_{s,r} = -\Omega^{22}_{s,r} + \frac{55}{8}\,\Omega^{11}_{s,r} - \frac{5}{2}\,\Omega^{12}_{s,r} + \frac{1}{2}\,\Omega^{13}_{s,r}.$$

The $\Omega^{ij}_{s,r}$ are the Chapman-Cowling integrals [8], which for the rigid sphere interaction read

$$\Omega^{ij}_{s,r} = \frac{(j+1)!}{8}\left[1 - \frac{1+(-1)^i}{2(i+1)}\right]\left(\frac{\pi\, k\, T}{2\,\mu^{sr}}\right)^{1/2}(d_s + d_r)^2$$

where d_s $(s = 1, 2)$ are the molecular diameters of every species s. Furthermore, the dimensionless macroscopic perturbed density (ρ^s), velocity ($\mathbf{v}^s$), temperature (τ^s), stress tensor ($\tilde{P}^s_{ij}$) and heat flux ($\mathbf{q}^s$), appearing in the McCormack model, are defined as follows

$$\rho^1 = \frac{1}{\pi^{3/2}}\int_{\mathbf{R}^3} h^1\, \mathrm{e}^{-|\mathbf{c}|^2}\, d\mathbf{c}, \qquad \rho^2 = \frac{1}{(\pi M_{12})^{3/2}}\int_{\mathbf{R}^3} h^2\, \mathrm{e}^{-\frac{|\mathbf{c}|^2}{M_{12}}}\, d\mathbf{c}$$

$$\mathbf{v}^1 = \frac{1}{\pi^{3/2}}\int_{\mathbf{R}^3} \mathbf{c}\, h^1\, \mathrm{e}^{-|\mathbf{c}|^2}\, d\mathbf{c}, \qquad \mathbf{v}^2 = \frac{1}{(\pi M_{12})^{3/2}}\int_{\mathbf{R}^3} \mathbf{c}\, h^2\, \mathrm{e}^{-\frac{|\mathbf{c}|^2}{M_{12}}}\, d\mathbf{c}$$

$$\tau^1 = \frac{1}{\pi^{3/2}}\int_{\mathbf{R}^3}\left(\frac{2}{3}\,|\mathbf{c}|^2 - 1\right)h^1\, \mathrm{e}^{-|\mathbf{c}|^2}\, d\mathbf{c}$$

$$\tau^2 = \frac{1}{(\pi M_{12})^{3/2}}\int_{\mathbf{R}^3}\left(\frac{2}{3M_{12}}\,|\mathbf{c}|^2 - 1\right)h^2\, \mathrm{e}^{-\frac{|\mathbf{c}|^2}{M_{12}}}\, d\mathbf{c}$$

$$\tilde{P}^1_{ij} = \frac{1}{\pi^{3/2}}\int_{\mathbf{R}^3}\left(c_i\, c_j - \frac{1}{3}|\mathbf{c}|^2\,\delta_{ij}\right) h^1\, \mathrm{e}^{-|\mathbf{c}|^2}\, d\mathbf{c}$$

$$\tilde{P}^2_{ij} = \frac{1}{(\pi M_{12})^{3/2}}\int_{\mathbf{R}^3}\left(c_i\, c_j - \frac{1}{3}|\mathbf{c}|^2\,\delta_{ij}\right) h^2\, \mathrm{e}^{-\frac{|\mathbf{c}|^2}{M_{12}}}\, d\mathbf{c}$$

$$\mathbf{q}^1 = \frac{1}{\pi^{3/2}}\int_{\mathbf{R}^3}\frac{1}{2}\mathbf{c}\left(|\mathbf{c}|^2 - \frac{5}{2}\right) h^1\, \mathrm{e}^{-|\mathbf{c}|^2}\, d\mathbf{c}$$

$$\mathbf{q}^2 = \frac{1}{(\pi M_{12})^{3/2}}\int_{\mathbf{R}^3}\frac{1}{2}\mathbf{c}\left(\frac{|\mathbf{c}|^2}{M_{12}} - \frac{5}{2}\right) h^2\, \mathrm{e}^{-\frac{|\mathbf{c}|^2}{M_{12}}}\, d\mathbf{c}$$

References

1. Andries, P., Aoki, K., Perthame, B.: A consistent BGK-type model for gas mixtures. J. Stat. Phys. **106**, 993–1018 (2002)
2. Bisi, M., Lorenzani, S.: High-frequency sound wave propagation in binary gas mixtures flowing through microchannels. Phys. Fluids **28**, 052003 (2016)
3. Bowler, J.R., Johnson, E.A.: Anomalous dispersion in disparate-mass gas mixtures. Phys. Rev. Lett. **54**, 329–332 (1985)
4. Cercignani, C.: Mathematical Methods in Kinetic Theory. Plenum, New York (1969)
5. Cercignani, C., Daneri, A.: Flow of a rarefied gas between two parallel plates. J. Appl. Phys. **34**, 3509–3513 (1963)
6. Cercignani, C., Sernagiotto, F.: The method of elementary solutions for time dependent problems in linearized kinetic theory. Ann. Phys. **30**, 154–167 (1964)

7. Cercignani, C., Lampis, M., Lorenzani, S.: Plane Poiseuille-Couette problem in micro-electro-mechanical systems applications with gas-rarefaction effects. Phys. Fluids **18**, 087102 (2006)
8. Chapman, S., Cowling, T.G.: The Mathematical Theory of Non-uniform Gases. Cambridge University Press, New York (1970)
9. Desvillettes, L., Lorenzani, S.: Sound wave resonances in micro-electro-mechanical systems devices vibrating at high frequencies according to the kinetic theory of gases. Phys. Fluids **24**, 092001 (2012)
10. Garcia, R.D.M., Siewert, C.E.: The linearized Boltzmann equation: sound-wave propagation in a rarefied gas. Z. Angew. Math. Phys. **57**, 94–122 (2006)
11. Gosse, L.: Computing Qualitatively Correct Approximations of Balance Laws. Springer, Milan (2013)
12. Grad, H.: Theory of rarefied gases. In: Devienne, F.M. (ed.) Rarefied Gas Dynamics, pp. 100–138. Pergamon, New York (1960)
13. Groppi, M., Spiga, G.: A Bhatnagar-Gross-Krook-type approach for chemically reacting gas mixtures. Phys. Fluids **16**, 4273–4284 (2004)
14. Huck, R.J., Johnson, E.A.: Possibility of double sound propagation in disparate-mass gas mixtures. Phys. Rev. Lett. **44**, 142–145 (1980)
15. Kalempa, D., Sharipov, F.: Sound propagation through a rarefied gas confined between source and receptor at arbitrary Knudsen number and sound frequency. Phys. Fluids **21**, 103601 (2009)
16. Landau, L.D., Lifshitz, E.M.: Fluid Mechanics. Pergamon, New York (1989)
17. Lighthill, J.: Waves in Fluids. Cambridge University Press, New York (1978)
18. Lorenzani, S., Gibelli, L., Frezzotti, A., Frangi, A., Cercignani, C.: Kinetic approach to gas flows in microchannels. Nanoscale Microscale Thermophys. Eng. **11**, 211–226 (2007)
19. Loyalka, S.K., Cheng, T.C.: Sound-wave propagation in a rarefied-gas. Phys. Fluids **22**, 830–836 (1979)
20. McCormack, F.J.: Construction of linearized kinetic models for gaseous mixtures and molecular gases. Phys. Fluids **16**, 2095–2105 (1973)
21. Naris, S., Valougeorgis, D., Kalempa, D., Sharipov, F.: Gaseous mixture flow between two parallel plates in the whole range of the gas rarefaction. Physica A **336**, 294–318 (2004)
22. Naris, S., Valougeorgis, D., Kalempa, D., Sharipov, F.: Flow of gaseous mixtures through rectangular microchannels driven by pressure, temperature, and concentration gradients. Phys. Fluids **17**, 100607 (2005)
23. Siewert, C.E., Valougeorgis, D.: Concise and accurate solutions to half-space binary-gas flow problems defined by the McCormack model and specular-diffuse wall conditions. Eur. J. Mech. B/Fluids **23**, 709–726 (2004)
24. Siewert, C.E., Valougeorgis, D.: The McCormack model: channel flow of a binary gas mixture driven by temperature, pressure and density gradients. Eur. J. Mech. B/Fluids **23**, 645–664 (2004)
25. Sirovich, L., Thurber, J.K.: Propagation of forced sound waves in rarefied gasdynamics. J. Acoust. Soc. Am. **37**, 329–339 (1965)

Convergent Lagrangian Discretization for Drift-Diffusion with Nonlocal Aggregation

Daniel Matthes and Benjamin Söllner

Abstract A Lagrangian discretization for nonlinear aggregation-diffusion equations in one space dimension is presented, and its convergence is rigorously analyzed. In comparison to related works by the first author and Osberger (ESAIM Math Model Numer Anal 48:697–726, 2014; Found Comput Math 1–54, 2015) on Lagrangian schemes for drift-diffusion equations, convergence is proven directly on the level on the Lagrangian maps, without passage through the density formulation.

1 Introduction

1.1 Drift-Diffusion Equations with Non-local Aggregation

A time-implicit discretization of the gradient flow for the energy functional,

$$\mathbf{E}(u) = \int_J \frac{u^m}{m-1}\,\mathrm{d}x + \frac{1}{2}\int_J u \cdot W * u\,\mathrm{d}x + \int_J Vu\,\mathrm{d}x \tag{1}$$

with respect to the L^2-Wasserstein distance W on the space $\mathscr{P}(J)$ of probability densities $u : J \to \mathbb{R}_{\geq 0}$ on the interval $J := [a, b]$ is studied. We shall assume:

1. *superlinear diffusion:* $m > 1$;
2. *attractive interaction:* $W : [a-b, b-a] \to \mathbb{R}$ C^2-smooth, symmetric and convex;
3. *confining potential:* $V : J \to \mathbb{R}$ is C^2-smooth, non-negative, and uniformly convex with modulus $\lambda_V \geq 0$, i.e., $V'' \geq \lambda_V$.

The first and the second integral term in the definition of $\mathbf{E}$ are in competition with each other in the following sense: the first penalizes large values of the density u, and favours an equal distribution of u on J; whereas the second one penalizes,

D. Matthes (✉) • B. Söllner
Technische Universität München, Zentrum Mathematik/M8, Boltzmannstr. 3, 85747 Garching, Germany
e-mail: matthes@ma.tum.de; b.soellner@ma.tum.de

L. Gosse, R. Natalini (eds.), *Innovative Algorithms and Analysis*,
Springer INdAM Series 16, DOI 10.1007/978-3-319-49262-9_12

roughly speaking, the spreading of u and is the smaller, the more of u's total mass is concentrated in the neighborhood of a single point. The third integral in (1) has a global confining effect on u too, although negligible with respect to the second term. That Wasserstein gradient flow of $\mathbf{E}$ is formally given as the solution operator to the following initial boundary value problem for a transport equation,

$$\partial_t u + \partial_x \big(u\, \mathbf{v}[u]\big) = 0, \tag{2}$$

$$\mathbf{v}[u](t,a) = \mathbf{v}[u](t,b) = 0, \tag{3}$$

$$u(0,x) = u^0(x), \tag{4}$$

where, at any $t \geq 0$, the velocity field $\mathbf{v}[u]$ depends on its respective density u via,

$$-\mathbf{v} = \partial_x \left(\frac{\delta \mathbf{E}}{\delta u} \right) = m\, u^{m-2} u_x + W' * u + V'. \tag{5}$$

Substitution of (5) into (2) produces a nonlinear drift-diffusion equation with a non-local aggregation term,

$$\boxed{\partial_t u = (u^m)_{xx} + \big(u\, W' * u\big)_x + \big(u\, V'\big)_x\, .} \tag{6}$$

A key feature of the dynamics is the competition between the spreading of mass induced by the diffusion and the concentration of mass due to the aggregation term related to W. Equations of that type were studied intensively in the recent past. Indeed, the celebrated parabolic-elliptic Keller-Segel model[1] is of the form (6).

One of the central questions about the qualitative behavior of (6) is whether the interaction force W is strong enough to defeat diffusion and produce concentration of positive mass in one (or several) points, either at finite or in infinite time. For instance, a well-known result for the Keller-Segel model in two space dimensions is the existence of a critical mass such that any solution with a total mass that is less than the critical value exists globally, and any solution with supercritical mass blows up in finite time. Lagrangian schemes are well suited for numerical approximation of solutions that might develop concentration singularities, since the moving spatial mesh automatically refines those zones in which mass accumulates. Just recently, a Lagrangian scheme [7] has been shown to produce highly accurate approximations of the blow-up time and the blow-up profile in a one-dimensional chemotaxis model. Here, we give an alternative

[1] However, our smoothness hypothesis on W excludes application to kernels like $W(z) = \log |z|$.

discretization for a similar class of Eq. (6), and we prove analytical properties like convergence.

1.2 Lagrangian Formulation

For our purposes, the transport form (2) is appropriate, as it directly gives rise to the Lagrangian formulation: consider some sufficiently smooth velocity field $\mathbf{v}[u]$ associated to a solution u of (2), and define a time-dependent Lagrangian map X:

$$\partial_t X(t,\xi) = \mathbf{v}[u(t,\cdot)]\big(X(t,\xi)\big). \tag{7}$$

This solution is unique up to the choice of the initial map $X(0,\cdot) : I \to J$. Thanks to the solution formula for the transport equation, the corresponding u reads,

$$u\big(t, X(t,\xi)\big) = \frac{\bar{u}(X(t,\xi))}{X_\xi(t,\xi)}, \tag{8}$$

for a suitable (reference) probability density $\bar{u}$. Following the idea of the pioneering contributions in [12, 13], we shall choose $\bar{u}$ as the unit density on the interval $I = [0,1]$, in which case $X(t,\cdot) : I \to J$ is simply the inverse cumulative distribution function of $u(t,\cdot)$, see Sect. 2, for which (8) simplifies to

$$\boxed{u\big(t, X(t,\xi)\big) = \frac{1}{X_\xi(t,\xi)}.} \tag{9}$$

Accordingly, the Lagrangian map $X^0 : I \to J$ at initial time $t = 0$ is the uniquely determined "càdlàg" (right continuous with left limits) function, for which,

$$\int_a^{X^0(\xi)} u^0(x)\,\mathrm{d}x = \xi \quad \text{at each } \xi \in I. \tag{10}$$

An autonomous equation for X follows by inserting (10) into (7), yielding:

$$\begin{aligned} \partial_t X(t,\xi) &= \mathbf{v}\left[\frac{1}{X_\xi(t,\cdot)}\right](\xi) \\ &= -\left(X_\xi^{-m}\right)_\xi - \int_I W'\big(X(t,\xi) - X(t,\eta)\big)\,\mathrm{d}\eta - V'\big(X(t,\xi)\big), \end{aligned} \tag{11}$$

where X satisfies the boundary and initial conditions

$$X(t,0) = a, \quad X(t,1) = b, \tag{12}$$

$$X(0,\xi) = X^0(\xi). \tag{13}$$

A remarkable—albeit easily verifiable—fact playing a key role is that (11) is again a gradient flow with respect to the scalar product in $L^2(I)$, for the functional,

$$\mathbb{E}(X) = \mathbf{E}\left(\frac{1}{X_\xi} \circ X^{-1}\right)$$
$$= \int_I \frac{X_\xi^{-(m-1)}}{m-1}\,\mathrm{d}\xi + \iint_{I\times I} W\big(X(\xi) - X(\eta)\big)\,\mathrm{d}(\xi,\eta) + \int_I V\big(X(\xi)\big)\,\mathrm{d}\xi. \tag{14}$$

1.3 Overview of the Discretization

We discretize (11) by approximating X using **piecewise affine elements in space and the implicit Euler method in time**. For simplicity, let a sequence of $(\tau^{(k)}, \delta^{(k)})_{k=1,2,\ldots}$ of time steps $\tau^{(k)}$ and "mass lattice" widths $\delta^{(k)} = 1/k$ be given. Introduce the respective equidistant mesh

$$\boldsymbol{\xi}^{(k)} = (\xi_0, \xi_1, \ldots, \xi_k) \quad \text{with } \xi_i = i\delta^{(k)}, \tag{15}$$

and the associated canonical hat functions $\theta_0^{(k)}, \ldots, \theta_k^{(k)} : I \to \mathbb{R}$ which are globally continuous, piecewise affine on the intervals $[\xi_{i-1}, \xi_i]$, and satisfy $\theta_j^{(k)}(\xi_i) = \delta_{ij}$. Our discrete approximation $X^{(k)}$ for the solution X to (11)–(13) is a piecewise constant in time and piecewise affine in space function $[0,T] \times I \to J$ such that

$$X^{(k)}(t,\cdot) = \mathbf{X}[x^n] := \sum_{i=0}^{k} x_i^n \theta_i^{(k)} \quad \text{for all } t \in ((n-1)\tau, n\tau], \tag{16}$$

where $x^n = (x_0^n, x_1^n, \ldots, x_k^n) \in \mathbb{R}^{k+1}$ stand for $x_i^n = X^{(k)}(t, \xi_i)$ at the nodes, so that

$$a = x_0^n < x_1^n < \ldots < x_k^n = b. \tag{17}$$

The set of all such "monotone" vectors in $\mathbb{R}^{k+1}$ is denoted by $\mathfrak{x}^{(k)}$. See Fig. 1 for the geometrical interpretation of the Lagrangian map $X = \mathbf{X}[x]$ and its associated density u from (9) for a given $x \in \mathfrak{x}^{(k)}$: X is the linear interpolation of the x_i's over the ξ_i's, and u is piecewise constant on the intervals (x_{i-1}, x_i). Our fully discrete scheme

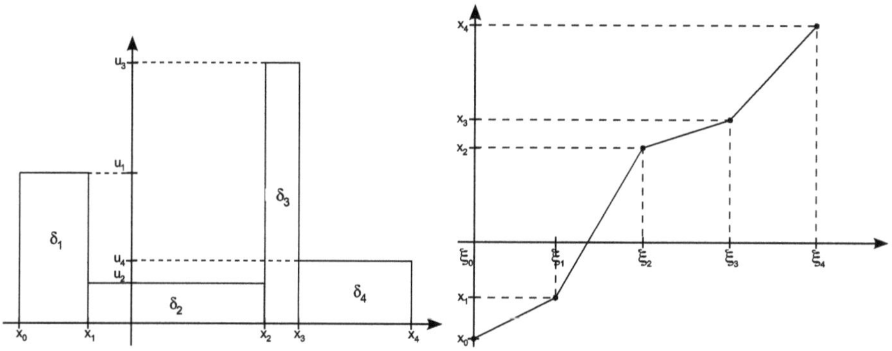

Fig. 1 A piecewise affine Lagrangian map (*right*) and its associated probability density (*left*)

amounts to solving inductively for $n = 1, 2, \ldots$ the following implicit difference equation for x^n in $\mathfrak{x}^{(k)}$:

$$\left[S^{(k)}\left(\frac{x^n - x^{n-1}}{\tau^{(k)}}\right)\right]_j = -\frac{1}{\delta^{(k)}}\left(\left(\frac{x_{j+1}^n - x_j^n}{\delta^{(k)}}\right)^{-m} - \left(\frac{x_j^n - x_{j-1}^n}{\delta^{(k)}}\right)^{-m}\right)$$
$$- \iint_{I\times I} W'\big(\mathbf{X}[x^n](\xi) - \mathbf{X}[x^n](\eta)\big)\big(\theta_j^{(k)}(\xi) - \theta_j^{(k)}(\eta)\big)\,\mathrm{d}\xi\,\mathrm{d}\eta$$
$$- \int_{\xi_{j-1}}^{\xi_{j+1}} V'\big(\mathbf{X}[x^n](\xi)\big)\theta_j^{(k)}(\xi)\,\mathrm{d}\xi, \quad j = 1, 2, \ldots, k-1, \tag{18}$$

where the integral expressions on the right-hand side can be calculated explicitly for various relevant choices of V and W. The stiffness matrix $S^{(k)} \in \mathbb{R}^{(k-1)\times(k-1)}$ is the standard one associated to the hat functions $\theta_i^{(k)}$, see (31).

Theorem 1 *Given a monotone vector $x^0 \in \mathfrak{x}^{(k)}$, then a solution $(x^n)_{n\geq 0}$ to system* (18) *is obtained by solving the uniformly convex minimization problem*

$$\frac{1}{2\tau}(x - x^{n-1})^T S^{(k)}(x - x^{n-1}) + \mathbb{E}\big(\mathbf{X}[x]\big) \longrightarrow \min \tag{19}$$

inductively for $n = 1, 2, \ldots$ on the set $\mathfrak{x}^{(k)}$, see (17). *In particular, the solution (x^n) is defined for all $n \geq 0$ and is uniquely determined by x^0. The corresponding Lagrangian map $X^{(k)}$ defined via* (16) *dissipates the entropy, that is*

$$\mathbb{E}\big(X^{(k)}(t,\cdot)\big) \leq \mathbb{E}\big(X^{(k)}(s,\cdot)\big) \quad \textit{for } 0 \leq s \leq t \leq T. \tag{20}$$

Moreover, any two discrete solutions $\hat{X}^{(k)}$ and $\check{X}^{(k)}$,—obtained for the same choice of $\tau^{(k)}$ and $\delta^{(k)}$, but possibly for different initial conditions $\hat{X}^0$ and $\check{X}^0$—contract *exponentially in time at rate λ: for $0 \le s \le t \le T$,*

$$\|\hat{X}^{(k)}(t,\cdot) - \check{X}^{(k)}(t,\cdot)\|_{L^2} \le (1+\lambda\tau)^{-\lfloor \frac{t-s}{\tau} \rfloor} \|\hat{X}^{(k)}(s,\cdot) - \check{X}^{(k)}(s,\cdot)\|_{L^2}. \tag{21}$$

Our discretization preserves the comparison principle, too, being understood in a weaker sense, since—unless $W \equiv 0$ and $V \equiv 0$—one cannot expect that solutions u to (6) remain between the upper and lower bounds of u^0. However, there is an explicit control on how quickly the bounds grow in time.

Theorem 2 *Let $\underline{z}, \overline{z} > 0$ be such that*

$$\underline{z} \le \partial_\xi \mathbf{X}[x^0] \le \overline{z}.$$

Introduce

$$\overline{K} := \sup W'' + \sup V'' \ge \lambda_V, \tag{22}$$

and assume the time step $\tau^{(k)} > 0$ and the spatial mesh width $\delta^{(k)} = \frac{1}{k}$ satisfy

$$\frac{\left(\delta^{(k)}\right)^2}{\tau^{(k)}} \le 9me^{-2\overline{K}T}\overline{z}. \tag{23}$$

Then, at any index n with $n\tau^{(k)} \le T$, the corresponding $(x^n)_{n=0,1,\ldots}$ is such that,

$$(1+\tau^{(k)}\overline{K})^{-n}\underline{z} \le \partial_\xi \mathbf{X}[x^n] \le (1+\tau^{(k)}\lambda_V)^n \overline{z}. \tag{24}$$

1.4 Main Convergence Result

To any given sequence $(x^n)_{n=0,1,\ldots}$ of monotone vectors $x^n \in \mathfrak{x}^{(k)}$, we associate the piecewise-constant-in-time and piecewise-affine-in-space Lagrangian map

$$X^{(k)} : [0,T] \times I \to J, \qquad \text{via (16).}$$

Theorem 3 *Assume that the initial condition in (10) is such that*

$$\overline{z} := 2 \sup_{\xi \in I} \partial_\xi X^0(\xi) < \infty.$$

- *Let a terminal time $T > 0$ and a sequence $(\tau^{(k)}, \delta^{(k)})_{k=1,2,\ldots}$ be given such that (23) is satisfied with $\overline{K}$ from (22), with $\tau^{(k)} \downarrow 0$, $\delta^{(k)} = \frac{1}{k}$;*
- *For each k, choose an initial datum $x^0_{(k)} \in \mathfrak{x}^{(k)}$ such that $\underline{z} \le \partial_\xi \mathbf{X}[x^0_{(k)}] \le \overline{z}$ for some $\underline{z} > 0$, and such that $\mathbf{X}[x^0_k] \to X^0$ in $L^2(I)$ as $k \to \infty$;*

- *Finally, for each k, let $X^{(k)} : [0,T]\times I \to J$ be the Lagrangian map corresponding to the solution $(x^n)_{n=0,1,\ldots}$ to* (18) *with initial condition $x^0 = x^0_{(k)}$;*

then, ***along a suitable sequence*** *$k \to \infty$*, ***the*** *$X^{(k)}$* ***converge strongly in*** $L^2([0,T]\times I)$ ***to a map*** $X_* \in L^2([0,T]; H^1(I))$ *with the following properties:*

1. *X_* is absolutely continuous w.r.t. t in $L^2(I)$, with $\partial_t X_* \in L^2([0,T]\times I)$;*
2. *each $X_*(t,\cdot)$ is monotonically increasing;*
3. *the derivative $(\partial_\xi X_*)^{-m}$ belongs to $L^1([0,T]\times I)$;*
4. *and finally, X_* is a weak solution to the initial boundary value problem* (11)–(13) *in the sense that $X_*(0,\cdot) = X^0$, at almost every $t \in [0,T]$,*

$$\lim_{\xi\downarrow 0} X_*(t,\xi) = a, \qquad \lim_{\xi\uparrow 1} X_*(t,\xi) = b$$

along with, for all test functions $\psi \in C_c^\infty\big((0,T)\times(0,1)\big)$,

$$\begin{aligned} -\int_0^T \int_I \partial_t\psi(t,\xi)\, X_*(t,\xi)\,\mathrm{d}\xi\,\mathrm{d}t &= \int_0^T \int_I \partial_\xi\psi(t,\xi)\big(\partial_\xi X_*(t,\xi)\big)^{-m}\,\mathrm{d}\xi\,\mathrm{d}t \\ &+ \frac{1}{2}\int_0^T \iint_{I\times I} \big(\psi(t,\xi)-\psi(t,\eta)\big)\, W'\big(X_*(t,\xi)-X_*(t,\eta)\big)\,\mathrm{d}(\xi,\eta)\,\mathrm{d}t \\ &+ \int_0^T \int_I \psi(t,\xi)\, V'\big(X_*(t,\xi)\big)\,\mathrm{d}\xi\,\mathrm{d}t. \end{aligned} \tag{25}$$

1.5 Related Results from the Literature

The idea to use Lagrangian methods for numerical approximation of diffusion equations dates back at least to a short note of MacCamy and Socolovsky [17]. A few years later, a more elaborate approach has been taken by Russo [23], which also works in multiple space dimensions. Still, it took a while until this approach gained momentum (see, however, [5] and references therein), which was when it was re-discovered in the context of Wasserstein gradient flows. After the "invention" of the JKO scheme [14] for analytical construction of such gradient flows, several theses [16, 21] of students from the Bonn school have been concerned with numerical exploration of the associated transport equations. About a decade ago, the practicability of the inverse distribution function for numerical solution of nonlinear diffusion and homogeneous kinetic equations has been demonstrated in two works of Gosse and Toscani [12, 13], where they also rigorous analyzed several properties of the resulting schemes. In the next years, the Gosse-Toscani discretization has been used on a variety of further PDEs, like the Keller-Segel model [4], the p-Laplace equation [1], the isentropic Euler equations [25], the Cahn-Hilliard equation [8], and the Quantum drift diffusion equation [9]. The convergence analysis for these schemes is rather recent [18, 19].

Nowadays, the focus is on **Lagrangian schemes that also work for diffusion equations in several space dimensions**. Theoretical bases were laid out in [10], where an evolution equation for the Lagrangian map in a "natural gauge" was derived. That one has been discretized using finite elements in [6, 7]. The basic principle is the same as in one dimension: a transport equation is derived for u,

$$\partial_t u = -\mathrm{div}\big(u\,\mathbf{v}[u]\big), \qquad \text{on } \Omega \subset \mathbb{R}^d,$$

along with the one associated to the vector field $\mathbf{v}$,

$$\partial_t X = \mathbf{v}[u] \circ X. \tag{26}$$

Its solution X plays the role of the *Lagrangian map*. By substituting $u = X_{\#}\bar{u}$ for an appropriate reference density $\bar{u}$ on Ω, (26) becomes an autonomous equation in X, which is again a gradient flow on an L^2-space of maps.

In contrast with the one-dimensional case, the resulting L^2-gradient flow is structurally much weaker than the original gradient flow in the Wasserstein metric. The flow's λ-uniform geodesic convexity in Wasserstein [20] is not seen at the level of the L^2-flow. The main cause for that loss of structure is the curvature of the Wasserstein space, which vanishes in dimension one, but is positive for higher-dimensional base spaces Ω. Alternative structure-preserving discretizations [3, 15] were designed based on geometric properties of optimal transport [24]. Their rigorous convergence analysis is ongoing research.

1.6 Hypotheses and Novelty of the Contribution

Apart from the inclusion of the non-local aggregation term related to W, the main difference of the work at hand in comparison to earlier convergence analyses [18, 19] by the first author with Osberger is that convergence is directly studied on the level of the Lagrangian map X, and not just for the resulting density $u = \frac{1}{X_\xi} \circ X^{-1}$.

This is more challenging since dynamics of the Lagrangian maps are less stable than the density's ones; this should be intuitively clear since the velocity field for the mass transport is not well-defined in regions of vanishing mass, and typically becomes singular near the edge of support. Therefore, in the realm of Theorem 3 we need to assume that **the density is initially (and hence, by the comparison principle, at *any* time, see Theorem 2) bounded away from zero**

on the entire spatial domain J. Two comments on our general hypotheses are as follow:

- Condition (23) seems to be essential for the proof of the comparison principle in Theorem 2, thus—indirectly—for the main convergence result, too. It is worthwhile to emphasize that it plays a completely different role than the CFL condition in the related scheme in [12] where it was needed to ensure stability of the explicit Euler iteration. Here, iterations are always well-defined thanks to the implicit time-integrator.
- We use equidistant mesh widths $\delta^{(k)}$ in "mass space": this restriction could be relaxed quite easily, admitting meshes in which the sizes of the largest and the smallest mass cells are bounded uniformly in k, but at the price of introducing more cumbersome notations and technical calculations. In fact, the results in [18] were proven for these more general meshes.

2 Background: The Inverse Distribution Function

In this short section, we discuss in more detail the relation between the functionals $\mathbf{E}$ from (1) and $\mathbb{E}$ from (14), and their respective gradient flows (6) and (11). Recall our notations $J = [a, b]$ and $I = [0, 1]$.

Definition 1 ((Inverse) Distribution Function) To each probability density $u \in \mathscr{P}(J)$, one associates the *distribution function* $U : J \to I$ and the *inverse distribution function* $X : I \to J$ as follows:

$$U(x) = \int_a^x u(x)\,\mathrm{d}x \quad \text{for each } x \in J,$$

and X is the uniquely determined cadlag function such that

$$U(X(\xi)) = \xi \quad \text{for all } \xi \in I. \tag{27}$$

Since an inverse distribution function X is monotonically increasing by definition, it is differentiable almost everywhere. Differentiation of (27) w.r.t. ξ and elementary manipulations yield the useful identities

$$u(x) = \frac{1}{X_\xi(U(x))}; \quad \text{and} \quad \partial_\xi X(\xi) = \frac{1}{u(X(\xi))}\,. \tag{28}$$

The importance of the inverse distribution function in connection with Wasserstein gradient flows can be understood from the following

Lemma 1 *Let $u_1, u_2 \in \mathscr{P}(J)$ be probability densities, and let X_1, X_2 be their respective inverse distribution functions. Then the L^2-Wasserstein distance between u_1 and u_2 equals to the L^2-distance between X_1 and X_2:*

$$\mathrm{W}(u_1, u_2) = \left(\int_a^b (X_1(\xi) - X_2(\xi))^2 \, \mathrm{d}\xi\right)^{1/2} = \|X_1 - X_2\|_{L^2} . \tag{29}$$

In other words, the correspondence between probability densities and their respective inverse distribution functions is actually an isometry between the metric spaces

- $(\mathscr{P}(J); \mathrm{W})$ of probability densities endowed with the L^2-Wasserstein metric W,
- and the affine subspace $\left(L^2_*(I), (\cdot,\cdot)_{L^2}\right)$ of the Hilbert space $L^2(I)$, consisting of strictly monotone maps $X : I \to J$ with $X(0) = a$ and $X(1) = b$.

Proof See [24, Theorem 2.18], or [2, Sect. 7.2].

Lemma 2 *For each $u \in \mathscr{P}(J)$ and its corresponding inverse distribution function $X : I \to J$, we have that $\mathbf{E}(u) = \mathbb{E}(X)$.*

Proof The result follows by a change of variables $x = X(\xi)$ inside the definition of $\mathbf{E}(u)$ in (1), and a straight-forward computation, using the identity (28) above:

$$\begin{aligned}
\mathbf{E}(u) &= \int_a^b \left[\frac{u^m}{m-1} + \frac{1}{2} u\, W * u + V \cdot u\right] \mathrm{d}x \\
&= \int_0^1 \Bigg[\frac{u(X(\xi))^m}{m-1} \\
&\quad + \frac{1}{2}\int_0^1 \Big(u(X(\xi)) W(X(\xi) - X(\eta)) u(X(\eta))\Big) \partial_\eta X(\eta)\, \mathrm{d}\eta \\
&\quad + V(X(\xi)) u(X(\xi))\Bigg] X_\xi(\xi)\, \mathrm{d}\xi \\
&= \int_0^1 \left[\frac{(\partial_\xi X(\xi))^{-(m-1)}}{m-1} + \frac{1}{2}\int_0^1 \Big(W(X(\xi) - X(\eta))\Big)\, \mathrm{d}\eta + V(X(\xi))\right] \mathrm{d}\xi \\
&= \mathbb{E}(X).
\end{aligned}$$

Finally, we relate the evolution equations (6) and (11). Recall that the first one is the gradient flow of $\mathbf{E}$ w.r.t. the L^2-Wasserstein distance on $\mathscr{P}(J)$, while the second is the gradient flow of $\mathbb{E}$ w.r.t. the inner product on $L^2(I)$.

Lemma 3 *Let a solution $X : [0, T] \times I \to J$ to the initial boundary value problem* (11)–(13) *be given. Assume that X is smooth and satisfies* $\inf \partial_\xi X > 0$.

Then,

$$u : [0, T] \times I \to \mathbb{R}, \qquad \textit{with} \qquad u(t, \cdot) := \frac{1}{X_\xi(t, \cdot)} \circ X(t, \cdot)^{-1} \tag{30}$$

is well-defined, smooth, positive, and solves (2)–(4).

Proof According to (28), the function $u(t, \cdot)$ defined in (30) is a probability density that has $X(t, \cdot)$ as its inverse distribution function. Further note that differentiating (27) *twice* w.r.t. ξ yields the formula

$$u_x \circ X = -\frac{X_{\xi\xi}}{X_\xi^3}.$$

To conclude preparations, take the total time derivative of $U(t, X(t, \xi)) = \xi$ to obtain

$$\partial_t U(t, X(t, \xi)) = -\frac{\partial_t X(t, \xi)}{X_\xi(t, \xi)}.$$

Multiply both sides of (11) with $\frac{1}{-X_\xi(t,\xi)}$ and use these identities to get,

$$\begin{aligned}\partial_t U(t, X(t, \xi)) &= m u^{m-1}(t, X(t, \xi)) u_x(t, X(t, \xi)) \\ &+ \int_0^1 u(t, X(t, \xi)) W'(X(t, \xi) - X(t, \eta)) \, \mathrm{d}\eta + u(t, X(t, \xi)) \cdot V'(X(t, \xi)).\end{aligned}$$

Let $x = X(t, \xi)$ and $y = X(t, \eta)$, and differentiate w.r.t. x. This gives (6).

3 Piecewise-Linear *P*1 Discretization in Space

For any k, discretization in space is done by restricting $X : I \to J$ to maps of the form

$$X = \mathbf{X}[x] = \sum_{i=0}^{k} x_i \theta_i^{(k)},$$

where $x \in \mathfrak{x}^{(k)}$ is an arbitrary monotone vector, see (17), and the $\theta_0^{(k)}, \theta_1^{(k)}, \ldots, \theta_k^{(k)} : I \to \mathbb{R}$ are the standard hat functions, i.e.,

$$\theta_i^{(k)}(\xi) := \begin{cases} \frac{\xi - \xi_{i-1}}{\delta^{(k)}} & \text{for } \xi \in [\xi_{i-1}, \xi_i), \\ \frac{\xi_{i+1} - \xi}{\delta^{(k)}} & \text{for } \xi \in [\xi_i, \xi_{i+1}], \\ 0 & \text{elsewhere.} \end{cases} \tag{31}$$

We denote the respective space of such maps $\mathscr{X}_k$ and endow it with the scalar product $(\cdot,\cdot)_{L^2}$ of the Hilbert space $L^2(I)$. Note that $\mathscr{X}_k$ is not linear but only affine, because each $X \in \mathscr{X}_k$ satisfies the boundary conditions $X(0) = a$ and $X(1) = b$, and the scalar product is evaluated on pairs of *differences* between elements, that is,

$$(\hat{X} - \check{X}, \hat{Y} - \check{Y})_{L^2} = \int_I \big(\hat{X}(\xi) - \check{X}(\xi)\big)\big(\hat{Y}(\xi) - \check{Y}(\xi)\big)\,\mathrm{d}\xi \tag{32}$$

for arbitrary $\hat{X}, \check{X}, \hat{Y}, \check{Y} \in \mathscr{X}_k$.

In this section, we shall collect some properties of $\mathscr{X}_k$, and in particular of the restriction $\mathbb{E}^{(k)}$ of $\mathbb{E}$ to $\mathscr{X}_k$. Let k be fixed throughout the following discussion.

3.1 A Scalar Product on Vectors

X establishes an affine-linear one-to-one correspondence between monotone vectors in $\mathfrak{x}^{(k)}$ and maps in $\mathscr{X}_k$. Hence, the $L^2(I)$-scalar product (32) induces a scalar product on $\mathfrak{x}^{(k)}$, given by a $(k-1)\times(k-1)$ real matrix $S^{(k)}$ such that

$$(\hat{x} - \check{x}, \hat{y} - \check{y})_S := (\hat{x} - \check{x})^T S^{(k)} (\hat{y} - \check{y}) = \big(\mathbf{X}[\hat{x}] - \mathbf{X}[\check{x}], \mathbf{X}[\hat{y}] - \mathbf{X}[\check{y}]\big)_{L^2}$$

for each $\hat{x}, \check{x}, \hat{y}, \check{y} \in \mathfrak{x}^{(k)}$. This matrix is well-known in the literature:

$$S^{(k)} = \frac{\delta^{(k)}}{6}\begin{pmatrix} 4 & 1 & & & \\ 1 & 4 & 1 & & \\ & \ddots & \ddots & \ddots & \\ & & 1 & 4 & 1 \\ & & & 1 & 4 \end{pmatrix}. \tag{33}$$

By an elementary estimate using the binomial theorem, we obtain:

Lemma 4 *For any $v \in \mathbb{R}^{k-1}$, we have*

$$v^T S^{(k)} v \geq \frac{\delta^{(k)}}{3}.$$

The difference vectors $\hat{x} - \check{x}$ and $\hat{y} - \check{y}$ are always zero in their zeroth and kth components, so it suffices to consider the $k-1$ remaining components in the scalar product.

3.2 Properties of the Restricted Entropy Functional

Recall that the functional $\mathbb{E} : \mathscr{X} \to \mathbb{R}$ is given by

$$\mathbb{E}(X) = \mathscr{A}(X) + \mathscr{B}(X) + \mathscr{C}(X)$$

with the components

$$\mathscr{A}(X) = \int_0^1 \frac{(\partial_\xi X(\xi))^{-(m-1)}}{m-1}\, \mathrm{d}\xi,$$
$$\mathscr{B}(X) = \frac{1}{2}\int_0^1 \int_0^1 W(X(\xi) - X(\eta))\, \mathrm{d}\eta\, \mathrm{d}\xi,$$
$$\mathscr{C}(X) = \int_0^1 V(X(\xi))\, \mathrm{d}\xi.$$

$\mathbb{E}$ induces a functional $\mathbb{E}^{(k)} : \mathfrak{x}^{(k)} \to \mathbb{R}$ for each k via

$$\mathbb{E}^{(k)}(x) = \mathbb{E}(\mathbf{X}[x]),$$

and we define accordingly $\mathscr{A}^{(k)}, \mathscr{B}^{(k)}, \mathscr{C}^{(k)} : \mathfrak{x}^{(k)} \to \mathbb{R}$.

Lemma 5 *Each $\mathbb{E}^{(k)}$ is bounded from below, and is λ_V-convex.*

Here the modulus of convexity is meant with respect to the scalar product $(\cdot, \cdot)_S$, that is, the inequality

$$\left(\partial_{x_i x_j} \mathbb{E}^{(k)}\right)_{\substack{i=1,\ldots,k-1\\ j=1,\ldots,k-1}} \geq \lambda_V\, S \tag{34}$$

holds according to the order of symmetric matrices.

Proof We start with showing the claim about convexity. Since the map $\mathbf{X} : \mathfrak{x}^{(k)} \to \mathscr{X}$ is affine, and since the scalar product on $\mathfrak{x}^{(k)}$ is the one induced from the L^2-scalar product on $\mathscr{X}$, it suffices to verify that $\mathbb{E}$ is λ_V-convex on $\mathscr{X}$. $\mathscr{A}$ is convex since $s \mapsto s^{-(m-1)}$ is convex on $s > 0$, and since differentiation and integration are linear operations. Similarly, since the map $X \mapsto X(\xi) - X(\eta)$ is affine and integration is linear, the convexity of W is inherited to $\mathscr{B}$. Finally, since $V'' \geq \lambda_V$ by hypothesis, it clearly follows that $\mathscr{C}$ is uniformly convex of modulus λ_V.

For the differentiable function $\mathbb{E}^{(k)}$, boundedness from below is now a direct consequence of uniform convexity.

Corollary 1 *For $\hat{x}, \check{x} \in \mathfrak{x}^{(k)}$, we have*

$$(\hat{x} - \check{x})^T \left(\partial_j \mathbb{E}^{(k)}(\hat{x}) - \partial_j \mathbb{E}^{(k)}(\check{x})\right)_{j=0,\ldots,k} \geq \lambda_V (\hat{x} - \check{x})^T S^{(k)} (\hat{x} - \check{x})\ . \tag{35}$$

Proof A Taylor expansion of $\mathbb{E}^{(k)}$ using (34) implies that

$$\mathbb{E}^{(k)}(x+\Delta x) \geq \mathbb{E}^{(k)}(x) + (\Delta x)^T \big(\partial_{x_j}\mathbb{E}^{(k)}(x)\big)_{j=0,\dots,k} + \frac{\lambda_V}{2}(\Delta x)^T S^{(k)}(\Delta x). \tag{36}$$

First, choose $x := \check{x}$ and $\Delta x := \hat{x} - \check{x}$ in (36); next, choose $x := \hat{x}$ and $\Delta x := \check{x} - \hat{x}$ in (36). Add the resulting inequalities to obtain (35).

The rest of this section is devoted to calculating the first and second derivatives of the components of $\mathbb{E}^{(k)}$.

Lemma 6 *The first derivatives of $\mathscr{A}^{(k)}$, $\mathscr{B}^{(k)}$ and $\mathscr{C}^{(k)}$ are given as follows:*

$$\partial_{x_j}\mathscr{A}^{(k)}(x) = \left(\frac{x_{j+1}-x_j}{\delta^{(k)}}\right)^{-m} - \left(\frac{x_j - x_{j-1}}{\delta_k}\right)^{-m}, \tag{37}$$

$$\partial_{x_j}\mathscr{B}^{(k)}(x) = \int_0^1\int_0^1 W'\big(\mathbf{X}[x](\xi) - \mathbf{X}[x](\eta)\big)\theta_j^{(k)}(\xi)\,\mathrm{d}\xi\,\mathrm{d}\eta, \tag{38}$$

$$\partial_{x_j}\mathscr{C}^{(k)}(x) = \int_0^1 V'\big(\mathbf{X}[x](\xi)\big)\theta_j^{(k)}(\xi)\,\mathrm{d}\xi, \tag{39}$$

for all $j = 1, 2, \dots, k-1$.

Proof In this proof, we abbreviate $\partial_j := \frac{\partial}{\partial x_j}$. To begin with, let us show that

$$\mathscr{A}^{(k)}(x) = \frac{\delta^{(k)}}{m-1}\sum_{i=1}^{k}\left(\frac{x_i - x_{i-1}}{\delta^{(k)}}\right)^{-(m-1)}. \tag{40}$$

Indeed, splitting up the integral into the intervals (ξ_{i-1}, ξ_i), on each of which we have that $\partial_\xi \mathbf{X}[x](\xi) = \frac{x_i - x_{i-1}}{\delta^{(k)}}$, it follows that

$$\begin{aligned}\mathscr{A}^{(k)}(\mathbf{X}[x]) &= \sum_{i=1}^{k}\int_{\xi_{i-1}}^{\xi_i}\frac{(\partial_\xi \mathbf{X}[x](\xi))^{-(m-1)}}{m-1}\,\mathrm{d}\xi \\ &= \sum_{i=0}^{k-1}\frac{\delta^{(k)}}{m-1}\left(\frac{x_i - x_{i-1}}{\delta^{(k)}}\right)^{-(m-1)}.\end{aligned}$$

From here, it follows that

$$\begin{aligned}\partial_j\mathscr{A}^{(k)}(x) &= \frac{\delta^{(k)}}{m-1}\sum_{i=1}^{k}\partial_j\left[\left(\frac{x_i - x_{i-1}}{\delta^{(k)}}\right)^{-(m-1)}\right] \\ &= \left(\frac{x_{j+1}-x_j}{\delta^{(k)}}\right)^{-m} - \left(\frac{x_j - x_{j-1}}{\delta_k}\right)^{-m}.\end{aligned}$$

To obtain the formula for the derivative of $\mathscr{B}$, it suffices to observe that

$$\partial_j \mathbf{X}[x](\xi) = \sum_{i=0}^{k} \partial_j x_i \theta_i^{(k)}(\xi) = \theta_j^{(k)}(\xi),$$

since then, using that W' is odd,

$$\begin{aligned}\partial_j \mathscr{B}^{(k)}(x) &= \frac{1}{2}\int_0^1\int_0^1 W'\big(\mathbf{X}[x](\xi) - \mathbf{X}[x](\eta)\big)(\theta_j^{(k)}(\xi) - \theta_j^{(k)}(\eta))\,\mathrm{d}\xi\,\mathrm{d}\eta \\ &= \int_0^1\int_0^1 W'(X(\xi) - X(\eta))\theta_j^{(k)}(\xi)\,\mathrm{d}\xi\,\mathrm{d}\eta.\end{aligned}$$

Finally, the calculation of $\mathscr{C}^{(k)}$'s derivative goes along the same lines.

Lemma 7 *The second derivatives of $\mathscr{A}^{(k)}$, $\mathscr{B}^{(k)}$ and $\mathscr{C}^{(k)}$ are given as follows:*

$$\partial_{x_j x_\ell}\mathscr{A}^{(k)}(x) = \frac{m}{\delta^{(k)}} \cdot \begin{cases} \left(\frac{x_{j+1}-x_j}{\delta^{(k)}}\right)^{-m-1} + \left(\frac{x_j - x_{j-1}}{\delta^{(k)}}\right)^{-m-1} & \text{if } j = \ell, \\ -\left(\frac{x_{j+1}-x_j}{\delta^{(k)}}\right)^{-m-1} & \text{if } \ell = j+1, \\ -\left(\frac{x_j - x_{j-1}}{\delta^{(k)}}\right)^{-m-1} & \text{if } \ell = j-1, \\ 0 & \text{in all other cases}, \end{cases} \tag{41}$$

$$\partial_{x_j x_\ell}\mathscr{B}^{(k)}(x) = \int_0^1\int_0^1 W''(X(\xi) - X(\eta))\left(\theta_\ell^{(k)}(\xi) - \theta_\ell^{(k)}(\eta)\right)\theta_j^{(k)}(\xi)\,\mathrm{d}\xi\,\mathrm{d}\eta, \tag{42}$$

$$\partial_{x_j x_\ell}\mathscr{C}^{(k)}(x) = \int_0^1 V''(X(\xi))\theta_j^{(k)}(\xi)\theta_\ell^{(k)}(\xi)\,\mathrm{d}\xi. \tag{43}$$

Despite a seeming asymmetry, $\partial_{x_j x_\ell}\mathscr{B}^{(k)} = \partial_{x_\ell x_j}\mathscr{B}^{(k)}$, because W'' is even.

Proof Calculations are elementary and follow the same lines as in Lemma 6.

4 Proof of Theorem 1: Discrete Evolution in Time

Throughout this section, k is a fixed index, with respective parameters $\tau^{(k)} > 0$ and $\delta^{(k)} = \frac{1}{k}$. However, certain estimates are independent of k.

4.1 The Minimizing Movement Scheme

The functional $\tilde{\mathbb{E}}_k : \mathfrak{x}^{(k)} \times \mathfrak{x}^{(k)} \to \mathbb{R}$ is the expression on the left hand side of (19):

$$\tilde{\mathbb{E}}_k(\hat{x};\check{x}) = \frac{1}{2\tau^{(k)}}(\hat{x}-\check{x})^T S^{(k)}(\hat{x}-\check{x}) + \mathbb{E}^{(k)}(\hat{x}).$$

Given the initial datum $x^0 \in \mathfrak{x}^{(k)}$, we define the sequence $(x^n)_{n=0,1,\dots}$ as described in Theorem 1: the successor of x^{n-1} is obtained as the minimizer of $\tilde{\mathbb{E}}_k(\cdot;x^{n-1})$ over $\mathfrak{x}^{(k)}$. Let us show that the x^n are well-defined this way.

Lemma 8 *For each $\check{x} \in \mathfrak{x}^{(k)}$, there is a unique minimizer of $\tilde{\mathbb{E}}_k(\cdot;\check{x})$ on $\mathfrak{x}^{(k)}$.*

Proof Clearly, $x \mapsto \tilde{\mathbb{E}}_k(x;\check{x})$ is a continuous function on the convex space $\mathfrak{x}^{(k)}$.

- (existence) We know from Lemma 5 above that $\mathbb{E}^{(k)}$ is a uniformly convex functional on $\mathfrak{x}^{(k)}$, with modulus of convexity λ_V. By the definition of the scalar product on $\mathfrak{x}^{(k)}$, the map $x \mapsto \frac{1}{2}(x-\check{x})^T S^{(k)}(x-\check{x})$ is uniformly convex with modulus unity. Hence, $x \mapsto \tilde{\mathbb{E}}_k(x;\check{x})$ is uniformly convex of modulus $\lambda_V + \frac{1}{\tau^{(k)}}$. By elementary convex analysis, the lower continuous extension of this functional to the closure of $\mathfrak{x}^{(k)}$ in $\mathbb{R}^{k-1}$ possesses precisely one minimizer $\hat{x}$.
- It remains show that $\hat{x} \in \mathfrak{x}^{(k)}$. Assume otherwise; then there is a sequence $(\hat{x}^{(\ell)})_{\ell=1,2,\dots}$ of elements $\hat{x}^{(\ell)} \in \mathfrak{x}^{(k)}$ such that $\tilde{\mathbb{E}}_k(\hat{x}^{(\ell)};\check{x})$ converges to the minimal value for $\ell \to \infty$, but $\hat{x}^{(\ell)}$ itself converges to "the boundary" of $\mathfrak{x}^{(k)}$. The latter implies that there is some index $i \in \{1,2,\dots,k\}$ such that $x_i^{(\ell)} - x_{i-1}^{(\ell)} \to 0$. However, since

$$\tilde{\mathbb{E}}_k(x^{(\ell)};\check{x}) \geq \frac{(\delta^{(k)})^m}{m-1}(x_i^{(\ell)} - x_{i-1}^{(\ell)})^{-(m-1)} + \frac{1}{2}W(0) + \underline{V}$$

diverges to infinity, $(\hat{x}^{(\ell)})_{\ell=1,2,\dots}$ cannot be a minimizing sequence.

We can now easily prove the first claim in Theorem 1, namely that the nth iterate x^n satisfies the system (18) of difference equations. Indeed, since x^n is the minimizer of $\tilde{\mathbb{E}}(\cdot;x^{n-1})$, one has that

$$0 = \partial_{x_j}\Big|_{x=x^n} \tilde{\mathbb{E}}(x;x^{n-1}) \tag{44}$$

for each index $j = 1,2,\dots,k-1$. On the one hand, clearly

$$\partial_{x_j}\Big|_{x=x^n}\left(\frac{1}{2\tau^{(k)}}(x-x^{n-1})^T S^{(k)}(x-x^{n-1})\right) = \left[S^{(k)}\frac{x^n-x^{n-1}}{\tau^{(k)}}\right]_j.$$

One the other hand, the derivatives of $\mathbb{E}^{(k)}$ have been calculated in Lemma 6. By combining these, it is seen that (44) is equivalent to (18).

4.2 Energy Estimates

Lemma 9 *For every* $\overline{n} > \underline{n} \geq 0$, *we have that*

$$\frac{\tau^{(k)}}{2}\sum_{n=\underline{n}+1}^{\overline{n}}\left\|\frac{x^n - x^{n-1}}{\tau^{(k)}}\right\|_S^2 \leq \mathbb{E}^{(k)}(x^{\underline{n}}) - \mathbb{E}^{(k)}(x^{\overline{n}}). \tag{45}$$

In particular, it follows that

- *the entropy values* $\mathbb{E}^{(k)}(x^n)$ *are monotonically decreasing, i.e.,*

$$\mathbb{E}^{(k)}(x^n) \leq \mathbb{E}^{(k)}(x^{n-1}); \tag{46}$$

- *the total action of the curve* $(x^n)_{n=0,1,\ldots}$ *is bounded:*

$$\frac{\tau^{(k)}}{2}\sum_{n=1}^{\infty}\left\|\frac{x^n - x^{n-1}}{\tau^{(k)}}\right\|_S^2 \leq \mathbb{E}^{(k)}(x^0). \tag{47}$$

Note that (46) immediately implies (20).

Proof Since x^n minimizes $\tilde{\mathbb{E}}_k(\cdot; x^{n-1})$, one has in particular that

$$\tau^{(k)}\left\|\frac{x^n - x^{n-1}}{\tau^{(k)}}\right\|_S^2 + \mathbb{E}(x^n) = \tilde{\mathbb{E}}_k(x^n; x^{n-1}) \leq \tilde{\mathbb{E}}_k(x^{n-1}; x^{n-1}) = \mathbb{E}^{(k)}(x^{n-1}).$$

Summation of these inequalities from $n = \underline{n} + 1$ to $n = \overline{n}$ yields (45). Then, (46) and (47) correspond to $\underline{n} = n - 1, \overline{n} = n$, and $\underline{n} = 1, \overline{n} \to \infty$, respectively.

Let us rephrase this result in terms of the interpolating function $X^{(k)}$.

Lemma 10 *The curve* $X^{(k)}$ *is "almost" Hölder-continuous w.r.t. time:*

$$\forall t, s \in [0, T], \quad \|X^{(k)}(t) - X^{(k)}(s)\|_{L^2} \leq \left(2\mathbb{E}(X^{(k)}(0,\cdot)\right)^{\frac{1}{2}}(|t-s| + \tau^{(k)})^{\frac{1}{2}} \tag{48}$$

Proof Assume $s < t$, and let $\underline{n}, \overline{n}$ be such that $s \in ((\underline{n} - 1)\tau^{(k)}, \underline{n}\tau^{(k)}]$, and $t \in ((\overline{n} - 1)\tau^{(k)}, \overline{n}\tau^{(k)}]$. Note that

$$\overline{n}\tau^{(k)} - \underline{n}\tau^{(k)} \leq t - s + \tau^{(k)}. \tag{49}$$

We estimate, using the inequality (47) above:

$$
\begin{aligned}
\|X^{(k)}(t,\cdot) - X^{(k)}(s,\cdot)\|_{L^2} &= \|x^{\overline{n}} - x^{\underline{n}}\|_S \\
&\leq \sum_{n=\underline{n}+1}^{\overline{n}} \|x^n - x^{n-1}\|_S \\
&= \tau^{(k)} \sum_{n=\underline{n}+1}^{\overline{n}} \left\| \frac{x^n - x^{n-1}}{\tau^{(k)}} \right\|_S \\
&\leq \left(\tau^{(k)} \sum_{n=\underline{n}+1}^{\overline{n}} \left\| \frac{x^n - x^{n-1}}{\tau^{(k)}} \right\|_S^2 \right)^{\frac{1}{2}} \left(\tau^{(k)} \sum_{n=\underline{n}+1}^{\overline{n}} 1 \right)^{\frac{1}{2}} \\
&\leq \left(2\mathbb{E}^{(k)}(x^0)\right)^{\frac{1}{2}} (\overline{n}\tau^{(k)} - \underline{n}\tau^{(k)})^{\frac{1}{2}}.
\end{aligned}
$$

Using (49), the claim (48) follows.

It remains to verify the contraction estimate (21). Let $(\hat{x}^n)_{n=0,1,\ldots}$ and $(\check{x}^n)_{n=0,1,\ldots}$ be the two solutions to (18) corresponding to $\hat{X}^{(k)}$ and $\check{X}^{(k)}$, respectively. According to (44), we have for every n that

$$
S^{(k)} \frac{\hat{x}^n - \hat{x}^{n-1}}{\tau^{(k)}} + \nabla_x \mathbb{E}^{(k)}(\hat{x}^n) = 0, \qquad S^{(k)} \frac{\check{x}^n - \check{x}^{n-1}}{\tau^{(k)}} + \nabla_x \mathbb{E}^{(k)}(\check{x}^n) = 0.
$$

Adding,

$$
S^{(k)}(\hat{x}^n - \check{x}^n) - S^{(k)}(\hat{x}^{n-1} - \check{x}^{n-1}) = \tau^{(k)}\big(\nabla\mathbb{E}^{(k)}(\hat{x}^n) - \nabla\mathbb{E}^{(k)}(\hat{x}^{n-1})\big).
$$

Rearranging terms and multiplying both sides on the left with $\big(S^{(k)}\big)^{-1/2}$,

$$
\big(S^{(k)}\big)^{1/2}(\hat{x}^n - \check{x}^n) + \tau^{(k)}\big(S^{(k)}\big)^{-1/2}(\nabla\mathbb{E}^{(k)}(x^n) - \nabla\mathbb{E}^{(k)}(\check{x}^n)) = \big(S^{(k)}\big)^{1/2}(\hat{x}^{n-1} - \check{x}^{n-1}) \,.
$$

Next, take the (usual euclidean) norm of both sides. On the left hand side,

$$
\begin{aligned}
&\left\|\hat{x}^n - \check{x}^n\right\|_S^2 \\
&+ 2\tau^{(k)}(\hat{x}^n - \check{x}^n)^T(\nabla\mathbb{E}^{(k)}(\hat{x}^n) - \nabla\mathbb{E}^{(k)}(\check{x}_n)) \\
&+ \big(\tau^{(k)}\big)^2 (\nabla\mathbb{E}^{(k)}(\hat{x}^n) - \nabla\mathbb{E}^{(k)}(\check{x}_n))^T S^{(k)}(\nabla\mathbb{E}^{(k)}(\hat{x}_n) - \nabla\mathbb{E}^{(k)}(\check{x}^n)) \\
&= \left\|\hat{x}^{n-1} - \check{x}^{n-1}\right\|_S^2.
\end{aligned}
$$

The term with $(\tau^{(k)})^2$ is non-negative and can thus be neglected. Further, we apply Corollary 1 to the $\tau^{(k)}$-term, which yields

$$(1+2\tau^{(k)}\lambda_V)\|\hat{x}^n-\check{x}^n\|_S^2 \le \|\hat{x}^{n-1}-\check{x}^{n-1}\|_S^2.$$

Iteration from $n=\underline{n}-1$ to $n=\overline{n}$ leads to

$$\|\hat{x}^{\overline{n}}-\check{x}^{\overline{n}}\|_S \le (1+2\tau^{(k)}\lambda_V)^{-\frac{\overline{n}-\underline{n}}{2}}\|\hat{x}^{\underline{n}}-\check{x}^{\underline{n}}\|_S.$$

The contraction estimate (21) now simply follows by recalling that

$$X^{(k)}(s,\cdot)=\mathbf{X}[x^{\underline{n}}] \quad \text{for all } s\in((\underline{n}-1)\tau^{(k)},\underline{n}\tau^{(k)}],$$
$$X^{(k)}(t,\cdot)=\mathbf{X}[x^{\overline{n}}] \quad \text{for all } t\in((\overline{n}-1)\tau^{(k)},\overline{n}\tau^{(k)}].$$

We finish this section by deriving another energy estimate that will play a crucial role in the convergence proof.

Lemma 11 *For each $N>0$, we have that*

$$\begin{aligned}&\tau^{(k)}\sum_{n=1}^{N}\delta^{(k)}\sum_{j=1}^{k-1}\left(\frac{1}{\delta^{(k)}}\left[\left(\frac{x_{j+1}^n-x_j^n}{\delta^{(k)}}\right)^{-m}-\left(\frac{x_j^n-x_{j-1}^n}{\delta^{(k)}}\right)^{-m}\right]\right)^2\\&\le 6\mathbb{E}^{(k)}(x^0)+\frac{8}{9}N\tau^{(k)}\big(\|V\|_{C^1}+\|W\|_{C^1}\big).\end{aligned} \tag{50}$$

Proof Arguing like in the proof of Corollary 1—see (36)—we conclude from the convexity of $\mathbb{E}^{(k)}$ that

$$\begin{aligned}\frac{1}{\tau^{(k)}}\big(\mathbb{E}^{(k)}(x^n)-\mathbb{E}^{(k)}(x^{n-1})\big) &\ge \left(\frac{x^n-x^{n-1}}{\tau^{(k)}}\right)^T\nabla\mathbb{E}^{(k)}(x^n)\\&=\nabla\mathbb{E}^{(k)}(x^n)^T\big(S^{(k)}\big)^{-1}\nabla\mathbb{E}^{(k)}(x^n)\\&\ge\frac{3}{\delta^{(k)}}\nabla\mathbb{E}^{(k)}(x^n)^T\nabla\mathbb{E}^{(k)}(x^n),\end{aligned}$$

where Euler-Lagrange equations (44) and the positivity estimate in Lemma 4 were used, respectively. Summing this inequality over $n=1,\dots,N$ yields

$$\tau^{(k)}\sum_{n=1}^{N}\frac{1}{\delta^{(k)}}\sum_{j=1}^{k-1}\big(\partial_{x_j}\mathbb{E}^{(k)}(x^n)\big)^2\le 3\big(\mathbb{E}(x^0)-\mathbb{E}(x^N)\big)\le 3\mathbb{E}(x^0).$$

Inside the sum on the left hand side, we estimate as follows:

$$\begin{aligned}\left(\partial_{x_j}\mathbb{E}^{(k)}(x^n)\right)^2 &= \left(\partial_{x_j}\mathscr{A}^{(k)}(x^n) + \partial_{x_j}\mathscr{B}^{(k)}(x^n) + \partial_{x_j}\mathscr{C}^{(k)}(x^n)\right)^2 \\ &\geq \frac{1}{2}\left(\partial_{x_j}\mathscr{A}^{(k)}(x^n)\right)^2 - 2\left(\partial_{x_j}\mathscr{B}^{(k)}(x^n)\right)^2 - 2\left(\partial_{x_j}\mathscr{C}^{(k)}(x^n)\right)^2.\end{aligned}$$

Derivatives are in Lemma 6. As $\left(\partial_{x_j}\mathscr{A}^{(k)}(x^n)\right)^2$ brings the desired term on the left hand side of (50), it remains to estimate the contribution of the other terms.

$$\begin{aligned}\frac{1}{\delta^{(k)}}\sum_{j=1}^{k-1}\left(\partial_{x_j}\mathscr{C}^{(k)}(x^n)\right)^2 &= \frac{1}{\delta^{(k)}}\sum_{j=1}^{k-1}\left(\int_0^1 V'(\mathbf{X}[x^n](\xi))\theta_j^{(k)}(\xi)\,\mathrm{d}\xi\right)^2 \\ &\leq \frac{\|V\|_{C^1}}{\delta^{(k)}}\sum_{j=1}^{k-1}\left(\int_0^1 \theta_j^{(k)}(\xi)^2\,\mathrm{d}\xi\right) \\ &= \frac{\|V\|_{C^1}}{\delta^{(k)}}\sum_{j=1}^{k-1}\frac{4\left(\delta^{(k)}\right)^2}{9} \leq \frac{4}{9}\|V\|_{C^1},\end{aligned}$$

since $\delta^{(k)} = \frac{1}{k}$, and, in a completely analogous manner,

$$\frac{1}{\delta^{(k)}}\sum_{j=1}^{k-1}\left(\partial_{x_j}\mathscr{C}^{(k)}(x^n)\right)^2 \leq \frac{4}{9}\|W\|_{C^1}.$$

Summarizing the contributions, we arrive at (50).

5 Proof of Theorem 2: Comparison Principles

5.1 *The Difference Quotients*

We begin with an alternative formulation of the fully discrete evolution. We pass from the formulation in terms of $x^n \in \mathfrak{x}^{(k)}$ to one in terms of $z^n \in \mathbb{R}^k$, with

$$z_i^n = \frac{x_i^n - x_{i-1}^n}{\delta^{(k)}}.$$

Note that

$$\partial_\xi X^{(k)}(t,\xi) = z_i^n \quad \text{for all } t \in ((n-1)\tau, n\tau] \text{and} \xi \in (\xi_{i-1}, \xi_i).$$

Definition 2 For $i = 1, \ldots, k$, define

$$\Theta_i^{(k)} : [0,1] \to \mathbb{R} \quad \text{with} \quad \Theta_i^{(k)}(\xi) := \int_0^\xi \left(\theta_{i-1}^{(k)}(\hat{\xi}) - \theta_i^{(k)}(\hat{\xi})\right) \, \mathrm{d}\hat{\xi} \tag{51}$$

which takes the explicit form for $i = 2, \ldots, k-1$ and $\xi^{(k)} = (\xi_0, \ldots, \xi_k)$:

$$\Theta_i^{(k)}(\xi) = \begin{cases} \frac{1}{2\delta^{(k)}}(\xi - \xi_{i-2})^2 & \text{for } \xi \in [\xi_{i-2}, \xi_{i-1}] \\ \frac{3}{4}\delta^{(k)} - \frac{1}{4\delta^{(k)}}(2\xi - (\xi_i + \xi_{i-1}))^2 & \text{for } \xi \in [\xi_{i-1}, \xi_i] \\ \frac{1}{2\delta^{(k)}}(\xi_{i+1} - \xi)^2 & \text{for } \xi \in [\xi_i, \xi_{i+1}] \\ 0 & \text{elsewhere} \end{cases} \tag{52}$$

and

$$\Theta_1^{(k)}(\xi) = \begin{cases} \delta^{(k)} - \frac{1}{2\delta^{(k)}}\xi^2 & \text{for } \xi \in [0, \xi_1] \\ \frac{1}{2\delta^{(k)}}(\xi_2 - \xi)^2 & \text{for } \xi \in [\xi_1, \xi_2] \\ 0 & \text{elsewhere} \end{cases} \tag{53}$$

as well as

$$\Theta_k^{(k)}(\xi) = \begin{cases} \frac{1}{2\delta^{(k)}}(\xi - \xi_{k-2})^2 & \text{for } \xi \in [\xi_{k-2}, \xi_{k-1}] \\ \delta^{(k)} - \frac{1}{2\delta^{(k)}}(1 - \xi)^2 & \text{for } \xi \in [\xi_{k-1}, 1] \\ 0 & \text{elsewhere.} \end{cases} \tag{54}$$

For $i = 2, \ldots, k-1$ we have $supp\Theta_i^{(k)} = [\xi_{k-2}, \xi_{k+1}]$ and for $i = 1, k$ we have $supp\Theta_1^{(k)} = [0, \xi_2]$ and $supp\Theta_k^{(k)} = [\xi_{k-2}, 1]$.

Hereafter, we denote maximal/minimal values of difference quotients as:

$$\overline{z}^n := \max_i z_i^n \quad \text{and} \quad \underline{z}^n := \min_i z_i^n. \tag{55}$$

Lemma 12 *The entries of the matrix,*

$$\tilde{S}_{i,j} := \int_{\xi_{j-1}}^{\xi_j} \Theta_i(\xi) \, \mathrm{d}\xi, \quad \textit{for} \quad i,j = 1, \ldots, k,$$

are explicitly given as follows:

$$\tilde{S}_{i,j} = \begin{cases} \frac{4}{6}\left(\delta^{(k)}\right)^2 & \text{for } 2 \leq i = j \leq k-1 \\ \frac{5}{6}\left(\delta^{(k)}\right)^2 & \text{for } i = j \in \{1, k\} \\ \frac{1}{6}\left(\delta^{(k)}\right)^2 & \text{for } i \in \{j \pm 1\} \\ 0 & \text{elsewhere.} \end{cases} \tag{56}$$

Proof For $2 \leq i \leq k-1$ we have

$$\begin{aligned} \tilde{S}_{i,i} &= \int_{\xi_{i-1}}^{\xi_i} \Theta_i(\xi)\, \mathrm{d}\xi \\ &= \int_{\xi_{i-1}}^{\xi_i} \frac{3}{4}\delta^{(k)} - \frac{1}{4\delta^{(k)}}\left(2\xi - (\xi_i + \xi_{i-1})\right)^2\, \mathrm{d}\xi \\ &= \frac{4}{6}\left(\delta^{(k)}\right)^2. \end{aligned}$$

The other cases are similar.

Lemma 13 *For* $i = 2, \ldots, k-1$

$$\begin{aligned} \frac{1}{\tau^{(k)}}[\tilde{S}(z^n - z^{n-1})]_i = &- (z_{i+1}^n)^{-m} + 2(z_i^n)^{-m} - (z_{i-1}^n)^{-m} \\ &- \sum_{j=i-1}^{i+1} z_j^n \int_0^1 \int_{\xi_{j-1}}^{\xi_j} W''(\mathbf{X}[x^n](\xi) - \mathbf{X}[x^n](\eta))\Theta_i(\xi)\, \mathrm{d}\xi\, \mathrm{d}\eta \\ &- \sum_{j=i-1}^{i+1} z_j^n \int_{\xi_{j-1}}^{\xi_j} V''\left(\mathbf{X}[x^n](\xi)\right)\Theta_i(\xi)\, \mathrm{d}\xi \end{aligned} \tag{57}$$

and for the boundaries we have

$$\begin{aligned} \frac{1}{\tau^{(k)}}[\tilde{S}(z^n - z^{n-1})]_1 = &- (z_2^n)^{-m} + (z_1^n)^{-m} \\ &- \sum_{j=1}^{2} z_j^n \int_0^1 \int_{\xi_{j-1}}^{\xi_j} W''(\mathbf{X}[x^n](\xi) - \mathbf{X}[x^n](\eta))\Theta_1(\xi)\, \mathrm{d}\xi\, \mathrm{d}\eta \\ &- \sum_{j=1}^{2} z_j^n \int_{\xi_{j-1}}^{\xi_j} V''(\mathbf{X}[x^n](\xi))\Theta_1(\xi)\, \mathrm{d}\xi \end{aligned} \tag{58}$$

as well as

$$\frac{1}{\tau^{(k)}}[\tilde{S}(z^n - z^{n-1})]_k = -\,(z^n_{k-1})^{-m} + (z^n_k)^{-m}$$
$$- \sum_{j=k-1}^{k} z^n_j \int_0^1 \int_{\xi_{j-1}}^{\xi_j} W''(\mathbf{X}[x^n](\xi) - \mathbf{X}[x^n](\eta))\Theta_k(\xi)\,\mathrm{d}\xi\,\mathrm{d}\eta$$
$$- \sum_{j=k-1}^{k} z^n_j \int_{\xi_{j-1}}^{\xi_j} V''(\mathbf{X}[x^n](\xi))\Theta_k(\xi)\,\mathrm{d}\xi\,. \tag{59}$$

Proof These elementary calculations were carried out explicitly in [18].

5.2 *Proving the Comparison Principle*

Only the proof for the maximum principle, i.e., for the second half of the double inequality in (24), is given explicitly; the proof for the minimum principle is very similar. We proceed by induction on n: **we assume that** (24) **holds for** $n-1$, **in order to conclude that it also holds for** n, **provided** $n\tau^{(k)} < T$.

- Since $\frac{\mathrm{d}}{\mathrm{d}s}(-s^{-m}) = ms^{-m-1}$ is strictly increasing for $s > 0$ and $\overline{z}^n \geq z^n_i$ for all i we get the estimates (note the $\pm$)

$$(z^n_i)^{-m} - (z^n_{i\pm1})^{-m} \leq -m(\overline{z}^n)^{-m-1}(z^n_i - z^n_{i\pm1})\,, \tag{60}$$

  along with,

$$[\tilde{S}z_{n-1}]_i \leq \left(\delta^{(k)}\right)^2 \overline{z}^{n-1} \tag{61}$$

  thanks to $\overline{z}^{n-1} \geq z_{i,n-1}$ for all i and the explicit representation of $\tilde{S}$ in (56).
- Yet, consider the Euler-Lagrange equation from Lemma 13, and assume that $\overline{z}^n = z^n_i$ at one of the sites $i \in \{2, \ldots, k-1\}$, i.e. the maximal value of z^n is attained at one of the "interior points". Multiply the Euler-Lagrange equation by $\tau^{(k)}$:

$$[\tilde{S}z^n]_i = [\tilde{S}z^{n-1}]_i + \tau^{(k)}\big[\big((z^n_i)^{-m} - (z^n_{i-1})^{-m}\big) + \big((z^n_i)^{-m} - (z^n_{i+1})^{-m}\big)\big]$$
$$- \tau^{(k)} \sum_{j=i-1}^{i+1} z^n_j \int_0^1 \int_{\xi_{j-1}}^{\xi_j} W''(\mathbf{X}[x^n](\xi) - \mathbf{X}[x^n](\eta))\Theta_i(\xi)\,\mathrm{d}\xi\,\mathrm{d}\eta$$
$$- \tau^{(k)} \sum_{j=i-1}^{i+1} z^n_j \int_{\xi_{j-1}}^{\xi_j} V''(\mathbf{X}[x^n](\xi))\Theta_i(\xi)\,\mathrm{d}\xi$$

$$
\begin{aligned}
&\leq [\tilde{S}z^{n-1}]_i + \tau^{(k)}\big[\big((z_i^n)^{-m} - (z_{i-1}^n)^{-m}\big) + \big((z_i^n)^{-m} - (z_{i+1}^n)^{-m}\big)\big] \\
&\qquad - \tau^{(k)}\underline{W}'' \sum_{j=i-1}^{i+1} z_j^n \int_{\xi_{j-1}}^{\xi_j} \Theta_i(\xi)\,\mathrm{d}\xi - \tau^{(k)}\underline{V}'' \sum_{j=i-1}^{i+1} z_j^n \int_{\xi_{j-1}}^{\xi_j} \Theta_i(\xi)\,\mathrm{d}\xi \\
&\leq [\tilde{S}z^{n-1}]_i - \tau^{(k)} m(\overline{z}^n)^{-m-1}[(z_i^n - z_{i-1}^n) + (z_i^n - z_{i+1}^n)] \\
&\qquad - \tau^{(k)}(\underline{W}'' + \underline{V}'') \sum_{j=i-1}^{i+1} \tilde{S}_{i,j} z_j^n \\
&\leq [\tilde{S}z^{n-1}]_i - \tau^{(k)} m(\overline{z}^n)^{-m-1}[(z_i^n - z_{i-1}^n) + (z_i^n - z_{i+1}^n)] - \tau^{(k)}\lambda_V[\tilde{S}z^n]_i.
\end{aligned} \tag{62}
$$

This yields

$$
(1 + \tau^{(k)}\lambda_V)[\tilde{S}z^n]_i \leq [\tilde{S}z^{n-1}]_i - \tau^{(k)} m(\overline{z}^n)^{-m-1}[(z_i^n - z_{i-1}^n) + (z_i^n - z_{i+1}^n)] \tag{63}
$$

If $\tau^{(k)}$ is so small that $(1 + \tau^{(k)}\lambda_V) \geq e^{\frac{1}{2}\tau^{(k)}\lambda_V}$, positivity of z yields,

$$
(1 + \tau^{(k)}\lambda_V)\frac{4}{6}\left(\delta^{(k)}\right)^2 z_i^n \leq (1 + \tau^{(k)}\lambda_V)[\tilde{S}z^n]_i \leq [\tilde{S}z^{n-1}]_i \leq \left(\delta^{(k)}\right)^2 \overline{z}^{n-1},
$$

which implies

$$
\begin{aligned}
\overline{z}^n &\leq \frac{6}{4}(1 + \tau^{(k)}\lambda_V)^{-1}\overline{z}^{n-1} \\
&\leq 3(1 + \tau^{(k)}\lambda_V)^{-(n-1)}\overline{z}^0 \leq 3\left(e^{\frac{1}{2}\tau\lambda_V}\right)^{-(n-1)}\overline{z}^0 \leq \underbrace{3e^{-\frac{1}{2}T\lambda_V}\overline{z}^0}_{=:\hat{Z}}.
\end{aligned} \tag{64}
$$

- We are left with

$$
\begin{aligned}
[\tilde{S}z^n]_i &= \left(\frac{\left(\delta^{(k)}\right)^2}{6} z_{i+1}^n + \frac{4\left(\delta^{(k)}\right)^2}{6} z_i^n + \frac{\left(\delta^{(k)}\right)^2}{6} z_{i-1}^n\right) \\
&= \left(\frac{\left(\delta^{(k)}\right)^2}{6}(z_i^n - z_i^n + z_{i+1}^n) + \frac{4\left(\delta^{(k)}\right)^2}{6} z_i^n + \frac{\left(\delta^{(k)}\right)^2}{6}(z_i^n - z_i^n + z_{i-1}^n)\right) \\
&= \left(\delta^{(k)}\right)^2 z_i^n - \frac{\left(\delta^{(k)}\right)^2}{6}[(z_i^n - z_{i+1}^n) + (z_i^n - z_{i-1}^n)].
\end{aligned}
$$

Elementary manipulations and recalling that $z_i^{n-1} \leq \overline{z}^{n-1}$ yields

$$(1 + \tau^{(k)}\lambda_V)\left(\delta^{(k)}\right)^2 \overline{z}^n \leq \left(\delta^{(k)}\right)^2 \overline{z}^{n-1}$$
$$+ \left(\frac{1 + \tau^{(k)}\lambda_V}{6}\left(\delta^{(k)}\right)^2 - \tau^{(k)} m(\hat{Z})^{-m-1}\right)\left[(z_i^n - z_{i-1}^n) + (z_i^n - z_{i+1}^n)\right]$$

where the right summand on the right hand side is non-positive by (23).

This closes the induction step. The proof for the minimum principle is almost identical: one needs to reverse the inequality signs, and replace λ_V by $\overline{K}$. The only difference is that in (64), an additional estimate $\underline{z}^n \leq \overline{z}^n$ is needed. This is the reason why $\overline{z}^0$ appears in (23) instead of the expected $\underline{z}^0$.

6 Proof of Theorem 3: Convergence Result

6.1 Convergence of the Lagrangian Map $X^{(k)}$

In the first step, we apply a standard argument to show:

Lemma 14 *$X^{(k)}$ converges uniformly w.r.t. $t \in [0, T]$ strongly in $L^2(I)$ to the limit function $X_* \in C^{\frac{1}{2}}([0, T]; L^2(I))$.*

Proof Recall that $X^{(k)}$ is "almost" Hölder-continuous thanks to the estimate (48). With that at hand, the strengthened form of the Arzela-Ascoli theorem given in [2, Proposition 3.3.1] immediately yields the claim.

Remark 1 As a consequence, we have that $X_*(0, \cdot)$ is the strong $L^2(I)$-limit of $X^{(k)}(0, \cdot) = X^0$, hence X_* attains the initial condition (13).

The difficult part is to **show strong convergence of $X^{(k)}$ to X_* in $L^2([0, T]; H^1(I))$**. Here we rely on a version of Aubin-Lions compactness lemma as given in [22, Theorem 2]. To this end, introduce the functional $\mathfrak{F} : H^1(I) \to \mathbb{R} \cup \{+\infty\}$ as follows:

$$\mathfrak{F}(Z) = \begin{cases} \|\partial_\xi Z\|_{\mathrm{TV}}^2 & \text{if } Z(0) = a, Z(1) = b, \text{and } \partial_\xi Z \text{ of bounded variation;} \\ +\infty & \text{otherwise.} \end{cases}$$

In the sense of [22], the functional $\mathfrak{F}$ is,

- *normal*: $\mathfrak{F}$ is sequentially lower semi-continuous with respect to convergence in $H^1(I)$. This is obvious since the (distributional) derivative is a linear operation on $H^1(I)$, and the total variation semi-norm is convex.

- *coercive*: sublevels of $\mathfrak{F}$ are relatively compact in $H^1(I)$. By classical results on the total variation, see e.g. [11], it follows from boundedness of $\|\partial_\xi Z_k\|_{\mathrm{TV}}$ for a given sequence $(Z_k)_{k=1,2,\ldots}$ in $H^1(I)$ that a subsequence of $\partial_\xi Z_k$ converges strongly in any $L^q(I)$. In combination with Z_k being essentially bounded between a and b, strong convergence of a subsequence of Z_k in $H^1(I)$ follows.

We are now going to verify the two main hypotheses from [22] for compactness of our sequence $(X^{(k)})_{k=1,2\ldots}$ in $L^2([0,T];H^1(I))$.

Lemma 15 *Our sequence* $(X^{(k)})_{k=1,2,\ldots}$ *has the following two properties:*

- Tightness w.r.t. $\mathfrak{F}$, *that is:*

$$\sup_{k=1,2,\ldots}\int_0^T \mathfrak{F}(X^{(k)}(t,\cdot))\,\mathrm{d}t < \infty, \tag{65}$$

- strong integral equicontinuity in t, *that is:*

$$\limsup_{h\downarrow 0}\left(\limsup_{k\to\infty}\int_0^{T-h}\|X^{(k)}(t+h,\cdot)-X^{(k)}(t,\cdot)\|^2_{L^2(I)}\,\mathrm{d}t\right)=0. \tag{66}$$

Proof • By Lemma 14, $X^{(k)}(t,\cdot)$ converge uniformly w.r.t. $t\in[0,T]$ in $L^2(I)$ towards a Hölder continuous limit, implying the integral equicontinuity (66).

- It remains to show (65): since $X^{(k)}(t,\cdot)$ is a piecewise affine function, and hence $\partial_\xi X^{(k)}$ is piecewise constant, its total variation semi-norm is simply the sum of the moduli of all jumps. Hence:

$$\int_0^T \mathfrak{F}(t,\partial_\xi X^{(k)}(t))\,\mathrm{d}t = \int_0^T \|\partial_\xi X^{(k)}(t)\|^2_{\mathrm{TV}}\,\mathrm{d}t$$

$$= \tau^{(k)}\sum_{n=1}^N\left(\sum_{i=1}^k|\partial_\xi X^{(k)}(n\tau^{(k)},\xi_i+\delta^{(k)}/2)-\partial_\xi X^{(k)}(n\tau^{(k)},\xi_{i-1}+\delta^{(k)}/2)|\right)^2\,\mathrm{d}t$$

$$\le \tau^{(k)}\sum_{j=1}^N\left[\left(\sum_{i=0}^{k-1}\delta^{(k)}\right)\left(\delta^{(k)}\sum_{i=0}^{k-1}\frac{1}{(\delta^{(k)})^2}\left|\left(\frac{x_{i+1}^n-x_i^n}{\delta^{(k)}}\right)-\left(\frac{x_i^n-x_{i-1}^n}{\delta^{(k)}}\right)\right|^2\right)\right]$$

$$\le \tau^{(k)}\sum_{j=1}^N\frac{\underline{z}^{2(m+1)}}{m}\delta^{(k)}\sum_{i=0}^{k-1}\frac{1}{(\delta^{(k)})^2}\left|\left(\frac{x_{i+1}^n-x_i^n}{\delta^{(k)}}\right)^{-m}-\left(\frac{x_i^n-x_{i-1}^n}{\delta^{(k)}}\right)^{-m}\right|^2,$$

which is k-uniformly bounded thanks to the energy estimate (50).

From [22, Theorem 2] one obtains the desired strong convergence of $X^{(k)}$ in $L^2([0,T];H^1(I))$. The integral equicontinuity (66) w.r.t. the L^2-norm is sufficient, since the L^2-norm is obviously a lower semi-continuous functional on $H^1(I)$.

Proposition 1 *1. A subsequence of* $(X^{(k)})_{k=1,2,\ldots}$ *converges in* $L^2([0,T];H^1(I))$ *to* X_*.

2. Along that subsequence, $(\partial_\xi X^{(k)})^{-m}$ *converges strongly in* $L^1([0,T]\times I)$ *to* $(\partial_\xi X_*)^{-m}$.

6.2 X_* is a Weak Solution of Our Inverse Distribution PDE

In the following, for a test function $\varphi \in C_c^\infty((0,T)\times I)$, we define $\varphi^{(k)} : [0,T]\times I \to \mathbb{R}$ as the function that is piecewise affine in ξ with $\phi^{(k)}(t,\xi_i) = \phi(t,\xi_i)$ for $i = 1,2,\ldots,k-1$. Note that $\phi^{(k)} \to \phi$ uniformly on $[0,T]\times I$, and that $\partial_\xi\phi^{(k)} \to \partial_\xi\phi$ in $L^\infty([0,T]\times I)$.

Lemma 16 *For every* $\varphi \in C_c^\infty((0,T)\times I)$,

$$\begin{aligned}
&\int_0^T\int_0^1 \frac{X^{(k)}(t,\xi)-X^{(k)}(t-\tau^{(k)},\xi)}{\tau^{(k)}}\varphi^{(k)}(t,\xi)\,\mathrm{d}\xi\,\mathrm{d}t \\
&= \int_0^T\int_0^1 (\partial_\xi X^{(k)}(t,\xi))^{-m}\partial_\xi\varphi^{(k)}(t,\xi)\,\mathrm{d}\xi\,\mathrm{d}t \\
&\quad - \int_0^T\int_0^1\int_0^1 W'(X(t,\xi)-X(t,\eta))\varphi^{(k)}(t,\xi)\,\mathrm{d}\eta\,\mathrm{d}\xi\,\mathrm{d}t \\
&\quad - \int_0^T\int_0^1 V'(X(t,\xi))\varphi^{(k)}(t,\xi)\,\mathrm{d}\xi\,\mathrm{d}t\,.
\end{aligned} \tag{67}$$

Proof This is but an alternative way to write down the discrete evolution (18). Indeed, multiply (18) with $\varphi(\xi_j)$, sum over $j = 1,2,\ldots,k-1$, use that $\phi^{(k)} = \sum_{j=1}^{k-1}\varphi(\xi_j)\theta_j^{(k)}$, and recall the definition of $S^{(k)}$ as L^2-scalar product of the $\theta_i^{(k)}$'s.

With this result at hand, we proceed to the proof of the weak formulation. Let $\varphi \in C_c^\infty((0,T)\times I)$ be fixed, and assume, without loss of generality, that $\tau^{(k)}$ is so small that $\mathit{supp}\varphi \subset (\tau^{(k)}, T-\tau^{(k)})\times(0,1)$.

1. First, to show that

$$\lim_{k\to\infty}\int_0^T\int_0^1 \frac{X^{(k)}(t+\tau^{(k)},\xi)-X^{(k)}(t,\xi)}{\tau^{(k)}}\varphi^{(k)}(t,\xi)\,\mathrm{d}\xi\,\mathrm{d}t$$

$$= -\int_0^T\int_0^1 \hat{X}(t,\xi)\partial_t\phi(t,\xi)\,\mathrm{d}\xi\,\mathrm{d}t\,,$$

we use

$$\int_0^T \int_0^1 \frac{X^{(k)}(t+\tau^{(k)},\xi) - X^{(k)}(t,\xi)}{\tau^{(k)}} \varphi^{(k)}(t,\xi)\, \mathrm{d}\xi\, \mathrm{d}t$$

$$= \int_0^T \int_0^1 X^{(k)}(t,\xi) \frac{\varphi^{(k)}(t-\tau^{(k)},\xi) - \varphi^{(k)}(t,\xi)}{\tau^{(k)}}\, \mathrm{d}\xi\, \mathrm{d}t.$$

Since $X^{(k)} \xrightarrow{L^2([0,T]\times I)} X_*$ and $\frac{\varphi^{(k)}(t-\tau^{(k)},\xi)-\varphi^{(k)}(t,\xi)}{\tau^{(k)}} \xrightarrow{unif.} -\partial_t \phi$, we are done.

2. Next we show

$$\lim_{k\to\infty} \int_0^T \int_0^1 \frac{\partial_\xi \varphi^{(k)}(t,\xi)}{\partial_\xi X^{(k)}(t,\xi)^m}\, \mathrm{d}\xi\, \mathrm{d}t = \int_0^T \int_0^1 \frac{\partial_\xi \phi(t,\xi)}{\partial_\xi \hat{X}(t,\xi)^m}\, \mathrm{d}\xi\, \mathrm{d}t$$

which can actually be seen quite quickly, just by considering

$$\left| \int_0^T \int_0^1 (\partial_\xi X^{(k)}(t,\xi))^{-m} \partial_\xi \varphi^{(k)}(t,\xi) - (\partial_\xi \hat{X}(t,\xi))^{-m} \partial_\xi \phi(t,\xi)\, \mathrm{d}\xi\, \mathrm{d}t \right|$$

$$\leq \int_0^T \int_0^1 \Big[\left| (\partial_\xi X^{(k)}(t,\xi))^{-m} \partial_\xi \varphi^{(k)}(t,\xi) - (\partial_\xi X^{(k)}(t,\xi))^{-m} \partial_\xi \phi(t,\xi) \right|$$

$$+ \left| (\partial_\xi X^{(k)}(t,\xi))^{-m} \partial_\xi \phi(t,\xi) - (\partial_\xi \hat{X}(t,\xi))^{-m} \partial_\xi \phi(t,\xi) \right| \Big]\, \mathrm{d}\xi\, \mathrm{d}t$$

$$= \int_0^T \int_0^1 \Big[\left| (\partial_\xi X^{(k)}(t,\xi))^{-m} \right| \left| \partial_\xi \varphi^{(k)}(t,\xi) - \partial_\xi \phi(t,\xi) \right|$$

$$+ \left| (\partial_\xi X^{(k)}(t,\xi))^{-m} - (\partial_\xi \hat{X}(t,\xi))^{-m} \right| \left| \partial_\xi \phi(t,\xi) \right| \Big]\, \mathrm{d}\xi\, \mathrm{d}t\,.$$

The claim follows from here right away by Proposition 1, and by the uniform boundedness of $(\partial_\xi X^{(k)})^{-m}$ thanks to the minimum principle.

3. We verify the claim for the last two summands simultaneously, again by using the trick above. But first we see that by Lipschitz-continuity of W' and V' we have that $W'(X^{(k)}(t,\xi) - X^{(k)}(t,\eta))$ and $V'(X^{(k)}(t,\xi))$ both converge

in $L^2([0,T] \times I)$ to $W'(\hat{X}(t,\xi) - \hat{X}(t,\eta))$ and $V'(\hat{X}(t,\xi))$ respectively. This yields

$$\int_0^T \int_0^1 \int_0^1 \left|W'(\hat{X}(\xi) - \hat{X}(\eta))\right| \left|\varphi^{(k)}(t,\xi) - \phi(t,\xi)\right|$$

$$+ |\phi(t,\xi)| \left|W'(X^{(k)}(t,\xi) - X^{(k)}(t,\eta)) - W'(\hat{X}(\xi) - \hat{X}(\eta))\right| \,\mathrm{d}\eta\,\mathrm{d}\xi\,\mathrm{d}t\,,$$

and

$$\int_0^T \int_0^1 \left|V'(\hat{X}(t,\xi))\right| \left|\varphi^{(k)}(t,\xi) - \phi(t,\xi)\right|$$

$$+ |\phi(t,\xi)| \left|V'(X^{(k)}(t,\xi)) - V'(\hat{X}(t,\xi))\right| \,\mathrm{d}\xi\,\mathrm{d}t\,.$$

By boundedness of W', V' and ϕ, and thanks to the convergence of $W'(X^{(k)}(t,\xi) - X^{(k)}(t,\eta))$ and $V'(X^{(k)}(t,\xi))$ in $L^2([0,T] \times I)$ as well as the uniform convergence of $\phi^{(k)}$ to ϕ. So the claim follows.

7 Numerical Results

As a general observation, the effort of our own discretization is comparable to a standard finite difference scheme for the original evolution equation in u: in both cases, we are numerically solving a parabolic equation in one space dimension, either (6) for u, or (11) for X. However, a big advantage of our scheme (apart from the structure preserving properties) is the **preservation of mass and positivity**, which does not come for free with standard finite difference solvers for (6). As usual with implicit schemes, the **highest costs—both in terms of implementation effort, and in terms of time consumption in the numerical experiments—are related to Newton iterations**, to be carried out in each time step. An explicit time stepping, like in [12, 13], is faster and far easier to implement, but requires a CFL condition of the type $\tau^{(k)} < K(\delta^{(k)})^2$ for stability. In particular, such a condition would be in opposition to our own inverse CFL condition 23, that we need to establish the discrete comparison principle and the eventually the convergence.

7.1 Implementation

7.1.1 Initial Data

Since our focus was on the inverse distribution functions in this paper, we chose to give an initial inverse distribution function $X^{(0)}$ rather than an initial density $u^{(0)}$. If one is given an initial density $u^{(0)}$, then one can, either analytically or numerically achieve the corresponding inverse distribution function, so this is no restriction.

As already pointed out, we used a equidistant grid

$$\xi^{(k)} = (0 = \xi_0^{(k)}, \xi_1^{(k)}, \dots, \xi_k^{(k)} = 1), \qquad \text{on} \qquad [0, 1],$$

to discretize initial conditions. To achieve the corresponding discrete initial data $X^{(k),0}$ from a $X^{(0)}$ we used the results from Chap. 3, i.e. we set $x_i^{(0)} = X^{(0)}(\xi_i)$. And the resulting $X^{(k),0}(\xi) = \sum_{i=0}^{k} \theta_i^{(k)}(\xi) x_i^{(0)}$ converge in L^2 to the initial $X^{(0)}$ as was shown. Finally we chose $X^{(0)}$ as a member of the set $\mathfrak{X}$, i.e. $X^{(0)}$ increases at least with some positive rate and is Lipschitz. In the last numerical experiments we take a short look on how to deal with initial conditions that do have vertical slopes (which correspond to intervals with $u^{(0)} = 0$).

7.1.2 Iteration

We have

$$0 = \nabla \tilde{\mathbb{E}}_k(x, x^{(n-1)}) = \frac{1}{\tau} S^{(k)}(x - x^{(n-1)}) + \nabla \mathbb{E}^{(k)}(x),$$

which becomes just $\nabla \mathbb{E}_k(x)$ for $x = x^{(n-1)}$ as in the pseudocode. The corresponding Jacobian is given by

$$\nabla^2 \tilde{\mathbb{E}}(x, x^{(n-1)}) = \frac{1}{\tau} S^{(k)} + \nabla^2 \mathbb{E}^{(k)}(x).$$

The explicit expressions for the first and second derivatives of $\mathbb{E}$ have been calculated in Lemmata 6 and 7. We use the following damped Newton method, where the damping parameter is adjusted in each step such that the next x lies again in $\mathfrak{x}^{(k)}$.

$x = x^{(n-1)};$
$t = 0;$
while $t < T$
    $dx = -(\frac{1}{\tau}S + \nabla^2\mathbb{E}^{(k)}(x))^{-1}\nabla\mathbb{E}^{(k)}(x);$
    while $x + dx \notin \mathfrak{x}^{(k)};$
        $dx = \frac{1}{2}dx;$
    end
    $x = x + dx;$
    $t = t + \tau;$
end
$x^{(n)} = x;$

For the evaluation of $\nabla\mathbb{E}_k$ and $\nabla^2\mathbb{E}_k$ above an explicit expression for the following integrals is needed:

$$[\nabla\mathscr{B}^{(k)}]_j(x) = \int_0^1\int_0^1 W'(X(\xi) - X(\eta))\theta_j^{(k)}(\xi)\,\mathrm{d}\xi\,\mathrm{d}\eta$$

$$[\nabla^2\mathscr{B}^{(k)}]_{i,j}(x) = \int_0^1\int_0^1 W''(X(\xi) - X(\eta))\left(\theta_i^{(k)}(\xi) - \theta_i^{(k)}(\eta)\right)\theta_j^{(k)}(\xi)\,\mathrm{d}\xi\,\mathrm{d}\eta$$

$$[\nabla\mathscr{C}^{(k)}]_j(x) = \int_0^1 V'(X(\xi))\theta_j^{(k)}(\xi)\,\mathrm{d}\xi$$

$$[\nabla^2\mathscr{C}^{(k)}]_{i,j}(x) = \int_0^1 V''(X(\xi))\theta_i^{(k)}(\xi)\theta_j^{(k)}(\xi)\,\mathrm{d}\xi\,.$$

The expressions for the last two integrals have been given in [18], so we will not give then here again. We will give the explicit expressions for the first and one summand of the second integral. At the core of these considerations lies the decomposition

$$\int_0^1\int_0^1 W'(X(\xi) - X(\eta))\theta_n^{(k)}(\xi)\,\mathrm{d}\xi\,\mathrm{d}\eta$$
$$= \sum_{i,j=1}^{k}\int_{\xi_{i-1}}^{\xi_i}\int_{\xi_{j-1}}^{\xi_j} W'(X(\xi) - X(\eta))\theta_n^{(k)}(\xi)\,\mathrm{d}\xi\,\mathrm{d}\eta\,,$$

and that on these intervals, the $\theta_n^{(k)}$ are affine linear, so we can integrate by parts. Doing this sufficiently often, one can get rid of the $\theta_n^{(k)}$ entirely and by assuming that one has the first and second antiderivative, $\mathscr{W}$ and $\hat{\mathscr{W}}$ of W, we can then write

$$\int_0^1\int_0^1 W'(X(\xi) - X(\eta))\theta_i^{(k)}(\xi)\,\mathrm{d}\xi\,\mathrm{d}\eta = \left(\delta^{(k)}\right)^2\Bigg[$$
$$\left(\frac{1}{x_{i+1} - x_i} - \frac{1}{x_i - x_{i-1}}\right)\sum_{j=1}^{k}\frac{1}{x_j - x_{j-1}}(\mathscr{W}(x_i - x_j) - \mathscr{W}(x_i - x_{j-1}))$$

$$+\frac{1}{(x_i-x_{i-1})^2}\sum_{j=1}^{k}\frac{1}{x_j-x_{j-1}}\Big((\hat{\mathscr{W}}(x_i-x_j)-\hat{\mathscr{W}}(x_i-x_{j-1}))$$

$$-(\hat{\mathscr{W}}(x_{i-1}-x_j)-\hat{\mathscr{W}}(x_{i-1}-x_{j-1}))\Big)$$

$$-\frac{1}{(x_{i+1}-x_i)^2}\sum_{j=1}^{k}\frac{1}{x_j-x_{j-1}}\Big((\hat{\mathscr{W}}(x_{i+1}-x_j)-\hat{\mathscr{W}}(x_{i+1}-x_{j-1}))$$

$$-(\hat{\mathscr{W}}(x_i-x_j)-\hat{\mathscr{W}}(x_i-x_{j-1}))\Big)\Bigg],$$

for the gradient. The Hessian can be split into two integrals

$$\int_0^1\int_0^1 W''(X(\xi)-X(\eta))\left(\theta_l^{(k)}(\xi)-\theta_l^{(k)}(\eta)\right)\theta_n^{(k)}(\xi)\,\mathrm{d}\xi\,\mathrm{d}\eta =$$

$$\int_0^1\int_0^1 W''(X(\xi)-X(\eta))\theta_l^{(k)}(\xi)\theta_n^{(k)}(\xi)\,\mathrm{d}\xi\,\mathrm{d}\eta$$

$$-\int_0^1\int_0^1 W''(X(\xi)-X(\eta))\theta_l^{(k)}(\eta)\theta_n^{(k)}(\xi)\,\mathrm{d}\xi\,\mathrm{d}\eta$$

and as the first integral strongly resembles the integral for the components of the Hessian of $\mathscr{C}$ with $V''=W''$ etc., we only give the explicit expression for the second integral here.

$$\int_0^1\int_0^1 W''(X(\xi)-X(\eta))\theta_l^{(k)}(\eta)\theta_n^{(k)}(\xi)\,\mathrm{d}\xi\,\mathrm{d}\eta =$$

$$=\delta^{(k)}\Bigg[\left(\frac{1}{x_i-x_{i-1}}-\frac{1}{x_{i+1}-x_i}\right)\Bigg[\left(\frac{1}{x_{j+1}-x_j}-\frac{1}{x_j-x_{j-1}}\right)W(x_i-x_j)$$

$$-\frac{1}{(x_j-x_{j-1})^2}\left(\mathscr{W}(x_i-x_j)-\mathscr{W}(x_i-x_{j-1})\right)$$

$$+\frac{1}{(x_{j+1}-x_j)^2}\left(\mathscr{W}(x_i-x_{j+1})-\mathscr{W}(x_i-x_j)\right)\Bigg]$$

$$+\frac{1}{(x_{i+1}-x_i)^2}\Bigg[\left(\frac{1}{x_{j+1}-x_j}-\frac{1}{x_j+x_{j-1}}\right)(\mathscr{W}(x_{i+1}-x_j)-\mathscr{W}(x_i-x_j))$$

$$-\frac{1}{(x_j-x_{j-1})^2}\Big(\hat{\mathscr{W}}(x_{i+1}-x_j)-\hat{\mathscr{W}}(x_{i+1}-x_{j-1})-(\hat{\mathscr{W}}(x_i-x_j)+\hat{\mathscr{W}}(x_i-x_{j-1})\Big)$$

$$+ \frac{1}{(x_{j+1}-x_j)^2}\Big(\hat{\mathscr{W}}(x_{i+1}-x_{j+1}) - \hat{\mathscr{W}}(x_{i+1}-x_j) - (\hat{\mathscr{W}}(x_i - x_{j+1}) + \hat{\mathscr{W}}(x_i - x_j)\Big)\Big]$$

$$+ \frac{1}{(x_i - x_{i-1})^2}\Bigg[\Bigg(\frac{1}{x_{j+1}-x_j} - \frac{1}{x_j + x_{j-1}}\Bigg)(\mathscr{W}(x_i - x_j) - \mathscr{W}(x_{i-1} - x_j))$$

$$- \frac{1}{(x_j - x_{j-1})^2}\Big(\hat{\mathscr{W}}(x_i - x_j) - \hat{\mathscr{W}}(x_i - x_{j-1}) - (\hat{\mathscr{W}}(x_{i-1} - x_j) + \hat{\mathscr{W}}(x_{i-1} - x_{j-1})\Big)$$

$$+ \frac{1}{(x_{j+1}-x_j)^2}\Big(\hat{\mathscr{W}}(x_i - x_{j+1}) - \hat{\mathscr{W}}(x_i - x_j) - (\hat{\mathscr{W}}(x_{i-1} - x_{j+1}) + \hat{\mathscr{W}}(x_{i-1} - x_j)\Big)\Big]\Bigg] .$$

7.2 *Numerical Experiments*

The following numerical experiments are performed with $m = 20$ which was chosen since it emphasizes the struggle between the interaction potential an the diffusion. The interaction potential was chosen to be $W(\xi) = \xi^2$ which possesses all of the assumed properties, including that we can give a first and second antiderivative for the numerical calculations.

7.2.1 Reference Solution

The reference solution was calculated with the following parameters. We chose a sufficiently large discretization of space, $k = 5000$ grid points as well as a sufficiently small time-step $\tau = 0.01$. Furthermore the **external potential was omitted as we wanted to concentrate on both diffusion and interactions**. Additionally, the external potential was already studied in [18]. We used an initial condition $X^{(0)}$ on the interval $J = [a, b] = [-1, 1]$ which gives rise to a density $u^{(0)}$ which is asymmetric, positive, possesses three accumulations of mass one of which is found at the right border of the interval and can be seen in Fig. 2 to the left. We use

$$x^{(0)} : [0, 1] \to [-1, 1];$$

$$X^{(0)}(\xi) = \frac{1}{5\pi}\big[10\xi(\xi - \tfrac{1}{2})(\xi - 1) + 1)\big]\sin(5\pi\xi) + 2\xi - 1 .$$

One can observe how the left spike moves to the right, driven by the interaction potential which wants to accumulate all of the mass. The middle mass

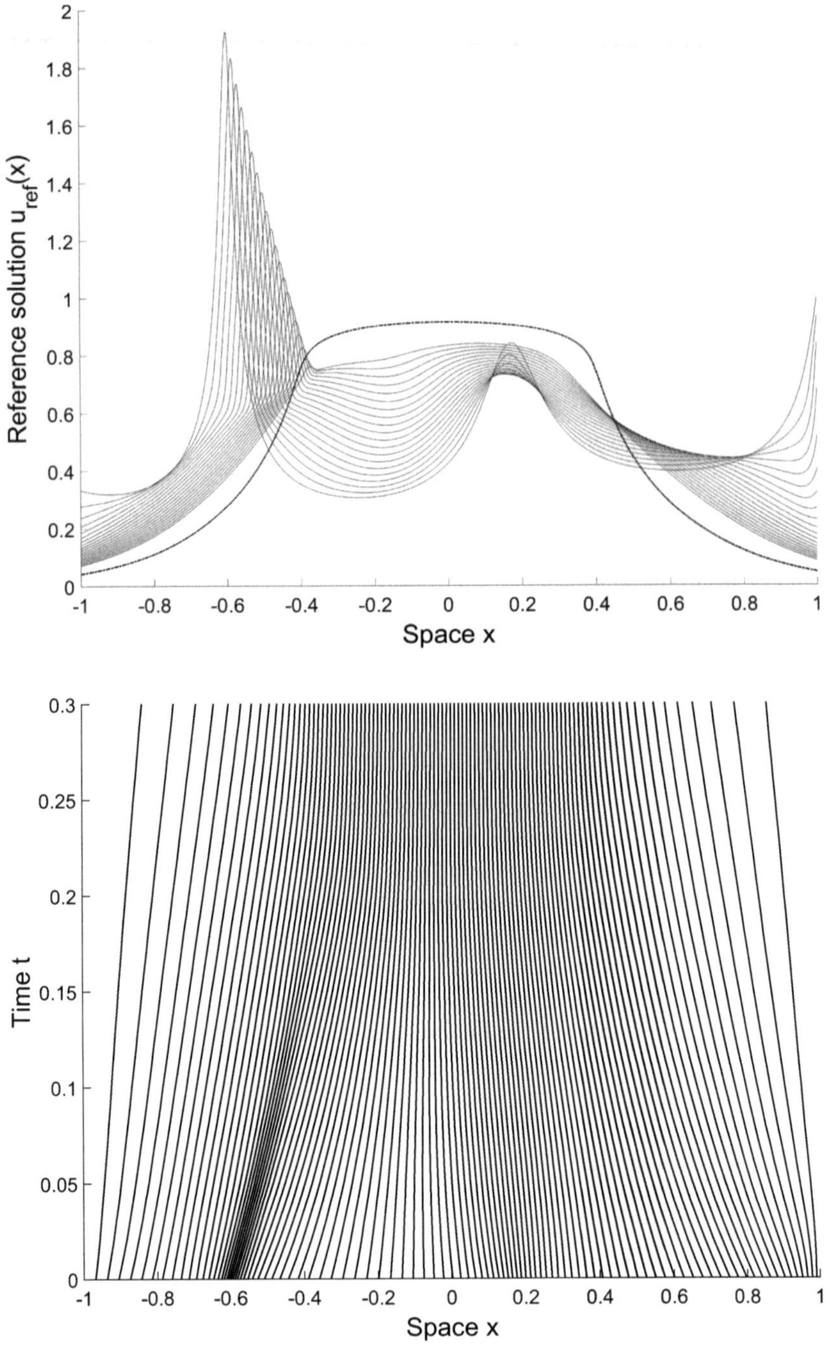

Fig. 2 *Left*: The evolution of the (reference) solution u with initial condition at times $0, \tau, \ldots, 20\tau$ where $\tau = 0.01$ and $k = 5000$ grid points. The *dotted line* (*top*) shows the calculated stationary solution. *Bottom*: the corresponding particle trajectories

concentration loses some of its shape due to diffusion, but keeps the shape assumed after 2 timesteps, since it is held together by the interaction potential. A similar behaviour can be seen at the left spike. While it collapses, a smaller rounder mass accumulation is build at the base of the spike, keeping shape and moving to the mass center. Finally the pile at the right border tries to stay together for three steps due to interaction, but is ultimately absorbed to the middle pile.

In the stationary solution one can see the influence of the diffusion and the interaction potential quite clearly. The flat top of the mesa-like distribution of mass is caused by diffusion, which is, by $m = 20$ rather localized. At the edge of the mesa the interaction potential takes over and accumulates the mass, creating cliffs. The picture of the particle trajectories shows how the x_i actually move over time. Especially interesting is again, the influence of diffusion, causing high densities of lines to spread and the ongoing accumulation of the lines towards the center.

7.2.2 Fixed τ

In this series of experiments we fixed a time step $\tau = 0.01$. Figure 3 shows to the left the L^1 and to the right the L^2 error w.r.t. the reference solution for the interpolation points (graphs from top to bottom) $k = 50, 100, 200, 400, 800$ and 1600 grid points.

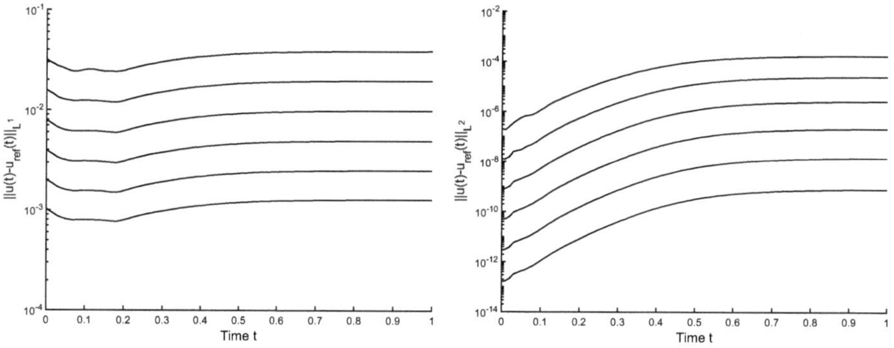

Fig. 3 Time-growth of numerical errors with fixed time step $\tau = 0.01$, using $k = 50, 100, 200, 400, 800$ and 1600 grid points. The curves in both pictures are 50, the *upmost*, counting through to 1600 the *lowest* in each picture

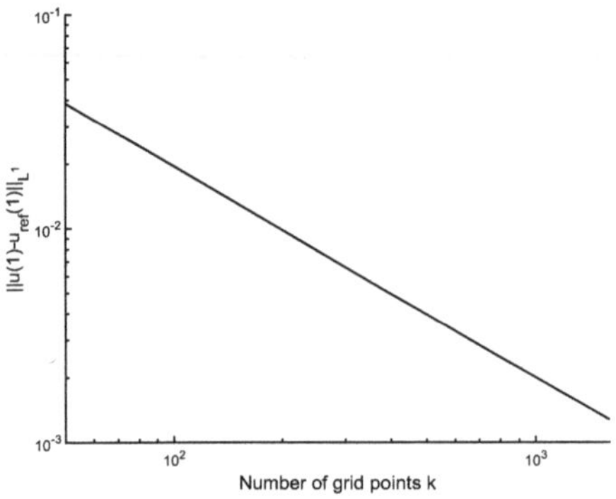

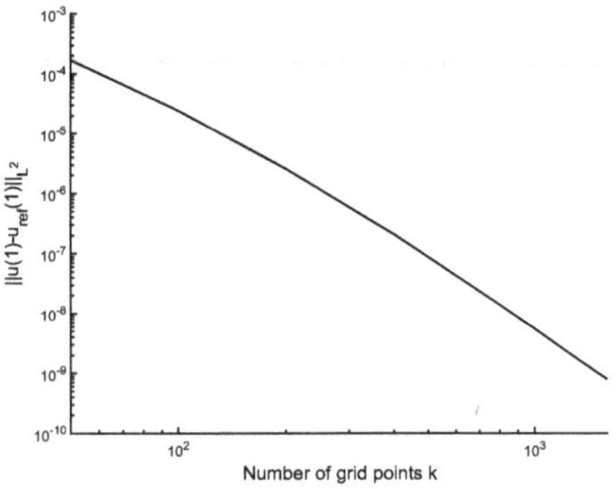

Fig. 4 *Left*: This plot shows the L^1-error of the densities u at $T = 1$ against the number of grid points. *Right*: The corresponding L^2-error of the inverse distribution functions X at $T = 1$

Note on the left hand side the counterintuitive drop of the error for small times which can be explained by the contractivity of the nonlinear diffusion in a neighbourhood of the initial condition. Figure 4 shows the L^1-error at the time horizon $T = 0.3$ of the calculated solutions w.r.t. the reference solution plotted with logarithmic scale on both axes. The convergence is of order k^{-1}.

7.2.3 Non-negative Initial Data

This series of experiments deals with merely non-negative initial densities (or initial inverse distributions which possess at least one vertical jump discontinuity). We use the initial inverse distribution function

$$X^{(0)} : [0, 1] \to [-\tfrac{1}{2}, \tfrac{1}{2}]; \quad X^{(0)}(\xi) = \frac{1}{\frac{9}{2}\pi} \sin((\tfrac{9}{4}\pi + \tfrac{\pi}{2})\xi) + \xi - \tfrac{1}{2}$$

which gives rise to a density with support $\text{supp}(u) = [-0.5, 0.5] \subset [-1, 1]$. Since our discretization by the inverse distribution functions discretizes the density isovolumetric, the intervals which do not lie in the support are simply absorbed into intervals $[x_i, x_{i+1}]$ with either x_i or x_{i+1} in the interior of the support of our density. Therefore the discrete density is always positive and the value of our discretization on these intervals decreases with increasing grid points k. This way the natural approach to approximate merely non-negative initial data by positive initial densities $u_k + \varepsilon_k$ with $\varepsilon_k \to 0$ and $\varepsilon_k > 0$ is taken care of by the discretization itself (Figs. 5 and 6).

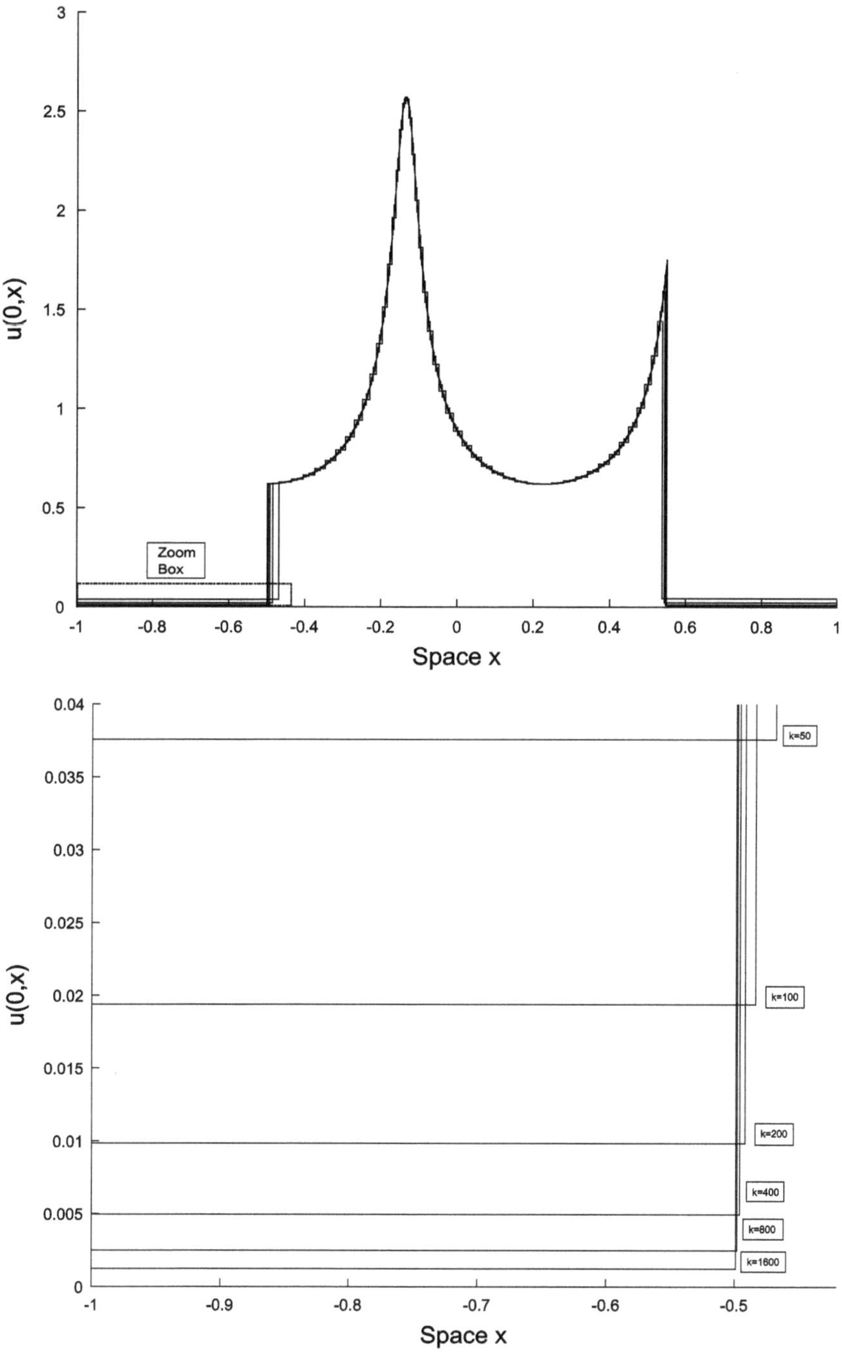

Fig. 5 *Upper figure*: The initial densities for $k = 50, 100, 200, 400, 800$ and 1600 grid points. *Lower figure*: Zoom on the behaviour at the left interval where $u^{(0)} = 0$. Note how the discretization takes care of the "$u + \varepsilon$"-approximation by itself

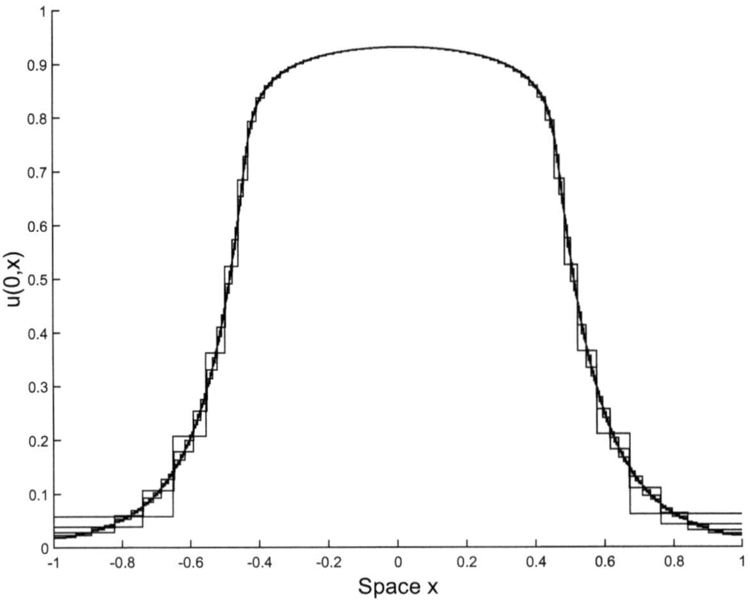

Fig. 6 Discrete solution corresponding to the initial condition from Fig. 5 at time $T = 1$

References

1. Agueh, M., Bowles, M.: One-dimensional numerical algorithms for gradient flows in the p-Wasserstein spaces. Acta Appl. Math. **125**, 121–134 (2013)
2. Ambrosio, L., Gigli, N., Savaré, G.: Gradient Flows in Metric Space of Probability Measures. Lectures in Mathematics. Birkäuser, Boston (2005)
3. Benamou, J.-D., Carlier, G., Mérigot, Q., Oudet, E.: Discretization of functionals involving the Monge–Ampère operator. Numer. Math. **134**, 1–26 (2015)
4. Blanchet, A., Calvez, V., Carrillo, J.A.: Convergence of the mass-transport steepest descent scheme for the subcritical Patlak-Keller-Segel model. SIAM J. Numer. Anal. **46**, 691–721 (2008)
5. Budd, C.J., Collins, G.J., Huang, W.Z., Russell, R.D.: Self-similar numerical solutions of the porous-medium equation using moving mesh methods. R. Soc. Lond. Philos. Trans. Ser. A Math. Phys. Eng. Sci. **357**, 1047–1077 (1993)
6. Carrillo, J.A., Moll, J.S.: Numerical simulation of diffusive and aggregation phenomena in nonlinear continuity equations by evolving diffeomorphisms. SIAM J. Sci. Comput. **31**, 4305–4329 (2009/2010)
7. Carrillo, J.A., Ranetbauer, H., Wolfram, M.-T.: Numerical simulation of nonlinear continuity equations by evolving diffeomorphisms (2016). Preprint. arXiv:1605.05921
8. Cavalli, F., Naldi, G.: A Wasserstein approach to the numerical solution of the one-dimensional Cahn-Hilliard equation. Kinet. Relat. Models **3**, 123–142 (2010)
9. Düring, B., Matthes, D., Milišić, J.P.: A gradient flow scheme for nonlinear fourth order equations. Discrete Contin. Dyn. Syst. Ser. B **14**, 935–959 (2010)
10. Evans, L.C., Savin, O., Gangbo, W.: Diffeomorphisms and nonlinear heat flows. SIAM J. Math. Anal. **37**, 737–751 (2005)

11. Giusti, E.: Minimal Surfaces and Functions of Bounded Variation. Monographs in Mathematics, vol. 80. Birkhäuser Verlag, Boston (1984)
12. Gosse, L., Toscani, G.: Identification of asymptotic decay to self-similarity for one-dimensional filtration equations. SIAM J. Numer. Anal. **43**, 2590–2606 (2006) (electronic)
13. Gosse, L., Toscani, G.: Lagrangian numerical approximations to one-dimensional convolution-diffusion equations. SIAM J. Sci. Comput. **28**, 1203–1227 (2006) (electronic)
14. Jordan, R., Kinderlehrer, D., Otto, F.: The variational formulation of the Fokker-Planck equation. SIAM J. Math. Anal. **29**, 1–17 (1998)
15. Junge, O., Matthes, D., Osberger, H.: A fully discrete variational scheme for solving nonlinear Fokker-Planck equations in higher space dimensions (2015). Preprint. arXiv:1509.07721
16. Leven, M.: Gradientenfluß-basierte Diskretisierung parabolischer Gleichungen, Diplomarbeit, Universität Bonn (2002)
17. MacCamy, R.C., Socolovsky, E.: A numerical procedure for the porous media equation, hyperbolic partial differential equations, II. Comput. Math. Appl. **11**, 315–319 (1985)
18. Matthes, D., Osberger, H.: Convergence of a variational Lagrangian scheme for a nonlinear drift diffusion equation. ESAIM Math. Model. Numer. Anal. **48**, 697–726 (2014)
19. Matthes, D., Osberger, H.: A convergent Lagrangian discretization for a nonlinear fourth order equation. Found. Comput. Math. 1–54 (2015)
20. McCann, R.J.: A convexity principle for interacting gases. Adv. Math. **128**, 153–179 (1997)
21. Roessler, T.: Discretizing the porous medium equation based on its gradient flow structure — a consistency paradox. Technical report, Available online at http://sfb611.iam.uni-bonn.de/uploads/150-komplett.pdf (2004)
22. Rossi, R., Savaré, G.: Tightness, integral equicontinuity and compactness for evolution problems in Banach spaces. Ann. Sc. Norm. Super. Pisa Cl. Sci. (5) **2**, 395–431 (2003)
23. Russo, G.: Deterministic diffusion of particles. Commun. Pure Appl. Math. **43**, 697–733 (1990)
24. Villani, C.: Topics in Optimal Transportation. Graduate Studies in Mathematics, vol. 58. American Mathematical Society, Providence, RI (2003)
25. Westdickenberg, M., Wilkening, J.: Variational particle schemes for the porous medium equation and for the system of isentropic Euler equations. M2AN Math. Model. Numer. Anal. **44**, 133–166 (2010)